Metalloenzymes in Denitrification
Applications and Environmental Impacts

RSC Metallobiology Series

Titles in the Series:
1: Mechanisms and Metal Involvement in Neurodegenerative Diseases
2: Binding, Transport and Storage of Metal Ions in Biological Cells
3: 2-Oxoglutarate-Dependent Oxygenases
4: Heme Peroxidases
5: Molybdenum and Tungsten Enzymes: Biochemistry
6: Molybdenum and Tungsten Enzymes: Bioinorganic Chemistry
7: Molybdenum and Tungsten Enzymes: Spectroscopic and Theoretical Investigations
8: Metal Chelation in Medicine
9: Metalloenzymes in Denitrification: Applications and Environmental Impacts

Metalloenzymes in Denitrification
Applications and Environmental Impacts

Edited by

Isabel Moura
Universidade Nova de Lisboa, Portugal
Email: isabel.moura@fct.unl.pt

José J. G. Moura
Universidade Nova de Lisboa, Portugal
Email: jose.moura@fct.unl.pt

Sofia R. Pauleta
Universidade Nova de Lisboa, Portugal
Email: srp@fct.unl.pt

and

Luisa B. Maia
Universidade Nova de Lisboa, Portugal
Email: luisa.maia@fct.unl.pt

ROYAL SOCIETY OF CHEMISTRY

THE QUEEN'S AWARDS FOR ENTERPRISE: INTERNATIONAL TRADE 2013

RSC Metallobiology Series No. 9

Print ISBN: 978-1-78262-334-2
PDF eISBN: 978-1-78262-376-2
EPUB eISBN: 978-1-78262-926-9
ISSN: 2045-547X

A catalogue record for this book is available from the British Library

Published by The Royal Society of Chemistry,
Thomas Graham House, Science Park, Milton Road,
Cambridge CB4 0WF, UK

Registered Charity Number 207890

For further information see our web site at www.rsc.org

Printed in the United Kingdom by CPI Group (UK) Ltd, Croydon, CR0 4YY, UK

Preface

The choice of the topic for this book was the easiest part. By itself, denitrification is a very important topic due to its wide range of applications and impacts. The topic has been also a key driver of the editors' research groups and we were aware of a need for an overview of new discoveries and results.

The book itself would not have been possible without the generous efforts of our co-authors. All of us are involved in a 'multitasking' world, and agreeing to sum up, share and reveal this new knowledge was an enormous effort that will improve our understanding of this enormous puzzle. We hope to reach a wide range of people and that the collected information will be useful not only for research, but also for pedagogic purposes.

A final word for a colleague who left us during this process: Simon DeVries, a long-time friend and a generous and imaginative scientist. He was unable to finish his assigned Chapter 6, but Yoshitsugu Shiro was able, at short notice, to step in and complete the assignment.

Scientific life involves groups of people researching in faraway places. During the preparation of this book, we were very impressed by the range of scientific information gathered and the competence and cooperation of our co-authors. We value, most of all, the friendship of this group of people.

Isabel Moura, José J. G. Moura, Sofia R. Pauleta and Luisa B. Maia.

RSC Metallobiology Series No. 9
Metalloenzymes in Denitrification: Applications and Environmental Impacts
Edited by Isabel Moura, José J. G. Moura, Sofia R. Pauleta and Luisa B. Maia

Published by the Royal Society of Chemistry, www.rsc.org

Contents

RSC Metallobiology Series No. 9
Metalloenzymes in Denitrification: Applications and Environmental Impacts
Edited by Isabel Moura, José J. G. Moura, Sofia R. Pauleta and Luisa B. Maia

Published by the Royal Society of Chemistry, www.rsc.org

CHAPTER 1

A Bird's Eye View of Denitrification in Relation to the Nitrogen Cycle

ISABEL MOURA*[a], LUÍSA B. MAIA[a], SOFIA R. PAULETA[a] AND JOSÉ J. G. MOURA[a]

[a]UCIBIO, REQUIMTE, Departamento de Química, Faculdade de Ciências e Tecnologia, Campus da Caparica, 2829-516 Caparica, Portugal
*E-mail: isabelmoura@fct.unl.pt

1.1 Introduction

Nitrate and nitrite are key players in the broad nitrogen biogeochemical cycle that is fundamental for life processes. These oxo-anions are involved in key pathways and in the turnover/recycling of nitrogen in the atmosphere, hydrosphere and geosphere. From a human perspective, nitrite and nitrate have a "Janus face": they are important as food preservatives,[1] but are also implied in oncogenic problems due to the formation of *N*-nitrosamines.[2–6] On the other hand, nitrite is now considered to be beneficial to health, whether formed endogenously or added therapeutically. The chapter will identify all of the components that are involved and put into context all of the diverse pathways that contribute to the nitrogen biogeochemical cycle (Figure 1.1). Emphasis will be given to the denitrification pathway—the actors, the mechanisms

RSC Metallobiology Series No. 9
Metalloenzymes in Denitrification: Applications and Environmental Impacts
Edited by Isabel Moura, José J. G. Moura, Sofia R. Pauleta and Luisa B. Maia

Published by the Royal Society of Chemistry, www.rsc.org

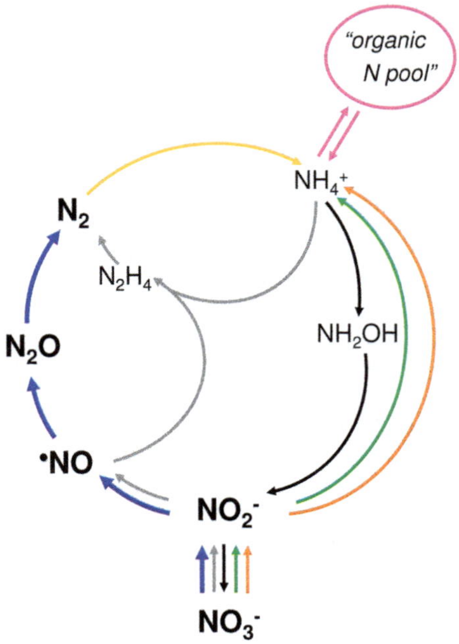

Figure 1.1 Biochemical cycle of nitrogen. Denitrification, blue arrows; dinitrogen fixation, yellow arrow; assimilatory ammonification, orange arrows; "organic nitrogen pool", pink arrows; dissimilatory nitrate reduction to ammonium, green arrows; nitrification, black arrows; anaerobic ammonium oxidation, grey arrows.

and the regulation of these "simple" chemical reactions—always keeping in mind the environmental and (bio)technological applications.

1.2 The Nitrogen Cycle

Hydrogen, oxygen, carbon and nitrogen are absolutely essential for life. Nitrogen is the fourth most abundant element in living organisms,[7] being used for the biosynthesis of *N*-containing compounds (amino acids, nucleosides and other fundamental compounds). Two nitrogen assimilatory pathways provide the reduced nitrogen that is recycled and stored in an "organic nitrogen pool". In addition, some organisms also use nitrogen compounds as substrates for "respiration" and, for that purpose, several nitrogen dissimilatory pathways have also evolved. Nitrogen is fixed mainly from the largest pool, atmospheric dinitrogen (containing about 70% of all nitrogen) and is recycled through the dissimilatory pathways.[8–12]

1.3 Classic and New Pathways

Two nitrogen *assimilatory* pathways provide the reduced nitrogen (ammonium) that is essential for biosynthetic purposes (Figure 1.1, yellow and orange arrows). Free-living archaea, bacteria and symbiotic bacteria are capable of dinitrogen fixation (yellow arrow) in a reaction that is catalysed by molybdenum/iron-dependent nitrogenases,[13–15] a reaction widely occurring in soils, oceans and the Earth's crust. Less commonly, vanadium/iron or iron-only enzymes are involved. Nitrate assimilation is dependent on the transformation of nitrate into nitrite, catalysed by molybdenum-dependent nitrate reductases,[16,17] and then on the reduction of nitrite to ammonium, catalysed by sirohaem-containing nitrite reductases.[18–25] The enzymes share similarities in prokaryotic and eukaryotic organisms.[26–28] In bacteria, fungi and plants, ammonium enters the "organic nitrogen pool" (Figure 1.1, pink arrows) in the form of two amino acids—glutamine and glutamate—through the concerted action of the enzymes glutamine synthase and glutamate synthase. Ammonium can also be directly incorporated into glutamate through the glutamate dehydrogenase that is present in all forms of life and is also relevant in amino acid catabolism.

The organic nitrogen, in the form of amino and amide groups, can be transferred, transformed and incorporated into other amino acids and several other nitrogen-containing biomolecules (Figure 1.1, pink arrows). The organic nitrogen is then recycled between all living organisms through food chains, waste products and organic decay. Mineralisation also takes part in the process of returning ammonium back to the environment.

The biogeochemical cycle of nitrogen continues with four well-understood nitrogen *dissimilatory* pathways, in which nitrogen compounds are used as electron donors/acceptors to derive energy *via* respiration.

1.3.1 Dissimilatory Nitrate Reduction to Ammonium and Nitrification

Dissimilatory reduction of nitrate to ammonium (DNRA; Figure 1.1, green arrows) is used for anaerobic growth. In this pathway, nitrite is directly reduced to ammonium using a multi-*c*-type haem-containing nitrite reductase.

Nitrifiers use the "opposite" aerobic oxidation of ammonium to nitrite (and to nitrate) for the chemoautotrophic fixation of inorganic carbon (nitrification; Figure 1.1, black arrows). Hence, DNRA and nitrification can be regarded as alternative short-circuits within the vast dinitrogen reservoir.[29,30] Aerobic ammonium-oxidising bacteria use ammonium almost exclusively as the electron donor in "respiration" and oxidise it to hydroxylamine (catalysed by ammonium monooxygenase) and then to nitrite (catalysed by hydroxylamine oxidoreductase; Figure 1.1, black arrows). In addition, archaea (from marine and soil environments) seem able to oxidise ammonium for chemolithoautotrophic growth.[31–34]

Nitrite-oxidising bacteria are responsible for the final oxidation of nitrite to nitrate through a reaction that is catalysed by a molybdenum-containing nitrite oxidoreductase (Figure 1.1, black arrow). These bacteria are chemolithoautotrophs that derive energy from the oxidation of nitrite to nitrate in a strictly aerobic process. Here, all of the carbon that is needed is satisfied by carbon dioxide assimilation (*via* the Calvin cycle).[35–41]

Nitrite oxidation is a relevant step of the nitrification pathway and probably the main source of nitrate under aerobic conditions. However, as described below, nitrite is also oxidised through other pathways.

1.3.2 Anaerobic Oxidation of Ammonium to Dinitrogen

Ammonium can also be oxidised anaerobically *via* the anaerobic ammonium oxidation (AnAmmOx) pathway (Figure 1.1, grey arrows) and is used for anaerobic chemoautotrophic growth of planctomycetes.[42–48] In this pathway, ammonium is first oxidised by nitric oxide (NO) to yield hydrazine in a reaction that is catalysed by the di-haem-*c*-containing enzyme hydrazine synthase. Hydrazine is then oxidised to dinitrogen by the octa-haem-*c*-containing enzyme hydrazine oxidoreductase.[44,49–51] The necessary NO is formed *via* nitrite reduction by a cd_1 haem-containing nitrite reductase, which is a well-described enzyme of the denitrification pathway. In this way, AnAmmOx is coupled to nitrite reduction. Recently, a bacterial strain KSU-1 that follows this pathway was hypothesised to contain a copper-containing nitrite reductase enzyme (which is also a well-known enzyme of the denitrification pathway), suggesting that the enzymatic machinery might be more flexible than initially thought.[52,53]

This AnAmmOx pathway is presently one of the challenging and unknown features of the nitrogen biogeochemical cycle, operating in marine, freshwater and terrestrial ecosystems.

1.3.3 Anaerobic Reduction of Nitrate to Dinitrogen—Denitrification

This book is devoted to the denitrification pathway, an anaerobic process that is used by a wide range of microorganisms for energy generation in which nitrate, present in the soil or water, is reduced to gaseous dinitrogen. Nitrate accumulation and the release of nitrous oxide into the atmosphere due to the excessive use of fertilisers in agriculture are examples of two environmental problems in which denitrification plays a central role.

The denitrification[54,55] metabolic pathway is found in a wide range of organisms, from archaea to eubacteria, and even in some eukaryotes (*e.g.*, fungi, protozoa and benthic *Foraminifera* and *Gromiida*).[55–57] It occurs in both autotrophic and heterotrophic organisms, coupled to the anaerobic oxidation of organic carbon.

The reduction of nitrate to dinitrogen gas (eqn (1.5)) is accomplished by five different types of metalloenzymes in four surprisingly simple steps (eqn (1.1)–(1.4)): nitrate is reduced to nitrite, then to NO, then to nitrous oxide and finally to dinitrogen. A battery of enzymes is utilised, namely the reductases of nitrate, nitrite, NO and nitrous oxide. There are also multiple electron donors that interact with these enzymes to transfer the electrons that are required in these stepwise reactions.[54,58–60] The importance of the pathway has inspired efforts to understand the structures of the participating enzymes and their enzymatic mechanisms.

$$NO_3^- + 2e^- + 2H^+ \rightarrow NO_2^- + H_2O \tag{1.1}$$
$$NO_2^- + e^- + 2H^+ \rightarrow {}^{\bullet}NO + H_2O \tag{1.2}$$
$$2{}^{\bullet}NO + 2e^- + 2H^+ \rightarrow N_2O + H_2O \tag{1.3}$$
$$N_2O + 2e^- + 2H^+ \rightarrow N_2 + H_2O \tag{1.4}$$
$$\text{Global reaction: } 2NO_3^- + 10e^- + 12H^+ \rightarrow N_2 + 6H_2O \tag{1.5}$$

The four sequential steps involve three one-oxygen atom abstractions (eqn (1.1), (1.2) and (1.4)) and a one-oxygen atom transfer coupled with N–N bond formation (eqn (1.3)). Each step is catalysed by specific metalloenzymes whose 3D structures have all been determined.[54,59]

The metal diversity detected in this pathway is acknowledged, being a delight in bioinorganic chemistry. A dependence upon molybdenum, iron and copper is observed. Two different types of nitrite reductases (eqn (1.2)) can be found: an iron-dependent enzyme, featuring *c*-type and d_1-type haem centres (cd_1 haem-containing nitrite reductase) and a copper-dependent enzyme (copper-containing nitrite reductase).[54,57,58,61,62] The initial step of nitrate transformation to nitrite (eqn (1.1)) is carried out by nitrate reductase, which is part of the bis-pyranopterin guanosine dinucleotide-containing family of mononuclear molybdo-enzymes.[63] The last two steps of denitrification use exquisite machineries. NO reductase (eqn (1.3)) contains both haem iron and non-haem iron, building up an active site that resembles cytochrome *c* oxidase (*i.e.*, is copper/iron dependent).[64–66] Nitrous oxide reductase (eqn (1.4)), the last enzyme of this pathway, contains a novel tetranuclear copper centre (Cu_Z), which is a unique and challenging structure in biology.[67,68]

All of the diverse pathways interact and close the nitrogen biogeochemical cycle (Figure 1.1). The fixed nitrogen is eventually lost to the atmosphere, soils, crust and oceans.

1.3.4 New Avenues/New Challenges

In addition to the well-recognised four nitrogen dissimilatory pathways, several new processes have been identified, including anaerobic phototrophic nitrite oxidation,[69,70] microaerobic nitrification/AnAmmOx coupling,[69–75] hyper-thermophilic dinitrogen fixation/methane production[76] and "denitrification/intra-aerobic methane oxidation".[77,78]

Over recent years, the discovery and addition of new organisms participating in the nitrogen cycle and the scrutiny of new ways to transform the nitrogen oxo-anions have added more complexity to this biogeochemical cycle, revealing an intricate network of pathways. The regulation and control of these complex pathways is still a challenging problem, and future research will certainly clarify some of the presently established dogmas.

1.4 Book Outlook

This book, devoted to metalloenzymes in denitrification, aims to give a concise update of the multiple and complex aspects of denitrification. Two initial overarching chapters (Chapter 1—A Bird's Eye View of Denitrification in Relation to the Nitrogen Cycle, by Isabel Moura, Luisa Maia, Sofia R. Pauleta and José J. G. Moura; and Chapter 2—Nitric Oxide Production, Damage and Management During Anaerobic Nitrate Reduction to Ammonia, by Jeffrey A. Cole) will provide a general overview of the subject, placing the denitrification pathway in the nitrogen cycle and connecting some of its substrates and products, such as nitrate and nitrite, with the other pathways of the nitrogen cycle. The role of NO as a by-product of these enzymes, and its role as a signalling molecule, as well as the new functions of proteins such as Hcp and YtfE, is reviewed.

The following chapters will cover the key enzymes and bioinorganic aspects of denitrification, including nitrate reductase (Chapter 3—Structure, Function and Mechanisms of Respiratory Nitrate Reductases, by Pablo J. Gonzalez, Maria G. Rivas and José J. G. Moura), iron-dependent nitrite reductase (Chapter 4—Nitrite Reductase—Cytochrome cd_1, by Serena Rinaldo, Giorgio Giardina and Francesca Cutruzzolà), copper-dependent nitrite reductase (Chapter 5—Structure and Function of Copper Nitrite Reductase, by Masaki Nojiri), NO reductase (Chapter 6—Structure and Function of Nitric Oxide Reductases, by Takehiko Tosha and Yoshitsugu Shiro) and nitrous oxide reductase (Chapter 7—Insights into Nitrous Oxide Reductase, by Sofia R. Pauleta, Cíntia Carreira and Isabel Moura). The book will also cover the efforts that have been made to model the active sites and the reactions involved (Chapter 8—Model Compounds for Molybdenum Nitrate Reductases, by Amit Majumdar and Sabyasachi Sarkar; Chapter 9—Model Compounds for Nitric Oxide Reductase, by Abhishek Dey and Somdatta Ghosh Dey; and Chapter 10—Model Compounds of Copper-Containing Enzymes Involved in Bacterial Denitrification, by Brittany J. Johnson and Neal P. Mankad). Electron transfer and molecular recognition in denitrification and nitrate dissimilatory pathways will be addressed in Chapter 11 by Rui M. Almeida, Simone Dell'Acqua, Sofia R. Pauleta, Isabel Moura and José J. G. Moura. Aspects on nitrate and nitrite transport will also be covered (Chapter 12—Channels and Transporters for Nitrogen Cycle Intermediates, by Susana L. A. Andrade and Oliver Einsle), as well as the regulation of denitrification (Chapter 13—Regulation of Denitrification, by Stephen Spiro). The denitrification pathway in fungi and Gram-positive bacteria will also be addressed (Chapter 14—Denitrification in Fungi, by Hirofumi Shoun and Shinya Fushinobu; and Chapter 15—Denitrification and Non-Denitrifier Nitrous Oxide Emission in Gram-Positive Bacteria, by Yi-hua Sun, Paul De Vos and Kim Heylen). Emphasis will also be given to the environmental impact, water treatment (biotechnology) and modelling/mitigating of its negative impacts (Chapter 16—Denitrification Processes for Wastewater Treatment, by Bing-Jie Ni, Yuting Pan, Jianhua Guo, Bernardino Virdis, Shihu Hu, Xueming Chen and Zhiguo Yuan). Finally, the novel human nitrate/nitrite/NO signalling pathway will be discussed within the nitrogen cycle scenario (Chapter 17—Lessons from Denitrification for the Human Metabolism of Signalling Nitric Oxide, by Luísa B. Maia and José J. G. Moura).

The book will update the state of the art and new aspects on the structure and function of the enzymes involved in denitrification, the regulation of this pathway, mechanistic revelations/implications and the efforts on the chemical modelling of the reaction centres. Applied aspects are an added value in respect to environment, regulation, measurement and mitigation.

Acknowledgements

This work was supported by the Unidade de Ciências Biomoleculares Aplicadas—UCIBIO, which is financed by national funds from FCT/MEC (UID/Multi/04378/2013) and co-financed by the ERDF under the PT2020 Partnership Agreement (POCI-01-0145-FEDER-007728). LBM thanks Fundacão para a Ciência e a Tecnologia (FCT), MEC, for a fellowship grant (SFRH/BPD/111404/2015, which is financed by national funds and co-financed by FSE). SRP is an IF fellow also supported by FCT, and IM thanks the financial support provided by FCT through PTDC/QUI-BIQ/116481/2010 and PTDC/BBB-BQB/0129/2014.

References

1. E. F. Binkerd and O. E. Kolari, *Food Cosmet. Toxicol.*, 1975, **13**, 661.
2. S. R. Tannenbaum, A. J. Sisnkey, M. Weisman and W. J. Bishop, *Cancer Inst.*, 1974, **53**, 79.
3. V. M. Craddock, *Nature*, 1983, **306**, 638.
4. H. Bartsch and R. Montesano, *Carcinogenesis*, 1984, **5**, 1381.
5. D. Forman, S. Al-Dabbagh and R. Doll, *Nature*, 1985, **313**, 620.
6. M. H. Ward, T. M. deKok, P. Levallois, J. Brender, G. Gulis, B. T. Nolan and J. VanDerslice, *Environ. Health Perspect.*, 2005, **113**, 1607.
7. J. J. R. Fraústo da Silva and R. J. P. Williams, *The Biological Chemistry of the Elements: The Inorganic Chemistry of Life*, Oxford University Press, Oxford, 2001.
8. D. J. Richardson and N. J. Watmough, *Curr. Opin. Chem. Biol.*, 1999, **3**, 207.
9. J. N. Galloway, F. J. Dentener, D. G. Capone, E. W. Boyer, R. W. Howarth, S. P. Seitzinger, G. P. Asner, C. C. Cleveland, P. A. Green, E. A. Holland, D. M. Karl, A. F. Michaels, J. H. Porter, A. R. Townsend and C. J. Vorosmarty, *Biogeochemistry*, 2004, **70**, 153.
10. H. Bothe, S. J. Ferguson and W. E. Newton, *Biology of the nitrogen cycle*, Elsevier, Amsterdam, 2007.
11. D. E. Canfield, A. N. Glazer and P. G. Falkowski, *Science*, 2010, **330**, 192.
12. R. M. Martínez-Espinosa, J. A. Cole, D. J. Richardson and N. J. Watmough, *Biochem. Soc. Trans.*, 2011, **39**, 175.

13. B. M. Hoffman, D. Lukoyanov, Z. Y. Yang, D. R. Dean and L. C. Seefeldt, *Chem. Rev.*, 2014, **114**, 4041.
14. Y. Hu and M. W. Ribbe, *JBIC, J. Biol. Inorg. Chem.*, 2015, **20**, 435.
15. R. Bjornsson, F. Neese, R. R. Schrock, O. Einsle and S. DeBeer, *JBIC, J. Biol. Inorg. Chem.*, 2015, **20**, 447.
16. M. Cerqueira, P. J. Gonzalez, C. D. Brondino, M. J. Romão, C. C. Romão, I. Moura and J. J. B. Moura, *J. Comput. Chem.*, 2009, **30**, 2466.
17. P. J. González, C. Correia, I. Moura, C. D. Brondino and J. J. G. Moura, *J. Inorg. Biochem.*, 2006, **100**, 1015.
18. E. J. Hewitt, *Annu. Rev. Plant Physiol.*, 1975, **26**, 73.
19. M. G. Guerrero, J. M. Vega and M. Losada, *Ann. Rev. Plant Physiol.*, 1981, **32**, 169.
20. D. B. Knaff and M. Hirasawa, *Biochim. Biophys. Acta*, 1991, **1056**, 93.
21. J. T. Lin and V. Stewart, *Adv. Microb. Physiol.*, 1997, **39**, 1.
22. K. Fischer, G. G. Barbier, H. J. Hecht, R. R. Mendel, W. H. Campbell and G. Schwarz, *Plant Cell*, 2005, **17**, 1167.
23. E. Flores, J. E. Frias, L. M. Rubio and A. Herrero, *Photosynth. Res.*, 2005, **83**, 117.
24. C. Moreno-Vivián and E. Flores, in *Biology of the nitrogen cycle*, ed. H. Bothe, S. J. Ferguson and W. E. Newton, Elsevier, Amsterdam, 2007, p. 263.
25. A. J. Gates, V. M. Luque-Almagro, A. D. Goddard, S. J. Ferguson, D. Roldán and D. J. Richardson, *Biochem. J.*, 2011, **435**, 743.
26. M. Hirasawa, K. Fukushima, G. Tamura and D. B. Knaff, *Biochim. Biophys. Acta*, 1984, **791**, 145.
27. C. Pino, F. Olmo-Mira, P. Cabello, M. Martínez-Luque, F. Castillo, M. D. Roldán and C. Moreno-Vivián, *Biochem. Soc. Trans.*, 2006, **34**, 127.
28. S.-W. Kim, S. Fushinobu, S. Zhou, T. Wakagi and H. Shoun, *Appl. Environ. Microbiol.*, 2009, **75**, 2652.
29. H. P. Koops and A. Pommerening-Roser, *FEMS Microbiol. Ecol.*, 2001, **37**, 1.
30. S. E. Vlaeminck, A. G. Hay, L. Maignien and W. Verstraete, *Environ. Microbiol.*, 2011, **13**, 283.
31. M. Könneke, A. E. Bernhard, J. R. de la Torre, C. B. Walker, J. B. Waterbury and D. A. Stahl, *Nature*, 2005, **437**, 543.
32. C. Schleper, G. Jurgens and M. Jonuscheit, *Nat. Rev. Microbiol.*, 2005, **3**, 479.
33. S. J. Hallam, K. T. Konstantinidis, N. Putnam, C. Schleper, Y. Watanabe, J. Sugahara, C. Preston, J. de la Torre, P. M. Richardson and E. F. DeLong, *Proc. Natl. Acad. Sci. U. S. A.*, 2006, **103**, 18296.
34. R. Hatzenpichler, E. V. Lebedeva, E. Spieck, K. Stoecker, A. Richter, H. Daims and M. Wagner, *Proc. Natl. Acad. Sci. U. S. A.*, 2008, **105**, 2134.
35. A. J. Smith and D. S. Hoare, *J. Bacteriol.*, 1968, **95**, 844.
36. E. Bock, *Arch. Microbiol.*, 1976, **108**, 305.
37. E. Bock, H. Sundermeyer-Klinger and E. Stackebrandt, *Arch. Microbiol.*, 1983, **136**, 281.

38. K. Kirstein and E. Bock, *Arch. Microbiol.*, 1993, **160**, 447.
39. A. Teske, E. Alm, J. M. Regan, S. Toze, B. E. Rittmann and D. A. Stahl, *J. Bacteriol.*, 1994, **176**, 6623.
40. J. A. Dodsworth, B. Hungate, J. R. Torre, H. Jiang and B. P. Hedlund, *Methods Enzymol.*, 2011, **486**, 171.
41. E. Spieck and A. Lipski, *Methods Enzymol.*, 2011, **486**, 109.
42. M. S. M. Jetten, M. Strous, K. T. van de Pas-Schoonen, J. Schalk, U. G. van Dongen, A. A. van de Graaf, S. Logemann, G. Muyzer, M. C. van Loosdrecht and J. G. Kuenen, *FEMS Microbiol. Rev.*, 1998, **22**, 421.
43. M. Strous, J. A. Fuerst, E. H. M. Kramer, S. Logemann, G. Muyzer, K. T. van de Pas-Schoonen, R. Webb, J. G. Kuenen and M. S. M. Jetten, *Nature*, 1999, **400**, 446.
44. M. G. Klotz and L. Y. Stein, *FEMS Microbiol. Lett.*, 2008, **278**, 146.
45. M. S. M. Jetten, L. van Niftrik, M. Strous, B. Kartal, J. T. Keltjens and H. J. M. Op den Camp, *Crit. Rev. Biochem. Mol. Biol.*, 2009, **44**, 65.
46. N. M. de Almeida, W. J. Maalcke, J. T. Keltjens, M. S. Jetten and B. Kartal, *Biochem. Soc. Trans.*, 2011, **39**, 303.
47. B. Kartal, W. J. Maalcke, N. M. de Almeida, I. Cirpus, J. Gloerich, W. Geerts, H. J. M. Op den Camp, H. R. Harhangi, E. M. Janssen-Megens, K.-J. Francoijs, H. G. Stunnenberg, J. T. Keltjens, M. S. M. Jetten and M. Strous, *Nature*, 2011, **479**, 127.
48. B. Kartal, L. van Niftrik, J. T. Keltjens, H. J. Op den Camp and M. S. Jetten, *Adv. Microb. Physiol.*, 2012, **60**, 211.
49. M. Strous, E. Pelletier, S. Mangenot, T. Rattei, A. Lehner, M. W. Taylor, M. Horn, H. Daims, D. Bartol-Mavel, P. Wincker, V. Barbe, N. Fonknechten, D. Vallenet, B. Segurens, C. Schenowitz-Truong, C. Medigue, A. Collingro, B. Snel, B. E. Dutilh, H. J. M. Op den Camp, C. van der Drift, I. Cirpus, K. T. van de Pas-Schoonen, H. R. Harhangi, L. van Niftrik, M. Schmid, J. Keltjens, J. van de Vossenberg, B. Kartal, H. Meier, D. Frishman, M. A. Huynen, H.-W. Mewes, J. Weissenbach, M. S. M. Jetten, M. Wagner and D. Le Paslier, *Nature*, 2006, **440**, 790.
50. M. Shimamura, T. Nishiyama, H. Shigetomo, T. Toyomoto, T. Kawahara, K. Furukawa and T. Fujii, *Appl. Environ. Microbiol.*, 2007, **73**, 1065.
51. M. Li, T. Ford, X. Li and J.-D. Gu, *Environ. Sci. Technol.*, 2011, **45**, 3547.
52. D. Hira, H. Toh, C. T. Migita, H. Okubo, T. Nishiyama, M. Hattori, K. Furukawa and T. Fujii, *FEBS Lett.*, 2012, **586**, 1658.
53. B. Kartal, M. M. Kuypers, G. Lavik, J. Schalk, H. J. Op den Camp, M. S. Jetten and M. Strous, *Environ. Microbiol.*, 2007, **9**, 635.
54. I. Moura and J. J. G. Moura, *Curr. Opin. Chem. Biol.*, 2001, **5**, 168.
55. L. B. Maia and J. J. G. Moura, *Chem. Rev.*, 2014, **114**, 5273.
56. E. Piña-Ochoa, S. Høgslund, E. Geslin, T. Cedhagen, N. Peter Revsbech, L. P. Nielsen, M. Schweizer, F. Jorissen, S. Rysgaard and N. Risgaard-Petersen, *Proc. Natl. Acad. Sci. U. S. A.*, 2010, **107**, 1148.
57. S.-W. Kim, S. Fushinobu, S. Zhou, T. Wakagi and H. Shoun, *Appl. Environ. Microbiol.*, 2009, **75**, 2652.
58. W. G. Zumft, *Microbiol. Mol. Biol. Rev.*, 1997, **61**, 533.

59. O. Einsle and P. M. Kroneck, *J. Biol. Chem.*, 2004, **385**, 875.
60. A. Vázquez-Torres and A. J. Bäumler, *Curr. Opin. Microbiol.*, 2015, **29**, 1.
61. N. Lehnert, T. C. Berto, M. G. I. Galinato and L. E. Goodrich, *The Handbook of Porphyrin Science*, World Scientific, New Jersey, 2011.
62. E. I. Solomon, D. E. Heppner, E. M. Johnston, J. W. Ginsbach, J. Cirera, M. Qayyum, M. T. Kieber-Emmons, C. H. Kjaergaard, R. G. Hadt and L. Tian, *Chem. Rev.*, 2014, **114**, 3659.
63. L. B. Maia, I. Moura and J. J. G. Moura, *Molybdenum and Tungsten Enzymes*, ed. R. Hille, C. Schulzke and M. L. Kirk, Royal Society of Chemistry, 2016, in press.
64. T. Tosha and Y. Shiro, *IUBMB Life*, 2013, **65**, 217.
65. S. de Vries and I. Schröder, *Biochem. Soc. Trans.*, 2002, **30**, 662.
66. A. L. Ducluzeau, B. Schoepp-Cothenet, R. van Lis, F. Baymann, M. J. Russell and W. Nitschke, *J. R. Soc., Interface*, 2014, **11**, 20140196.
67. L. K. Schneider, A. Wüst, A. Pomowski, L. Zhang and O. Einsle, *Met. Ions Life Sci.*, 2014, **14**, 177.
68. S. R. Pauleta, S. Dell'Acqua and I. Moura, *Coord. Chem. Rev.*, 2013, **2**, 332.
69. A. K. Boal and A. C. Rosenzweig, *Science*, 2012, **337**, 1617.
70. S. Demaneche, L. Philippot, M. M. David, E. Navarro, T. M. Vogel and P. Simonet, *Appl. Environ. Microbiol.*, 2009, **75**, 534.
71. M. Könneke, A. E. Bernhard, J. R. de la Torre, C. B. Walker, J. B. Waterbury and D. A. Stahl, *Nature*, 2005, **437**, 543.
72. C. A. Francis, J. M. Beman and M. M. Kuypers, *ISME J.*, 2007, **1**, 19.
73. P. Lam, M. M. Jensen, G. Lavik, D. F. McGinnis, B. Muller, C. J. Schubert, R. Amann, B. Thamdrup and M. M. M. Kuypers, *Proc. Natl. Acad. Sci. U. S. A.*, 2007, **104**, 7104.
74. J. van de Vossenberg, J. E. Rattray, W. Geerts, B. Kartal, L. van Niftrik, E. G. van Donselaar, J. S. Sinninghe Damsté, M. Strous and M. S. Jetten, *Environ. Microbiol.*, 2008, **10**, 3120.
75. J. Yan, H. J. Op den Camp, M. S. Jetten, Y. Y. Hu and S. C. Haaijer, *Syst. Appl. Microbiol.*, 2010, **33**, 407.
76. M. P. Mehta and J. A. Baross, *Science*, 2006, **314**, 1783.
77. A. A. Raghoebarsing, A. Pol, K. T. van de Pas-Schoonen, A. J. Smolders, K. F. Ettwig, W. I. Rijpstra, S. Schouten, J. S. Damsté, H. J. Op den Camp, M. S. Jetten and M. Strous, *Nature*, 2006, **440**, 918.
78. K. F. Ettwig, M. K. Butler, D. Le Paslier, E. Pelletier, S. Mangenot, M. M. Kuypers, F. Schreiber, B. E. Dutilh, J. Zedelius, D. de Beer, J. Gloerich, H. J. Wessels, T. van Alen, F. Luesken, M. L. Wu, K. T. van de Pas-Schoonen, H. J. Op den Camp, E. M. Janssen-Megens, K. J. Francoijs, H. Stunnenberg, J. Weissenbach, M. S. Jetten and M. Strous, *Nature*, 2010, **464**, 543.

CHAPTER 2

Nitric Oxide Production, Damage and Management During Anaerobic Nitrate Reduction to Ammonia

JEFFREY A. COLE[a]

[a]School of Biosciences and the Institute of Microbiology and Infection, University of Birmingham, Birmingham, B15 2TT, UK
*E-mail: j.a.cole@bham.ac.uk

2.1 "Things Ain't What They Used to Be!"

Forty years ago, the scientific literature recognised only two types of nitrate reduction: nitrate assimilation and denitrification. With data then available from only a few types of microorganisms, denitrification could be defined simply as the anaerobic conversion of nitrate *via* nitrite to dinitrogen in a process that conserves energy for anaerobic growth. As more data accumulated, it was realised that some bacteria are able to "denitrify", even though they lack a functional nitrate reductase.[1] At the other end of the pathway, organisms were found that reduce nitrite to nitric oxide (NO) and N_2O, but they lack a functional nitrous oxide reductase.[2] Only in 1980 was it finally accepted that many bacteria living in electron-rich anaerobic environments reduce nitrite to ammonia, rather than to N_2.[3,4] In contrast to denitrification,

RSC Metallobiology Series No. 9
Metalloenzymes in Denitrification: Applications and Environmental Impacts
Edited by Isabel Moura, José J. G. Moura, Sofia R. Pauleta and Luisa B. Maia

Published by the Royal Society of Chemistry, www.rsc.org

less energy is conserved in this process. As six electrons are required to reduce nitrite to ammonia, but only one is required to convert it to NO, nitrite reduction to ammonia provides a selective advantage compared to denitrification when electron donors are abundant but electron acceptors are scarce. This situation occurs typically, but not exclusively, during anaerobic growth. Thus, its primary physiological function is to maintain redox balance during the oxidation of reduced carbon substrates and to optimise their oxidation to a level that enables them to be used to support growth. Until recently, there were no examples of anaerobic bacteria that catalyse both denitrification and nitrate reduction to ammonia. However, as documented below, the genomics era has provided at least one exception to this dogma.

We can blame the Kuenen laboratory in Delft for exposing the next layers of complexity in the denitrification story! Before 1984, almost all of the literature had focused on the membrane-bound nitrate reductase, now referred to as NarG. Controversies had focused on whether the active site was located in the cytoplasm or in the periplasm,[5–9] the stoichiometry of energy conservation[10,11] and the functions of the NarH, NarJ and NarI subunits.[12] Robertson and Kuenen[13] then upset the denitrification community by showing that some bacteria, in particular what was then known as *Thiosphaera pantotrophus*, could denitrify in the presence of excess oxygen. Ferguson, Richardson and their colleagues confirmed this new heresy,[14–16] showing that in some bacteria the nitrate reductase is a soluble protein located in the periplasm of Gram-negative bacteria. Therefore, no proton motive force is generated during electron transfer from the quinol pool to the periplasmic nitrate reductase. Twenty years later, diehards were still maintaining that denitrification was a purely anaerobic process, despite excellent data to the contrary even in the original paper, as well as subsequent reports.[13]

The next seminal advance from the Kuenen laboratory was the discovery of AnAmmOx: the anaerobic oxidation of ammonia by nitrite in a process in which almost (but not quite) all of the nitrogen is released as N_2, with water and small quantities of nitrate being the other final products.[17] Subsequently, the Jetten laboratory beautifully demonstrated the biochemistry and diversity of this process, a fascinating story that is beyond the scope of this chapter. However, its relevance to denitrification is two-fold: first, the final product is N_2, so it is another process in which "fixed" nitrogen is removed from the biosphere; second, as in denitrification, NO is an obligate intermediate in the biochemical pathway.[18]

NO also features as an intermediate in another process that was documented either by the Jetten laboratory or by former students who were involved in these seminal studies. This was anaerobic oxidation of methane by nitrate or by nitrite. This research resulted in the remarkable demonstration that some strictly anaerobic bacteria can generate their own oxygen from NO.[19] The nitrogen atoms from two molecules of NO are released as dinitrogen, and it has been suggested that this reaction predates oxygenic photosynthesis as the original source of oxygen in the atmosphere.[19]

A further complication to the denitrification story came from studies of what has previously been considered to be the opposite process: nitrification. Early studies of the aerobic oxidation of ammonia to nitrate focused on just two bacterial genera: the ammonia oxidising *Nitrosomonas* and the nitrite oxidising *Nitrobacter*. *Nitrosomonas europaea* and *Nitrosomonas eutropha* cannot oxidise nitrite to nitrate. Conversely, *Nitrobacter winogradskyi* cannot oxidise ammonia. Despite the realisation even 20 years ago that nitrification is widely distributed amongst almost all major groups of bacteria, Teske *et al.*[20] stated that "No organism that has been described is capable of fully oxidising ammonia to nitrate." This statement was still held to be true when the first draft of this review was completed, but two seminal papers published in 2015 destroyed this remaining dogma.[21,22] In a concerted collaboration, two *Nitrospira* strains were shown to oxidise ammonia completely to nitrate in a process that was called Comammox.

A much earlier twist in the story came with the realisation that many nitrifying bacteria require oxygen-limited conditions in order to thrive, and that under microaerobic conditions, many of them can denitrify as well as nitrify.[23–25] Thus, like the AnAmmOx process in Planctomycetes genera such as *Brocadia*, *Jettenia*, *Kuenia* and *Scalindula*, NO is produced even by more conventional ammonia-oxidising bacteria.

There have been many subsequent twists and complications in the story of how fixed nitrogen is converted to gaseous products that are unavailable to productive agriculture. However, this review will focus on a process that is still emerging: the production of NO by enteric and other bacteria that primarily catalyse the dissimilatory reduction of nitrite to ammonia. Key questions, some of which remain unanswered, concern the range of processes by which bacteria such as *Salmonella enterica* and *Escherichia coli* generate NO; the relative importance of these various pathways; their physiological relevance; how these bacteria protect themselves from nitrosative stress; the secondary consequences of failure to do so; and the relative importance of at least four different biochemical mechanisms of NO reduction.

In the context of NO, all of the above examples lead us to recall the song that was made famous by Ella Fitzgerald and Duke Ellington: "Things ain't what they use to be!".

2.2 Overview of Enzymes that Catalyse Denitrification or Nitrite Reduction to Ammonia

Figure 2.1 summarises the main differences between denitrification and nitrate reduction to ammonia. The major enzymes involved in denitrification are reviewed in the next chapters of this book, so only a simplified summary of the best-documented players is shown. The main purpose of the figure for the current chapter is to emphasise the differences, especially in the nitrite reductases, between the two competing processes. However, in order to understand how NO is produced and how its effects are managed by

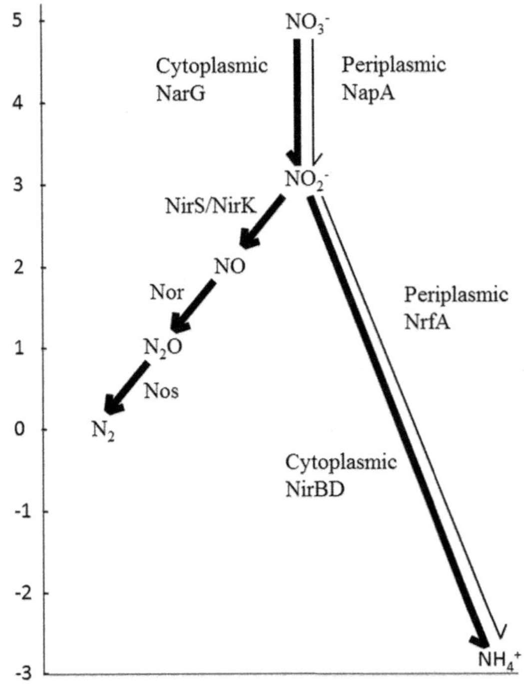

Figure 2.1 Major differences between denitrification and the dissimilatory reduction of nitrate to ammonia. The *y*-axis shows the redox state of the nitrogen in the various species involved. This highly simplified figure is designed to emphasise: (i) that similar nitrate reductases located in either or both the cytoplasmic membrane or in the periplasm catalyse the two-electron reduction of nitrate to nitrite; (ii) that there are two complete pathways for nitrate and nitrite reduction to ammonia, but only a single denitrification pathway; and (iii) that denitrification of nitrite involves a series of three one-electron reduction reactions, whereas the six-electron reduction of nitrite to ammonia occurs in a single step without the stoichiometric production of intermediate species.

bacteria that reduce nitrite directly to ammonia, it is essential first to review the role of the enzymes involved. The following sections will briefly summarise why different bacteria catalyse nitrate reduction to ammonia rather than denitrification, the cellular locations of the enzymes involved and how their synthesis is regulated.

The first obvious difference is that, in contrast to denitrification, which requires four enzymes, nitrate reduction to ammonia requires only two. Both processes start with the two-electron reduction of nitrate to nitrite, and both of them involve similar but alternative nitrate reductases with active sites located either in the cytoplasm or in the periplasm. Some bacteria encode only one of the two types of nitrate reductase, but there are many species of α-, β- and γ-proteobacteria that contain both proteins. In some bacteria, such as *Rhodobacter sphaeroides* and different species of *Shewanella*, there is even greater complexity: there are two different types of periplasmic nitrate reductase. One of the two Nap enzymes in *Rhodobacter sphaeroides* strain 2.4.3 is expressed mainly during aerobic growth to support aerobic denitrification; the second is expressed anaerobically to support nitrate respiration.[26] A different type of variation occurs in *Shewanella* strains, in which Nap-α, encoded by the *napEDABC* gene cluster, appears to support denitrification, while the alternative Nap-β, encoded by *napDAGHB*, supports nitrate reduction to ammonia.[27]

In a few species, such as *E. coli* and *S. enterica* serovar Typhimurium, there are also two alternative nitrite reductases that generate large quantities of ammonia, and again, one is located in the cytoplasm and the other in the periplasm. These bacteria therefore effectively have two independent pathways for the complete reduction of nitrate to ammonia—one located in the cytoplasm and the other in the periplasm—and their synthesis is regulated at least in part independently (see Figure 2.2). Note, however, that while both enzymes of the cytoplasmic pathway are induced in response to excess nitrate, both of the periplasmic enzymes are co-ordinately repressed by excess nitrate.

Why have both the cytoplasmic and the periplasmic pathways been retained by some anaerobic bacteria? Three important clues led to the design of competition experiments that revealed a plausible answer. First, some free-living bacteria reduce nitrate to ammonia, but lack the periplasmic pathway. An example is *Streptococcus carnosus*, which is a generally recognised as safe bacterium that is used for the maturation of German salami to give

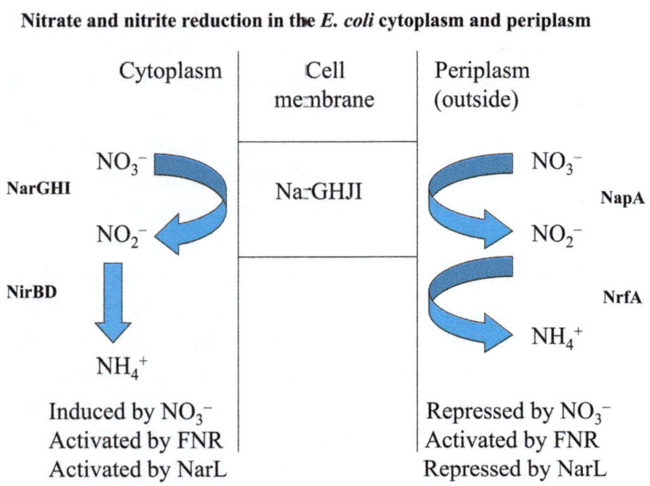

Figure 2.2 The cytoplasmic and periplasmic pathways of nitrate reduction to ammonia in enteric bacteria and their regulation by oxygen and nitrate. Although the nitrate reductase NarG is attached to the cytoplasmic membrane, its active site is located in the periplasm. The nitrite formed is either exported by the membrane protein NarK or reduced to ammonia by the cytoplasmic nitrite reductase NirBD.

the sausage its characteristic pink colour.[28] These bacteria are exposed to high concentrations of nitrate and nitrite in the absence of excess electron donors. In contrast, some enteric bacteria such as *Haemophilus parainfluenzae* and the rumen bacterium *Wolinella succinogenes* lack the cytoplasmic nitrate and nitrite reductases. They are found where nitrate and nitrite are scarce, but electron donors, such as carbohydrates, are abundant.[29,30] The ability to scavenge limited quantities of terminal electron acceptors would provide a selective advantage in environments dominated by fermentative bacteria. This led us to propose that the cytoplasmic pathway is critical for survival where nitrate is abundant but growth might be electron limited.[3,4] Soil and waste water treatment plants provide examples of such environments.[31] In contrast, the periplasmic pathway provides a selective advantage in the gastro-intestinal tracts of warm-blooded animals. It also dominates nitrate reduction in sulphide-rich thermal vents.[32] Further evidence that this might be the explanation came from detailed studies of how expression of the operons that encode the cytoplasmic and periplasmic nitrate reduction pathways is regulated in bacteria such as *E. coli* that retain both systems. In brief, higher concentrations of nitrate are required to induce expression of the cytoplasmic than the periplasmic nitrate reductase, and expression of the *nap* genes is repressed by excess nitrate.[33] This proposal was supported by results from competition experiments in which a chemostat was seeded with a mixture of two *E. coli* mutants, one deleted for the periplasmic nitrate reductase genes and the other deleted for the cytoplasmic nitrate reductase operon.[34] As predicted, the strain that retained the periplasmic pathway had the selective advantage in a nitrate-limited chemostat supplied with an excess of the non-fermentable carbon source, glycerol. In contrast, the strain with the cytoplasmic nitrate reductase rapidly dominated glycerol-limited cultures in which excess nitrate was available. Note that although more energy is conserved during nitrate reduction in the cytoplasm by NarG than by Nap in the periplasm, the bacteria have to pay a penalty. The product, nitrite, will also be formed in the cytoplasm, where it can potentially cause most damage. The relevance of this will be explained in subsequent sections of the review.

Figure 2.1 is far from comprehensive as far as the range of nitrate reductases is concerned. For example, nitrate assimilation by the Nas system is omitted. In Archaea, the nitrate reductase is clearly related to NarG from α-, β- and γ-proteobacteria, with an aspartate residue as a ligand to the catalytic molybdenum atom. However, the apoprotein is synthesised as a precursor with a leader peptide that includes a conserved SRRxFLK motif in its signal sequence that is typical of redox proteins that are transported across the membrane by the "twin arginine" translocation and secretion pathway. The active site of this nitrate reductase is therefore located outside the cytoplasm.[35]

To recap the details presented in the previous chapter, the first dedicated step in denitrification requires the one-electron reduction of nitrite to NO. Two types of nitrite reductase that catalyse this reaction have never yet been found in the same bacterium.[36] One of them is a Cu-containing nitrite reductase designated NirK; the other, designated NirS, is a cytochrome cd_1 nitrite reductase that receives electrons from a haem d_1 subunit (reviewed in Zumft[36] and in Chapter 4 of this book). Both types of nitrite reductase are located in the periplasm of Gram-negative bacteria, or in the case of NirK from *Neisseria* species, it is attached to the outer membrane.[37]

Wide variations are found in the amino acid sequences of NirS.[38] In contrast, NirK sequences fall more tightly into two subgroups. Heylen *et al.*[39] reported that NirK sequences are more similar in strains isolated from similar habitats than in strains of closely related bacteria from different habitats. This phenomenon typically results from horizontal gene transfer within an ecosystem.

As in denitrification, two alternative types of nitrite reductase catalyse the second step in nitrate reduction to ammonia (or, more correctly, the ammonium ion).[33] However, there the similarity ends. In denitrification, the alternative nitrite reductases catalyse the first of three one-electron reduction reactions that form what is essentially a single pathway on the periplasmic side of the cytoplasmic membrane or, in Gram-positive bacteria, attached to the outer side of the membrane. In contrast, one of the alternative nitrite reductases that catalyses the six-electron reduction of nitrite to ammonia, NirBD, is located in the cytoplasm. The other, NrfAB, is secreted into the periplasm (Figure 2.2).

2.3 Dissimilatory Reduction of Nitrate to Ammonia in the Cytoplasm of Gram-Negative Bacteria

The first enzyme in the cytoplasmic pathway for nitrate reduction to ammonia is the membrane-associated NarG. In most bacteria studied so far, the catalytic site for nitrate reduction faces the cytoplasm, so nitrate must first be transported across the cytoplasmic membrane. In enteric bacteria, the nitrate transport protein for respiratory nitrate reduction is the membrane protein, NarK. In *Paracoccus denitrificans*, NarK is a fused protein with two transport domains designated NarK1 and NarK2.[40] The NarK1 domain appears to function as a high-affinity nitrate–H^+ symporter, while NarK2 is proposed to be a lower-affinity antiport system.[41] In *E. coli*, two independent proteins, NarK and NarU, function as nitrate–nitrite antiporters,[42–44] while NirC is a nitrite channel.[42,44] Both liposome experiments and crystal structure data have shown that NarK is an antiporter operating by a rocker mechanism (more details are given in Chapter 12 of this book).[42,43,45] Whether the two sub-groups of NarK, found in some α- and β-proteobacteria, utilise separate transport mechanisms—one forming a channel and the other functioning as a true antiporter—remains to be confirmed.[40]

Electrons for nitrate reduction in the cytoplasm by NarG originate from the cytochrome bc_1 complex in α- and β-proteobacteria, but from NarI in the γ-proteobacteria. Most of the nitrate reduced in the cytoplasm of

Enterobacteriaceae is initially exported to the surrounding medium.[46] However, some of it is also reduced in the cytoplasm directly to ammonia by the NirBD nitrite reductase. NADH is the electron donor, and the catalytic site for cytoplasmic nitrite reduction is siroheme.[47,48] The NirB polypeptide also binds both an iron–sulphur centre and flavin adenine dinucleotide,[47] so the single 93 kDa polypeptide provides a complete electron transfer chain and sufficient electron storage capacity for the six-electron reduction reaction.[47–49] Hydroxylamine might be formed transiently at the NirB active site, but free hydroxylamine is neither a physiologically relevant intermediate nor an alternative substrate for NirBD.[49]

Siroheme is also the catalytic site of nitrite reduction by the assimilatory nitrite reductases, so genome databases often incorrectly annotate *nirB* sequences as genes for assimilatory nitrite reductases. *E. coli* and *S. enterica* serovar Typhimurium can indeed use nitrite as the only source of nitrogen for growth,[3,4] but only during anaerobic growth or during growth under conditions in which the supply of oxygen is so limited that the transcription factor, FNR (the regulator of fumarate and nitrate reduction), is in its active, reduced form.[2,3]

2.4 Nitrate Reduction to Ammonia in the Periplasm of Gram-Negative Bacteria

In *E. coli* and related bacteria, the periplasmic pathway for nitrate reduction involves two enzyme complexes: Nap (for nitrate reduction in the periplasm) and Nrf (for nitrite reduction by formate). Electrons for nitrate reduction are transferred by a variety of electron donors to quinones, which in turn transfer them *via* enzyme-specific quinol dehydrogenases to nitrate or nitrite. The six electrons donated to the periplasmic Nrf complex originate from the menaquinol rather than the ubiquinol pool. As several different electron donors other than formate can also reduce menaquinone, compounds other than formate, especially NADH, can also serve as electron donors for periplasmic nitrite reduction.

The catalytic subunit of the periplasmic nitrate reductase NapA is a molybdoprotein that is associated in many bacteria with a mono-haem *c*-type cytochrome, NapB.[50–52] The *E. coli* NapAB complex receives electrons from two sources: a tetra-haem *c*-type cytochrome, NapC,[53,54] and a membrane-associated quinol dehydrogenase, NapGH.[55] Many variations in subunit composition and electron transfer pathways are found in unrelated bacteria: increasing diversity is being revealed with the increasing number of completely sequenced bacteria genomes. Some of the most striking examples of this variation between genera include the detection of a nitrate reductase that is closely related to NarG with an active site facing the periplasm rather than the cytoplasm;[25] wide variations in electron transfer components amongst different *Shewanella* species;[26] and the range of effects of oxygen on *nap* gene expression.[15]

Less variation has so far been found in the periplasmic nitrite reductases, which are penta-haem *c*-type cytochromes designated NrfA. Electrons are delivered to NrfA from the menaquinol pool *via* the quinol dehydrogenase, a second penta-haem cytochrome, NrfB, or, in *Wolinella succinogenes*, by the *c*-type cytochrome, NrfH.[56]

Although no proton motive force is generated during electron transfer from the quinol pools to NapA or NrfA, periplasmic nitrate and nitrite reduction can still be described as respiratory processes when energy is conserved during electron transfer from the physiological electron donor, especially formate and NADH, to the quinone pool.[57]

2.5 Regulation of Nitrate Reduction to Ammonia by Enteric Bacteria

Intensive studies of the regulation of the transcription of the genes involved in nitrate reduction have revealed a totally unpredictable level of complexity. Superficially, there are two main levels of regulation: one by oxygen and the other by nitrate (see Figure 2.2). None of the four main enzymes, NarG, NapA, NrfAB or NirBD, is synthesised during oxygen-sufficient growth. This is because their expression is totally dependent upon a functional FNR protein,[58,59] which contains an oxygen-sensitive [4Fe–4S] iron–sulphur cluster. Oxygen degrades this cluster, converting the active dimeric protein into inactive monomers.[60–62] The rare reports that aerobic cultures of *E. coli* can assimilate nitrate were almost certainly based upon cultures in which the biomass density was sufficient to ensure that oxygen was consumed as rapidly as it was supplied, thus allowing the FNR protein to remain at least partially active.[63] However, such dogmatic statements must be viewed with caution, because single-base changes in the *fnr* gene can result in amino acid substitutions such as S24F and L28H that stabilise the [4Fe–4S] iron–sulphur centre from degradation by oxygen.[62,64,65] There is also evidence that FNR in *Neisseria meningitidis* is more oxygen stable than its *Neisseria gonorrhoeae* counterpart.[66] This correlates with the different exposure levels of these pathogenic bacteria to oxygen in their different locations within the human body.

Nitrate and, to a lesser extent, nitrite also regulate gene expression. There are two environmental nitrate sensors, NarX and NarQ, and two response regulators, NarL and NarP, which at least in part are interchangeable.[67,68] However, while NarX preferentially phosphorylates NarL, NarQ is equally able to phosphorylate both NarL and NarP.[69] Both NarX and NarQ are extremely sensitive to nitrate, but less responsive to nitrite.[70,71] Nitrate activates the kinase activity of NarX and NarQ and the phosphate groups are transferred to NarL and NarP, which in turn regulate gene expression. Phosphorylated NarL activates transcription at the NarG and NirB promoters. However, there are two binding sites for phosphorylated NarL in the *nap* and *nrf* regulatory regions (see Figure 2.3).[72] Low levels of phosphorylated NarL activate transcription at both promoters, but high levels of NarL-P bind to the lower-affinity sites to repress transcription. This complexity provides the mechanism by which the periplasmic pathway for nitrate reduction to ammonia is preferentially

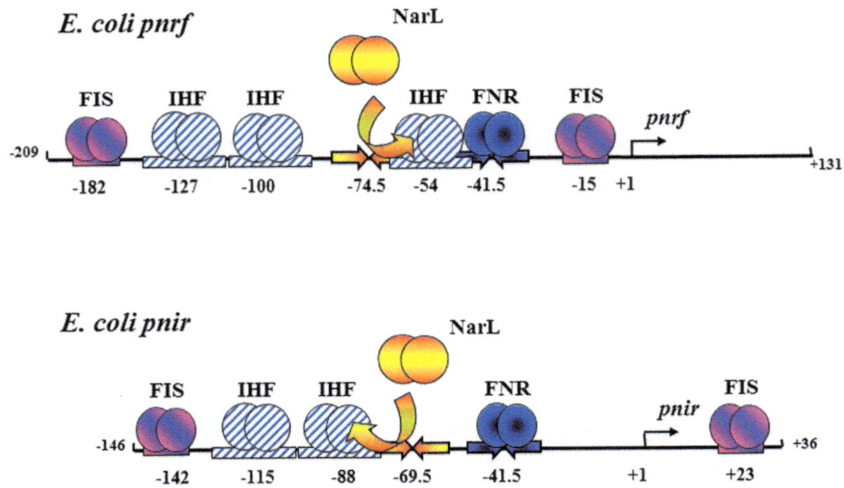

Figure 2.3 Transcription control of the genes for the *E. coli* periplasmic and cytoplasmic nitrite reductases, NrfAB and NirBD. Only some of the major factors that regulate gene expression are shown. For a fuller account, readers are directed to the EcoCyc website. For simplicity, binding sites are not shown for the nitric oxide-sensitive repressor NsrR or for NarP, the response regulator of the alternative two-component nitrate sensing system. The figure was prepared by D. Browning based upon his published data.[75–77,79] FIS: the nucleoid protein "factor for inversion stimulation"; IHF: integration host factor; FNR: the regulator of fumarate and nitrate reduction; NarL: the response regulator of the NarX–NarL two-component regulatory system.

expressed when only a limited concentration of nitrate is available, but the cytoplasmic pathway dominates when nitrate is abundant.[33,34] To add to this complexity, nitrate activates the phosphorylation of NarX, but nitrite promotes its dephosphorylation.[73] Thus, nitrate reduction by NarG in the cytoplasm tends to be suppressed when nitrite accumulates, but this effect is not sufficient to prevent the quantitative conversion of nitrate to nitrite, which can accumulate to extremely high levels.

Further layers of gene regulation are provided by proteins that various authors regard either as global transcription regulators or chromosomal structural proteins. The first of these to be documented was the "integration host factor", IHF, which was shown to bend DNA upstream of *narG* to facilitate the formation of an effective transcription activation complex between NarL, FNR and RNA polymerase.[74] Subsequently, IHF was found to insert similar bends in the *nirB* and *nrfA* regulatory regions. At the *nirB* promoter, IHF binds to two sites: an activation site centred 88 bases upstream of the transcription start and an activation site located 115 bases upstream (referred to as positions −8 and −115, respectively).[75,76] There is also a binding site for repression by FIS (the "factor for inversion stimulation") at position −145 (see Figure 2.3, lower panel). At the *nrf* promoter, IHF and

FIS both repress transcription, with IHF binding at a site centred 54 bases upstream of the transcription start site (see Figure 2.3, upper panel).[77] Repression by IHF is relieved when phosphorylated NarL or NarP bind to a site centred 71 bases upstream of the start of transcription, the opposite of the situation at the *narG* regulatory region. The binding site for FIS 15 bases upstream of the transcription start interferes with RNA polymerase binding. FIS is responsible for the catabolite repression of *nrf* expression in rich media, especially when glucose is available. In contrast, the catabolite repressor–activator protein, Cra, mediates catabolite repression at the *nirB* promoter.[78] The *nrf* promoter is also repressed by the NO-sensing repressor, NsrR, which binds to a site that overlaps the IHF binding site.[79] Repression is relieved under conditions of nitrosative stress due to the accumulation of nitrite. This requires a complex restructuring of the regulatory region, with some of the transcription factors effectively flipping their roles (see Figure 2.4). The resulting increase in nitrite reductase activity accelerates nitrite clearance and hence results in a decrease in NO generation. This complexity

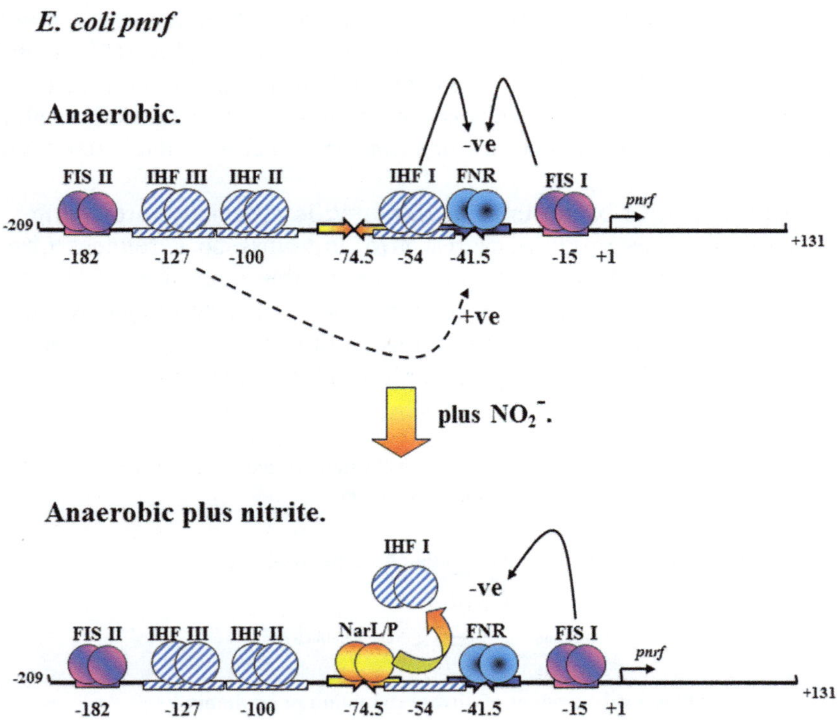

Figure 2.4 Restructuring of the regulatory region of the *nrfA* nitrite reductase operon in response to the availability of nitrite. The figure illustrates how the nucleoprotein complex at the promoter for the periplasmic nitrite reductase operon is restructured in response to the availability of nitrite. It was prepared by D. Browning based upon his published data.[75–77,79]

explains how various metabolic signals can be integrated by bacteria to regulate gene expression. It invites the question of whether for every transcription activator there is a counteracting transcription repressor.

2.6 "Denitrification" by Enteric Bacteria that Mainly Reduce Nitrate to Ammonia

Genes for NirK and NirS are absent from bacteria that reduce nitrite to ammonia. Unlike denitrifying bacteria, they are unable to convert nitrite quantitatively to NO. It was therefore significant when three laboratories independently reported partial denitrification by *E. coli* and other species, including soil bacteria that reduce nitrate to ammonia.[80–83] These reports originated in the observation that small quantities of nitrous oxide are released during anaerobic nitrite reduction to ammonia. Smith then demonstrated that the key step in this process—the reduction of nitrite to NO—is catalysed mainly, but not exclusively, by the membrane-associated nitrate reductase, NarG (see Figure 2.5).[84] Many subsequent studies confirmed the importance of NarG in the generation of NO from nitrite.[85–87] Nevertheless, there have been several reports that some NO is generated even in the absence of NarG (shown in brackets in Figure 2.5). Typically less than 1% of the nitrite reduced is converted *via* NO to N_2O,[82,84] though this can be increased substantially to above 20% under extreme growth conditions that bacteria are unlikely to encounter naturally.[88]

The Gram-positive bacterium *Bacillus vireti* is especially interesting.[89] It synthesises all three proteins, NarG, NrfA and NirB, so it reduces nitrate mainly to ammonia. It also generates considerable quantities of NO: 10% of the nitrite formed from nitrate reduction was converted to NO when the concentration of nitrate in the medium was limiting, but this increased to 50% when excess nitrate was supplied. Although *Bacillus vireti* lacks NirS and

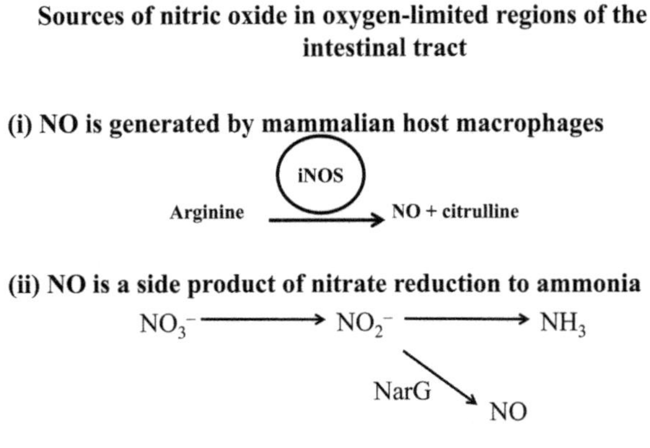

Figure 2.5 Sources of nitric oxide (NO) encountered by enteric bacteria in the lower gastro-intestinal tract.

NirK, it has a Cu-Nor NO reductase and a conventional nitrous oxide reductase, so any NO that is formed as a side product of nitrite reduction to ammonia is completely denitrified to N_2.

2.6.1 Nitrite Reductases Other than NirS and NirK Implicated in NO Formation

Various authors have demonstrated that the two nitrite reductases, Nir and Nrf, are also minor contributors to the generation of NO by *E. coli.* First it was shown that more NO is formed by a parent strain than a mutant that lacks the periplasmic nitrite reductase.[90] In contrast, Weiss concluded that the cytoplasmic nitrite reductase, NirBD, was the key player.[91] Their relative contributions remain unresolved primarily because of the reasons discussed in the previous three sections. The main reason for the apparent contradictions in the literature is that the various experiments were completed with different strains of bacteria, different growth conditions and in different media. With the complexity of the regulation of the enzymes involved (see Figures 2.3 and 2.4), it is not surprising that different groups have reached contrasting conclusions concerning the relative importance of different pathways for the conversion of nitrite to NO. It is clear, however, at least for *S. enterica* serovar Typhimurium, that the periplasmic nitrate reductase NapA is not a major source of NO.[88] Maximum conversion of nitrite to NO occurs when nitrate is more abundant than the supply of electrons required to reduce it to ammonia.

Expression of *narGHJI*, *nrfA-G* and *nirBDC* is regulated by many different transcription factors, each of which responds to different environmental signals. As briefly discussed above, the relevant environmental signals include not only the availability of nitrate or nitrite, but also their relative concentrations, the availability of a readily utilisable carbon source and the richness of the growth medium, all of which can be critical in determining the relative levels of expression of the *narG*, *nirB* and *nrfA* operons. A major problem in assessing how bacteria respond to nitrosative stress in that most of the experiments described in the literature refer to pure cultures of model organisms grown under severe conditions that are never found in natural environments. Such studies are useful because they show what is possible, but they are limited because they tell us little about how microbes living in mixed communities compete successfully to survive under conditions of stress. Is any natural environment similar to life in a conical flask or bench top fermenter containing Luria broth supplemented with 100 mM nitrate? Is any natural environment similar to a solution of minimal salts such as M9 medium prepared from Analar-grade purified components? Many publications (including some from the laboratory of the author!) are unable to distinguish between chemical and biochemical reactions. This problem has been highlighted in one of the most perceptive reviews of how the bacterial responses to nitrosative stress are regulated.[92]

2.7 Defence Against NO Toxicity in Bacteria that Reduce Nitrate to Ammonia

Although NarG is an insoluble cytoplasmic membrane protein, its active site for nitrate reduction is located on the cytoplasmic side of the membrane. Thus, it generates NO in the cytoplasm that, unless it is removed, will damage many types of cell component. Under anaerobic conditions, NO nitrosylates metal ions, especially iron atoms of iron–sulphur proteins and cytochromes. The [4Fe–4S] iron–sulphur centres of dehydratases are exposed and especially vulnerable to nitrosylation, so if NO accumulates, growth stops rapidly because enzymes such as aconitase, fumarase and the dihydroxyacid dehydratase IlvD are unable to function.[89] During anaerobic growth, the citric acid cycle cannot function as an energy-generating pathway and is blocked by repression of the genes required for 2-oxo-glutarate dehydrogenase synthesis. Nevertheless, the iron–sulphur protein aconitase B is required for the synthesis of glutamate and the amino acids derived from glutamate, such as glutamine, proline and tryptophan. The three *E. coli* fumarate hydratases also fulfil different roles during aerobic and anaerobic growth. The FumA and FumB isoenzymes both contain exposed [4Fe–4S] iron–sulphur centres that are exposed to solvent access.[93,94] They are closely related. One isoenzyme, FumA, is synthesised during aerobic growth in order to generate malate in the oxidative citric acid cycle, while FumB is synthesised anaerobically in order to convert malate to fumarate, which is then available for electrogenic reduction to succinate. Consequently, central metabolism is a critical target for nitrosative damage and bacteria have developed mechanisms to minimise damage, or to reverse it when it occurs.

During anaerobic growth, many types of damage are secondary consequences that first require the binding of NO to metal ions. *Trans*-nitrosylation of –SH groups of cysteine, peptides such as glutathione and proteins is metal dependent,[95] so under conditions of severe nitrosative stress, multiple pathways are required in order to protect against such damage and to repair it as rapidly as it occurs. It is well established that bacterial flavohaemoglobins such as Hmp catalyse rapid oxidation of NO to nitrate,[96–98] but in the absence of oxygen, such oxygen-dependent repair is unavailable. One or both of two alternative strategies must therefore be exploited: the NO can either be reduced to a harmless product; or it can be sequestered by an NO sink. However, only the first of these strategies completely removes the problem, as an NO sink (*e.g.*, glutathione) can release NO back into the cytoplasm.[99] The problem is delayed, but not resolved. Nevertheless, the sequestering strategy is well documented for some bacteria. A striking example is *Vibrio cholerae*, which relies almost entirely on Hmp to remove NO by oxidation to nitrate.[100] In the absence of the co-substrate oxygen, *Vibrio cholerae* was shown to be especially sensitive to NO, and iron–sulphur proteins of the dehydratase family were the prime targets for inactivation. A protein called NnrS is essential for some protection against nitrosative stress during anaerobic growth. NnrS is not a NO reductase, but apparently sequesters NO in order to minimise nitrosative damage. A similar role is played by cytochrome *c*′ in pathogenic *Neisseria*.[101,102]

As NO is far more stable in the absence of oxygen than in its presence, anaerobic cultures have developed a series of mechanisms to protect themselves against NO toxicity, or to repair the damage caused by NO accumulation. However, their relative importance has been a long-running source of controversy.

2.7.1 Three Pathways for NO Reduction by Enteric Bacteria

Three proteins synthesised by bacteria that reduce nitrate to ammonia have been reported to catalyse NO reduction, but the physiological relevance of these activities remains controversial.

First, Hmp can both oxidise NO and reduce it to N_2O.[103] Although Hmp is a cytoplasmic protein and therefore is correctly located to protect cytoplasmic proteins from inactivation, the rate of NO reduction is extremely slow, with a turnover number of only 0.02 s^{-1}. This led to the conclusion that Hmp does not provide physiologically relevant protection to anaerobic cultures.[104] This conclusion fully explains why transcription of the *hmp* gene is repressed during anaerobic growth by the transcription factor, FNR.[105,106] Evidence against an alternative possibility—that FNR is itself sensitive to NO inactivation and therefore can function as a NO-sensitive repressor molecule—will be presented later in this review.

The second NO reductase to be discovered was the flavorubredoxin, NorV.[107] In various bacteria, NorV is encoded by an operon that includes an NADH-dependent NorV reductase, the flavoprotein NorW. In *E. coli* and *S. enterica*, expression of *norVW* requires an RpoN-dependent transcription factor, the enhancer binding protein NorR.[108] In the absence of NO, NorR cannot bind to the *norVW* regulatory region, so the σ^{54} form of RNA polymerase cannot form an effective transcription complex. The regulatory domain of NorR contains a mononuclear non-heam iron centre, which reversibly binds NO.[109] Binding of NO stimulates the ATPase activity of NorR, which now binds to its DNA target, enabling the activation of transcription by RNA polymerase. As the expression of the Nor system is not regulated by FNR or by other oxygen-responsive transcription factors, NorVW can provide protection during both aerobic or anaerobic growth.

The third NO reductase in enteric bacteria is the periplasmic nitrite reductase, NrfAB, which can reduce NO to ammonia.[110] Regulation of transcription of the *nrf* operon is extremely complex, but one of the many transcription factors involved is NsrR, which is a NO-sensitive repressor of the Rrf2 family of repressor molecules. The sensor domain of NsrR is an iron–sulphur centre, variously characterised as a [2Fe–2S][111] or a [4Fe–4S] centre[112] that binds NO with high affinity. Despite a high V_{max} for NO reduction by the Nrf system, the K_m for NO is also very high.[113] Furthermore, its location precludes a role in the protection of cytoplasmic proteins from nitrosative damage. This led to the proposal that NrfAB provides a first line of defence against exogenous NO from external sources.[113] As NO is a freely diffusible gas that can equilibrate across membranes, a NO reductase located in the periplasm might be able to decrease the concentration of NO that is available to cross into the cytoplasm, where NorVW would "mop up" residual NO.[113]

The concentrations of NO used in all of the previous studies of bacterial NO reduction have been extremely high—far higher than those reported to accumulate in natural environments. The concentration of NO that accumulates in the bacterial cytoplasm during nitrate and nitrite reduction has not been reported, but it is likely to be far lower than the 10–200 μM of NO used in various laboratory experiments.[103,105,110,114,115] If the concentration that is sufficient to cause nitrosative stress is below 1 μM, neither Hmp nor NrfAB would provide effective protection. Furthermore, it has been shown that NsrR responds to far lower concentrations of NO than either NorR or NsrR.[116] This raises the fascinating question of whether sufficient NO accumulates to activate NorR, which would be essential for NorVW to play any role in protection. The alternative possibility is that NsrR-regulated genes of unknown function might provide a previously undiscovered but highly sensitive protection mechanism. This possibility will be discussed in the next section. If correct, the role of NorVW and NrfAB would be to provide auxiliary defence against severe NO attack. In contrast, a currently unidentified NsrR-regulated system would provide protection under normal conditions of nitrate and nitrite reduction. Note, however, that the NO dissociation constant for NorR was reported to be extremely low (50 ± 10 nM), being well within the expected physiologically relevant range.

This raises the previously unanswered question: what concentration of NO accumulates in the bacterial cytoplasm during nitrite reduction? No answer to this question has been reported, but it is known that only a minority of nitrite is reduced to NO rather than to ammonia, and the residual rate of NO reduction by strains that lack both of the nitrite reductases is less than 1% of that of the parent, in the order of 2 nmol min^{-1} (mg of dry cell mass)$^{-1}$. If one assumes that NO is reduced at the same rate as it is formed from nitrite by an enzyme with a K_m of 100 nM and a V_{max} of only 10 μmol min^{-1} (mg of dry cell mass)$^{-1}$, the cytoplasmic NO concentration would be less than 25 nM. This [NO] would be greater than 10-fold lower if the V_{max} for the hypothetical enzyme is 100 μmol min^{-1} (mg of dry cell mass)$^{-1}$. This calculation, based upon Michaelis–Menton kinetics, demonstrates that the physiological relevance of virtually all kinetic data for NO reduction in the literature should be viewed with caution.

2.7.2 Is There a High-Affinity But Previously Undiscovered NO Reductase in Anaerobic Bacteria?

Claire Vine in the author's laboratory constructed a quadruple mutant of *E. coli* K-12 that lacks both of the nitrite reductases, NirBD and NrfAB, as well as Hmp and NorV.[87,114,117] This strain not only lacks the three previously documented NO reductase activities, but also lacks the ability to reduce nitrite formed from nitrate reduction to ammonia. Although this mutant was far more sensitive to nitrosative stress than the parent strain, it was still moderately resistant to NO and retained the ability to reduce it. This strongly suggested that enteric bacteria synthesise another NO reductase that so far has evaded discovery. Several laboratories have identified a possible candidate for such an activity. This is the hybrid cluster protein, Hcp.

The name "hybrid cluster protein" originated from its unique structure. Although Hcp contains two iron–sulphur clusters, one of them is neither a conventional [2Fe–2S] nor a [4Fe–4S] cluster, but a hybrid [4Fe–2S–2O] cluster.[118–120] Unique structure implies an unique function, but until recently, the biochemical function of Hcp remained unresolved. NO is far more stable in the absence of oxygen than in its presence, and a series of reports has implicated Hcp in protecting anaerobic bacteria against nitrosative stress. Transcriptomic studies have revealed that *hcp* expression is strongly up-regulated in various bacteria under conditions of nitrosative stress.[121–124] Two copies of the *hcp* gene are found in the strict anaerobe *Desulfovibrio vulgaris* (Hildenborough strain) and in some other *Desulfovibrio* species. Inactivation of the *hcp2* gene resulted in a strain that was far more sensitive than its parent to nitrosative stress from NO generated by macrophages in response to infection.[125] An *S. enterica hcp* mutant generated less N_2O than its parent and was also sensitive to nitrosative stress.[88] Similar results have been reported for *Porphyromonas gingivalis*.[126] However, three other attempts to use *hcp* mutants to define a function were unproductive,[88,116,127] and concerted attempts to demonstrate NO reduction by purified Hcp to N_2O were unsuccessful.[120]

In enteric bacteria, transcription at the *hcp* promoter is totally dependent upon a functional FNR protein. Consequently, the *hcp–hcr* operon is expressed only during anaerobic or oxygen-limited growth. Transcription is also repressed by NsrR,[121,128–131] and it was shown that NsrR responds to much lower concentrations of NO than NorV, both of which are far lower than the concentrations that are sufficient to damage the iron–sulphur centre of FNR.[116] If, as discussed above, the concentration of NO in the bacterial cytoplasm is always very low, it would be logical for an NsrR-regulated gene to encode a protection mechanism that is capable of functioning under physiological conditions. This led us to investigate whether Hcp and its reductase, Hcr, constitute a high-affinity but low-activity NO reduction pathway. As Hcp is an iron–sulphur protein, we also were aware that its catalytic activity might be extremely unstable, and that special conditions might be required to retain this activity during purification. These properties—high affinity but low activity and sensitivity to substrate inhibition—would explain why laboratory experiments with concentrations of NO greatly in excess of those encountered naturally have failed, because any such Hcp-dependent activity with both a low K_m and a low V_{max} would have been masked by the lower affinities but higher activities of NorVW and NrfAB (and possibly even by the NO reductase activity of Hmp). Demonstration of such an activity would then require the construction of a strain that was deleted for all of the competing activities. Furthermore, many iron–sulphur proteins are unstable when exposed to air.

This hypothesis was recently confirmed by demonstrating that the *E. coli* Hcp is a high-affinity NO reductase that is the major enzyme for reducing NO stoichiometrically to N_2O under physiologically relevant conditions.[132] Deletion of *hcp* results in extreme sensitivity to NO during anaerobic growth and inactivation of the iron–sulphur proteins aconitase and fumarase by accumulated cytoplasmic NO. The residual NO reductase activity of the quadruple mutant mentioned above was lost upon deletion of the *hcp* gene, and during anaerobic growth, the resulting strain was extremely sensitive to both nitrite and NO. Site-directed mutagenesis confirmed that the hybrid cluster is essential for this NO reductase activity and for protection against nitrosative stress. The second gene of the *hcp* operon, *hcr*, codes for an NADH-dependent Hcp reductase, Hcr. First we showed that Hcp interacts tightly with Hcr both *in vivo* and *in vitro*. A surprising result was found when the *hcr* gene was deleted. Although Hcp and Hcr purified individually were unable to reduce NO either alone or when recombined, a co-purified preparation reduced NO *in vitro* with a K_m for NO of 500 nM. In an *hcr* mutant, Hcp was inactivated by NO concentrations above 100 nM, indicating that Hcr protects Hcp from nitrosylation by its substrate, NO. In summary, Hcp is indeed a high-affinity NO reductase that, physiologically, is the most effective enzyme for protecting bacteria from nitrosative stress in an anaerobic environment. The role of Hcp is especially important when small quantities of NO are generated in the bacterial cytoplasm during nitrate reduction catalysed by the cytoplasmic nitrate reductase, NarG.

2.8 Controversial Claims that the Transcription Factors FNR, OxyR and Fur are Physiologically Relevant NO Sensors

There has been a series of reports showing that transcription factors whose primary roles are to enable bacteria respond to oxygen, oxidative stress or iron limitation can also function as NO sensors. These reports were immediately followed by the question: are they physiologically relevant NO sensors? Examples from enteric bacteria include the oxygen-responsive FNR, the peroxide-responsive OxyR and the iron-sensitive repressor Fur. These reports have two things in common: first, each of the transcription factors is an iron-containing protein; and second, the concentrations of NO used in the experiments were far higher (in some cases up to three orders of magnitude higher) than those encountered by bacteria in their natural environments. A further complication is that some of these reports were published before the discovery of highly sensitive NO sensors such as NsrR and NorR. This topic was thoroughly reviewed by Spiro,[92] who emphasised the need to distinguish between physiologically relevant environmental sensing and secondary consequences of chemical damage, especially to iron-containing proteins.

The claim that FNR can function as a NO sensor arose from experiments in which anaerobic cultures of *E. coli* were exposed to 10–25 μM of NO, which resulted in a two-fold induction of an FNR-activated promoter or a two-fold repression of a FNR-repressed promoter.[90,106] These effects should be compared with the 10-fold increase in the activity of these promoters when the *nsrR* gene is deleted, and are comparable to the NsrR-dependent induction during anaerobic growth in the presence of nitrite.[114,116] High-level expression from a FNR-activated promoter in the absence of NsrR clearly proves that NsrR, rather than FNR, is the critical factor in regulating levels of gene expression in response to NO. Similarly, NO binds to ferrous ions and can therefore simulate iron starvation, resulting in increased transcription of Fur-repressed genes.[128] Both *S. enterica* Fur and *Bacillus subtilis* PerR have been reported to respond to NO,[126,129,130] but the effects are almost certainly due to secondary consequences of iron starvation in rich media or the effects of NO on other transcription factors that are much more sensitive to NO.

Reports that OxyR can serve as a NO sensor have not been substantiated independently, but were based upon the observation that high levels of NO lead to *S*-nitrosylation of OxyR.[124] However, virtually all (but not quite all!) of the data reported can be explained as indirect effects of chemical damage to FNR in a strain that is deficient in protection from reactive oxygen species.[118] Most laboratories grow inocula for experiments aerobically under conditions in which FNR is inactive and reactive oxygen species are abundant. The iron–sulphur centre of FNR would not be incorporated to yield the active dimer until all reactive oxygen species are removed, and hence anaerobic growth on nitrate will be prevented or delayed. Failure to activate protection mechanisms that repair nitrosylation damage to metalloproteins would lead to increased *trans*-nitrosylation of –SH groups, including the cysteines at the active site of OxyR. In support of this explanation, we have shown that an *oxyR* mutation has no effect on adaptation to nitrate-dependent growth when an anaerobic inoculum is used to start the culture, and once growth has commenced, there is no loss of NO-dependent induction of a NO response in the *oxyR* mutant compared with the parent strain. In summary, it is highly unlikely that S-nitrosylated OxyR is a physiologically relevant transcription activator of genes that provide protection against nitrosative stress.[125]

2.9 Regulation of Gene Expression Under Conditions of Nitrosative Stress

Whole-genome transcriptomic analyses of enteric bacteria have consistently identified three genes that are strongly induced by nitrosative stress during anaerobic growth. These are *hmp*, encoding a bacterial flavohaemoglobin, Hmp; *ytfE*, which encodes a protein that is implicated in the repair of damaged iron, and especially iron–sulphur, centres; and *hcp*, the gene for the enigmatic hybrid cluster protein. In enteric bacteria, these genes are regulated by NsrR, a transcription factor of the Rrf2 family that functions mainly as a repressor in the absence of NO.[115,121–123,128] NsrR has a high affinity for NO, which inactivates its repressor function and therefore de-represses *hmp*, *ytfE* and *hcp–hcr* expression.[111] A range of transcription factors regulates the response to nitrosative stress in other bacteria, many of which are related by the *E. coli* Crp–FNR family of proteins.[131] For example, in *Porphyromonas gingivalis*, HcpR regulates the synthesis of the hybrid cluster protein Hcp, and HcpR is essential for survival in mammalian host cells.[127]

The biochemical function of Hmp as an oxygen-dependent NO oxygenase is well established. In contrast, the physiological significance of a catalytically less efficient NO reduction function is more controversial. Corker and Poole[90] argued that because *hmp* expression is repressed by FNR and FNR contains a [4Fe–4S] iron–sulphur centre that is damaged by severe exposure to NO, Hmp protects bacteria against the NO damage generated during anaerobic growth. In contrast, Gardner and Gardner[104] showed that the catalytic efficiency of Hmp as a NO reductase is extremely low, with a k_{cat} of only 20 s^{-1} and a relatively high K_m for NO.

The third protein that is induced by nitrosative stress, YtfE, has been extensively implicated in the repair of the iron–sulphur centres of proteins damaged by nitrosation.[133–137] A series of other microarray studies with *Staphylococcus aureus*,[138] *Neisseria gonorrhoeae*[139] and *Neisseria meningitidis*[140] all identified YtfE or its homologues as proteins that are synthesised in response to nitrosative stress. The protein from *E. coli* was purified and shown to be a di-iron protein that can itself be nitrosylated.[134,141] Furthermore, the purified protein was shown to restore the activity of an iron–sulphur protein, fumarase, which has been inactivated by nitrosylation. In more recent studies, it has been shown that YtfE releases iron that can be incorporated into other iron-deficient proteins. This led the Saraiva group to conclude that the likely function of YtfE is to store iron and donate it to iron-deficient iron–sulphur centres damaged by either oxidative or nitrosative stress.[134]

Why, then, is there any remaining doubt about the primary role of YtfE? There are at least five reasons for this. First, it is conceptually difficult to imagine how a single purified protein can catalyse both the release of a nitrosylated iron atom and then replace it with a functional iron atom. Second, the rate of iron transfer from YtfE to other proteins is extremely slow,[136] being less than 0.01 s^{-1}. This rate would seem to be inadequate to repair damage sufficiently rapidly to be physiologically useful. Third, the concentrations of both the iron-donating YtfE and the acceptor protein used in these experiments were extremely high, suggesting that the observed iron transfer might simply be due to the slow dissociation of iron and its adsorption by the "accepting" molecule;[134,136] chemistry, yes, but not biochemistry! Fourth, the Fe–S centres of proteins vary greatly in both structure and accessibility, so it is difficult to envisage how YtfE can recognise such a very wide range of target molecules. Finally, as stated in the next section, there might be a more plausible role for YtfE that has yet to be tested, namely that it catalyses the simple release of NO, or even the NO–Fe complex, from the damaged protein. This would imply that YtfE is an enzyme that releases NO into the bacterial cytoplasm.

2.10 Repair of Nitrosylation Damage by the YtfE Protein, also Known as RIC

The first reports that *ytfE* expression is induced by NO included experiments with a *ytfE* deletion mutant and the complementation of some of its defects by a *ytfE*$^+$ plasmid. The mutant showed a range of severe growth defects in anaerobic media.[115] It was also more sensitive than the parent to NO. Subsequently, it was realised that there had been an unlinked, 124-gene deletion during the construction of the mutant.[142] The deletion included genes for both molybdopterin guanine dinucleotide synthesis and the *hcp–hcr* operon.[27] Molybdoproteins are required for nitrate reduction and formate oxidation, as well as for the reduction of TMAO and DMSO, amongst others. This fully explained the anaerobic growth defects of the mutant. As stated above, the hybrid cluster protein Hcp plays a critical role in protecting bacteria against nitrosative stress, especially during anaerobic growth. This led us to conclude that the precise role of YtfE remains, at least in part, controversial, and the repair mechanism is unknown.

The proposal that YtfE is an enzyme that catalyses the release of NO from cytoplasmic proteins damaged by nitrosylation originated in the observation that a *ytfE* mutation significantly suppressed the growth defects of a strain that was defective in other mechanisms for reducing cytoplasmic NO. Critical to this discovery was our recent demonstration that the hybrid cluster protein Hcp is a high-affinity NO reductase that is essential for the anaerobic growth of strains that lack NorVW, NrfAB or Hmp under conditions of nitrosative stress. A further mutation in *ytfE* partially suppressed this growth defect, suggesting that YtfE is at least partially responsible for the release of NO into the cytoplasm. Transformation with a *ytfE*$^+$ expression plasmid restored extreme sensitivity to nitrosative stress, confirming that YtfE was directly responsible for the observed growth inhibition. Independent evidence for this proposal came from the demonstration that the growth inhibition of the *hcp* mutant was reversed by the addition of PTIO, which stoichiometrically degrades NO as it is formed. Deletion of *narG* results in the production of far less cytoplasmic NO from nitrite than an isogenic *narG*$^+$ strain, and also suppressed the growth defects of the Hcp$^-$ YtfE$^+$ strain in the presence of nitrate or nitrite. Finally, less de-repression of *hcp::lacZ* transcription was found in a *ytfE* mutant than in the parent strain, again indicating that YtfE contributes significantly to the release of cytoplasmic NO.

These combined observations led us to propose the model for how very low concentrations of NO released into the *E. coli* cytoplasm are detoxified (see Figure 2.6). The model proposes that NO binds rapidly to metalloproteins such as aconitase, fumarase, IlvD and probably many others, inactivating their metabolic functions. I suggest that the function of YtfE is to reverse this initial damage, allowing bacterial growth to continue under conditions of mild nitrosative stress. According to this model, the role of Hcp is to reduce the resulting very low concentration of NO as rapidly as it is formed

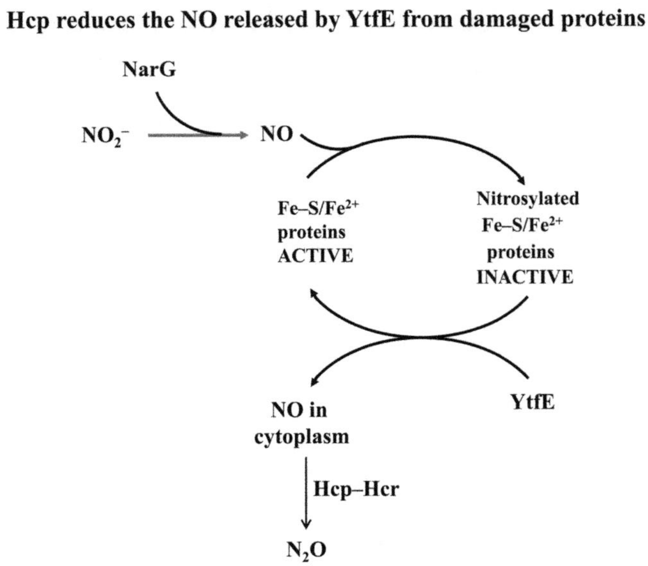

Figure 2.6 Hypothetical scheme showing a possible mechanism for the reactivation of Fe–S proteins damaged by nitrosative stress. According to this hypothesis, the function of YtfE (for repair of iron centres and also known as RIC) is to recycle the nitric oxide (NO) trapped by the nitrosylation of key target proteins. This generates NO in the cytoplasm, which must then be scavenged by the high-affinity NO reductase, Hcp.

(see Figure 2.6). In summary, I suggest that YtfE and Hcp coordinate the repair of nitrosative damage and prevent NO accumulation. In the absence of YtfE, secondary reactions, such as *trans*-nitrosylation of the –SH groups of proteins and peptides such as glutathione, would result in damage that is more difficult to repair.

Acknowledgements

The author is grateful to Douglas Browning for preparing Figures 2.3 and 2.4 of this chapter; to David Richardson, Gary Rowley and Ligia Saraiva for sharing data in advance of publication; to Ligia Saraiva for gifts of plasmids and sharing protocols; and to Robert Poole, Mariana Tinajero Trejo, Lars Bakken and Linda Bergaust for collaboration with the hybrid cluster protein project.

References

1. J. S. Knapp and V. L. Clark, *Infect. Immunol.*, 1984, **46**, 176.
2. L. Hart, A. D. Larson and C. S. McCleskey, *J. Bacteriol.*, 1965, **89**, 1104.
3. J. A. Cole, *FEMS Microbiol. Lett.*, 1978, **4**, 327.
4. J. A. Cole and C. M. Brown, *FEMS Microbiol. Lett.*, 1980, **7**, 65.

5. P. B. Garland, A. J. Downie and B. A. Haddock, *Biochem. J.*, 1975, **152**, 547.
6. P. John, *J. Gen. Microbiol.*, 1978, **98**, 231.
7. R. W. Jones, W. J. Ingeldew, A. Graham and P. B. Garland, *Biochem. Soc. Trans.*, 1978, **6**, 1287.
8. J. K. Kristjansson and T. C. Hollocher, *J. Bacteriol.*, 1979, **137**, 1227.
9. P. R. Alefounder and S. J. Ferguson, *Biochem. J.*, 1980, **192**, 231.
10. H. W. van Verseveld, E. M. Meijer and A. H. Stouthamer, *Arch. Microbiol.*, 1977, **111**, 17.
11. E. M. Meijer, J. W. Van Der Zwaan, R. Wever and A. H. Stouthamer, *Eur. J. Biochem.*, 1979, **Y6**, 69.
12. F. Blasco, C. Iobbi, G. Giordano, M. Chippaux and V. Bonnefoy, *Mol. Gen. Genet.*, 1989, **218**, 249.
13. L. A. Robertson and J. G. Kuenen, *J. Gen. Microbiol.*, 1983, **129**, 2847.
14. A. G. McEwan, J. B. Jackson and S. J. Ferguson, *Arch. Microbiol.*, 1984, **137**, 344.
15. L. C. Bell, D. J. Richardson and S. J. Ferguson, *FEBS Lett.*, 1990, **265**, 85.
16. B. C. Berks, D. J. Richardson, C. Robinson, A. Reilly, R. T. Aplin and S. J. Ferguson, *Eur. J. Biochem.*, 1994, **220**, 117.
17. A. Mulder, A. A. van de Graaf, L. A. Robertson and J. G. Kuenen, *FEMS Microbiol. Ecol.*, 1995, **16**, 177.
18. M. S. M. Jetten, L. van Niftrik, M. Strous, B. Kartal, J. T. Keltjens and H. J. M. Op den Camp, *Crit. Rev. Biochem. Mol. Biol.*, 2009, **44**, 65.
19. K. F. Ettwig, M. K. Butler, D. Le Paslier, E. Pelletier, S. Mangenot, M. M. M. Kuypers, F. Schreiber, B. E. Dutilh, J. Zedelius, D. de Beer, J. Gloerich, H. J. C. T. Wessels, T. van Alen, F. Luesken, M. L. Wu, K. T. van de Pas-Schoonen, H. J. M. Op den Camp, E. M. Janssen-Megens, K.-J. Francoijs, H. Stunnenberg, J. Weissenbach, M. S. M. Jetten and M. Strous, *Nature*, 2010, **464**, 543.
20. A. Teske, E. Alm, A. Regan, S. Toze, B. E. Rittmann and D. A. Stahl, *J. Bacteriol.*, 1964, **176**, 6623.
21. M. A. H. J. van Kessel, D. R. Speth, M. Albertsen, P. H. Nielsen, H. J. M. Op den Camp, B. Kartal, M. S. M. Jetten and S. Lücker, *Nature*, 2015, **528**, 555.
22. H. Dalms, E. V. Lebedeva, P. Pjevac, P. Han, C. Herbold, M. Albertsen, N. Jehmlich, M. Palatinszky, J. Vierheilig, A. Bulaev, R. H. Kirkgaard, M. von Bergen, T. Rattei, B. Bendinger, P. H. Nielsen and M. Wagner, *Nature*, 2015, **528**, 504.
23. T. J. Goreau, W. A. Kaplan, S. C. Wofsy, M. B. McElroy, F. W. Valois and S. W. Watson, *Appl. Environ. Microbiol.*, 1980, **40**, 526.
24. F. Lipschultz, O. C. Zafiriou, S. C. Wofsy, M. D. McElroy, F. W. Valois and W. Watson, *Nature*, 1981, **294**, 641.
25. P. J. Martikainen, *Appl. Environ. Microbiol.*, 1985, **50**, 1519.
26. A. Hartsock and J. P. Shapleigh, *J. Bacteriol.*, 2011, **193**, 6483.
27. P. J. L. Simpson, D. J. Richardson and R. Codd, *Microbiology*, 2010, **156**, 302.

28. H. Neubauer and G. Götz, *J. Bacteriol.*, 1996, **178**, 2005.
29. H. U. Liepe, *Forum Mikrobiol.*, 1982, **5**, 10.
30. J. Simon, M. Sänger, S. C. Schuster and R. Gross, *Mol. Microbiol.*, 2003, **49**, 69.
31. D. A. Flanagan, L. G. Gregory, J. P. Carter, A. Karakas-Sen, D. J. Richardson and S. Spiro, *FEMS Microbiol. Lett.*, 1999, **177**, 263.
32. H. Felbeck, J. J. Childress and G. N. Somero, *Nature*, 1983, **293**, 291.
33. L. Page, L. Griffiths and J. A. Cole, *Arch. Microbiol.*, 1990, **154**, 349.
34. L. Potter, P. Millington, G. Thomas and J. Cole, *Biochem. J.*, 1999, **344**, 77.
35. R. M. Martinez-Espinosa, E. J. Dridge, M. J. Bonete, J. N. Butt, C. S. Butler, F. Sargent and D. J. Richardson, *FEMS Microbiol. Lett.*, 2007, **276**, 129.
36. W. G. Zumft, *Microbiol. Mol. Biol. Rev.*, 1997, **61**, 533.
37. G. T. Hoehn and V. L. Clark, *Infect. Immunol.*, 1992, **60**, 4695.
38. G. Braker, J. Zhou, L. Wu, A. H. Devol and J. M. Tiedje, *Appl. Environ. Microbiol.*, 2000, **66**, 2096.
39. K. Heylen, D. Gevers, B. Vanparys, L. Wittebolle, J. Geets, N. Boon and P. De Vos, *Environ. Microbiol.*, 2006, **8**, 2012.
40. A. D. Goddard, J. W. Moir, D. J. Richardson and S. J. Ferguson, *Mol. Microbiol.*, 2008, **70**, 667.
41. S. Clegg, Y. Feng, W. Jia and J. A. Cole, *Mol. Microbiol.*, 2002, **44**, 143.
42. W. Jia, N. Tovell, S. Clegg, M. Trimmer and J. Cole, *Biochem. J.*, 2009, **417**, 295.
43. H. Zheng, G. Wisedchaisri and T. Gonen, *Nature*, 2013, **497**, 647.
44. W. Lü, N. J. Schwarzer, J. Du, E. Gerbig-Smentek, S. L. A. Andrade and O. Einsle, *Proc. Natl. Acad. Sci. U. S. A.*, 2013, **109**, 18395.
45. M. Fukuda, H. Takeda, H. E. Kato, S. Doki, K. Ito, A. D. Maturana, R. Ishitani and O. Nureki, *Nat. Commun.*, 2015, **6**, 7097.
46. J. A. DeMoss and P.-Y. Hsu, *J. Bacteriol.*, 1991, **173**, 3303.
47. R. Jackson, A. Cornish-Bowden and J. A. Cole, *Biochem. J.*, 1981, **193**, 861.
48. H. Macdonald and J. A. Cole, *Mol. Gen. Genet.*, 1985, **200**, 328.
49. R. H. Jackson, J. A. Cole and A. Cornish-Bowden, *Biochem. J.*, 1981, **199**, 171.
50. B. C. Berks, D. J. Richardson, C. Robinson, A. Reilly, R. T. Aplin and S. J. Ferguson, *Eur. J. Biochem.*, 1994, **220**, 117.
51. B. C. Berks, S. J. Ferguson, J. W. Moir and D. J. Richardson, *Biochim. Biophys. Acta*, 1995, **1232**, 97.
52. D. J. Richardson, B. C. Berks, D. A. Russell, S. Spiro and C. J. Taylor, *Cell. Mol. Life Sci.*, 2001, **58**, 165.
53. F. Reyes, M. D. Roldán, W. Klipp, F. Castillo and C. Moreno-Vivián, *Mol. Microbiol.*, 1996, **19**, 1307.
54. L. C. Potter and J. A. Cole, *Biochem. J.*, 1999, **344**, 69.
55. T. H. C. Brondijk, A. Nilavongse, N. Filenko, D. J. Richardson and J. A. Cole, *Biochem. J.*, 2004, **379**, 47.
56. J. Simon, R. Gross, O. Einsle, P. M. Kroneck, A. Kröger and O. Klimmek, *Mol. Microbiol.*, 2000, **35**, 686.

57. D. Richardson and G. Sawers, *Science*, 2002, **295**, 1842.
58. T. Peakman, J. Crouzet, J. F. Mayaux, S. Busby, S. B. Mohan, J. Wootton, R. Nicolson and J. A. Cole, *Eur. J. Biochem.*, 1990, **191**, 315.
59. P.-S. Jayaraman, T. C. Peakman, S. J. W. Busby, R. V. Quincey and J. A. Cole, *J. Mol. Biol.*, 1987, **196**, 781.
60. N. Khoroshilova, C. Popescu, E. Munck, H. Beinert and P. J. Kiley, *Proc. Natl. Acad. Sci. U. S. A.*, 1997, **94**, 6087.
61. J. C. Crack, A. A. Gaskell, M. R. Cheeseman, N. E. Le Brun and A. J. Thomson, *J. Am. Chem. Soc.*, 2008, **130**, 1749.
62. A. Volbeda, C. Darnault, O. Renoux, Y. Nicolet and J. C. Fontecilia-Camps, *Sci. Adv.*, 2015, **1**, e1501086.
63. H. J. Brons and A. J. B. Zehnder, *Arch. Microbiol.*, 1990, **153**, 531.
64. A. J. Jervis, J. C. Crack, G. White, P. J. Artymiuk, M. R. Cheesman, A. J. Thomson, N. E. Le Brun and J. Green, *Proc. Natl. Acad. Sci. U. S. A.*, 2009, **106**, 4659.
65. D. M. Bates, C. V. Popescu, N. Khoroshilova, K. Vogt, H. Beinert, E. Münck and P. J. Kiley, *J. Biol. Chem.*, 2000, **275**, 6234.
66. J. D. Rock, M. J. Thomson, R. C. Read and J. W. Moir, *J. Bacteriol.*, 2007, **89**, 1138.
67. R. S. Rabin and V. Stewart, *J. Bacteriol.*, 1993, **175**, 3259.
68. R. Cavicchioli, I. Schröder, M. Constanti and R. P. Gunsalus, *J. Bacteriol.*, 1995, **177**, 2416.
69. C. E. Noriega, H.-Y. Lin, L.-L. Chen, S. B. Williams and V. Stewart, *Mol. Microbiol.*, 2010, **75**, 394.
70. H. Wang, C. P. Tseng and R. P. Gunsalus, *J. Bacteriol.*, 1999, **181**, 5303.
71. H. Wang and R. P. Gunsalus, *J. Bacteriol.*, 2000, **182**, 5813.
72. A. J. Darwin, K. L. Tyson, S. J. W. Busby and V. Stewart, *Mol. Microbiol.*, 1997, **25**, 583.
73. S. B. Williams and V. Stewart, *J. Bacteriol.*, 1997, **179**, 721.
74. X. Zhang and J. A. DeMoss, *J. Bacteriol.*, 1996, **178**, 3971.
75. D. F. Browning, J. A. Cole and S. J. W. Busby, *Mol. Microbiol.*, 2000, **37**, 1258.
76. D. F. Browning, J. A. Cole and S. J. W. Busby, *J. Bacteriol.*, 2008, **190**, 7258.
77. D. F. Browning, D. C. Grainger, C. M. Beatty, A. J. Wolfe, J. A. Cole and S. J. W. Busby, *Mol. Microbiol.*, 2005, **57**, 496.
78. K. Tyson, S. Busby and J. Cole, *Arch. Microbiol.*, 1997, **168**, 240.
79. D. F. Browning, D. J. Lee, S. Spiro and S. J. W. Busby, *J. Bacteriol.*, 2010, **192**, 3824.
80. M. S. Smith and K. Zimmerman, *Soil Sci. Soc. Am. J.*, 1981, **45**, 865.
81. T. Satoh, S. S. M. Hom and K. T. Shanmugam, in *Genetic engineering of symbiotic nitrogen fixation and conservation of fixed nitrogen*, ed. J. H. Lyons, R. C. Valentine, D. A. Phillips, D. W. Rains, and R. C. Huffaker, Plenum Publishing Corp., New York, 1981, p. 473.
82. M. S. Smith, *Appl. Environ. Microbiol.*, 1982, **43**, 854.
83. B. H. Bleakley and J. M. Tiedje, *Appl. Environ. Microbiol.*, 1982, **44**, 1342.

84. M. S. Smith, *J. Bacteriol.*, 1983, **45**, 1545.
85. P. Taverna and B. Sedgwick, *J. Bacteriol.*, 1996, **178**, 5105.
86. R. Metheringham and J. Cole, *Microbiology*, 1997, **143**, 2647.
87. C. E. Vine and J. A. Cole, *FEMS Microbiol. Lett.*, 2011, **325**, 99.
88. G. Rowley, D. Hensen, H. Felgate, A. Arkenberg, C. Appia-Ayme, K. Prior, C. Harrington, S. J. Field, J. N. Butt, E. Baggs and J. Richardson, *Biochem. J.*, 2012, **441**, 755.
89. D. Mania, K. Heylen, R. J. M. van Spanning and Å. Frostegård, *Environ. Microbiol.*, 2014, **16**, 3196.
90. H. Corker and R. K. Poole, *J. Biol. Chem.*, 2003, **278**, 31584.
91. B. Weiss, *J. Bacteriol.*, 2006, **188**, 829.
92. S. Spiro, *FEMS Microbiol. Rev.*, 2007, **31**, 193.
93. X. Duan, J. Yang, B. Ren, G. Tang and H. Ding, *Biochem. J.*, 2009, **417**, 783.
94. D. H. Flint, M. H. Emtage and J. R. Guest, *Biochem.*, 1992, **31**, 10331.
95. Y. Zhang and N. Hogg, *Free Radical Biol. Med.*, 2005, **38**, 831.
96. P. R. Gardner, A. M. Gardner, L. A. Martin and A. L. Salzman, *Proc. Natl. Acad. Sci. U. S. A.*, 1998, **95**, 10378.
97. J. S. Bang, L. Liu, A. Vazquez-Torres, M. L. Crouch, J. S. Stamler and F. C. Fang, *J. Biol. Chem.*, 2006, **281**, 28039.
98. T. M. Stevanin, R. K. Poole, E. A. Demoncheaux and R. C. Read, *Infect. Immunol.*, 2002, **70**, 4399.
99. D. R. Noble, H. R. Swift and D. L. H. Williams, *Chem. Commun.*, 1999, **22**, 2317.
100. A. M. Stern, B. Liu, L. R. Bakken, J. P. Shapleigh and J. Zhu, *J. Bacteriol.*, 2013, **195**, 4702.
101. J. W. Moir, *Biochim. Biophys. Acta*, 1999, **1430**, 65.
102. S. M. Turner, J. W. B. Moir, L. Griffiths, T. W. Overton, H. Smith and J. A. Cole, *Biochem. J.*, 2005, **388**, 545.
103. S. O. Kim, Y. Orii, D. Lloyd, M. N. Hughes and R. K. Poole, *FEBS Lett.*, 1999, **445**, 389.
104. A. M. Gardner and P. R. Gardner, *J. Biol. Chem.*, 2002, **277**, 8166.
105. R. K. Poole, M. F. Anjum, J. Membrillo-Hernández, S. O. Kim, M. N. Hughes and V. Stewart, *J. Bacteriol.*, 1996, **178**, 5487.
106. H. Cruz-Ramos, J. Crack, G. Wu, M. N. Hughes, C. Scott, A. J. Thomson, J. Green and R. K. Poole, *EMBO J.*, 2002, **21**, 3235.
107. A. M. Gardner, R. A. Helmick and P. R. Gardner, *J. Biol. Chem.*, 2002, **277**, 8172.
108. A. M. Gardner, C. R. Gessner and P. R. Gardner, *J. Biol. Chem.*, 2003, **278**, 10081.
109. B. D'Autréaux, N. P. Tucker, R. Dixon and S. Spiro, *Nature*, 2005, **437**, 769.
110. S. R. Poock, E. R. Leach, J. W. B. Moir, J. A. Cole and D. J. Richardson, *J. Biol. Chem.*, 2002, **277**, 23664.
111. N. P. Tucker, M. G. Hicks, T. A. Clarke, J. C. Crack, G. Chandra, N. E. Le Brun, R. Dixon and M. I. Hutchings, *PLoS One*, 2008, **3**, e3623.

112. J. C. Crack, J. Munnoch, E. L. Dodd, F. Knowles, M. M. Al Bassam, S. Kamali, A. A. Holland, S. P. Cramer, C. J. Hamilton, M. K. Johnson, A. J. Thomson, M. I. Hutchings and N. E. Le Brun, *J. Biol. Chem.*, 2015, **290**, 12689.
113. J. H. van Wonderen, B. Burlat, D. J. Richardson, M. R. Cheeseman and J. N. Butt, *J. Biol. Chem.*, 2008, **283**, 9587.
114. C. E. Vine, S. K. Purewal and J. A. Cole, *FEMS Microbiol. Lett.*, 2011, **325**, 108.
115. M. C. Justino, J. B. Vicente, M. Teixeira and L. M. Saraiva, *J. Biol. Chem.*, 2005, **280**, 2636.
116. J. E. Karlinsey, I.-S. Bang, L. A. Becker, E. R. Frawley, S. Porwollik, H. F. Robbins, V. C. Thomas, R. Urbano, M. McClelland and F. C. Fang, *Mol. Microbiol.*, 2012, **85**, 1179.
117. C. E. Vine and J. A. Cole, *Biochem. Soc. Trans.*, 2011, **39**, 213.
118. A. F. Arendsen, J. Hadden, G. Card, A. S. McAlpine, S. Bailey, V. Zaitsev, E. H. M. Duke, P. F. Lindley, M. Krockel, A. X. Trautwein, M. C. Feiters, J. M. Charnock, C. D. Garner, S. J. Marritt, A. J. Thomson, I. M. Kooter, M. K. Johnson, W. A. M. Van den Berg, W. M. A. M. Van Dongen and W. R. Hagen, *JBIC, J. Biol. Inorg. Chem.*, 1998, **3**, 81.
119. W. A. M. van den Berg, W. R. Hagen and W. M. A. M. van Dongen, *Eur. J. Biochem.*, 2000, **267**, 666.
120. D. Aragão, E. P. Mitchell, C. F. Frazão and P. F. Lindley, *Acta Cryst.*, 2008, **D64**, 665.
121. N. Filenko, S. Spiro, D. Browning, D. Squire, T. Overton, J. Cole and C. Constantinidou, *J. Bacteriol.*, 2007, **189**, 4410.
122. C. C. Constantinidou, J. L. Hobman, M. D. Patel, C. W. Penn, J. A. Cole and T. W. Overton, *J. Biol. Chem.*, 2006, **281**, 4802.
123. D. M. Bodenmiller and S. Spiro, *J. Bacteriol.*, 2006, **188**, 874.
124. D. Seth, A. Hausladen, Y.-J. Wang and J. S. Stamler, *Science*, 2012, **336**, 470.
125. M. C. O. Fugueiredo, S. A. L. Lobo, S. H. Sousa, F. P. Pereira, J. D. Wall, L. S. Nobre and L. M. Saraiva, *J. Bacteriol.*, 2013, **195**, 2684.
126. J. P. Lewis, S. S. Yanamandra and C. Anaya-Bergman, *Infect. Immunol.*, 2012, **80**, 3319.
127. C. C. Kim, D. Monack and S. Falkow, *Infect. Immunol.*, 2003, **71**, 3196.
128. P. Mukhopadhyay, M. Zheng, L. A. Bedzyk, R. A. LaRossa and G. Storz, *Proc. Natl. Acad. Sci. U. S. A.*, 2004, **101**, 745.
129. B. Troxell, R. C. Fink, S. Porwollik, M. McClelland and H. M. Hassan, *BMC Microbiol.*, 2011, **11**, 236.
130. C. M. Moore, M. M. Nakano, T. Wang, R. W. Ye and J. D. Helmann, *J. Bacteriol.*, 2004, **186**, 4655.
131. D. A. Rodionov, I. L. Dubchak, A. P. Arkin, E. J. Alm and M. S. Gelfand, *PLoS Comput. Biol.*, 2005, **1**, e55.
132. J. Wang, C. E. Vine, B. K. Balasiny, J. Rizk, C. L. Bradley, M. Tinajero Trejo, R. K. Poole, L. L. Bergaust, L. R. Bakken and J. A. Cole, *Mol. Microbiol.*, 2016, **100**, 877.

133. M. C. Justino, C. C. Almeida, V. L. Gonclaves, M. Teixeira and L. M. Saraiva, *FEMS Microbiol. Lett.*, 2006, **257**, 278.
134. M. C. Justino, C. C. Almeida, M. Teixeira and L. M. Saraiva, *J. Biol. Chem.*, 2007, **282**, 10352.
135. T. W. Overton, M. C. Justino, Y. Li, J. M. Baptista, A. M. Melo, J. A. Cole and L. M. Saraiva, *J. Bacteriol.*, 2008, **190**, 2004.
136. L. S. Nobre, R. Garcia-Serres, S. Todorovix, P. Hildebrandt, M. Teixeira, J. M. Latour and L. M. Saraiva, *PloS One*, 2014, **9**, e95222.
137. M. C. Justino, J. M. Baptista and L. M. Saraiva, *Biometals*, 2009, **22**, 99.
138. A. R. Richardson, P. M. Dunman and F. C. Fang, *Mol. Microbiol.*, 2006, **61**, 927.
139. T. Overton, R. Whitehead, Y. Li, L. A. S. Snyder, N. J. Saunders, H. Smith and J. A. Cole, *J. Biol. Chem.*, 2006, **281**, 33115.
140. M. J. Thomson, R. C. Read and J. W. B. Moir, *J. Bacteriol.*, 2007, **189**, 1138.
141. S. Todorovic, M. C. Justino, G. Wellenreuther, P. Hildebrandt, D. H. Murgida, W. Meyer-Klaucke and L. M. Saraiva, *JBIC, J. Biol. Inorg. Chem.*, 2008, **13**, 765.
142. C. E. Vine, M. C. Justino, L. Saraiva and J. A. Cole, *J. Microbiol. Methods*, 2010, **81**, 77.

CHAPTER 3

Structure, Function and Mechanisms of Respiratory Nitrate Reductases

PABLO J. GONZALEZ[a], MARIA G. RIVAS[a] AND JOSÉ J. G. MOURA*[b]

[a]Departamento de Física, Facultad de Bioquímica y Ciencias Biológicas, Universidad Nacional del Litoral, Paraje El Pozo, S3000ZAA Santa Fe, Argentina; [b]UCIBIO, REQUIMTE, Departamento de Química, Faculdade de Ciências e Tecnologia, Universidade Nova de Lisboa, Campus da Caparica, 2829-516 Caparica, Portugal

*E-mail: jose.moura@fct.unl.pt

3.1 Introduction

Denitrification is a naturally occurring process of the biogeochemical nitrogen cycle that is carried out mainly by heterotrophic facultative bacteria.[1,2] This process starts with the two-electron reduction of nitrate into nitrite (see eqn (3.1)).

$$NO_3^- + 2H^+ + 2e^- \rightarrow NO_2^- + H_2O \quad E^\circ = 420\ \text{mV vs. SHE} \tag{3.1}$$

The latter is a key redox reaction driven by living organisms expressing the oxidoreductase respiratory nitrate reductase (Nar; EC 1.7.99.4).[1–5] Nars

RSC Metallobiology Series No. 9
Metalloenzymes in Denitrification: Applications and Environmental Impacts
Edited by Isabel Moura, José J. G. Moura, Sofia R. Pauleta and Luisa B. Maia

Published by the Royal Society of Chemistry, www.rsc.org

are heterotrimeric membrane-anchored enzymes containing the second-row element molybdenum ($_{42}$Mo) at the active site and iron ($_{26}$Fe) in the form of iron–sulphur (FeS) clusters and heams as additional redox cofactors.[4–7]

Mo is a trace element that is widespread in nature and can be found as an essential piece in the active sites of many different enzymes from eukaryotic (excepting *Protozoa*) and prokaryotic (*Archaea* and *Bacteria*) cells.[8–11] In contrast to the first-row transition metals commonly found in metalloenzymes as cofactors (Fe, Cu, Zn, Co, Ni), the biochemistry of Mo is not based on cationic species Mo^{n+} (n = 1–3), but instead, it is determined by the chemistry of its higher oxidation states (IV, V and VI).[12,13] Thanks to its higher valence chemistry, which resembles that of non-metals oxoanions, Mo has the capability to catalyse O-atom transfer reactions, which are essential in the nitrogen (nitrate reductase and nitrite oxidoreductase), sulphur (sulphite oxidase [SO] and thiosulfate reductase), carbon (CO dehydrogenase), chlorine (perchlorate reductase and chlorate reductase) and arsenic (arsenite oxidase and arsenate reductase) cycles. In order to be catalytically competent in the context of a living cell, Mo depends on the biosynthesis of sophisticated enzyme cofactors such as pyranopterin (pyranopterin monophosphate [PMP], pyranopterin cytidine dinucleotide [PCD] and pyranopterin guanosine dinucleotide [PGD]) for mononuclear Mo enzymes[14,15] or FeMoco for nitrogenase.[16–18]

Taking into account structural data, primary sequence alignments and molecular taxonomy studies, Mo-containing enzymes can be classified into two main groups.[8–10,23,24] The first group includes enzymes containing the heteronuclear FeMoco cluster (*e.g.*, nitrogenase), while all enzymes harbouring the Mo (or W) atom in mononuclear form belong to the second group. In the latter, the metal is coordinated by one or two *cis*-dithiolene groups from pyranopterin molecules. The coordination sphere is usually completed by oxygen, sulphur or selenium ligands.[8–10,24,25] The mononuclear Mo(W)-containing enzymes are capable of catalysing redox and non-redox reactions and, on the basis of structural and phylogenetic analysis, they can be divided into four families (Figure 3.1): the xanthine oxidase (XO) family, the SO family, the W-aldehyde oxidoreductase [AOR] family and the DMSO reductase (DMSOr) family.[†]

The XO family of mononuclear Mo enzymes is probably the best characterized.[26,27] These enzymes contain the Mo ion coordinated to one PMP or PCD molecule. While PMP is present in all eukaryotic Mo enzymes, PCD is present only in those from prokaryotic sources (*e.g.*, AOR from *Desulfovibrio* species,[28–31] CO dehydrogenase,[32] nicotinate dehydrogenase,[33] isoquinoline oxidoreductase[34] and 4-hydroxybenzoyl-CoA reductase[35]). The Mo ion

[†]The classification of tungstoenzymes has been a matter of some controversy. Some authors gather all the tungsten-containing enzymes into a single family, and others divide the tungstoenzymes between the DMSOr family and an aldehyde:ferredoxin oxidoreductase group. This latter classification is based on the fact that the great majority of the prokaryotic molybdo- and tungsto-enzymes harbour at the active site two PGDs (which is not the case for the aldehyde:-ferredoxin oxidoreductases, whose pyranopterin cofactor molecules are in the mononucleotide form). This classification is useful when one aims to highlight the structural/functional similarities within the family.

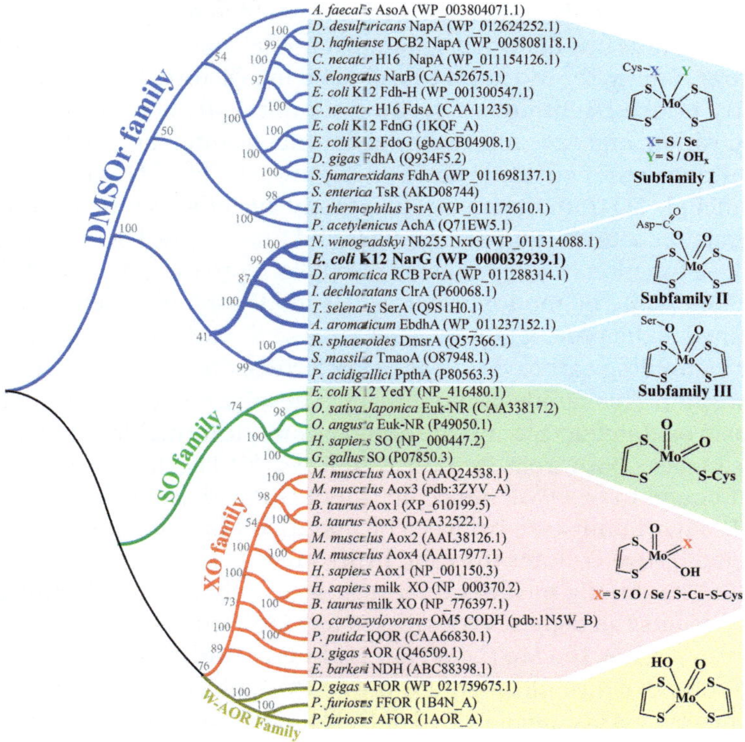

Figure 3.1 Molecular phylogenetic analysis of the mature catalytic subunits of mononuclear Mo(W)-containing enzymes. The evolutionary history was inferred using the neighbour-joining method.[19] The bootstrap consensus tree inferred from 10 000 replicates is taken to represent the evolutionary history of the taxa analysed.[20] Branches corresponding to partitions reproduced in fewer than 50% bootstrap replicates are collapsed. The percentages of replicate trees in which the associated taxa clustered together in the bootstrap test (10 000 replicates) are shown next to the branches. The evolutionary distances were computed using the p-distance method and are in the units of the number of amino acid differences per site.[21] The analysis involved 44 amino acid sequences. All positions with less than 50% site coverage were eliminated; that is, fewer than 50% alignment gaps, missing data and ambiguous bases were allowed at any position. There were a total of 924 positions in the final dataset. Evolutionary analyses were conducted in MEGA6.[22] Sequence accession numbers are in parentheses. AsoA: arsenite oxidase; NapA: periplasmic nitrate reductase; NarB: assimilatory nitrate reductase; Fdh/Fdn/Fdo/Fds: formate dehydrogenase; TsR: thiosulfate reductase; PsrA: polysulfide reductase; AchA: acetylene hydratase; NxrG: nitrite oxidase–reductase; NarG: respiratory nitrate reductase; PcrA: perchlorate reductase; ClrA: chlorate reductase; SerA: selenite reductase; EbdhA: ethylbenzene dehydrogenase; DmsrA: dimethyl sulphoxide reductase; TmaoA: trimethylamine oxide reductase; PpthA: pyrogallol phloroglucinol transhydrolase; Euk-NR: eukaryotic nitrate reductase; SO: sulphite oxidase; Aox: aldehyde oxidase; XO: xanthine oxidase; CODH: carbon monoxide dehydrogenase; IQOR: isoquinoline oxidase–reductase; AOR: aldehyde oxidase–reductase; NDH: nicotinate dehydrogenase; AFOR: aldehyde ferredoxin oxidase–reductase; FFOR: formaldehyde ferredoxin oxidase–reductase.

coordination sphere is completed with oxygen (oxo or hydroxo), sulphur or selenium ligands in a distorted square-pyramidal geometry (Figure 3.1).

In the SO family, the Mo ion is coordinated in a distorted square-pyramidal geometry to one *cis*-dithiolene group from one PMP, one oxo group at the apical position, one oxo group at the equatorial position and an S-atom from a Cys residue in the equatorial position (Figure 3.1).[11,36–38] Members of the SO family include SO from mammals, plants and bacteria, as well as the eukaryotic nitrate reductases from plants, algae and fungi.[39] The bacterial YedY protein whose function is still unknown also belongs to the SO family.[40–42]

Another family of mononuclear Mo- and W-containing enzymes to be established is the W-AOR family. They catalyse the oxidation of aldehydes to the corresponding carboxylic acids using ferredoxin as a physiological electron acceptor. This family of proteins includes the homodimeric aldehyde:-ferredoxin oxidoreductase and the homotetrameric formaldehyde:ferredoxin oxidoreductase, both from *Pyrococcus furiosus*.[43,44] The active site comprises a W ion bound to two PMP molecules (not PGD) with a similar arrangement to the DMSOr family (see below). The coordination sphere of the W ion is completed by two O atoms (oxo and hydroxo).

The DMSOr family includes enzymes exclusively from prokaryotes. This family presents the highest variability in terms of active site composition. As a common feature, the Mo (or W) ion is always coordinated by four sulphur ligands from two dithiolene groups of two PGD molecules. The fifth coordination position is occupied by O, S or Se from amino acid side chains (O-Ser, O-Asp, S-Cys, Se-Cys), while the sixth is occupied by oxygen (oxo and hydroxo) or sulphur atoms (one exception to this classification is provided by arsenite oxidase, in which the fifth and sixth positions are occupied by two oxo or one oxo plus one hydroxo groups).[8–10,13,23,24] In the DMSOr family, the Mo (or W) is hexacoordinated in a distorted trigonal-prismatic geometry (Figure 3.1). Although DMSOr family members present a high degree of similarity in the overall polypeptide fold of their catalytic subunits, variations at the active site in terms of metal ligands and surrounding amino acid residues result in remarkably different specificities and reactivities, which dictate the variety of functions of the enzymes belonging to this family. Taking into account that the structural properties of the active site and biochemical data are in line with primary sequence alignments and molecular taxonomy studies, the DMSOr family could be split into three subfamilies (Figure 3.1).[7,13,45,46] Considering Mo(W) enzymes with reported crystallographic structures, subfamily I is represented by periplasmic nitrate reductases (Nap),[47–51] polysulphide reductases (Psr)[52] and formate dehydrogenases (Fdh).[53–56] The Mo at the active sites is coordinated by a Cys residue in the case of Nap and Psr and by a Se-Cys or Cys residue in the case of Fdh. Subfamily III is exemplified by DMSOr[57,58] and trimethylamine *N*-oxide reductase.[59,60] In these enzymes, a Ser side chain occupies the fifth coordination position of the metal ion. Mo enzymes of subfamily II with reported structures include the respiratory Nar and ethylbenzene dehydrogenase (Ebdh).[61–63] In both enzymes, the coordination sphere of the Mo ion is completed by one or two O atoms from the carboxylate side chain of an Asp residue. Though structural data have not been reported at present, this subfamily also includes the nitrite oxidoreductase (Nxr), as well as the perchlorate, chlorate and selenate reductases (see Figure 3.1; Pcr, Clr and Ser, respectively).

In the next sections, we will review the molecular and spectroscopic properties of Nars, as well as the regulation of *nar* structural genes and those involved in cofactor biosynthesis and enzyme maturation. Also, a brief description of the role of Nar in the energy-conserving respiratory process called denitrification occurring within the bacterial cell is presented, as well as some key aspects of the current knowledge of the catalytic mechanisms.

3.2 Structural and Spectroscopic Properties of Respiratory Nars

Nars isolated from organisms capable of performing denitrification are oligomeric enzymes composed of three different subunits.[64–73] The large catalytic α-subunit (NarG: 110–140 kDa) and the electron transfer (ET) β-subunit (NarH: 50–65 kDa) are located in the cell cytoplasm and are anchored to the cell membrane through a hydrophobic patch, and the quinol-oxidising γ-subunit (NarI: 19–25 kDa), which is entirely buried in the phospholipid bilayer (Figure 3.2).[4] At present, only the X-ray crystallographic structure of Nar from *Escherichia coli* K12 (*Ec*) has been reported. Eleven Protein Data Bank (PDB) files of *Ec* Nar corresponding to the wild-type (WT), site-directed variants

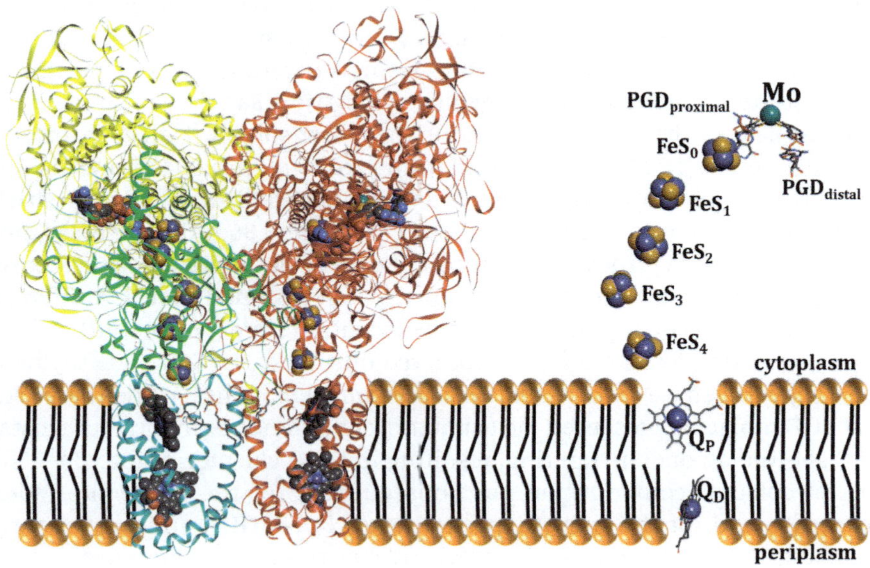

Figure 3.2 Biological assembly $(\alpha\beta\gamma)_2$ of *Escherichia coli* NarGHI in the bacterial cell membrane. The NarGHI heterotrimer on the left is coloured by chain (α, yellow; β, green; γ, light blue/teal). Atoms are coloured as follows: Mo, blue/teal ball; Fe, blue balls; S, yellow balls or sticks; C, grey sticks; N, blue sticks; O, red sticks.

and complexes with small molecules have been deposited in the PDB (codes: 1Q16, 1R27, 3EGW, 1SIW, 1Y4Z, 1Y5I, 1Y5L, 1Y5N, 3IR5, 3IR6 and 3IR7). In the bacterial cell membrane, Nars exist as a dimer of heterotrimers $(\alpha\beta\gamma)_2$, having a flower-like arrangement (Figure 3.2, left).[74] Each NarGHI heterotrimer harbours eight redox cofactors, which are separated by 12–14 Å, describing an ET pathway of ~90 Å (Figure 3.2, right).

3.2.1 The Catalytic Subunit NarG

The first X-ray structure reported in 2003 by Bertero *et al.* (PDB code: 1Q16[61]) was solved at 1.9 Å resolution using a detergent-solubilised enzyme, which was purified as NarGHI heterotrimer. The *Ec* NarG subunit has 1244 amino acid residues and a molecular mass of ~140 kDa. This polypeptide chain, which is larger than most of the α-subunits of the Mo enzymes of the DMSOr family, presents an αβ-type fold that can be divided into five domains: four involved in Mo-bisPGD binding, and a fifth domain that gives shape to the funnel-like cavity where substrate and product molecules diffuse towards/from the active site.[61,62,75]

3.2.1.1 The Mo-bisPGD Cofactor

As in the rest of mononuclear Mo(W) enzymes of the DMSOr family, the large α-subunit contains the Mo ion coordinated to four dithiolene sulphurs from two PGD molecules, forming the Mo-bisPGD. The Mo ion coordination sphere is completed with two O atoms from the carboxylate side chain of Asp_{222}, which is bound in a bidentate fashion determining a metal centre with distorted trigonal-prismatic geometry (Figure 3.3a).

Nevertheless, the X-ray structure of the water-soluble NarGH heterodimer (PDB code: 1R27) published by Jormakka *et al.* in 2004 showed a Mo ion in which the Asp_{222} carboxylate is a monodentate ligand,[62] and the sixth coordination position is occupied by an O atom at 1.86 Å (Figure 3.3b), not observed in the heterotrimer.[61] Since crystals of both NarGHI and NarGH were obtained under aerobic conditions, it was assumed that the different active site forms correspond to the same oxidation state (*i.e.*, Mo[VI]).

An interesting feature that was observed only in *Ec* NarGHI and ethylbenzene dehydrogenase from *Aromatoleum aromaticum* is that, while the proximal PGD molecule presented the usual tricyclic pyranopterin form, the distal PGD showed an open pyran ring, resulting in a bicyclic pterin state.[61,63] This special cofactor configuration has consequences for the active site redox properties and for the catalytic properties of *Ec* NarGHI, and was recently studied using kinetic, electrochemical and spectroscopic techniques.[76]

The spectroscopic properties of the Mo-bisPGD cofactor of Nars from different bacteria have been addressed mainly through electron paramagnetic resonance (EPR) spectroscopy.[67,70,77–81] As in the rest of mononuclear Mo enzymes, in Nars, the Mo ion can be found in three possible oxidation

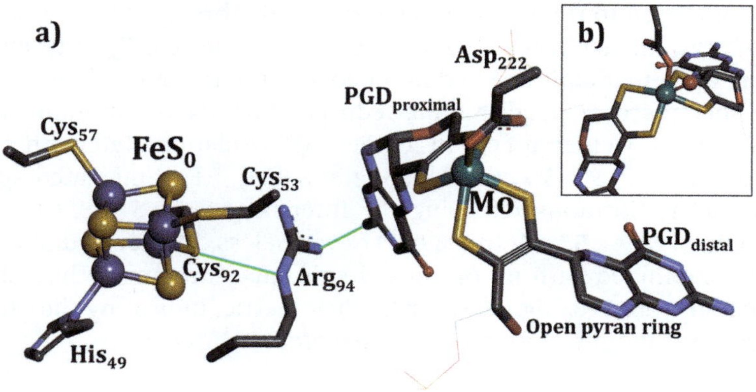

Figure 3.3 The redox cofactors of the catalytic (NarG) subunit from NarGHI (Protein Data Bank code: 1Q16). (a) From left to right: the FeS_0 cluster with the His and Cys ligands, the Arg_{94} putatively involved in electron transfer from FeS_0 to the proximal pyranopterin guanosine dinucleotide (PGD), and the Mo-bisPGD cofactor showing the Asp_{222} bidentate coordination mode. (b) Mo site showing the monodentate coordination mode of Asp_{222} (Protein Data Bank code: 1R27). Atoms colours as in Figure 3.2.

states—+IV, +V and +VI—although only Mo(V) has an open electronic shell (paramagnetic), therefore being EPR detectable. The stabilisation of Mo(V) species in Nars is possible thanks to the gap between the reduction potentials of the $Mo^{6+/5+}$ (E_m = 190 mV) and $Mo^{5+/4+}$ (E_m = 95 mV) redox couples.[80] In the absence of exogenous ligands (*i.e.*, oxoanions, halides, *etc.*), two different Mo(V) EPR signals can be observed. These paramagnetic species are in a pH-dependent equilibrium. The low-pH form shows a low-anisotropy rhombic signal (g_1 = 2.001, g_2 = 1.986, g_3 = 1.964), with resonance lines split by a solvent-exchangeable proton, which is clearly visible through X-band CW-EPR (A_{av} = 9.5 G). In the high-pH form (g_1 = 1.987, g_2 = 1.981, g_3 = 1.962), the hyperfine interaction is not evident (A_{av} = 3.5 G). Although there is no OH/OH_2 ligand in the reported X-ray structures of *Ec* Nar,[61,62] taking into account the hyperfine coupling constant values measured through EPR spectroscopy and computational (DFT) simulations of the EPR parameters of Mo sites,[82] it can be assumed that the sixth ligand of Mo(V) is an OH_x. Therefore, the bidentate coordination should be opened upon reduction to Mo(V) to bind an OH_x ligand, which gives rise to the solvent-exchangeable proton observable in the low- and high-pH Mo(V) species.

3.2.1.2 The FeS_0 Cluster

Besides the Mo-bisPGD cofactor, the NarG subunit also harbours the redox centre FeS_0 (Figure 3.3a, left), which is a [4Fe–4S] cluster that presents (similar to that observed in bacterial [Ni–Fe] and [Fe-only] hydrogenases)[83–86] one Fe atom coordinated by a His residue (His_{49}), and the remaining three coordinated by Cys residues (Cys_{53}, Cys_{57} and Cys_{92}).[61,62] X-ray data showed that

FeS_0 is connected to the proximal PGD molecule through the guanidinium group of Arg_{94}. In fact, recent reports showed that the FeS_0 E_m value and EPR signature are remarkably affected in the R94S site-directed variant.[87]

EPR spectroscopy of dithionite-reduced Nar (NarG mutant lacking the Mo-bisPGD) yielded a FeS_0 in a $[4Fe–4S]^{1+}$ oxidation state with a high-spin ground state ($S = 3/2$, $g_1 = 5.556$, $g_2 = 5.023$).[88] EPR-mediated spectropotentiometric titrations following the intensity of the $S = 3/2$ high-spin signal yielded $E_m = -55$ mV (pH 8.0).[88] Nevertheless, further potentiometric titrations combined with an optimised spin quantitation method showed that the high-spin FeS_0 signal was substoiciometric, indicating that another FeS_0 electronic ground state might be also present.[89]

3.2.2 The ET Subunit NarH

The β-subunit (NarH) serves, essentially, as a wire connecting NarI to NarG. It contains three [4Fe–4S] clusters (FeS_1, FeS_2 and FeS_3) and one [3Fe–4S] (FeS_4), the latter located at the NarH–NarI interface (Figure 3.4). In the case of the *Ec* NarGHI, the reduction potentials of the FeS centres present in NarH, as well as those in NarG and NarI, have been determined by UV-vis- and EPR-mediated potentiometric titrations.[88,90,91]In the case of NarH, it was proposed that this subunit can be divided in two domains (A and B), each one harbouring one low- and one high-potential cluster. Domain A contains FeS_1 (E_m = 130 mV) and FeS_2 (E_m = −420 mV), while domain B contains FeS_3 (E_m = −55 mV) and FeS_4 (E_m = 180 mV) (Figure 3.4). Taking into account these values, the literature previous to the report of the Nar X-ray structure suggested that, owing to its very low E_m value, FeS_2 would not take part in the ET pathway, having only a structural role. Naturally, this hypothesis was discarded when the NarGHI

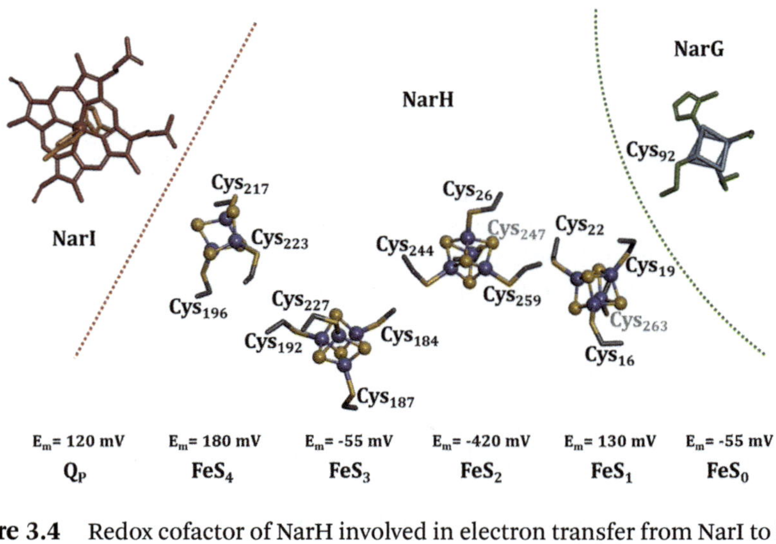

Figure 3.4 Redox cofactor of NarH involved in electron transfer from NarI to NarG. Atoms colours as in Figure 3.2.

X-ray data showed that FeS_2 was positioned as part of the pathway, acting as an ET relay.[61] Subsequent studies proposed that the low reduction potentials of FeS_3 and mainly FeS_2 would set an ET barrier that may serve to match the rates of ET and substrate reduction.[75]

3.2.3 The QH_2-Oxidising Subunit NarI

Membrane-soluble quinols are oxidised at the γ-subunit (NarI) to gather the electrons needed to reduce nitrate into nitrite. Early biochemical studies showed that *Ec* NarGHI is capable of catalysing the oxidation of both ubiquinol (UQH_2) and menaquinol (MQH_2) pools.[92] Moreover, recent biochemical and spectroscopic studies showed that demethylmenaquinol ($DMQH_2$, E_m = 36 mV), a third quinone produced by *Ec* and having an intermediate reduction potential between MQH_2 (E_m = −70 mV) and UQH_2 (E_m = 100 mV), is also a substrate of *Ec* NarGHI.[93]

NarI from *Ec* has a molecular mass of ~25 kDa and is composed mainly of α-helices that form five transmembrane segments that surround two b-type haems that present bis-His axial coordination.[94] NarI is completely immersed in the cell membrane and is associated to the water-soluble NarGH dimer thanks to a hydrophobic patch present in NarH (Figure 3.5). Additional attachment is provided by an N-terminal α-helix of NarG that passes through NarH (Figure 3.5, red dotted box) and forms a loop that contacts the NarI C-terminal fragment, and finally an α-helix that interacts with the cytoplasmic edge of the first transmembrane α-helix of NarI (Figure 3.5, blue dotted boxes).[61] All of these non-bonding interactions contribute to the correct anchoring and orientation of the water-soluble NarGH dimer to the membrane.

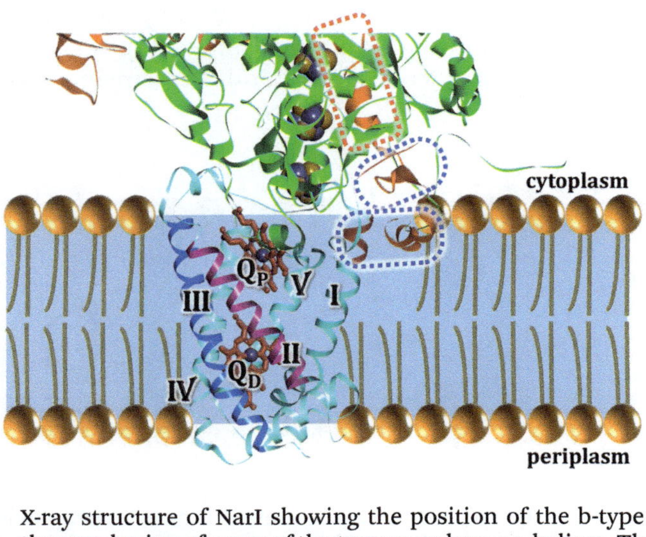

Figure 3.5 X-ray structure of NarI showing the position of the b-type haems and the numbering of some of the transmembrane α-helices. The red dotted box highlights the NarG α-helix that crosses through NarH, while the blue dotted boxes highlight the NarG–NarI contact sites.

When the NarGHI X-ray structure was reported, it was assumed that both haems would be involved in the oxidation of the membrane-soluble quinols (QH_2) coupled to the proton (H^+) translocation to the periplasm.[61] Later, X-ray structural studies showed that both in WT and K86A-NarI forms, the pentachlorophenol molecule (PCP, an UQH_2 analogue and potent inhibitor of quinol:nitrate oxidoreductase activity) binds in a cavity formed by highly conserved nonpolar residues of helices II (Figure 3.6, purple) and III (Figure 3.6, blue) at the Q_D site.[95] Besides these hydrophobic interactions, the PCP–OH group is H-bonded to one propionate group of haem b_D and to the imidazole side chain of His_{66} (Figure 3.6), the latter also being used as one of the axial ligands to the haem Fe atom.

The short edge-to-edge distance between PCP and haem b_D (~2.8 Å), the notable effect on the haem b_D EPR spectrum only and the competitive inhibitory behaviour (K_{iC} = 57 nm and K_{iU} = 490 nm) against plumbagin (a MQH_2 analogue) suggested that this compound would occupy the physiological electron donor binding site (Q-site).[95] Though no X-ray data exist, a similar effect was observed with the menaquinol analog 2-*n*-heptyl-4-hydroxyquinoline-*N*-oxide (HOQNO) through kinetic and spectroscopic studies.[95] *In silico* molecular models supported the hypothesis that both UQH_2 and MQH_2 would bind at the same cleft as PCP and HOQNO.[95]

In spite of PCP being observed only in one position (cleft B), on the basis of kinetic studies and visual inspection of the X-ray structure showing a larger pocket (known as cleft A) where both haems b_D and b_P are exposed, some authors hypothesised that there probably exists a second QH_2 binding site that has no ionisable groups (cleft A would be used for QH_2 oxidation without proton translocation).[95] However, pulse EPR spectroscopy studies showed that haem b_D is the only QH_2 oxidation site.[96–100] The spectroscopic studies specifically showed that menasemiquinone-8 (produced by *Ec* under anaerobic conditions), ubisemiquinone-8 (produced by *Ec* under aerobic and semiaerobic conditions) and demethylmenasemiquinone

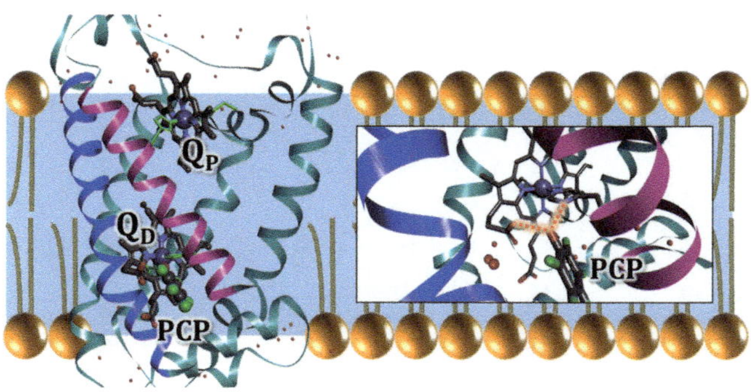

Figure 3.6 X-ray structure of NarI in complex with pentachlorophenol (PCP; Protein Data Bank code: 1Y4Z). Inset: close-up view of the PCP molecule at the Q_D site showing H-bond interactions with His_{66} and the propionate group from haem b_D.

intermediates can be stabilised at the Q_D site, forming an H-bond between the semiquinone O atom and the imidazole of His_{66}.[93,98] This interaction might be essential to reducing the Q_D site, while the H-bond with the haem propionate (as seen in the X-ray data of NarGHI–PCP complexes) would serve to release protons into the cell periplasm in order to generate a proton motive force (PMF).[95]

3.3 Gene Expression Control and Maturation of Respiratory Nars

3.3.1 Regulation at the Transcriptional and Translational Levels

In *Ec*, the *narGHJI* operon is located in the chlC locus of the chromosome. This operon encodes the NarGHI complex and NarJ. The latter is an exclusive chaperone that is essential for nitrate reductase activity as it participates in Mo-bisPGD cofactor insertion into NarG prior to the NarGH complex's attachment to NarI in the membrane.[4,75,101–104] Nar expression control is performed at different levels according to the cell requirements. Mo is translocated as molybdate (MoO_4^{2-}) *via* the ModABC transport system. When Mo is already available inside the cell, it forms a complex with a ModE protein and represses the expression of the transporter.[105] In addition, the ModE–Mo complex induces the expression of the structural *nar* genes, as well as the *moaABCDE* operon. The latter encodes enzymes involved in the Mo-bisPGD biosynthesis.[106,107] The *narGHJI* operon expression is also regulated by the global transcription factor FNR (fumarate–nitrate regulation). The FNR protein tightly controls the transition from aerobic to anaerobic conditions, repressing the synthesis of enzymes required in aerobic respiration[108,109] and increasing the expression of proteins needed for anaerobic metabolism.[110–112] The induction of *nar* genes by FNR is a complex process that occurs under anaerobic conditions and in the presence of nitrate. FNR binds to the promoter P_{narG} together with NarL and IHF.[112] The specific binding of NarL is promoted by its nitrate-dependent phosphorylation, which is catalysed by the sensor kinases NarX. The induction of the synthesis of *narGHJI* by the NarXL two-component system (TCS) is achieved under high nitrate concentrations. A similar TCS called NarQP was described in some organisms.[113] In contrast to NarXL, NarQP acts under low nitrate concentration and in the presence of nitrite.[114,115] NarL also controls the expression of *frdABCD* (fumarate reductase), *narK*, *nrfABCDEFG* (nitrite reductase) and *napFDAGHBC* (periplasmic nitrate reductase), being the last two enzyme complexes that are also regulated by NarP.[116] This complementary regulatory system detects the nitrate:nitrite ratio and inactivates NarL at low nitrate levels, favouring nitrite consumption. The gene expression regulation of Naps (as well as the phylogenetic, biochemical, spectroscopic and structural properties) have been recently reviewed.[117]

The discovery of highly conserved RNA sequences that respond to small molecules and are located upstream of genes encoding Mo enzymes, Mo transport and Mo cofactor biosynthesis shows a new aspect of Mo enzyme regulation. These segments of mRNA are known as riboswitches and control the translation of genes/operons at a translational level.[118] A clear example of this kind of regulation is constituted by the *moaABCDE* operon, which is regulated by the availability of Mo cofactors in the cell.[119–121] Moreover, some riboswitches can discriminate between Mo and W cofactors, and this fact could be key to answering the question of how homologous enzymes specifically incorporate Mo or W.[120] Besides the *narGHJI* operon, *Ec* encodes the *narZWYV* operon in the chlZ locus. This operon is expressed constitutively at low levels and its role is associated with adaptation processes in the transition from aerobic conditions to anoxia.[122,123]

3.3.2 NarGHI Maturation and Assembly

Maturation of the NarGHI complex (Figure 3.7) is a widely studied process in which NarJ has a crucial role.[75,124] NarJ prevents the attachment of the NarGH dimer to NarI at the inner face of the lipid bilayer until full NarGH maturation.[104] This exclusive chaperone also binds to undefined sites in the catalytic subunit and orchestrates the sequential incorporation of the metal cofactors.[103] Mutagenesis and X-ray crystallography studies showed that insertion

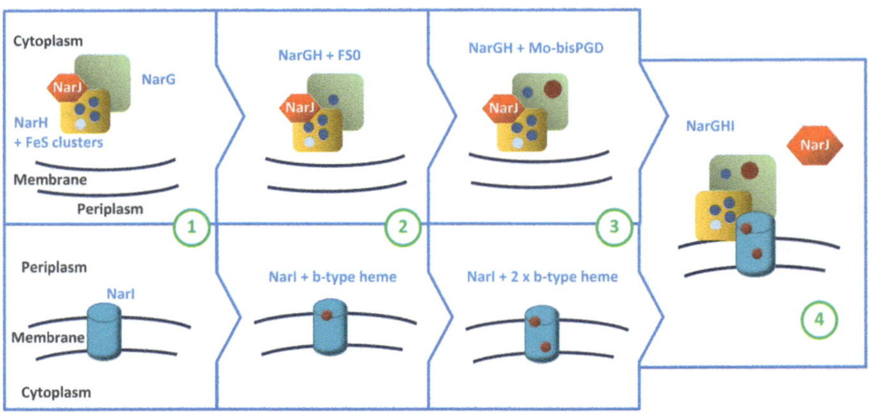

Figure 3.7 Maturation of the NarGHI complex. The maturation of the NarGH complex occurs in the cytoplasm. NarJ interacts with the NarG subunit of the NarGH complex, driving the sequential insertion of FeS_0 and Mo-bisPGD (images 1–3, top). A conformational change of the NarGH complex occurs after cofactor incorporation, triggering NarJ dissociation and attachment to the membrane through NarI (image 4). FeS biosynthesis machinery incorporates the FeS clusters in NarH during translation. NarI maturation is an inner membrane process where the b-type haems (b_D and b_P) are sequentially incorporated (images 1–3, bottom).

of the proximal FeS_0 cluster must occur previous to the Mo-bisPGD insertion.[87,88,103] The Mo-bisPGD delivery and insertion involves not only NarJ, but also several enzymes from the Mo cofactor biosynthesis pathway.[125,126] Cofactor insertion into NarGH produces conformational changes that trigger NarJ dissociation and subsequent membrane anchoring through the interactions described in Section 3.2.3 and Figure 3.5.[75] Insertion of one [3Fe–4S] and three [4Fe–4S] clusters into the NarH subunit occurs during translation and is assisted by FeS cluster biosynthetic machinery.[127] The NarI maturation is a process that occurs at the cytoplasmic side of the cell membrane and involves the sequential incorporation of the two b-type haems.[103]

3.4 Metabolic Role

The reduction of nitrate to nitrite catalysed by Nars is an energy-conserving process (*i.e.*, coupled to the translocation of protons to the cell periplasm in order to generate a PMF necessary for ATP synthesis).[128,129] To accomplish this, the MQH_2 pool produced under denitrifying conditions is oxidised at the periplasmic side of the lipid bilayer by the Q_D site. Two protons are released into the periplasm and two electrons are conducted through the ET pathway to the Mo ion at the active site, where nitrate is reduced, consuming two cytoplasmic protons. This results in a net balance of $4H^+/2e^-$ (Figure 3.8).

Given that the catalytic subunit NarG is cytoplasmic and nitrate cannot diffuse across the cell membrane, the latter is introduced into the cell by two transporters. According to studies performed in *Ec* K12, at the beginning of denitrification, nitrate is translocated by the proton:nitrate symporter $NarK_1$.

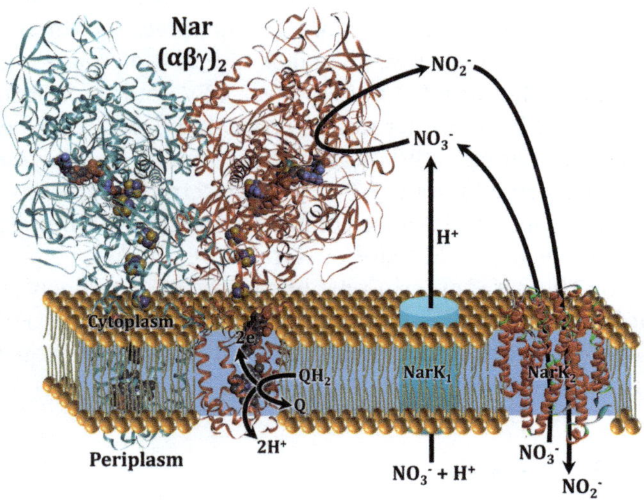

Figure 3.8 Schematic representation of the respiratory nitrate reduction process catalysed by nitrate reductases. Protein Data Bank codes: 3EGW and 4JR9 for Nar-$(\alpha\beta\gamma)_2$ and $NarK_2$, respectively.

During this period, the co-translocated protons are consumed in the nitrate reduction. Once nitrite is accumulated in the cytoplasm, $NarK_1$ is "switched-off" and nitrate is translocated by the nitrate:nitrite antiporter $NarK_2$, maintaining the steady state.[130] The first X-ray structure of a nitrate:nitrite antiporter was recently published.[131] The structural data of *Ec* $NarK_2$ revealed a positively charged translocation pathway without protonable residues, confirming that $NarK_2$ is a nitrate:nitrite exchanger that cannot co-transport protons.[131,132]

3.5 Catalytic Mechanism

In contrast to Naps, the catalytic mechanism of nitrate reduction by Nars was not studied at a molecular level using a computational approach.[133–135] Nevertheless, extensive biochemistry, electrochemistry and mutagenesis combined with X-ray crystallography and spectroscopic techniques were performed in order to understand QH_2 oxidation (see Section 3.2.3) and NO_3^- reduction by Nars.

Early kinetic studies performed in NarGHI from *Ec* and *Paracoccus denitrificans* showed that when viologen dyes are used as electron donors, Nar follows a Theorell–Chance mechanism in which nitrate binding is required in order to promote the ET from the donor to the active site.[68,136] On the other hand, when quinols were used as electron donors, the substrate's addition sequence is the same, but a two-site enzyme substitution mechanism with accumulation of ternary complexes is followed. In contrast with this observations, on the basis of protein film voltammetry (PFV) studies performed on Nars from *Ec* K12 and *Paracoccus pantotrophus* GB17, it was proposed that the Mo site would be reduced prior to nitrate binding.[137,138] In this sense, only Mo(v) and Mo(iv) could bind the substrate, with Mo(v) having a higher affinity for nitrate than Mo(iv). This model was used to explain the substrate concentration-dependent voltammetric response (wave-shape) of several Mo enzymes (see ref. 137 and 138, and page 2422 of ref. 139). However, PFV studies performed on the water-soluble NarGH heterodimer from *Marinobacter hydrocarbonoclasticus* 617 supported the early kinetic studies,[68,136] showing that the enzyme–substrate complex would be formed prior to ET from the electrode to the Mo site.[140]

At present, there is no absolute consensus on the fine details of nitrate reduction processes at the Mo site of Nars. Nevertheless, very recent studies showed that the pyranopterin portions of the PGD molecules are key in defining the redox properties and reactivity of the Mo ion.[76] Further studies will be needed in order to fully understand how this key enzyme of the denitrification pathway performs its task.

Acknowledgements

This work was supported by the Unidade de Ciências Biomoleculares Aplicadas-UCIBIO, which is financed by national funds from FCT/MEC (UID/Multi/04378/2013) and co-financed by the ERDF under the PT2020 Partnership Agreement (POCI-01-0145-FEDER-007728).

References

1. W. G. Zumft, *Microbiol. Mol. Biol. Rev.*, 1997, **61**, 533.
2. D. J. Richardson and N. J. Watmough, *Curr. Opin. Chem. Biol.*, 1999, **3**, 207.
3. B. C. Berks, S. J. Ferguson, J. W. Moir and D. J. Richardson, *Biochim. Biophys. Acta*, 1995, **1232**, 97.
4. P. J. Gonzalez, C. Correia, I. Moura, C. D. Brondino and J. J. Moura, *J. Inorg. Biochem.*, 2006, **100**, 1015.
5. C. Moreno-Vivian, P. Cabello, M. Martinez-Luque, R. Blasco and F. Castillo, *J. Bacteriol.*, 1999, **181**, 6573.
6. P. Tavares, A. S. Pereira, J. J. Moura and I. Moura, *J. Inorg. Biochem.*, 2006, **100**, 2087.
7. H. Dobbek, *Coord. Chem. Rev.*, 2011, **255**, 1104.
8. R. Hille, *Chem. Rev.*, 1996, **96**, 2757.
9. R. Hille, *Trends Biochem. Sci.*, 2002, **27**, 360.
10. R. Hille, J. Hall and P. Basu, *Chem. Rev.*, 2014, **114**, 3963.
11. G. Schwarz, R. R. Mendel and M. W. Ribbe, *Nature*, 2009, **460**, 839.
12. R. J. Williams and J. J. Frausto da Silva, *Biochem. Biophys. Res. Commun.*, 2002, **292**, 293.
13. P. J. Gonzalez, M. G. Rivas, C. S. Mota, C. D. Brondino, I. Moura and J. J. G. Moura, *Coord. Chem. Rev.*, 2013, **257**, 315.
14. S. Leimkühler, M. M. Wuebbens and K. V. Rajagopalan, *Coord. Chem. Rev.*, 2011, **255**, 1129.
15. G. Schwarz and R. R. Mendel, *Coord. Chem. Rev.*, 2011, **255**, 1145.
16. O. Einsle, *JBIC, J. Biol. Inorg. Chem.*, 2014, **19**, 737.
17. O. Einsle, F. A. Tezcan, S. L. Andrade, B. Schmid, M. Yoshida, J. B. Howard and D. C. Rees, *Science*, 2002, **297**, 1696.
18. Y. Hu and M. W. Ribbe, *Coord. Chem. Rev.*, 2011, **255**, 1218.
19. N. Saitou and M. Nei, *Mol. Biol. Evol.*, 1987, **4**, 406.
20. J. Felsenstein, *Evolution*, 1985, **39**, 783.
21. M. Nei and S. Kumar, *Molecular Evolution and Phylogenetics*, Oxford University Press, New York, 2000.
22. K. Tamura, G. Stecher, D. Peterson, A. Filipski and S. Kumar, *Mol. Biol. Evol.*, 2013, **30**, 2725.
23. C. D. Brondino, M. G. Rivas, M. J. Romao, J. J. Moura and I. Moura, *Acc. Chem. Res.*, 2006, **39**, 788.
24. M. J. Romao, *Dalton Trans.*, 2009, 4053.
25. M. K. Johnson, D. C. Rees and M. W. Adams, *Chem. Rev.*, 1996, **96**, 2817.
26. C. D. Brondino, M. J. Romao, I. Moura and J. J. Moura, *Curr. Opin. Chem. Biol.*, 2006, **10**, 109.
27. R. Hille, *Arch. Biochem. Biophys.*, 2005, **433**, 107.
28. R. Huber, P. Hof, R. O. Duarte, J. J. G. Moura, I. Moura, M. Y. Liu, J. LeGall, R. Hille, M. Archer and M. J. Romao, *Proc. Natl. Acad. Sci. U. S. A.*, 1996, **93**, 8846.
29. J. M. Rebelo, J. M. Dias, R. Huber, J. J. G. Moura and M. J. Romao, *JBIC, J. Biol. Inorg. Chem.*, 2001, **6**, 791.

30. M. J. Romao, M. Archer, I. Moura, J. J. G. Moura, J. Legall, R. Engh, M. Schneider, P. Hof and R. Huber, *Science*, 1995, **270**, 1170.
31. T. Santos-Silva, F. Ferroni, A. Thapper, J. Marangon, P. J. Gonzalez, A. C. Rizzi, I. Moura, J. J. G. Moura, M. J. Romao and C. D. Brondino, *J. Am. Chem. Soc.*, 2009, **131**, 7990.
32. H. Dobbek, L. Gremer, R. Kiefersauer, R. Huber and O. Meyer, *Proc. Natl. Acad. Sci. U. S. A.*, 2002, **99**, 15971.
33. N. Wagener, A. J. Pierik, A. Ibdah, R. Hille and H. Dobbek, *Proc. Natl. Acad. Sci. U. S. A.*, 2009, **106**, 11055.
34. I. Bonin, B. M. Martins, V. Purvanov, S. Fetzner, R. Huber and H. Dobbek, *Structure*, 2004, **12**, 1425.
35. M. Unciuleac, E. Warkentin, C. C. Page, M. Boll and U. Ermler, *Structure*, 2004, **12**, 2249.
36. C. Feng, G. Tollin and J. H. Enemark, *Biochim. Biophys. Acta*, 2007, **1774**, 527.
37. U. Kappler and S. Bailey, *J. Biol. Chem.*, 2005, **280**, 24999.
38. C. Kisker, H. Schindelin, A. Pacheco, W. A. Wehbi, R. M. Garrett, K. V. Rajagopalan, J. H. Enemark and D. C. Rees, *Cell*, 1997, **91**, 973.
39. K. Fischer, G. G. Barbier, H. J. Hecht, R. R. Mendel, W. H. Campbell and G. Schwarz, *Plant Cell*, 2005, **17**, 1167.
40. S. J. Brokx, R. A. Rothery, G. Zhang, D. P. Ng and J. H. Weiner, *Biochemistry*, 2005, **44**, 10339.
41. L. Loschi, S. J. Brokx, T. L. Hills, G. Zhang, M. G. Bertero, A. L. Lovering, J. H. Weiner and N. C. Strynadka, *J. Biol. Chem.*, 2004, **279**, 50391.
42. G. J. Workun, K. Moquin, R. A. Rothery and J. H. Weiner, *Microbiol. Mol. Biol. Rev.*, 2008, **72**, 228.
43. M. K. Chan, S. Mukund, A. Kletzin, M. W. Adams and D. C. Rees, *Science*, 1995, **267**, 1463.
44. Y. Hu, S. Faham, R. Roy, M. W. Adams and D. C. Rees, *J. Mol. Biol.*, 1999, **286**, 899.
45. C. A. McDevitt, P. Hugenholtz, G. R. Hanson and A. G. McEwan, *Mol. Microbiol.*, 2002, **44**, 1575.
46. R. A. Rothery, G. J. Workun and J. H. Weiner, *Biochim. Biophys. Acta*, 2008, **1778**, 1897.
47. P. Arnoux, M. Sabaty, J. Alric, B. Frangioni, B. Guigliarelli, J. M. Adriano and D. Pignol, *Nat. Struct. Biol.*, 2003, **10**, 928.
48. C. Coelho, P. J. Gonzalez, J. G. Moura, I. Moura, J. Trincao and M. Joao Romao, *J. Mol. Biol.*, 2011, **408**, 932.
49. J. M. Dias, M. E. Than, A. Humm, R. Huber, G. P. Bourenkov, H. D. Bartunik, S. Bursakov, J. Calvete, J. Caldeira, C. Carneiro, J. J. Moura, I. Moura and M. J. Romao, *Structure*, 1999, **7**, 65.
50. B. J. N. Jepson, S. Mohan, T. A. Clarke, A. J. Gates, J. A. Cole, C. S. Butler, J. N. Butt, A. M. Hemmings and D. J. Richardson, *J. Biol. Chem.*, 2007, **282**, 6425.
51. S. Najmudin, P. J. Gonzalez, J. Trincao, C. Coelho, A. Mukhopadhyay, N. M. Cerqueira, C. C. Romao, I. Moura, J. J. Moura, C. D. Brondino and M. J. Romao, *JBIC, J. Biol. Inorg. Chem.*, 2008, **13**, 737.

52. M. Jormakka, K. Yokoyama, T. Yano, M. Tamakoshi, S. Akimoto, T. Shimamura, P. Curmi and S. Iwata, *Nat. Struct. Mol. Biol.*, 2008, **15**, 730.
53. J. C. Boyington, V. N. Gladyshev, S. V. Khangulov, T. C. Stadtman and P. D. Sun, *Science*, 1997, **275**, 1305.
54. M. Jormakka, S. Tornroth, B. Byrne and S. Iwata, *Science*, 2002, **295**, 1863.
55. H. Raaijmakers, S. Macieira, J. M. Dias, S. Teixeira, S. Bursakov, R. Huber, J. J. Moura, I. Moura and M. J. Romao, *Structure*, 2002, **10**, 1261.
56. H. C. Raaijmakers and M. J. Romao, *JBIC, J. Biol. Inorg. Chem.*, 2006, **11**, 849.
57. H. Schindelin, C. Kisker, J. Hilton, K. V. Rajagopalan and D. C. Rees, *Science*, 1996, **272**, 1615.
58. F. Schneider, J. Lowe, R. Huber, H. Schindelin, C. Kisker and J. Knablein, *J. Mol. Biol.*, 1996, **263**, 53.
59. M. Czjzek, J. P. Dos Santos, J. Pommier, G. Giordano, V. Mejean and R. Haser, *J. Mol. Biol.*, 1998, **284**, 435.
60. L. Zhang, K. J. Nelson, K. V. Rajagopalan and G. N. George, *Inorg. Chem.*, 2008, **47**, 1074.
61. M. G. Bertero, R. A. Rothery, M. Palak, C. Hou, D. Lim, F. Blasco, J. H. Weiner and N. C. Strynadka, *Nat. Struct. Biol.*, 2003, **10**, 681.
62. M. Jormakka, D. Richardson, B. Byrne and S. Iwata, *Structure*, 2004, **12**, 95.
63. D. P. Kloer, C. Hagel, J. Heider and G. E. Schulz, *Structure*, 2006, **14**, 1377.
64. G. R. Chaudhry, I. M. Chaiken and C. H. MacGregor, *J. Biol. Chem.*, 1983, **258**, 5828.
65. G. R. Chaudhry and C. H. MacGregor, *J. Biol. Chem.*, 1983, **258**, 5819.
66. G. R. Chaudhry and C. H. MacGregor, *J. Bacteriol.*, 1983, **154**, 387.
67. C. Godfrey, C. Greenwood, A. J. Thomson, R. C. Bray and G. N. George, *Biochem. J.*, 1984, **224**, 601.
68. A. Craske and S. J. Ferguson, *Eur. J. Biochem.*, 1986, **158**, 429.
69. A. L. Ballard and S. J. Ferguson, *Eur. J. Biochem.*, 1988, **174**, 207.
70. C. Correia, S. Besson, C. D. Brondino, P. J. Gonzalez, G. Fauque, J. Lampreia, I. Moura and J. J. Moura, *JBIC, J. Biol. Inorg. Chem.*, 2008, **13**, 1321.
71. B. Lledo, R. M. Martinez-Espinosa, F. C. Marhuenda-Egea and M. J. Bonete, *Biochim. Biophys. Acta*, 2004, **1674**, 50.
72. L. I. Hochstein and F. Lang, *Arch. Biochem. Biophys.*, 1991, **288**, 380.
73. F. M. Ferroni, M. G. Rivas, A. C. Rizzi, M. E. Lucca, N. I. Perotti and C. D. Brondino, *BioMetals*, 2011, **24**, 891.
74. A. Parkin, C. F. Blanford, R. A. Rothery, R. Macey, M. Bertero, N. C. J. Strynadka, F. A. Armstrong and J. H. Weiner, *to be published*.
75. A. Magalon, J. G. Fedor, A. Walburger and J. H. Weiner, *Coord. Chem. Rev.*, 2011, **255**, 1159.
76. S. Y. Wu, R. A. Rothery and J. H. Weiner, *J. Biol. Chem.*, 2015, **290**, 25164.
77. S. P. Vincent and R. C. Bray, *Biochem. J.*, 1978, **171**, 639.
78. G. N. George, R. C. Bray, F. F. Morpeth and D. H. Boxer, *Biochem. J.*, 1985, **227**, 925.
79. N. Turner, A. L. Ballard, R. C. Bray and S. J. Ferguson, *Biochem. J.*, 1988, **252**, 925.

80. A. Magalon, M. Asso, B. Guigliarelli, R. A. Rothery, P. Bertrand, G. Giordano and F. Blasco, *Biochemistry*, 1998, **37**, 7363.
81. T. Hettmann, S. Anemuller, H. Borcherding, L. Mathe, P. Steinrucke and S. Diekmann, *FEBS Lett.*, 2003, **534**, 143.
82. F. Biaso, B. Burlat and B. Guigliarelli, *Inorg. Chem.*, 2012, **51**, 3409.
83. J. W. Peters, W. N. Lanzilotta, B. J. Lemon and L. C. Seefeldt, *Science*, 1998, **282**, 1853.
84. Y. Higuchi, T. Yagi and N. Yasuoka, *Structure*, 1997, **5**, 1671.
85. P. M. Matias, C. M. Soares, L. M. Saraiva, R. Coelho, J. Morais, J. Le Gall and M. A. Carrondo, *JBIC, J. Biol. Inorg. Chem.*, 2001, **6**, 63.
86. A. Volbeda, M. H. Charon, C. Piras, E. C. Hatchikian, M. Frey and J. C. Fontecilla-Camps, *Nature*, 1995, **373**, 580.
87. R. A. Rothery, M. G. Bertero, T. Spreter, N. Bouromand, N. C. Strynadka and J. H. Weiner, *J. Biol. Chem.*, 2010, **285**, 8801.
88. R. A. Rothery, M. G. Bertero, R. Cammack, M. Palak, F. Blasco, N. C. Strynadka and J. H. Weiner, *Biochemistry*, 2004, **43**, 5324.
89. P. Lanciano, A. Savoyant, S. Grimaldi, A. Magalon, B. Guigliarelli and P. Bertrand, *J. Phys. Chem. B*, 2007, **111**, 13632.
90. R. A. Rothery, F. Blasco, A. Magalon, M. Asso and J. H. Weiner, *Biochemistry*, 1999, **38**, 12747.
91. R. A. Rothery, A. Magalon, G. Giordano, B. Guigliarelli, F. Blasco and J. H. Weiner, *J. Biol. Chem.*, 1998, **273**, 7462.
92. B. J. Wallace and I. G. Young, *Biochim. Biophys. Acta*, 1977, **461**, 84.
93. J. Rendon, E. Pilet, Z. Fahs, F. Seduk, L. Sylvi, M. Hajj Chehade, F. Pierrel, B. Guigliarelli, A. Magalon and S. Grimaldi, *Biochim. Biophys. Acta*, 2015, **1847**, 739.
94. A. Magalon, D. Lemesle-Meunier, R. A. Rothery, C. Frixon, J. H. Weiner and F. Blasco, *J. Biol. Chem.*, 1997, **272**, 25652.
95. M. G. Bertero, R. A. Rothery, N. Boroumand, M. Palak, F. Blasco, N. Ginet, J. H. Weiner and N. C. Strynadka, *J. Biol. Chem.*, 2005, **280**, 14836.
96. S. Grimaldi, P. Lanciano, P. Bertrand, F. Blasco and B. Guigliarelli, *Biochemistry*, 2005, **44**, 1300.
97. P. Lanciano, A. Magalon, P. Bertrand, B. Guigliarelli and S. Grimaldi, *Biochemistry*, 2007, **46**, 5323.
98. R. Arias-Cartin, S. Lyubenova, P. Ceccaldi, T. Prisner, A. Magalon, B. Guigliarelli and S. Grimaldi, *J. Am. Chem. Soc.*, 2010, **132**, 5942.
99. S. Grimaldi, R. Arias-Cartin, P. Lanciano, S. Lyubenova, B. Endeward, T. F. Prisner, A. Magalon and B. Guigliarelli, *J. Biol. Chem.*, 2010, **285**, 179.
100. S. Grimaldi, R. Arias-Cartin, P. Lanciano, S. Lyubenova, R. Szenes, B. Endeward, T. F. Prisner, B. Guigliarelli and A. Magalon, *J. Biol. Chem.*, 2012, **287**, 4662.
101. D. C. Bay, C. S. Chan and R. J. Turner, *BMC Evol. Biol.*, 2015, **15**, 110.
102. F. Blasco, J.-P. Dos Santos, A. Magalon, C. Frixon, B. Guigliarelli, C.-L. Santini and G. Giordano, *Mol. Microbiol.*, 1998, **28**, 435.
103. P. Lanciano, A. Vergnes, S. Grimaldi, B. Guigliarelli and A. Magalon, *J. Biol. Chem.*, 2007, **282**, 17468.

104. A. Vergnes, J. Pommier, R. Toci, F. Blasco, G. Giordano and A. Magalon, *J. Biol. Chem.*, 2006, **281**, 2170.
105. W. R. Hagen, *Coord. Chem. Rev.*, 2011, **255**, 1117.
106. A. Hasona, W. T. Self, R. M. Ray and K. T. Shanmugam, *FEMS Microbiol. Lett.*, 1998, **169**, 111.
107. P. M. McNicholas, S. A. Rech and R. P. Gunsalus, *Mol. Microbiol.*, 1997, **23**, 515.
108. P. A. Cotter, V. Chepuri, R. B. Gennis and R. P. Gunsalus, *J. Bacteriol.*, 1990, **172**, 6333.
109. S. Spiro, R. E. Roberts and J. R. Guest, *Mol. Microbiol.*, 1989, **3**, 601.
110. P. A. Cotter and R. P. Gunsalus, *J. Bacteriol.*, 1989, **171**, 3817.
111. G. Unden, S. Becker, J. Bongaerts, G. Holighaus, J. Schirawski and S. Six, *Arch. Microbiol.*, 1995, **164**, 81.
112. M. S. Walker and J. A. DeMoss, *Mol. Microbiol.*, 1991, **5**, 353.
113. V. Stewart, *Biochem. Soc. Trans.*, 2003, **31**, 1.
114. C. E. Noriega, H.-Y. Lin, L.-L. Chen, S. B. Williams and V. Stewart, *Mol. Microbiol.*, 2010, **75**, 394.
115. R. S. Rabin and V. Stewart, *J. Bacteriol.*, 1993, **175**, 3259.
116. A. J. Darwin and V. Stewart, *J. Mol. Biol.*, 1995, **251**, 15.
117. C. Sparacino-Watkins, J. F. Stolz and P. Basu, *Chem. Soc. Rev.*, 2014, **43**, 676.
118. T. D. Ames and R. R. Breaker, *The Chemical Biology of Nucleic Acids*, John Wiley & Sons, Ltd, 2010, p. 433.
119. L. A. Anderson, E. McNairn, T. Leubke, R. N. Pau and D. H. Boxer, *J. Bacteriol.*, 2000, **182**, 7035.
120. E. E. Regulski, R. H. Moy, Z. Weinberg, J. E. Barrick, Z. Yao, W. L. Ruzzo and R. R. Breaker, *Mol. Microbiol.*, 2008, **68**, 918.
121. K. P. Baker and D. H. Boxer, *Mol. Microbiol.*, 1991, **5**, 901.
122. F. Blasco, C. Iobbi, J. Ratouchniak, V. Bonnefoy and M. Chippaux, *Mol. Gen. Genet.*, 1990, **222**, 104.
123. V. Bonnefoy and J. A. Demoss, *Antonie van Leeuwenhoek*, 1994, **66**, 47.
124. F. Blasco, J. Pommier, V. Augier, M. Chippaux and G. Giordano, *Mol. Microbiol.*, 1992, **6**, 221.
125. T. Palmer, C.-L. Santini, C. Iobbi-Nivol, D. J. Eaves, D. H. Boxer and G. Giordano, *Mol. Microbiol.*, 1996, **20**, 875.
126. A. Vergnes, K. Gouffi-Belhabich, F. Blasco, G. Giordano and A. Magalon, *J. Biol. Chem.*, 2004, **279**, 41398.
127. M. S. Kesawat, B. K. Das, M. Kumar, G. R. Bhaganagare and Manorama, *Biological Nitrogen Fixation*, John Wiley & Sons, Inc, 2015, p. 57.
128. M. Jormakka, B. Byrne and S. Iwata, *FEBS Lett.*, 2003, **545**, 25.
129. D. Richardson and G. Sawers, *Science*, 2002, **295**, 1842.
130. N. J. Wood, T. Alizadeh, D. J. Richardson, S. J. Ferguson and J. W. Moir, *Mol. Microbiol.*, 2002, **44**, 157.
131. H. Zheng, G. Wisedchaisri and T. Gonen, *Nature*, 2013, **497**, 647.
132. M. Fukuda, H. Takeda, H. E. Kato, S. Doki, K. Ito, A. D. Maturana, R. Ishitani and O. Nureki, *Nat. Commun.*, 2015, **6**, 7097.

133. N. M. Cerqueira, P. A. Fernandes, P. J. Gonzalez, J. J. Moura and M. J. Ramos, *Inorg. Chem.*, 2013, **52**, 10766.
134. N. M. Cerqueira, P. J. Gonzalez, C. D. Brondino, M. J. Romao, C. C. Romao, I. Moura and J. J. Moura, *J. Comput. Chem.*, 2009, **30**, 2466.
135. N. M. Cerqueira, P. J. Gonzalez, P. A. Fernandes, J. J. Moura and M. J. Ramos, *Acc. Chem. Res.*, 2015, **48**, 2875.
136. F. F. Morpeth and D. H. Boxer, *Biochemistry*, 1985, **24**, 40.
137. L. J. Anderson, D. J. Richardson and J. N. Butt, *Biochemistry*, 2001, **40**, 11294.
138. S. J. Elliott, K. R. Hoke, K. Heffron, M. Palak, R. A. Rothery, J. H. Weiner and F. A. Armstrong, *Biochemistry*, 2004, **43**, 799.
139. C. Leger and P. Bertrand, *Chem. Rev.*, 2008, **108**, 2379.
140. J. Marangon, P. M. Paes de Sousa, I. Moura, C. D. Brondino, J. J. Moura and P. J. Gonzalez, *Biochim. Biophys. Acta*, 2012, **1817**, 1072.

CHAPTER 4

Nitrite Reductase – Cytochrome cd$_1$

SERENA RINALDO[a], GIORGIO GIARDINA[a] AND FRANCESCA CUTRUZZOLÀ*[a]

[a]Department of Biochemical Sciences "A. Rossi Fanelli" and Istituto Pasteur Italia Fondazione Cenci Bolognetti, Sapienza University of Rome, P.le Aldo Moro 5, 00185 Rome, Italy
*E-mail: francesca.cutruzzola@uniroma1.it

4.1 The Haem Nitrite Reductase Gene *nirS*

Nitrite reductase (NIR) is a key enzyme in the dissimilative denitrification chain,[1] catalysing the reduction of nitrite to nitric oxide (NO) and representing the major known source of NO in bacteria.[2]

Two different classes of dissimilatory (NO-producing) NIRs have been characterised, containing either copper (CuNIR) or haem (cd_1NIR) as a cofactor, encoded by the *nirK* and *nirS* genes, respectively; the haem-containing enzyme, bearing two different haems (*i.e.*, *c*-haem and the d_1-haem), is found more frequently.[3] The two enzymes fulfil the same role *in vivo* as published by Zumft and coworkers, who demonstrated that the lack of cd_1NIR in *Pseudomonas stutzeri* can be complemented by CuNIR from *Pseudomonas aureofaciens.*[4]

The *nirS* gene is found in bacterial genomes in operons that usually also include the genes coding for ancillary functions, such as those required to

RSC Metallobiology Series No. 9
Metalloenzymes in Denitrification: Applications and Environmental Impacts
Edited by Isabel Moura, José J. G. Moura, Sofia R. Pauleta and Luisa B. Maia

Published by the Royal Society of Chemistry, www.rsc.org

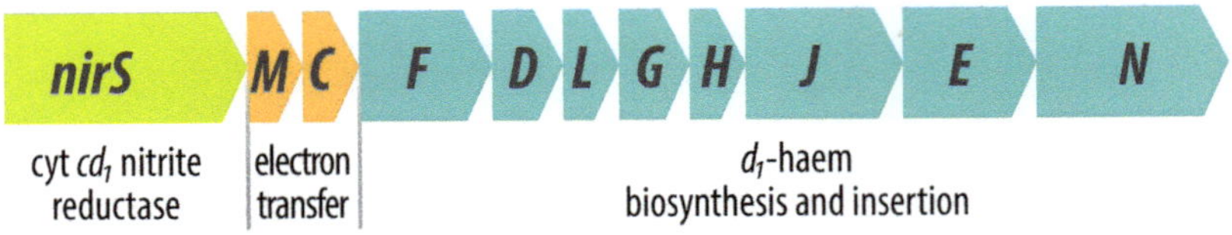

Figure 4.1 Organisation of the *nir* operon in *Pseudomonas aeruginosa*. The approximate sizes and transcriptional directions of the genes are given by the arrow boxes. Different colours refer to different biological function (*i.e.*, nitrite reductase, electron transfer and d_1-haem biosynthesis).

synthesise its specialised cofactor, the d_1-haem. The organisation of the genes in the *nir* operon from *P. aeruginosa* is shown in Figure 4.1 as an example; together with the *nirS* gene, the genes coding for the (putative) electron donor(s) are also found in this and other operons, such as *nirM*, coding for the small soluble *c*-type cytochrome c_{551}. In other bacteria, the *nirS* gene is found in more complex loci, such as in *P. stutzeri*, which also include regulatory factors.[5]

The nitrite-reducing capability can be located in selected bacterial species on transferable genetic units, as shown in *Thermus thermophilus*, in which denitrification is a strain-specific property.[6,7] In this species, the nitrate respiration capability is conferred by a conjugative element (NCE), whereas nitrite respiration requires a gene cluster (nitrite respiration cluster, NIC), located on a megaplasmid and encoding genes for NIR and ancillary proteins (*nirSJM*) and NO reductase (*nor*). These traits are easily transferred by natural competence or by a conjugation-like process and may be lost upon continuous aerobic growth, suggesting that their episomal location is a metabolic adaptation strategy.

The expression patterns of *nir* clusters from different bacteria are variable. For example, in *Paracoccus pantotrophus*, *nirS* is transcribed as a monocistronic operon from its own promoter only under anaerobic conditions with nitrate.[8] In contrast, in *P. stutzeri*, two transcripts are produced from a *nirSTB* operon, a monocistronic one for *nirS* and a tricistronic one including all genes.[9] The *Thermus thermophilus nirSJM* genes are expressed from two promoters, one producing a tricistronic mRNA under aerobic and anaerobic conditions and the other producing a bicistronic mRNA only under anoxic conditions plus a nitrogen oxide.[6]

In *P. aeruginosa*, the whole *nirSMCFDLGHJEN* cluster is co-transcribed from a single promoter.[10] Transcription of the *P. aeruginosa nir* operon is mainly controlled by the hierarchical action of two regulators, namely ANR (anaerobic regulator of arginine deiminase and nitrate reductase) and DNR (dissimilative nitrate respiration regulator), sensing oxygen and NO, respectively.[11,12] Under low-oxygen conditions, the *nir* operon is transcribed at low levels in an ANR-dependent manner; the *nir* promoter contains a FNR box, which was shown to be important for transcriptional regulation by ANR.[13] As suggested by Arai and coworkers,[13] the low-level expression of NIR might be necessary to initially provide NO for activation of the denitrification regulator DNR in the presence of nitrite. However, as the concentration of NO increases, ANR-mediated transcriptional regulation is inhibited by NO, which disrupts the $[4Fe–4S]^{2+}$ cluster in the ANR protein, resulting in the inability to bind to its cognate promoter DNA sequences.[14]

DNR is a haem-based gas sensor of the CRP–FNR superfamily, which positively responds to NO.[15–19] Other putative haem-containing transcription factors such as DNRD from *P. stutzeri* and NNR (nitrite and NO reductase regulator) from *Paracoccus denitrificans* were shown to be involved in the NO-dependent activation of their cognate *nirS* operons.[20,21] The affinity of the DNR–NO complex for its target DNA was recently determined to be in the nanomolar range.[22] Together with *nirS*, DNR activates other denitrification genes like *nirQ*, *norCB* (coding for NO reductase) and *nosZ* (coding for nitrous oxide reductase) in response to NO.[12,23] Taken together, all available evidence suggests that the expression of the *nirS* operon in *P. aeruginosa* involves a fine-tuned regulatory network mediated by ANR and DNR in response to oxygen and NO. Although the details of the regulation described in *P. aeruginosa* may differ from those that are operative in other bacterial species, which, to date, have been poorly characterised, it is likely that both oxygen and N-oxides do represent major factors in controlling *nirS* expression.

4.2 The *cd*$_1$NIR Protein

cd$_1$NIRs have been purified from many sources, including *P. aeruginosa*, *Pseudomonas nautica*, *P. stutzeri*, *P. pantotrophus*, *Paracoccus halodenitrificans*, *Roseobacter denitrificans*, *Azospirillum brasilense*, *Magnetospirillum magnetotacticum*, *Alcaligenes faecalis*, *Thiobacillus denitrificans*, *Ralstonia eutropha* and *Hydrogenobacter thermophilus*, and are generally located in the periplasmic space.[3]

cd$_1$NiR from *P. aeruginosa* (Pa-*cd*$_1$NIR) was initially characterised as a cytochrome oxidase for its tetraelectronic oxygen reductase activity;[24] later, Yamanaka's group[25] showed that the enzyme is also capable of catalysing the monoelectronic reduction of nitrite to NO, a reaction that is now accepted as being the only one that is physiologically relevant. This assignment has been later proved by genetic evidence in different biological backgrounds.[8,26,27] Recent data also assigned a role to Pa-*cd*$_1$NIR in mediating protein–protein interaction of the flagellar structural protein FliC and the protein chaperone DnaK in order to control flagellum-mediated motility.[28] The *NirS* mutant is defective in the formation of flagella and therefore in swimming motility; nevertheless, the NIR protein is required as a scaffold rather than for its NIR activity, given that mutations altering the sole *d*$_1$-haem biogenesis do not affect swimming ability.[28]

Among the purified *cd*$_1$NIR proteins, those from *P. aeruginosa* (Pa-*cd*$_1$NIR) and from *P. pantotrophus* (Pp-*cd*$_1$NIR) have been deeply characterised both structurally and functionally, and therefore, in this chapter, we will refer mainly to the properties of these two enzymes.

cd$_1$NIRs are homodimers containing one *c*-haem and one d_1-haem (3,8-dioxo-17-acrylate-porphyrindione) group in each subunit (Figure 4.2a). The electrons necessary to reduce the substrate(s) come from periplasmic macromolecular electron donors such as cytochrome c_{551} for Pa-*cd*$_1$NIR[29] or pseudoazurin and cytochrome c_{550} for Pp-*cd*$_1$NIR.[30] These electrons are accepted by the *c*-haem moiety of the enzyme and then internally transferred to the ferric d_1-haem, where catalysis occurs.[31] The d_1-haem (Figure 4.2a) belongs to the modified tetrapyrrole group and its chemical structure indicates that it falls within the iron-containing dioxo-isobacteriochlorin subgroup;[32] this cofactor is unique to *cd*$_1$NiRs[33] and it is responsible for the peculiar reactivity of *cd*$_1$NiRs with haem–iron ligands such as anions, NO and carbon monoxide (CO).[34] The uncommon structure of this cofactor requires a specialised biosynthetic pathway, which, up to now, has only been found in *cd*$_1$NiR-containing denitrifiers.[32]

The presence and the conservation of this peculiar cofactor and its dedicated biosynthetic pathway among denitrifiers clearly indicate that the

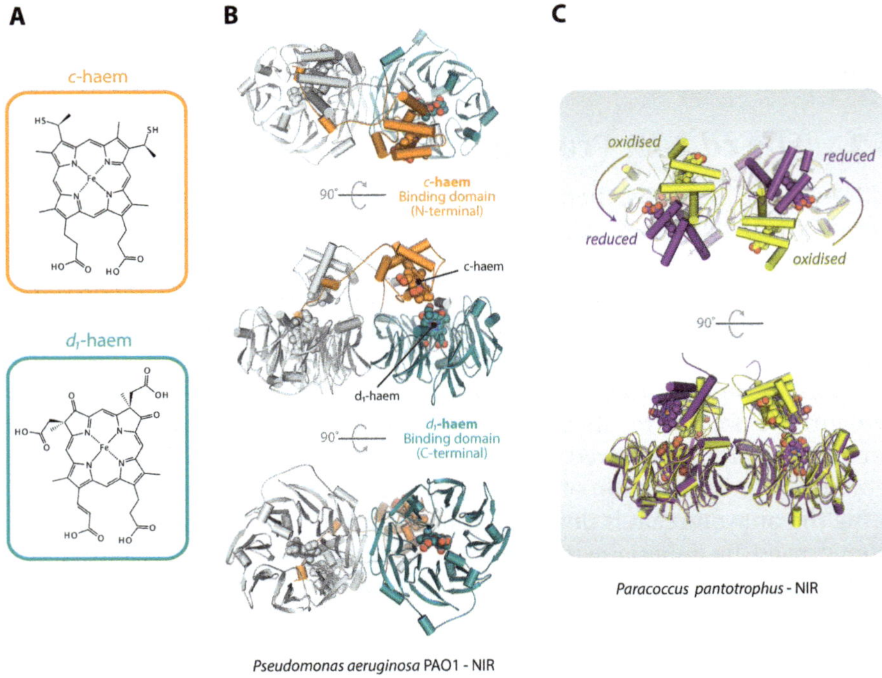

Figure 4.2 Structure of *cd*$_1$NIR and its cofactors. (a) Structure of the *c*-haem (upper frame) and d_1-haem (lower frame). (b) Structure of Pa-*cd*$_1$NIR in three different views (Protein Data Bank code: 1NIR). Colour code: *c*-domain, orange; d_1-domain, blue. (c) Superposition of oxidised "as isolated" Pp-*cd*$_1$NIR (in yellow; Protein Data Bank code: 1H9X) with the reduced derivative and crystallised under anaerobic conditions (in violet; Protein Data Bank code: 1QKS). The rotation of the *c*-domain in the two structures is clearly visible.

d_1-haem is designed to cope with nitrite reduction and NO dissociation in cd_1NiR. More intriguingly, the d_1-haem is also embedded in a complex protein environment that has evolved to efficiently couple the catalytic event (nitrite to NO) to an effective electron supply by tightly regulating the electron flow from the *c*-haem to the d_1-haem through redox-dependent conformational changes.

The state-of-the-art of the current knowledge on the structure and reactivity of the two cofactors (*c*-haem and d_1-haem) is reviewed in the following paragraphs, together with the structural features of the protein, including the redox-dependent conformational changes. Moreover, given the importance of these cofactors in determining the cd_1NIR activity, a paragraph describing their biogenesis is also included.

4.3 Structure and Reactivity of cd_1NIR

4.3.1 Structure Overview

The protein sequence of mature cd_1NIR, as compared to that translated from *nirS* genes, indicates that the protein is synthesised as a pre-protein, with a leader peptide responsible for the delivery into the periplasmic space. Pa-cd_1NIR and Pp-cd_1NIR mature proteins are 543 and 567 amino acids long, respectively.

As previously mentioned, the protein is a homodimer of two identical subunits. Within each subunit, the C-terminal domain, containing the d_1-haem cofactor (named the d_1-domain), presents an higher degree of sequence similarity between cd_1NIRs from different species than the N-terminal *c*-haem domain (*c*-domain), while sequence conservation is particularly low at the N-terminus region, outside of the *c*-domain fold.[26]

Structural information came from the high-resolution structure of both Pp-cd_1NIR (ferric, up to 1.2 Å resolution[35,36]) and Pa-cd_1NIR (ferric, 2.15 Å resolution[37]); the structures of the reduced and nitrite/NO-bound derivatives are also available.[38,39]

Pa-cd_1NIR and Pp-cd_1NIR share the same overall architecture, with each monomer being organised into two structurally distinct domains, one carrying the *c*-haem and the other the d_1-haem (Figure 4.2b and c). The dimeric state is stabilised by non-covalent interactions between monomers mainly involving hydrophobic interactions in the d_1-domain.[36,37] The distances between the haems within a subunit are about 11 Å edge to edge and 20 Å iron to iron; those between subunits are much larger (>40 Å).[36,37]

In Pa-cd_1NIR, the α-helical *c*-domain has a topology that is similar to that of class I cytochrome c (Figure 4.2b),[40] whereas a different connectivity of the helices is observed in Pp-cd_1NiR.[36] In both proteins, an N-terminal segment from the *c*-domain extends towards the d_1-domain and is inserted into the active site pocket. In Pp-cd_1NIR, this exchange occurs within the same monomer, but in Pa-cd_1NIR, a "domain swapping" occurs, since the tail of one monomer contacts the d_1-pocket of the neighbouring monomer (Figure

4.2b). The role of this N-terminal arm is still elusive; cd_1NIRs from other sources, such as *P. stutzeri*, lack this portion.[41] It has been proposed that this region could be involved in controlling the conformational changes occurring upon d_1-haem reduction (see below for details), but this hypothesis has been put into question by laser photolysis experiments on *P. stutzeri* cd_1NIR, whose kinetic behaviour suggests that a conformational rearrangement similar to that of Pa-cd_1NIR is taking place, despite the absence of the N-terminal arm.[41] Moreover, mutations of a conserved tyrosine falling in the d_1-haem pocket and coming from this arm in both Pa-cd_1NIR and Pp-cd_1NIR do not alter the overall enzyme activity.[42,43]

The d_1-domain forms a C-terminal rigid β-propeller bearing an eight-fold repeat of a four-antiparallel sheet motif (the "blade" of the propeller); the resulting central pocket accommodates the d_1-haem cofactor (Figure 4.2b).[36,37] The β-propeller of the enzymes from the two sources are superimposable and both are stabilised by a molecular "Velcro",[44] a sort of zipper where the C-terminal strand of the eighth blade interacts (*via* H-bonding) with three antiparallel strands coming from the N-terminal portion of such a domain (Figure 4.2b and c).

4.3.2 Haem Pockets and Reactivity

Structural and spectroscopic studies show that the identity of the protein ligands of the *c*-haem iron differs in cd_1NIRs from the two sources, being His_{51}–Met_{88} in Pa-cd_1NIR and His_{17}–His_{69} for Pp-cd_1NIR in the ferric form (Figure 4.3).[38,39,45–47] Interestingly, the coordination pattern of the ferric Pp-cd_1NIR changes in the ferrous state into His_{69}–Met_{106} in response to a large conformational change occurring upon reduction (Figure 4.3b).[39] Upon reduction of Pp-cd_1NIR, the loop of the *c*-domain holding Met_{106} shifts to the class I cytochrome c fold position, thus populating the canonical His/Met haem iron coordination.

The cytochrome c portion of cd_1NIR is crucial to delivering electrons from external donors to the active site; this domain is characterised by an unusual plasticity of the protein moiety, which controls the redox-dependent exchange of the *c*-haem protein ligands observed in Pp-cd_1NIR and confers a peculiar reactivity to the *c*-haem iron.

The reaction of cytochrome c with ligands such as NO usually occurs *in vitro* only under particular conditions,[48–51] involving either dissociation of the proximal methionine or its removal.[52,53] Nevertheless, it is clear that *c*-haem nitrosylation represents a signalling strategy in eukaryotes for controlling the release of mitochondrial cytochrome c in the cytoplasm during early apoptotic events.[51,54]

In Pa-cd_1NIR, the *c*-haem is hexacoordinated and low spin in both reduction states; contrary to expectations, this haem is able to form a complex with some ligands under relatively physiological conditions (NO and CN^-; see Table 4.1).[55–57] Previous literature data indicate that partial nitrosylation of the *c*-haem occurs very slowly (~10^{-4} s^{-1} at pH 6.4, upon addition of saturating

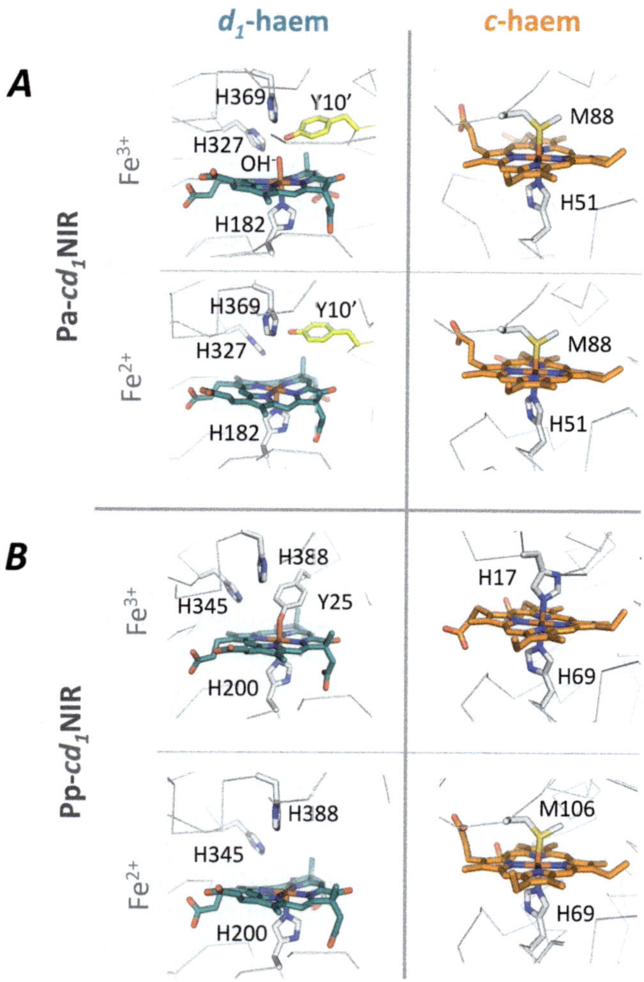

Figure 4.3 Structure of the d_1-haem (in blue) and the *c*-haem (in orange) pockets in Pa-*cd*$_1$NIR (a) and Pp-*cd*$_1$NIR (b) *cd*$_1$NIRs. Residues coordinating the haem iron are represented as sticks and numbered according to the corresponding protein sequence; in the case of Pa-*cd*$_1$NIR, Tyr_{10}, coming from the neighbouring monomer, is in yellow and numbered as Y10′. For each source, the corresponding haem coordination is reported both in the ferric and the ferrous states (upper and lower frames of each panel, respectively). Protein Data Bank codes: ferric Pa-*cd*$_1$NIR, 1NIR; ferrous Pa-*cd*$_1$NIR, 1BL9; ferric Pp-*cd*$_1$NIR, 1QKS; ferrous Pp-*cd*$_1$NIR, 1AOF.

atmosphere of NO gas) mainly at acidic pH, but the physiological relevance of such a reaction has not been determined.[55,57] It has been recently proposed that *c*-haem nitrosylation of *cd*$_1$NIR could represent a regulatory strategy in *P. aeruginosa*, as it occurs in eukaryotes.[57] When Pa-*cd*$_1$NIR is incubated with substrates under turnover conditions (excess of electrons and nitrite), the

Table 4.1 Ligand binding properties of the two haem in cd_1NIR. The dissociation constant (or the Michaelis constant in the case of NO_2^-) is reported when available (in Molar). N.D.: not determined.

Ligand	c^{3+} (M)	d_1^{3+}	c^{2+}	d_1^{2+}	References
O_2	No	No	No	10^{-4} Pa-cd_1NIR	77 and 82
CO	No	No	No	Cooperative binding, 6×10^{-6} Pa-cd_1NIR, 80×10^{-6} Pp-cd_1NIR	74 and 76
CN^-	2.3×10^{-3} Pa-cd_1NIR	Yes	No	9.5×10^{-6} Pa-cd_1NIR, 123×10^{-6} H369A Pa-cd_1NIR, 0.7×10^{-6} Pp-cd_1NIR	34,56,64
NO_2^-	N.D.	N.D.	N.D.	6×10^{-6} Pa-cd_1NIR, 12×10^{-6} Pp-cd_1NIR	85 and 91
NO	Yes	Yes	Yes	5.7×10^{-7} Pa-cd_1NIR, 0.9×10^{-7} H369A Pa-cd_1NIR, 1.8×10^{-7} Pp-cd_1NIR	34,57,86

NO produced *via* nitrite reduction binds to the *c*-haem, thus yielding fully nitrosylated *c*-haem simultaneously with the observed turnover ($k_{obs} = {\sim}10^{-2}$ s^{-1} at pH 6.2). This event occurs at two orders of magnitude more rapidly than with externally added NO gas and is also observed at pH 7.0 under conditions in which gaseous NO is unable to bind to the *c*-haem.[57] These data indicate that *c*-haem nitrosylation occurs mainly under conditions in which catalysis is favoured and product inhibition avoided (*i.e.*, balanced excess of both substrates); therefore, nitrosylation of the *c*-haem moiety might be considered a hallmark of a "healthy" cd_1NIR that is fully able to undergo turnover. A possible structural interpretation can be proposed, considering that the coordination bond of the *c*-haem iron with the sixth ligand (*i.e.*, Met88) is likely to be weakened during productive turnover (and therefore electron transfer), thus allowing NO to bind, in agreement with the observation that cd_1NIR conformations largely depend on the redox state of the protein. In *P. aeruginosa* cd_1NIR, *c*-haem reduction triggers a cascade of movements (see below) that could represent the "tip of the iceberg" of all possible conformations sampled by the *c*-haem domain during catalysis, possibly controlling and being controlled by *c*-haem nitrosylation.[57]

The d_1-haem in Pa-cd_1NIR is a low-spin hexacoordinate species in the ferric state and a high-spin pentacoordinate species in the ferrous state, as shown by EPR, NMR and magnetic circular dichroism measurements.[47,58,59] Structural data indicate that the so-called proximal ligand is provided by His_{182} and His_{200} in Pa-cd_1NIR and Pp-cd_1NIR, respectively,[37,39] while the sixth ligand in the low-spin (ferric) state is an hydroxide ion in Pa-cd_1NIR[37] and Tyr_{25}, a residue belonging to the N-terminal segment and thus to the *c*-domain, in Pp-cd_1NIR (Figure 4.3).[39] The His/Tyr Pp-cd_1NIR species is catalytically inert (resting, often referred to as the "as isolated" state of the protein[60]) and needs to be activated by reduction[61] or nitrite binding, but at a rate that is several orders of magnitude slower than turnover.[62]

As shown in Figure 4.4, the ligation and the oxidation states of the *c*- and d_1-haem iron confer to both Pa-cd_1NIR and Pp-cd_1NIR characteristic UV-vis absorbance spectra, in which the relative contribution of the *c*-haem and of the d_1-haem can be ascribed to specific absorption peaks.

Two conserved histidines, namely His_{327} and His_{369} in Pa-cd_1NIR and His_{345} and His_{388} in Pp-cd_1NIR, are located in the d_1-haem pocket (Figure 4.3); both of them are relevant for catalysis, as demonstrated by site-directed mutagenesis experiments.[36,63] These residues are likely to be involved in

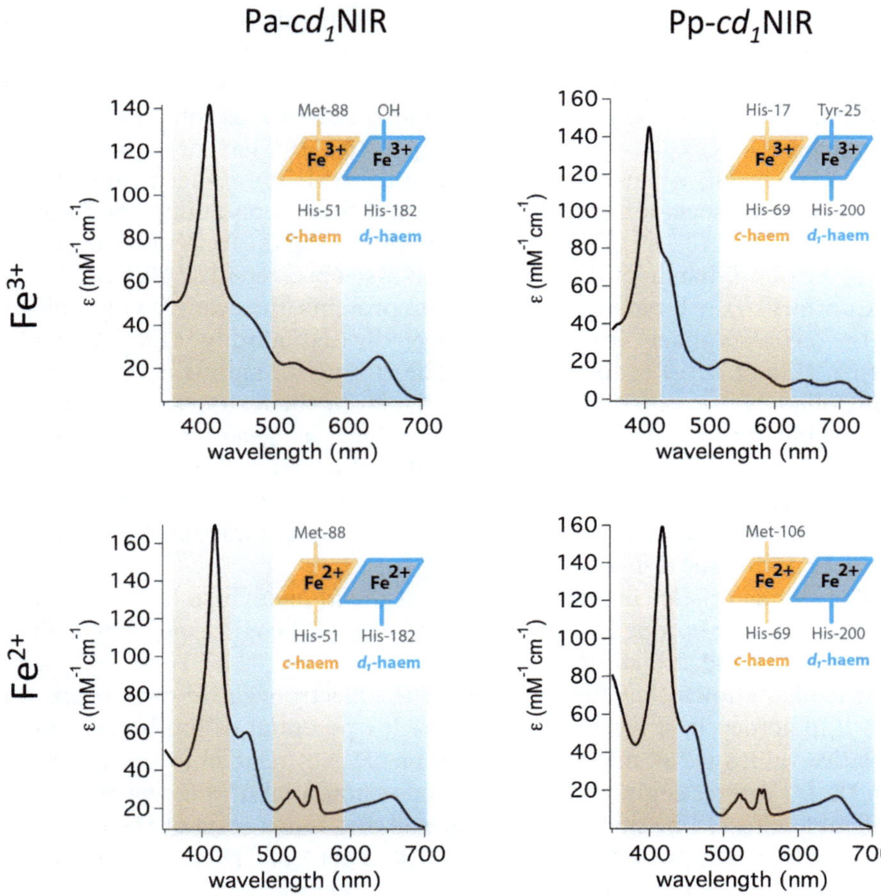

Figure 4.4 UV-vis absorbance spectra of Pa-cd_1NIR and Pp-cd_1NIR (left and right panels, respectively) both in the ferric and ferrous state (upper and lower panels, respectively). The regions of the spectrum where each cofactor mainly contributes are labelled in blue or orange, indicative of d_1-and *c*-haem, respectively. Ferric "as isolated" Pp-cd_1NIR shows the following maxima: 406, 525, 644 and 792 nm; ferric Pa-cd_1NiR shows the following maxima: 411, 520 and 640 nm. The reduced proteins display virtually identical spectra, thus reflecting their greater structural similarity. The ferrous peaks are 418, 460 and 521 nm, the α-band (split into a doublet whose peaks are 547 and 553 nm) and a broad peak at 655 nm.

H-bonding-mediated stabilisation of substrate bound to the haem iron, as well as providing the protons necessary to form a water molecule, produced as a by-product of nitrite reduction.[63–65] The structure of the nitrite-bound form of Pp-cd_1NIR indeed shows that one of the oxygens of nitrite is suitably positioned for H-bonding with the ε-nitrogen of both histidines.[39]

The peculiar chemical traits of the d_1-haem, such as the oxo groups and the acrylate side chain, are relevant for explaining its unusual reactivity with the classical haem iron ligands in cd_1NIR (see Table 4.1).[66] This peculiar reactivity is also observed in studies with the free cofactor or when the d_1-haem is reconstituted *in vitro* into other protein environments (myoglobin).[34,66–69] As an example, in the ferrous state, both the d_1-haem extracted from cd_1NIR and the d_1-haem in the holoprotein react with CO and NO quite weakly ($K_{D\,CO}$ = 6–80 μM and $K_{D\,NO}$ = 0.2–0.6 μM for cd_1NIRs; see also Table 4.1) as compared to other b-type haems ($K_{D\,CO}$ = ~10^{-2} μM and $K_{D\,NO}$ = ~10^{-3}–10^{-4} μM for globins).[34,70–76] Myoglobin itself, when reconstituted *in vitro* with the d_1-haem, displays a more rapid dissociation rate of NO than its b-haem containing counterpart ($k_{off\,NO}$ = 2 s^{-1} *vs.* 10^{-4} s^{-1}),[34] thus indirectly confirming the (relatively) low affinity of ferrous d_1-haem for this ligand that is also observed in cd_1NIRs.

Contrary to the b-type-containing haem proteins, the cyanide anion binds to the ferrous d_1-haem iron with unexpectedly high affinity (K_{DCN} = 1–10 μM for cd_1NIRs *vs.* 0.1–1 M for globins).[34] The affinity for anions is significantly controlled by the protein moiety, in particular by the residues located in the distal side of the d_1-haem pocket (Figure 4.3), as suggested by mutations of the conserved distal histidines in Pa-cd_1NIR, which strongly affect the affinity for cyanide or the substrate nitrite, and in turn the catalytic properties.[63,64] It is also known that while holo-Pa-cd_1NIR binds sodium metabisulphite upon reduction, the free d_1-haem does not.[68,69]

The characteristic chemical structure of the d_1-haem also leads to anomalous vibrational properties (high frequencies) of the corresponding Fe–CO and Fe–NO bond stretching in Pa-cd_1NIR.[75]

It is also known that the presence of the electronegative oxo groups on the d_1 macrocycle shifts the midpoint redox potential of the iron to more positive values (+287 mM in Pa-cd_1NIR at pH 7.0) relative to protoporphyrin IX.[77,78] Interestingly, the midpoint redox potential of the c-haem is considerably affected by the presence, ligation and redox state of the d_1-haem, ranging from +288 to +380 mV in Pa-cd_1NIR at pH 7.0.[77] This phenomenon is indicative of cross-talk between the two redox centres and suggests that the electron transfer pathway is affected by the electronic (and conformational) rearrangements occurring during reduction and catalysis.

4.3.3 Redox-Dependent Conformational Changes

A wealth of data suggests that the interplay between the two cofactors of cd_1NIR involves redox-controlled conformational rearrangements occurring in the surrounding protein environment. These data include static and dynamic structural studies by X-ray crystallography on different redox and ligation states, coupled to spectroscopic studies (mainly by UV-vis absorbance) monitoring the haem environment in the presence of different substrates/ligands of the two cofactors.

First of all, the structures of the two redox states of Pp-*cd*$_1$NIR show significant differences: the first residue of the *c*-domain is the Asp$_9$ in both subunits in the ferric state, but in the reduced state, obtained by soaking the oxidised "as isolated" crystals with reductants, it becomes Asp$_{36}$ and Glu$_{26}$ for subunits A and B, respectively. Moreover, both His$_{17}$ and Tyr$_{25}$ are disordered in the crystals, and the large part of the H-bond network occurring between the *c*- and the *d*$_1$-domain is disrupted.[39] Catalysis can occur in the crystal soaked with reductants and nitrite: in the crystallographic structure of the reaction intermediates, the His/His *c*-haem iron coordination species can be populated from the His/Met ferrous intermediate of the nitrite reduction reaction.[39] Although observed in the crystal, this redox-dependent *c*-haem ligand transition is unlikely to be involved in the catalytic cycle in solution: the estimated His/His ligation redox potential is lower than that of His/Met (estimated $E^{\circ\prime} \leq +50$ mV *vs.* ~+250 mV) and therefore the His/His *c*-haem is not a suitable electron acceptor from external donors.[61] In agreement with this prediction, kinetic and spectroscopic data indicate that the oxidised form populated during turnover and productively reduced from external electron donors such as pseudoazurin presents the His/Met coordination of the *c*-haem iron.[61] Moreover, when this Met is mutated into His, the corresponding protein loses its electron transfer capability in the *c*-domain.[79]

Redox-dependent conformational changes are not confined to the haem pockets. A more dramatic rearrangement of the *c*-domain than that observed in the aforementioned crystallographic studies was shown in a study carried out by Sjogren and Hajdu.[80] These authors solved the crystal structure of the reduced Pp-*cd*$_1$NIR, crystallised directly under anaerobic conditions (differently from those discussed above, which were obtained by soaking of the oxidised crystals). Surprisingly, the *c*-domain was found to be rotated by 60° and shifted by approximately 20 Å (Figure 4.2c). In this alternative conformation, the electron transfer pathway from the *c*-haem to the *d*$_1$-haem is dramatically changed.[80] As demonstrated by spectroscopic evidence, re-oxidation of these crystals does not restore the His/His ligation of the *c*-haem, thus confirming that this haem environment of Pp-*cd*$_1$NIR protein refers to an intermediate of the kinetic pathway.[80] Contrary to the "as isolated" form, this re-oxidised protein is catalytically capable of re-entering the nitrite reduction pathway.[61]

Also in Pa-*cd*$_1$NIR, some conformational changes occur upon reduction, though to a lesser extent. The Tyr$_{10}$ side chain, falling within the *d*$_1$-haem pocket and interacting with the axial ligand of ferric *d*$_1$-haem (*i.e.*, the hydroxyl ion; Figure 4.3a), is rotated around the Cβ–Cγ bond away from the active site and the hydroxide ion is dissociated.[38] The structures of the oxidised crystals of Pa-*cd*$_1$NIR soaked with reductants, followed by cryoquenching of the reaction at different times, indicate that the structural modifications at the level of the two haem pockets are absent when the sole *c*-haem is reduced; this result clearly indicates that the conformational changes are not triggered by the *c*-haem reduction, but rather by the internal electron transfer from the *c*-haem to the d_1-haem.[81]

To summarise, the conformational changes occurring during the internal electron transfer represent the switching of redox energy into conformational energy, which serves to "time and tune the NIR reaction."[80] Interestingly, the structure of the NO-bound ferrous derivative of the H327A and H369A mutants of Pa-cd_1NIR displays the same *c*-domain rotation observed in the reduced Pp-cd_1NIR (see above), thus showing that the redox-associated conformational changes are somehow coupled to the protonation step required for nitrite reduction.[65]

4.4 Mechanistic Aspects of cd_1NIR Catalysis

Nitrite reduction to NO is the physiologically relevant activity of cd_1NIR, even though reduction of oxygen and (to a lesser extent) of hydroxylamine are also reported.[82–85] In both Pa-cd_1NIR and Pp-cd_1NIR, NO is efficiently produced in a pH-dependent fashion, with the pH optima set below 6.5, where turnover reaches 242 s^{-1} in the case of Pp-cd_1NIR or 11 s^{-1} in the case of Pa-cd_1NIR.[85,86] The K_M for nitrite is dependent on the nature of the electron donor, which provides electrons for the nitrite reduction, ranging from ~6 to ~50 μM for Pa-cd_1NIR[63,87] and from ~7 to ~70 μM for Pp-cd_1NIR.[85] Although less efficient, the reduction of oxygen into water has been characterised kinetically and was found to be optimal at acidic pH as in the case of nitrite reduction (K_M = ~30 μM and k_{cat} = ~2 s^{-1} at pH 7.0 for Pa-cd_1NIR and K_M = ~150 μM and k_{cat} = ~3-6 s^{-1} at pH 7.0 for Pp-cd_1NIR).[82,83,85]

Steady-state and pre-steady-state kinetic studies indicate that the catalytic cycle of nitrite reduction involves the following steps: substrate binding to the d_1-haem iron of the fully reduced enzyme; reduction by electrons coming from the d_1-haem iron and conversion to NO by a protonation/dehydration step; and product release and electron transfer to populate the starting fully reduced enzyme (summarised in Figure 4.5 and described in more detail below).

For many years, the final steps of the catalytic cycle (*i.e.*, product release and repopulation of the fully reduced enzyme[88]) were considered to be mechanistically puzzling. The main working hypothesis suggested that re-reduction of the d_1-haem might only occur after NO dissociation in order to avoid trapping of the enzyme in a stable ferrous–NO adduct (the so-called "dead end"). To achieve this goal, the enzyme should tightly control the internal electron transfer rate, coupling it to the large redox-dependent conformational rearrangements seen in the crystal structure. However, more recent data overturn the idea of the obligate "dead end" formation upon NO binding to ferrous d_1-haem, thus opening up novel scenarios regarding the mechanistic details of cd_1NIR's catalytic activity and, more generally, the reactivity of modified haems with NO.[34]

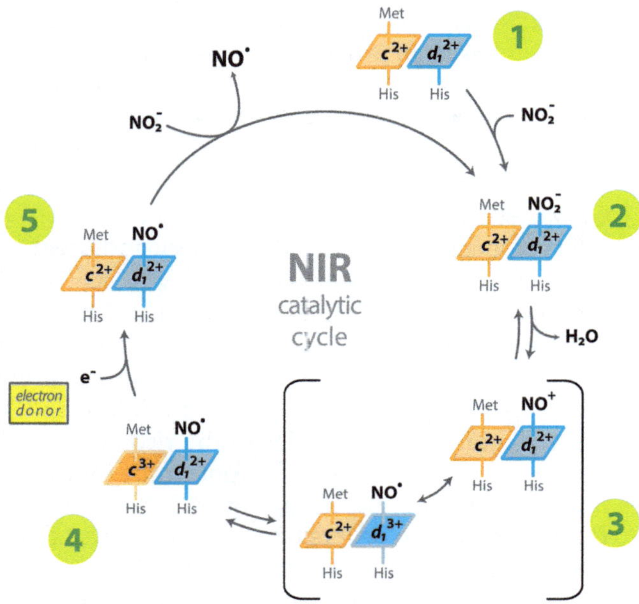

Figure 4.5 Proposed reaction mechanism for the catalytic cycle of cytochrome cd_1 nitrite reductases. Starting from the fully reduced enzyme (species 1, $c^{2+}d^{2+}$), nitrite binds to the ferrous d_1-haem, yielding the Michaelis complex (species 2, $c^{2+}d^{2+}NO_2^-$). Electron transfer from the d_1-haem iron to nitrite produces nitric oxide (NO) bound to the oxidised haem iron (species 3, $c^{2+}d^{3+}$NO) and a water molecule, which is released. All available data suggest that NO dissociates from the fully reduced NO-bound enzyme (species 5, $c^{2+}d^{2+}$NO) in both Pa-cd_1NIR[86] and Pp-cd_1NIR;[34] formation of species 5 implies that species 4 is populated by the internal electron transfer step (*i.e.*, reduction of the d_1-haem by the *c*-haem cofactor), followed by reduction by an external electron donor (yielding species 5). The physiological macromolecular reductants are small c-type cytochromes (such as *Pseudomonas aeruginosa* cytochrome c_{551} or *Paracoccus pantotrophus* cytochrome c_{550}) or copper proteins (such as pseudoazurin in *P. pantotrophus*), which are able to transfer the electron only to the *c*-haem. In principle, upon NO dissociation from species 5, the fully reduced enzyme should be reformed (species 1) and enter a new catalytic cycle. However, experimental data indicate that it is likely that dissociation of NO preferentially occurs upon nitrite binding (*i.e.*, a displacement reaction in the presence of excess nitrite takes place rather than species 1 being populated).[107] The scheme refers to a single monomer of the enzyme.

4.4.1 Substrate Binding

Binding of nitrite occurs to the Fe^{2+} of the d_1-haem, as demonstrated by kinetic and structural data;[39,89–91] the nitrite molecule binds *via* the N-atom in Pp-cd_1NIR, forming the so-called nitro-complex.[39] This binding mode agrees well with data on other haem proteins[92,93] and on synthetic iron porphyrin/nitrite complexes in which the nitro-complex was observed, regardless of the iron oxidation state.[94–96] This binding mode is also compatible with the catalytic mechanism that has been proposed to date, which involves the double protonation of a terminal oxygen atom of the nitrite molecule, mediated by the two conserved His residues in the d_1-haem pocket.[63]

The formation of the Michaelis–Menten complex occurs very rapidly, in the dead-time of a classical stopped-flow apparatus,[90,91] even at a suboptimal pH (8.0); the k_{on} was estimated to be $\geq 10^8\ M^{-1}\ s^{-1}$ in the case of Pa-cd_1NIR.[91] This value is orders of magnitude higher than the second-order rate constant for the bimolecular reaction with oxygen ($k_{on} = 3.3 \times 10^4\ M^{-1}\ s^{-1}$).[82,83]

As mentioned above, cd_1NIRs display unusual high affinity for anionic molecules in the ferrous state, including the inhibitor cyanide.[64,97] This peculiar behaviour could be ascribed to the d_1-haem bearing two electron-withdrawing carbonyl groups on the tetrapyrrole ring (Figure 4.2a), as well as to protein moiety, in particular to the two conserved histidines in the active site pocket (Figure 4.3).[36,37] Upon mutagenesis of the latter His residues into Ala in Pa-cd_1NIR, the affinities for nitrite and for cyanide drop (the k_{on} for cyanide is two orders of magnitude lower than that of the wild-type enzyme) and the turnover rate is reduced 100-fold.[63,64] The residue His_{369} in Pa-cd_1NIR is also crucial for stabilising negatively charged intermediates during the reduction of oxygen to water: replacement of His_{369} with Ala allows these intermediates to cause d_1-haem bleaching in the absence of excess reductants.[83]

Information on the very early events upon nitrite binding came from rapid freeze quenching experiments carried out on Pp-cd_1NIR where a species was populated within the first 130 μs after mixing the enzyme (reduced; species 1 in Figure 4.5) with the substrate nitrite; the UV-vis spectrum of this species is characterised by an unchanged c-haem contribution (still fully reduced), by the disappearance of the Soret peak of the d_1-haem at 460 nm (indicative of ligand binding at the d_1-haem) and by the formation of a peak at 620 nm. Given that this timeframe is significantly lower than the half time of the reaction (as inferred by the k_{cat} value; see above) and that no electrons have yet been delivered from the c-haem to the d_1-haem, it is likely that this is the Michaelis–Menten complex (species 2 in Figure 4.5), which decays within 1–2 ms into a mixed species containing both the $Fe^{2+}NO^+$ and Fe^{2+} species at the d_1-haem (peaking at 630 and 660 nm, respectively) and the Fe^{2+}/Fe^{3+} mixed state at the c-haem.[90] Since these experiments have been carried out without an excess of reductants, the enzyme can enter only one catalytic cycle. Thus, under these conditions, slow oxidation of the c-haem is observed, counterbalanced by a back reduction of the c-haem itself by the d_1-haem, yielding, within 11 ms, a species displaying ~55% of c-haem oxidation (in agreement with previous kinetic data).[89] This species is unlikely to be a genuine intermediate, but rather a resting state waiting for electron supply from external donors.[90] On the other hand, pre-steady-state experiments on Pa-cd_1NIR have been carried out in the presence of excess reductants to reach a stationary phase, whose features are further discussed below.

4.4.2 Electron Transfer From the *c*-Haem to the d_1-Haem

After the reduction of nitrite (species 3 in Figure 4.5), the d_1-haem is re-reduced by intramolecular electron transfer from the *c*-haem; this step is required to yield the nitrosylated mixed-valence species (species 4 in Figure 4.5), which in turn is reduced by external electron donors *via* the *c*-haem (species 5 in Figure 4.5). Studies carried out on cd_1NIRs from different sources indicate that this step occurs at considerably different rates: 3–4 s^{-1} in *P. aeruginosa*,[98,99] 1400 s^{-1} in *P. pantotrophus*,[100] 33 s^{-1} in *P. nautica*[101] and 23 s^{-1} in *P. stutzeri*.[102] Interestingly, in Pa-cd_1NIR, the internal electron transfer rate is very close to the overall turnover rate at pH 7.0 (6 s^{-1}),[86] and thus it has been proposed to be the rate-limiting step of the reaction for this specific enzyme.

All available data indicate that cd_1NIRs display cooperative behaviour. Redox titrations indicate that the *c*- to d_1-haem electron transfer process is cooperative, implying the existence of a strong thermodynamic driving force for a second electron transfer to follow the first one. In Pa-cd_1NIR, this cooperativity has been attributed to subunit–subunit interactions,[103] while in Pp-cd_1NIR, cooperativity resides in the same subunit, thus implying cross-talk between the *c*- and the d_1-haem moiety.[104]

More recently, negative cooperativity between the two d_1-haem sites in the Pa-cd_1NIR dimer has also been observed using pulse radiolysis experiments.[99] In more detail, the internal electron transfer rate decreases (by more than one order of magnitude) as the number of electrons introduced in the enzyme is increased.[99] This kinetic profile may be associated with a decrease in the electronic coupling between donor and acceptor sites as a consequence of the aforementioned conformational changes observed between reduced and oxidized Pa-cd_1NIR.[38] In agreement with this structural model, the internal electron transfer is faster (about 10-fold) when the conserved His_{369} residue is substituted with Ala.[99] The decrease of the positive charge density on the distal side of the d_1-haem due to the mutation destabilises the hydroxide ligand, lowering the energy barrier for electron transfer and thus increasing the rate.

The cooperative behaviour of cd_1NIRs has also been observed upon the binding of several ligands to the d_1-haem iron, thus further confirming the presence of cross-talk between the two redox centres and highlighting the heterogeneity of the monomers in the dimer.[73–76]

4.4.3 Catalysis and Product Release

After the internal electron transfer step, the ferrous nitrosylated d_1-haem is populated; as discussed above, in line with the other haem proteins characterised so far, this adduct could be very stable and could represent a "dead end" inhibited species (species 4 in Figure 4.5),[105] given that NO binds tightly to ferrous haem proteins.[71,106] A previous hypothesis on the catalytic mechanism envisaged the possibility that NO dissociates from the ferric d_1-haem prior to the delivery of the electron from the *c*-haem to the d_1-haem in order to yield the unbound reduced species.[63] However, as discussed above, NO dissociation from reduced cd_1NIR is fast (up to 200 s^{-1}),[34] several orders of magnitude greater than that measured from other b-haem-containing proteins, and the fully reduced, NO-bound species is not firmly inhibited by NO.[86,107] In light of this evidence, the catalytic scheme reported in Figure 4.5, which is also the simplest reaction cycle that one can draw for these enzymes, represents the most plausible pathway for nitrite reduction.

The observation that, contrary to expectations, the d_1-haem needs to be reduced in order to allow effective product release may explain why this class of enzymes requires two redox centres, with the second ensuring the proper (kinetically controlled) electron supply to the first.[34]

Even though fast NO dissociation from d_1-haem is a common feature of cd_1NIRs and, more generally, is a peculiar trait of the d_1-haem, this kinetic process represents the rate-limiting step of nitrite reduction only in Pp-cd_1NIR, in which it occurs at 200 s^{-1} and 65 s^{-1} ($k_{cat} = 72\ s^{-1}$).[34] Interestingly, the biphasic NO dissociation observed in Pp-cd_1NIR could mirror the heterogeneity of the two monomers in the dimer, as observed for other phenomena discussed previously. On the other hand, in Pa-cd_1NIR, NO dissociates at 70 s^{-1}, which is unlikely to be the rate-limiting step, due to the turnover rate being 6 s^{-1} (pH 7.0), a value that is closer to the electron transfer rate, as previously mentioned.[34,86]

Thus, the reaction scheme shown in Figure 4.5 is in agreement with all of the available data, including other experiments carried out on Pp-cd_1NIR showing that: (i) the intramolecular *c*-haem to d_1-haem electron transfer triggers product release;[108] (ii) the ferric–NO d_1-haem species is a very long-lived species in the absence of excess reductants;[89,90,109] and (iii) the protein works efficiently only in the presence of substrate and electron donor proteins (*i.e.*, it cannot release NO in the absence of reducing equivalents).[79,85,89,90]

The conclusion that NO is not an inhibitor of cd_1NIRs in the presence of excess substrates allows us to explain the linear rate of NO production observed even at tens of μM.[110] Indeed, the efficient binding of nitrite to the ferrous NO-bound adduct prevents enzyme trapping or inhibition, as demonstrated with displacement experiments, allowing the enzyme to enter a new catalytic cycle.[34,107]

The peculiar high affinity of ferrous d_1-haem for nitrite (see above), together with its unforeseen "low" affinity for NO, is therefore crucial to promoting NO dissociation and turnover. When the affinity of Pa-cd_1NIR for nitrite is decreased by site-directed mutagenesis, as in the case of the H369A and H327A mutants, the fully reduced, NO-bound derivative accumulates and nitrite fails to displace NO.[63,107] Other residues such as Tyr_{10}, located in the distal part of the d_1-haem pocket in the ferrous–NO-bound state, are less relevant for catalysis and NO stabilisation.[43,111]

ENDOR-DFT calculations reveal that the protein environment in cd_1NIRs plays a minor (but not null) role in controlling the reactivity of d_1-haem with NO:[111,112] these investigations indicate that, in Pa-cd_1NIR, the triad of distal residues (Tyr_{10}, His_{327} and His_{369}) assembles a highly mobile and flexible H-bonding network, which contributes to lowering the extent of the back-bonding effect of ferrous haem iron on NO. In summary, although the effect is not as relevant as that of d_1-haem, it plays a significant role in lowering the tightness of the Fe–NO interaction. This H-bonding network also shows a considerable degree of plasticity, since it can compensate for the loss of one residue (by mutagenesis), leading to unchanged Fe–N–O angles and bond lengths in the mutants as compared to the wild-type protein.[112,113]

The kinetic scheme described here clearly implies that electrons coming from external donors are readily provided in order to populate species 5 of Figure 4.5. The macromolecules involved in this reduction reaction are most likely (but possibly not exclusively) pseudoazurin and/or cytochrome c_{550} for Pp-cd_1NIR and cytochrome c_{551}, whose gene is found in the *nir* operon (Figure 4.1), for Pa-cd_1NIR. *In vitro* reduction of cd_1NIR can also be obtained with other macromolecules, such as horse heart cytochrome c and the *P. aeruginosa* copper protein azurin.[85,114–116]

Regardless the macromolecular partners involved, it is clear that the high affinity for nitrite, the allosteric control of the internal electron transfer rate and the exceptionally fast NO dissociation rate fit with the proposed (and simplest) catalytic mechanism (Figure 4.5), making these cd_1NIR traits unique among haem proteins.

4.5 Biogenesis of the *c*-Haem and the d_1-Haem

The biogenesis of cd_1NIR requires several steps, starting from the translocation of the protein across the inner membrane, followed by the delivery of the prosthetic groups into the periplasm. To obtain the mature cd_1NIR, covalent binding of *c*-haem (as observed for other periplasmic *c*-type cytochromes), insertion of the noncovalent d_1-haem and folding of the protein are required.[1,117] While the order of these events is not exactly known, the biosynthetic pathways of these prosthetic groups and some mechanistic details of their maturation have been reported. The biogenesis of prosthetic groups involves specialised enzymatic pathways, in the case of the d_1-haem found only in denitrifiers, finely regulated by environmental stimuli, such as anaerobic or O_2-limited growth and the presence of nitrates/nitrites.[1] More generally, denitrifying conditions demand a more abundant haem protein content and therefore a steady supply of porphyrins along the anaerobic pathways of synthesis.[1]

The biosynthesis of the haem precursor δ-aminolevulinic acid (ALA) is one of the checkpoints used by denitrifiers to switch to anaerobic and N-oxide-dependent metabolism. The *hemA* transcript (glutamyl-tRNA reductase) of *P. aeruginosa*, together with the transcript of *hemN* involved in the downstream O_2-independent protohaem IX (or b-haem) biogenesis, is increased during denitrification by the coordinated action of the transcriptional regulators ANR and DNR, sensing O_2 and NO, respectively.[118,119] ALA biosynthesis leads to the production of the haem precursor uroporphyrinogen III, which is at the crossroads between the biosynthesis of the d_1-haem and c-type cytochromes (*via* protohaem IX), with both pathways being necessary for cd_1NIR assembly.[1]

4.5.1 *c*-Haem

In *c*-type cytochromes, the apoprotein is covalently bound to a b-haem (protohaem IX) cofactor. *c*-Type cytochromes present two thioether bonds between the vinyls of the haem moiety at positions 2 and 4 of the tetrapyrrole ring and the thiols of two cysteines of a conserved haem-binding motif (C1XXC2H, where "X" denotes any residue). In Pa-cd_1NIR, the two conserved cysteines correspond to residues 47 and 50, while in Pp-cd_1NIR, they are Cys_{65} and Cys_{68}.

In Gram-negative bacteria, the *c*-type cytochromes exert their function in the periplasmic space: in this cellular compartment, the protohaem is covalently attached to the cognate apoprotein by the appropriate maturation apparatus. This mechanism was also confirmed for cd_1NIR by a genetic approach: in a *P. denitrificans* mutant lacking the *c*-type cytochrome biosynthetic machinery, the cd_1NIR polypeptide is found in the periplasmic space without both cofactors (fully apo-cd_1NIR).[120,121] Although the precise details of the machinery involved in maturation of the *c*-haem domain of cd_1NIR have not been clarified, the available data are briefly summarised here.

The apoprotein is usually translocated across the inner membrane as an unfolded protein *via* the Sec apparatus.[122] Among Gram-negative bacteria, in the α-, γ- and δ-proteobacteria (including *P. aeruginosa* and *P. pantotrophus*), cytochrome c maturation involves up to nine membrane-bound proteins belonging to system I, named CcmABCDEFGH(I) (see ref. 123 and 124 for up-to-date reviews). Given that the denitrification process requires the *de novo* synthesis of a substantial number of *c*-type cytochromes under anaerobic conditions, including cd_1NIR, it is highly likely—and to some extent proved—that the *ccm*-based system controls the maturation of the cytochrome c pool involved in denitrification. Genetic evidence indicates that cytochrome c-deficient mutants of *Bradyrhizobium japonicum*[125] and *P. aeruginosa*[126] lose the capability to anaerobically grow by nitrate respiration. On the other hand, the availability of the entire *ccm* operon in a single plasmid (pEC86)[127] allows the heterologous overexpression of semi-apocd_1NIR (*i.e.*, lacking the sole d_1-haem) in a recombinant, non-denitrifier background such as *Escherichia coli.*[128]

c-Type cytochrome biogenesis can be divided into three distinct steps: (i) transport and relay of haem b; (ii) preparation and chaperoning of ligation-competent *c*-type apocytochromes; and (iii) ligation of haem-apocytochrome c to yield holocytochrome c.

The haem b translocation step should mainly involve CcmA and CcmB, which show sequence conservation typical of the ABC transporter family;[123] however, an alternative route has been hypothesised given that periplasmic *b*-type cytochromes (but not *c*-type cytochromes) can be produced in the absence of Ccm proteins.[129] Once in the periplasm, haem is bound to CcmC only as a ternary complex with CcmE, to which haem is shuttled.[130] Release of holo-CcmE is then accomplished thanks to the coordinate activity of CcmD and the ATPase activity of CcmA. A recent hypothesis assigned to the haem-containing CcmF protein the role of a specific haem oxidoreductase, which could reduce the haem iron, although this is still a matter of debate.[131] Ferrous haem is required for thioether formation with the apocytochrome,[132] and this redox state favours haem release from holo-CcmE.[131] The same group demonstrated that the CcmABCDE pathway might not be strictly necessary for cytochrome c biogenesis (rather, it is involved in haem delivery) and that the "core" of the system I holocytochrome c synthetase is the CcmFH complex and CcmG (see below).[133]

The system I holocytochrome c synthetase is crucial for the thioreduction step of cytochrome c biogenesis, where the oxidised internal disulphide of the apocytochrome is reduced into two free thiols, leading to the competent species for haem ligation.[123] It has been shown that CcmF forms an integral membrane complex with CcmH where the formation of the thioether bond between haem and apocytochrome should occur.[134,135] CcmH and CcmG are membrane-tethered thioredox-active proteins that likely maintain the thiol groups of the apocytochrome in the reduced state, possibly *via* the formation of a mixed disulphide between CcmH and oxidised apocytochrome, which is subsequently resolved by CcmG.[136,137]

The final step involving apocytochrome chaperoning and haem attachment to the reduced haem-binding motif requires CcmF and, if present, CcmI, possibly together with CcmH;[123] the *ccmI* gene is present only in some Ccm operons, while in others, the corresponding ORF is present within the *ccmH* gene (as in *E. coli*). CcmI contains a TPR motif, which is known to provide a scaffold for protein–protein interactions or multiprotein complex assembly *via* the chaperoning of unfolded proteins[138]; this motif may provide a support for the unfolded apocytochrome, chaperoning it to the haem attachment site,[139] presumably located on CcmF,[133] although the mechanistic details are still unknown.[123]

4.5.2 d_1-Haem

A functional cytochrome cd_1NIR carrying both types of haem groups cannot be heterologously expressed from the *nirS* coding sequence alone. As an example, the expression of *nirS* from *P. aeruginosa* in non-denitrifying host such as *Pseudomonas putida* yields semi-apo cytochrome cd_1NIR, carrying only the *c*-haem cofactor.[140] The d_1-haem is an iron-containing dioxo-isobacteriochlorin with structural modifications, as compared to other "common" haems (Figure 4.2), and this protein has been found only in cd_1NIR. The main structural differences found in the d_1-haem, as compared to the others, are the saturation of two of the pyrrole rings and the presence of methyl groups at C2 and C7, methyl groups in the place of acetate groups at C12 and C18, oxo groups in the place of propionate groups at C3 and C8 and an acrylate group oxidised from a propionate group at C17 (Figure 4.2). Therefore, synthesis of the d_1-haem starting from uroporphyrinogen III requires: (i) methylation at rings I and II; (ii) decarboxylation at rings III and IV; (iii) introduction of the oxo groups at rings I and II; and (iv) dehydrogenation of the propionate sidechain on ring IV (Figure 4.6). It has been previously assumed that the last step should be the insertion of iron by a chelatase and the transport across the membrane so that it can be inserted into an apocytochrome, along with the *c*-haem.[1]

It has been proposed that modified haems, such as the d_1-haem, have been selected for their specific chemico-physical properties that allow them to cope with peculiar biochemical reactions.[141,142] This is the case for the d_1-haem; in the ferrous state, this cofactor shows high affinity for anions and an unexpected "low" affinity for NO as compared to other haems, thus allowing *cd*$_1$NIR to productively bind and reduce nitrite into NO, finally releasing this compound without being trapped.[34,64,107] These "*ad hoc*" properties could justify the conservation of a dedicated (and costly) biosynthetic pathway found in the denitrifiers' genomes alongside the *nirS* gene (*nir* operon[s]; see above and Figure 4.1).[1] It is known from mutagenesis studies that approximately eight gene products are needed, and while the mechanistic details have been reported very recently, many questions are still unanswered.

Most of the steps of d_1-haem biosynthesis are believed to take place in the bacterial cytoplasm, with the exception of the last step, which was suggested to be the dehydrogenation of one of the propionate side chains to the corresponding acrylate side chain of d_1-haem, occurring in the periplasm.[143,144]

The first step involves the conversion of uroporphyrinogen III into precorrin-2 (*i.e.*, methylation at rings I and II; Figure 4.6) by the action of the NirE protein, which is an S-adenosyl-L-methionine-dependent uroporphyrinogen III methyltransferase.[145,146] The route of reaching the d_1-haem cofactor from this intermediate was found to be partially shared with other modified haems, which requires the sirohaem tetrapyrrole as a precursor.[147]

Conversion of precorrin-2 into sirohaem requires two enzymatic reactions catalysed by a precorrin-2 dehydrogenase and a sirohydrochlorin ferrochelatase,[148] encoded by genes located in the *nir* operon of *P. aeruginosa*. The requirement of sirohaem as an actual intermediate of d_1-haem biosynthesis indicates that the ferrochelatase activity, which is needed for iron incorporation into the tetrapyrrole moiety, is not coded by the *nir* operon and possibly works in concert with a sirohaem synthesis pathway.[147] Moreover, this suggests that the chelatase activity is not the final step of d_1-haem biosynthesis, as it was assumed in the past.[1] Sirohaem is then decarboxylated to 12,18-didecarboxy-sirohaem by the action of sirohaem decarboxylase,[149] which is composed of the four proteins: NirD, NirL, NirG and NirH (*i.e.*, decarboxylation at rings III and IV; Figure 4.4).

The penultimate biosynthetic step to obtain d_1-haem involves the removal of the two propionate side chains on pyrrole rings I and II, with the

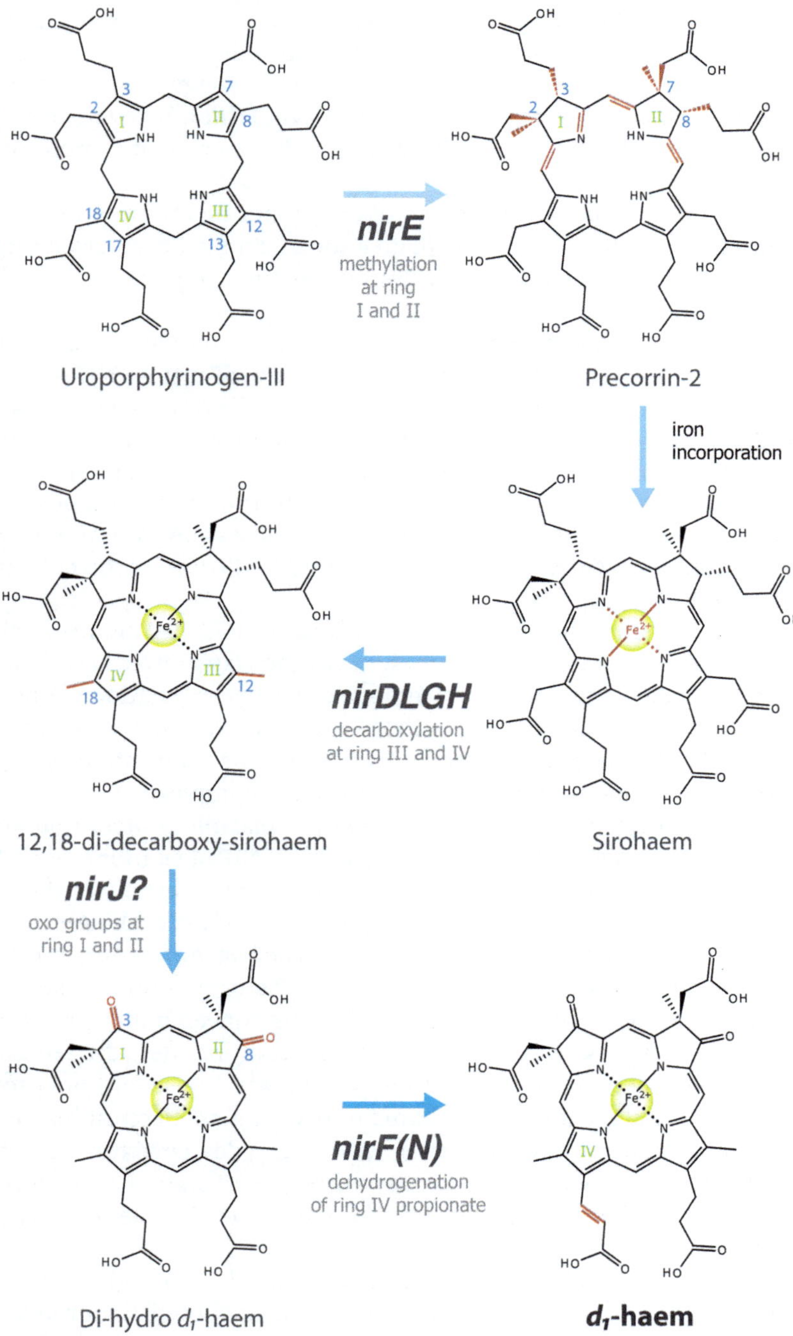

Figure 4.6 Main steps of d_1-haem biosynthesis starting from the uroporphyrinogen III intermediate. The intermediates are reported as chemical structures, while the enzymatic reactions are reported in grey. The genes directly involved in d_1-haem biogenesis and belonging to the *nir* operon(s) are represented as italic captions.

concomitant introduction of the carbonyl functions at these positions yielding the dihydro-d_1-haem (or pre-d_1-haem) (*i.e.*, introduction of the oxo groups at rings I and II; Figure 4.6); a current hypothesis suggests that this transformation is catalysed by the radical *S*-adenosylmethionine enzyme NirJ, an iron–sulphur-containing protein,[32] even though experimental proof of this is needed.

Pre-d_1-haem is then transported across the inner membrane *via* an as-yet unknown mechanism, possibly through a specific transporter, given that the proteins that are assumed to be involved in the final step are found to be in the periplasm.[128]

The latter step—the dehydrogenation of the propionate side chain on ring IV (Figure 4.6) to yield the d_1-haem to be "loaded" into cd_1NIR—involves NirF and NirN proteins, but the mechanistic details of this are still a matter of debate. NirF from *P. pantotrophus* is a soluble, periplasmic protein that is able to bind d_1-haem, and therefore it has been assumed that this protein catalyses the last reaction[143]; the same group also proposed that NirN could be involved in d_1-haem transfer and insertion into cd_1NIR, given that *P. pantotrophus* NirN was also able to bind and transfer the cofactor to cd_1NIR *in vitro*.[150]

On the other hand, the fact that *P. aeruginosa* NirF is membrane-anchored suggests that this protein could interact with the transporter and then take up the pre-d_1-haem and potentially catalyse the dehydrogenation of one of its propionate side chains.[128] These authors demonstrated that NirF, NirN and semi-apo cd_1NIR interact with each other during the maturation of cd_1NIR, after which the holo-enzyme dissociates from the complex.[128,151]

The NirN protein is a *c*-haem-containing periplasmic protein that shares about 24% amino acid sequence identity with cd_1NIR. The monomeric NirN from *P. aeruginosa* was able to bind d_1-haem *in vitro* and transfer the cofactor to cd_1NIR when the two proteins were mixed.[144] Indeed, the Δ*nirN* strain yields a holo-cd_1NIR protein whose cofactor type was different from that of the *P. aeruginosa* wild-type strain, being the cofactor of a pre-d_1-haem.[128,151] The fact that NirN is able to catalyse *in vitro* the conversion of pre-d_1-haem to d_1-haem strongly suggests that NirN is the terminal dehydrogenase catalysing the last step of the d_1-haem biosynthesis.[144] These authors propose a novel and unusual catalytic mechanism involving an electron bifurcation strategy starting from a one-electron accepting redox cofactor (*c*-haem) in order to catalyse a two-electron oxidation reaction. Such a mechanism is possible for NirN because its substrate/product itself is able to accept one electron.[144]

In the same study, NirF was found to bind pre-d_1-haem; given that NirF from *P. aeruginosa* is a periplasmic, membrane-anchored lipoprotein that interacts with both NirN and cd_1NIR,[128] the current hypothesis is that NirF participates in the pre-d_1-haem uptake from a still unknown membrane transporter and then further transfers the precursor to NirN. The latter protein then converts the tetrapyrrole into d_1-haem, which is finally transferred to cd_1NIR, yielding the active holo-form of the enzyme.[144]

4.6 Role of cd_1NIR in Biofilm and Quorum Sensing

It is well known that bacteria are able to live both as free-floating, planktonic organisms as well as organised bacterial communities (biofilms) attached to abiotic or biotic surfaces. The formation of biofilms is a crucial event in bacterial physiology; biofilms are abundant in environmental, industrial and medical settings, and they are becoming a relevant concern given their increased resistance to antibiotics and disinfectants to which planktonic bacteria are susceptible. To better appreciate the impact of biofilms, it should be considered, for example, that they are responsible for more than 70% of all nosocomial infections and that biofilm-related biofouling remains a key challenge in membrane-based water treatment systems. The biofilm lifecycle involves the formation of mushroom-like structures, known as microcolonies, in which bacterial cells are embedded in a self-produced matrix; motile cells from microcolonies can move to colonise new locations (biofilm dispersal). Active dispersal is initiated by the bacteria themselves and can be specifically triggered by several environmental cues, including NO, in various bacterial species. Dispersed cells can be more easily detected and removed and can be targeted by antimicrobial compounds.

Several studies suggest the involvement of the NO produced by cd_1NIR in the biofilm dynamics in *P. aeruginosa*, which is considered to be a model organism for studying biofilms and will be described in detail here.

Under aerobic conditions, nitrate metabolism controls swarming motility and affects biofilm architecture, showing a central role for NIR-derived NO in these events.[152,153] cd_1NIR activity and NO production are also required for biofilm dispersal under aerobic conditions.[152,154,155]

Under strict anaerobic conditions, NO accumulation may ultimately favour biofilm formation.[156] Very recently, cd_1NIR was also shown to control flagella production and swimming motility under anaerobic conditions, independently of its ability to produce NO, by forming a complex with the chaperone DnaK and the flagellar protein FliC in the periplasm.[28] An intriguing possibility suggested by these studies, which will require further investigation, is that cd_1NIR may promote motility in different ways depending on oxygen availability, either by producing NO to increase rhamnolipid synthesis or to activate other signalling pathways, as well as by directly controlling flagella formation *via* protein–protein interactions.

Another interesting link between denitrification and biofilm modulation in *P. aeruginosa* involves the cell-to-cell communication signalling systems collectively known as quorum sensing (QS) systems. *P. aeruginosa* possesses at least three different QS systems: two N-acyl-L-homoserine lactone (AHL) signals, represented by the LasR–LasI (*las*) and RhlR–RhlI (*rhl*) systems[157] and a third signal molecule, 2-heptyl-3-hydroxy-4-quinolone, referred to as the *Pseudomonas* quinolone signal (PQS).[158] The denitrification pathway (including the *nir* operon) is regulated by both the AHL and PQS systems *via* different mechanisms.[154,159–163] ANR, the oxygen-responsive transcription factor controlling the expression of denitrification genes, also controls QS in biofilms under low oxygen conditions (1%).[164] Regulation of *nirS* expression and, more generally, of denitrification by QS may thus be important in both aerobically and anaerobically grown biofilms. Interestingly, PQS addition to the growth medium specifically promotes NO accumulation,[161] thus suggesting a possible mechanism of endogenous regulation of biofilm dispersal.[154]

4.7 Haem NIR as a Bioresource

4.7.1 Nitrite Biosensing

As discussed above, the main reaction catalysed by cd_1NIR is nitrite reduction to NO. Due to their activity, NIRs are considered to be promising biocatalysts for building electrochemical nitrite biosensors.[165]

Previous electrochemical studies of cd_1NIR showed that the enzyme is capable of nitrite reduction only in the presence of partner electron donor proteins, since the usage of other electron donors results in poor activity. Enzymatic activity could be measured with cd_1NIR and its physiological electron donor in solution, incorporated into polymeric films or within a dialysis membrane on the electrode surface.[101,166,167] However, despite all of these efforts, including recent attempts to co-immobilise cd_1NIR from *Marinobacter hydrocarbonoclasticus* with its electron donor cytochrome c_{552},[168] the critical dependence of electrocatalysis upon interaction with the electron donor(s) has so far hampered the construction of an efficient nitrite biosensor based on cd_1NIR.

4.7.2 Functional Markers in Metagenomic Analysis

Together with the *nirK* gene coding for the copper-containing NIR and *nosZ* coding for nitrous oxide reductase, the *nirS* gene is widely used as a functional marker to identify the factors driving the variations in the abundance and structure of communities of denitrifying bacteria and their responses to the changing of environmental factors. The presence of the *nirS* gene is used to analyse not only *bona fide* denitrifiers, but also other communities, such as bacteria that are able to couple ammonia oxidation to nitrite reduction (AnAmmOx). The AnAmmOx bacterial *nirS* gene was shown to be a suitable functional biomarker for the investigation of AnAmmOx bacteria in the environment.[169]

Communities of denitrifiers with *nirS* respond differently to environmental gradients than those with *nirK*, suggesting that the two communities occupy different ecological niches.[170] The *nirS* and *nirK* genes are commonly used to characterise the community structure in specialised bioreactors, such as wastewater treatment plants[171] or other bioreactors.[172]

The modular structure of the denitrification pathway was previously suggested[1,26] and recently confirmed by a systematic analysis.[173] Based on the gene co-occurrence patterns, the majority of organisms with a *nirS* NIR are more likely to perform complete denitrification to N_2 than the *nirK* types under favourable environmental conditions. Thus, in the environment, as these authors suggest, the presence of *nor* genes in the genome of an organism also harbouring *nir* or *nosZ* serves as an indicator of whether the organism is a denitrifier *sensu stricto* or merely a nitrite or nitrous oxide reducer. Indeed, the ratio of *nirS*:*nirK*-type denitrifiers was recently shown to have a negative relationship with the potential for soil N_2O reduction.[174] On the other hand, among animal host-associated organisms, those with *nirK* are more abundant than the *nirS*-containing ones, suggesting that the copper-containing reductase might have played a more important role during animal host adaptation and pathogenicity. Interestingly, the assumption that the two NIRs are mutually exclusive[175] is also being contradicted by the finding that organisms harbouring both a *nirK* and a *nirS* gene in the same genome do exist,[173] opening up new scenarios regarding the relative importance of the two proteins in driving adaptation in environmental and human settings.

References

1. W. G. Zumft, *Microbiol. Mol. Biol. Rev.*, 1997, **61**, 533.
2. N. Castiglione, S. Rinaldo, G. Giardina, V. Stelitano and F. Cutruzzolà, *Antioxid. Redox Signal.*, 2012, **17**, 684.
3. S. Rinaldo and F. Cutruzzolà, in *Biology of the Nitrogen Cycle*, ed. H. Bothe, S. Ferguson and W. E. Newton, Elsevier, Amsterdam, 2007, p. 37.
4. W. G. Zumft, C. Braun and H. Cuypers, *Eur. J. Biochem.*, 1994, **219**, 481.
5. A. B. Glockner and W. G. Zumft, *Biochim. Biophys. Acta*, 1996, **1277**, 6.
6. L. Alvarez, C. Bricio, A. Hidalgo and J. Berenguer, *J. Bacteriol.*, 2014, **196**, 1350.
7. L. Alvarez, C. Bricio, A. Blesa, A. Hidalgo and J. Berenguer, *Appl. Environ. Microbiol.*, 2014, **80**, 19.
8. N. F. Saunders, S. J. Ferguson and S. C. Baker, *Microbiology*, 2000, **146**, 509.
9. E. Hartig and W. G. Zumft, *J. Bacteriol.*, 1999, **181**, 161.
10. S. Kawasaki, H. Arai, T. Kodama and Y. Igarashi, *J. Bacteriol.*, 1997, **179**, 235.
11. H. Arai, Y. Igarashi and T. Kodama, *FEBS Lett.*, 1995, **371**, 73.
12. H. Arai, T. Kodama and Y. Igarashi, *Mol. Microbiol.*, 1997, **25**, 1141.
13. M. Kuroki, Y. Igarashi, M. Ishii and H. Arai, *Environ. microbiol. Rep.*, 2014, **6**, 792.
14. S. S. Yoon, A. C. Karabulut, J. D. Lipscomb, R. F. Hennigan, S. V. Lymar, S. L. Groce, A. B. Herr, M. L. Howell, P. J. Kiley, M. J. Schurr, B. Gaston, K. H. Choi, H. P. Schweizer and D. J. Hassett, *EMBO J.*, 2007, **26**, 3662.
15. G. Giardina, S. Rinaldo, K. A. Johnson, A. Di Matteo, M. Brunori and F. Cutruzzolà, *J. Mol. Biol.*, 2008, **378**, 1002.
16. G. Giardina, S. Rinaldo, N. Castiglione, M. Caruso and F. Cutruzzolà, *Proteins*, 2009, **77**, 174.
17. N. Castiglione, S. Rinaldo, G. Giardina and F. Cutruzzolà, *Microbiology*, 2009, **155**, 2838.

18. G. Giardina, N. Castiglione, M. Caruso, F. Cutruzzolà and S. Rinaldo, *Biochem. Soc. Trans.*, 2011, **39**, 294.
19. S. Rinaldo, N. Castiglione, G. Giardina, M. Caruso, A. Arcovito, S. D. Longa, P. D'Angelo and F. Cutruzzolà, *Antioxid. Redox Signaling*, 2012, **17**, 1178.
20. K. U. Vollack and W. G. Zumft, *J. Bacteriol.*, 2001, **183**, 2516.
21. Y. Y. Lee, N. Shearer and S. Spiro, *Microbiology*, 2006, **152**, 1461.
22. L. Lobato, L. Bouzhir-Sima, T. Yamashita, M. T. Wilson, M. H. Vos and U. Liebl, *J. Biol. Chem.*, 2014, **289**, 26514.
23. H. Arai, M. Mizutani and Y. Igarashi, *Microbiology*, 2003, **149**, 29.
24. T. Horio, T. Higashi, H. Matsubara, K. Kusai, M. Nakai and K. Okunuki, *Biochim. Biophys. Acta.*, 1958, **29**, 297.
25. T. Yamanaka, A. Ota and K. Okunuki, *Biochim. Biophys. Acta*, 1961, **53**, 294.
26. W. G. Zumft and H. Korner, *Antonie Van Leeuwenhoek*, 1997, **71**, 43.
27. E. Zennaro, I. Ciabatti, F. Cutruzzolà, R. D'Alessandro and M. C. Silvestrini, *FEMS Microbiol. Lett.*, 1993, **109**, 243.
28. J. M. Borrero-de Acuna, G. Molinari, M. Rohde, T. Dammeyer, J. Wissing, L. Jansch, S. Arias, M. Jahn, M. Schobert, K. N. Timmis and D. Jahn, *J. Bacteriol.*, 2015, **197**, 3066.
29. F. Cutruzzolà, M. Arese, G. Ranghino, G. van Pouderoyen, G. Canters and M. Brunori, *J. Inorg. Biochem.*, 2002, **88**, 353.
30. I. V. Pearson, M. D. Page, R. J. M. van Spanning and S. J. Ferguson, *J. Bacteriol.*, 2003, **185**, 6308.
31. S. Rinaldo, A. Arcovito, G. Giardina, N. Castiglione, M. Brunori and F. Cutruzzolà, *Biochem. Soc. Trans.*, 2008, **36**, 1155.
32. A. A. Brindley, R. Zajicek, M. J. Warren, S. J. Ferguson and S. E. Rigby, *FEBS Lett.*, 2010, **584**, 2461.
33. J. W. A. Allen, P. D. Barker, O. Daltrop, J. M. Stevens, E. J. Tomlinson, N. Sinha, Y. Sambongi and S. J. Ferguson, *Dalton Trans.*, 2005, 3410, DOI: 10.1039/b508139b.
34. S. Rinaldo, K. A. Sam, N. Castiglione, V. Stelitano, A. Arcovito, M. Brunori, J. W. Allen, S. J. Ferguson and F. Cutruzzolà, *Biochem. J.*, 2011, **435**, 217.
35. S. C. Baker, N. F. Saunders, A. C. Willis, S. J. Ferguson, J. Hajdu and V. Fulop, *J. Mol. Biol.*, 1997, **269**, 440.
36. V. Fulop, J. W. Moir, S. J. Ferguson and J. Hajdu, *Cell*, 1995, **81**, 369.
37. D. Nurizzo, M. C. Silvestrini, M. Mathieu, F. Cutruzzolà, D. Bourgeois, V. Fülöp, J. Hajdu, M. Brunori, M. Tegoni and C. Cambillau, *Structure*, 1997, **5**, 1157.
38. D. Nurizzo, F. Cutruzzolà, M. Arese, D. Bourgeois, M. Brunori, C. Cambillau and M. Tegoni, *Biochemistry*, 1998, **37**, 13987.
39. P. A. Williams, V. Fulop, E. F. Garman, N. F. Saunders, S. J. Ferguson and J. Hajdu, *Nature*, 1997, **389**, 406.
40. G. R. Moore and G. W. Pettigrew, *Cytochromes c. Evolutionary, Structural and Physicochemical Aspects*, Springer-Verlag, Berlin, 1990.

41. E. K. Wilson, A. Bellelli, F. Cutruzzolà, W. G. Zumft, A. Gutierrez and N. S. Scrutton, *Biochem. J.*, 2001, **355**, 39.
42. E. H. Gordon, T. Sjögren, M. Löfqvist, C. D. Richter, J. W. Allen, C. W. Higham, J. Hajdu, V. Fülöp and S. J. Ferguson, *J. Biol. Chem.*, 2003, **278**, 11773.
43. F. Cutruzzolà, M. Arese, S. Grasso, A. Bellelli and M. Brunori, *FEBS Lett.*, 1997, **412**, 365.
44. T. Juhasz, Z. Szeltner, V. Fulop and L. Polgar, *J. Mol. Biol.*, 2005, **346**, 907.
45. M. R. Cheesman, S. J. Ferguson, J. W. Moir, D. J. Richardson, W. G. Zumft and A. J. Thomson, *Biochemistry*, 1997, 16267.
46. J. Sutherland, C. Greenwood, J. Peterson and A. J. Thomson, *Biochem. J.*, 1986, **233**, 893.
47. R. Timkovich and M. S. Cork, *Biochim. Biophys. Acta*, 1983, 742, 162.
48. P. Ascenzi, M. Coletta, R. Santucci, F. Polizio and A. Desideri, *J. Inorg. Biochem.*, 1994, **53**, 273.
49. Y. Orii and H. Shimada, *J. Biochem.*, 1978, **84**, 1542.
50. M. A. Sharpe and C. E. Cooper, *Biochem. J.*, 1998, **332**, 9.
51. G. Silkstone, S. M. Kapetanaki, I. Husu, M. H. Vos and M. T. Wilson, *J. Biol. Chem.*, 2010, **285**, 19785.
52. A. Schejter and P. George, *Nature*, 1965, **206**, 1150.
53. G. Silkstone, G. Stanway, P. Brzezinski and M. T. Wilson, *Biophys. Chem.*, 2002, **98**, 65.
54. C. M. Schonhoff, B. Gaston and J. B. Mannick, *J. Biol. Chem.*, 2003, **278**, 18265.
55. M. C. Silvestrini, A. Colosimo, M. Brunori, T. A. Walsh, D. Barber and C. Greenwood, *Biochem. J.*, 1979, **183**, 701.
56. D. Barber, S. R. Parr and C. Greenwood, *Biochem. J.*, 1978, **175**, 239.
57. S. Rinaldo, G. Giardina and F. Cutruzzolà, *Biochem. Biophys. Res. Commun.*, 2014, **451**, 449.
58. B. B. Muhoberac and D. C. Wharton, *J. Biol. Chem.*, 1983, **258**, 3019.
59. T. A. Walsh, M. K. Johnson, C. Greenwood, D. Barber, J. P. Springall and A. J. Thomson, *Biochem. J.*, 1979, **177**, 29.
60. J. W. Allen, N. J. Watmough and S. J. Ferguson, *Nat. Struct. Biol.*, 2000, **7**, 885.
61. J. W. A. Allen, M. R. Cheesman, C. W. Higham, S. J. Ferguson and N. J. Watmough, *Biochem. Biophys. Res. Commun.*, 2000, **279**, 674.
62. J. H. van Wonderen, C. Knight, V. S. Oganesyan, S. J. George, W. G. Zumft and M. R. Cheesman, *J. Biol. Chem.*, 2007, **282**, 28207.
63. F. Cutruzzolà, K. Brown, E. K. Wilson, A. Bellelli, M. Arese, M. Tegoni, C. Cambillau and M. Brunori, *Proc. Natl. Acad. Sci. U. S. A.*, 2001, **98**, 2232.
64. W. Sun, M. Arese, M. Brunori, D. Nurizzo, K. Brown, C. Cambillau, M. Tegoni and F. Cutruzzolà, *Biochem. Biophys. Res. Commun.*, 2002, **291**, 1.
65. K. Brown, V. Roig-Zamboni, F. Cutruzzolà, M. Arese, W. Sun, M. Brunori, C. Cambillau and M. Tegoni, *J. Mol. Biol.*, 2001, **312**, 541.
66. C. K. Chang, *Ciba Foundation Symposium*, 1994, **180**, 228.

67. M. B. Steup and B. B. Muhoberac, *J. Inorg. Biochem.*, 1989, **37**, 233.
68. S. R. Parr, M. T. Wilson and C. Greenwood, *Biochem. J.*, 1974, **139**, 273.
69. T. A. Walsh, M. K. Johnson, D. Barber, A. J. Thomson and C. Greenwood, *J. Inorg. Biochem.*, 1981, **14**, 15.
70. E. Antonini and M. Brunori, *Hemoglobin and Myoglobin in their reactions with ligands*, North Holland Pub., Amsterdam, 1971.
71. E. G. Moore and Q. H. Gibson, *J. Biol. Chem.*, 1976, **251**, 2788.
72. B. A. Springer, S. G. Sligar, J. S. Olson and G. N. Phillips Jr., *Chem. Rev.*, 1994, **94**, 699.
73. S. R. Parr, M. T. Wilson and C. Greenwood, *Biochem. J.*, 1975, **151**, 51.
74. E. K. Wilson, A. Bellelli, S. Liberti, M. Arese, S. Grasso, F. Cutruzzolà, M. Brunori and P. Brzezinski, *Biochemistry*, 1999, **38**, 7556.
75. T. K. Das, E. K. Wilson, F. Cutruzzolà, M. Brunori and D. L. Rousseau, *Biochemistry*, 2001, **40**, 10774.
76. T. Sjogren, M. Svensson-Ek, J. Hajdu and P. Brzezinski, *Biochemistry*, 2000, **39**, 10967.
77. M. C. Silvestrini, S. Falcinelli, I. Ciabatti, F. Cutruzzolà and M. Brunori, *Biochimie*, 1994, **76**, 641.
78. D. L. Compton and J. A. Laszlo, *J. Electroanal. Chem.*, 2002, **520**, 71.
79. R. S. Zajicek, M. L. Cartron and S. J. Ferguson, *Biochemistry.*, 2006, **45**, 11208.
80. T. Sjogren and J. Hajdu, *J. Biol. Chem.*, 2001, **276**, 29450.
81. D. Nurizzo, F. Cutruzzolà, M. Arese, D. Bourgeois, M. Brunori, C. Cambillau and M. Tegoni, *J. Biol. Chem.*, 1999, **274**, 14997.
82. C. Greenwood, D. Barber, S. R. Parr, E. Antonini, M. Brunori and A. Colosimo, *Biochem. J.*, 1978, **173**, 11.
83. F. Centola, S. Rinaldo, M. Brunori and F. Cutruzzolà, *FEBS J.*, 2006, **273**, 4495.
84. J. Singh, *Biochim. Biophys. Acta*, 1974, **333**, 28.
85. C. D. Richter, J. W. Allen, C. W. Higham, A. Koppenhofer, R. S. Zajicek, N. J. Watmough and S. J. Ferguson, *J. Biol. Chem.*, 2002, **277**, 3093.
86. S. Rinaldo, A. Arcovito, M. Brunori and F. Cutruzzolà, *J. Biol. Chem.*, 2007, **282**, 14761.
87. T. Yamanaka, *Ann. Rep. Sci. Works, Fac. Sci., Osaka Univ.*, 1963, **11**, 77.
88. I. Moura and J. J. Moura, *Curr. Opin. Chem. Biol.*, 2001, **5**, 168.
89. S. J. George, J. W. Allen, S. J. Ferguson and R. N. Thorneley, *J. Biol. Chem.*, 2000, **275**, 33231.
90. K. A. Sam, M. J. Strampraad, S. de Vries and S. J. Ferguson, *J. Biol. Chem.*, 2008, **283**, 27403.
91. M. C. Silvestrini, M. G. Tordi, G. Musci and M. Brunori, *J. Biol. Chem.*, 1990, **265**, 11783.
92. O. Einsle, A. Messerschmidt, R. Huber, P. M. Kroneck and F. Neese, *J. Am. Chem. Soc.*, 2002, **124**, 11737.
93. B. R. Crane, L. M. Siegel and E. D. Getzoff, *Biochemistry*, 1997, **36**, 12120.
94. G. R. Wyllie and W. R. Scheidt, *Chem. Rev.*, 2002, **102**, 1067.
95. H. Nasri, M. K. Ellison, M. Shang, C. E. Schulz and W. R. Scheidt, *Inorg. Chem.*, 2004, **43**, 2932.

96. L. Cheng, D. R. Powell, M. A. Khan and G. B. Richter-Addo, *Chem. Commun*, 2000, 2301.
97. A. Jafferji, J. W. Allen, S. J. Ferguson and V. Fulop, *J. Biol. Chem.*, 2000, **275**, 25089.
98. K. Kobayashi, A. Koppenhofer, S. J. Ferguson, N. J. Watmough and S. Tagawa, *Biochemistry*, 2001, **40**, 8542.
99. O. Farver, M. Brunori, F. Cutruzzolà, S. Rinaldo, S. Wherland and I. Pecht, *Biophys. J.*, 2009, **96**, 2849.
100. K. Kobayashi, A. Koppenhofer, S. J. Ferguson and S. Tagawa, *Biochemistry*, 1997, **36**, 13611.
101. H. Lopes, S. Besson, I. Moura and J. J. G. Moura, *J. Biol. Inorg. Chem.*, 2001, **6**, 55.
102. O. Farver, P. M. Kroneck, W. G. Zumft and I. Pecht, *Biophys. Chem.*, 2002, **98**, 27.
103. Y. Blatt and I. Pecht, *Biochemistry*, 1979, **18**, 2917.
104. A. Koppenhöfer, K. L. Turner, J. W. Allen, S. K. Chapman and S. J. Ferguson, *Biochemistry*, 2000, **39**, 4243.
105. B. A. Averill, *Chem. Rev.*, 1996, **96**, 2951.
106. V. G. Kharitonov, V. S. Sharma, D. Magde and D. Koesling, *Biochemistry*, 1997, **36**, 6814.
107. S. Rinaldo, M. Brunori and F. Cutruzzolà, *Biochem. Biophys. Res. Commun.*, 2007, **363**, 662.
108. K. A. Sam, S. A. Fairhurst, R. N. Thorneley, J. W. Allen and S. J. Ferguson, *J. Biol. Chem.*, 2008, **283**, 12555.
109. K. A. Sam, J. D. Tolland, S. A. Fairhurst, C. W. Higham, D. J. Lowe, R. N. Thorneley, J. W. Allen and S. J. Ferguson, *Biochem. Biophys. Res. Commun.*, 2008, **371**, 719.
110. G. J. Carr, M. D. Page and S. J. Ferguson, *Eur. J. Biochem.*, 1989, **179**, 683.
111. M. Radoul, F. Centola, S. Rinaldo, F. Cutruzzolà, I. Pecht and D. Goldfarb, *Inorg. Chem.*, 2009, **48**, 3913.
112. M. Radoul, Y. Barak, S. Rinaldo, F. Cutruzzolà, I. Pecht and D. Goldfarb, *Biochemistry*, 2012, **51**, 9192.
113. M. Radoul, D. Bykov, S. Rinaldo, F. Cutruzzolà, F. Neese and D. Goldfarb, *J. Am. Chem. Soc.*, 2011, **133**, 3043.
114. M. C. Silvestrini, M. G. Tordi, A. Colosimo, E. Antonini and M. Brunori, *Biochem. J.*, 1982, **203**, 445.
115. M. van de Kamp, M. C. Silvestrini, M. Brunori, J. Van Beeumen, F. C. Hali and G. W. Canters, *Eur. J. Biochem.*, 1990, **194**, 109.
116. J. W. Moir, D. Baratta, D. J. Richardson and S. J. Ferguson, *Eur. J. Biochem.*, 1993, **212**, 377.
117. D. A. Mavridou, S. J. Ferguson and J. M. Stevens, *IUBMB life*, 2013, **65**, 209.
118. R. Krieger, A. Rompf, M. Schobert and D. Jahn, *Mol. Genet. Genomics*, 2002, **267**, 409.
119. A. Rompf, C. Hungerer, T. Hoffmann, M. Lindenmeyer, U. Romling, U. Gross, M. O. Doss, H. Arai, Y. Igarashi and D. Jahn, *Mol. Microbiol.*, 1998, **29**, 985.

120. M. D. Page and S. J. Ferguson, *Mol. Microbiol.*, 1989, **3**, 653.
121. M. D. Page and S. J. Ferguson, *Mol. Microbiol.*, 1990, **4**, 1181.
122. L. Thony-Meyer and P. Kunzler, *Eur. J. Biochem.*, 1997, **246**, 794.
123. C. Travaglini-Allocatelli, *Scientifica*, 2013, **2013**, 505714.
124. A. F. Verissimo and F. Daldal, *Biochim. Biophys. Acta*, 2014, **1837**, 989.
125. T. M. Ramseier, H. V. Winteler and H. Hennecke, *J. Biol. Chem.*, 1991, **266**, 7793.
126. A. Ray and H. D. Williams, *FEMS Microbiol. Lett.*, 1996, **135**, 123.
127. E. Arslan, H. Schulz, R. Zufferey, P. Kunzler and L. Thony-Meyer, *Biochem. Biophys. Res. Commun.*, 1998, **251**, 744.
128. T. Nicke, T. Schnitzer, K. Münch, J. Adamczack, K. Haufschildt, S. Buchmeier, M. Kucklick, U. Felgenträger, L. Jänsch, K. Riedel and G. Layer, *Biosci. Rep.*, 2013, **33**, e00048.
129. M. Throne-Holst, L. Thony-Meyer and L. Hederstedt, *FEBS Lett.*, 1997, **410**, 351.
130. C. Richard-Fogal and R. G. Kranz, *J. Mol. Biol.*, 2010, **401**, 350.
131. B. San Francisco, E. C. Bretsnyder, K. R. Rodgers and R. G. Kranz, *Biochemistry*, 2011, **50**, 10974.
132. D. W. Nicholson and W. Neupert, *Proc. Natl. Acad. Sci. U. S. A.*, 1989, **86**, 4340.
133. B. San Francisco, M. C. Sutherland and R. G. Kranz, *Mol. Microbiol.*, 2014, **91**, 996.
134. Q. Ren, U. Ahuja and L. Thony-Meyer, *J. Biol. Chem.*, 2002, **277**, 7657.
135. C. Sanders, S. Turkarslan, D. W. Lee, O. Onder, R. G. Kranz and F. Daldal, *J. Biol. Chem.*, 2008, **283**, 29715.
136. A. Di Matteo, N. Calosci, S. Gianni, P. Jemth, M. Brunori and C. Travaglini-Allocatelli, *Proteins*, 2010, **78**, 2213.
137. A. Di Matteo, S. Gianni, M. E. Schinina, A. Giorgi, F. Altieri, N. Calosci, M. Brunori and C. Travaglini-Allocatelli, *J. Biol. Chem.*, 2007, **282**, 27012.
138. R. K. Allan and T. Ratajczak, *Cell Stress Chaperones*, 2011, **16**, 353.
139. E. Di Silvio, A. Di Matteo, F. Malatesta and C. Travaglini-Allocatelli, *Biochim. Biophys. Acta*, 2013, **1834**, 1554.
140. M. C. Silvestrini, F. Cutruzzolà, R. D'Alessandro, M. Brunori, N. Fochesato and E. Zennaro, *Biochem. J.*, 1992, **285**, 661.
141. J. W. Allen, P. D. Barker, O. Daltrop, J. M. Stevens, E. J. Tomlinson, N. Sinha, Y. Sambongi and S. J. Ferguson, *Dalton Trans.*, 2005, **7**, 3410.
142. S. Rinaldo, M. Brunori and F. Cutruzzolà, *Plant Signal. Behavior*, 2008, **3**, 135.
143. S. Bali, M. J. Warren and S. J. Ferguson, *FEBS J.*, 2010, **277**, 4944.
144. J. Adamczack, M. Hoffmann, U. Papke, K. Haufschildt, T. Nicke, M. Bröring, M. Sezer, R. Weimar, U. Kuhlmann, P. Hildebrandt and G. Layer, *J. Biol. Chem.*, 2014, **289**, 30753.
145. S. Storbeck, J. Walther, J. Muller, V. Parmar, H. M. Schiebel, D. Kemken, T. Dulcks, M. J. Warren and G. Layer, *FEBS J.*, 2009, **276**, 5973.
146. S. Storbeck, S. Saha, J. Krausze, B. U. Klink, D. W. Heinz and G. Layer, *J. Biol. Chem.*, 2011, **286**, 26754.

147. S. Bali, A. D. Lawrence, S. A. Lobo, L. M. Saraiva, B. T. Golding, D. J. Palmer, M. J. Howard, S. J. Ferguson and M. J. Warren, *Proc. Natl. Acad. Sci. U. S. A.*, 2011, **108**, 18260.
148. S. Bali, S. Rollauer, P. Roversi, E. Raux-Deery, S. M. Lea, M. J. Warren and S. J. Ferguson, *Mol. Microbiol.*, 2014, **92**, 153.
149. K. Haufschildt, S. Schmelz, T. M. Kriegler, A. Neumann, J. Streif, H. Arai, D. W. Heinz and G. Layer, *J. Mol. Biol.*, 2014, **426**, 3272.
150. R. S. Zajicek, S. Bali, S. Arnold, A. A. Brindley, M. J. Warren and S. J. Ferguson, *FEBS J.*, 2009, **276**, 6399.
151. Y. Li, S. Bali, S. Borg, E. Katzmann, S. J. Ferguson and D. Schuler, *J. Bacteriol.*, 2013, **195**, 4297.
152. N. E. Van Alst, K. F. Picardo, B. H. Iglewski and C. G. Haidaris, *Infect. Immun.*, 2007, **75**, 3780.
153. C. de la Fuente-Nunez, F. Reffuveille, K. E. Fairfull-Smith and R. E. Hancock, *Antimicrob. Agents Chemother.*, 2013, **57**, 4877.
154. N. Barraud, D. J. Hassett, S. H. Hwang, S. A. Rice, S. Kjelleberg and J. S. Webb, *J. Bacteriol.*, 2006, **188**, 7344.
155. N. Barraud, D. Schleheck, J. Klebensberger, J. S. Webb, D. J. Hassett, S. A. Rice and S. Kjelleberg, *J. Bacteriol.*, 2009, **191**, 7333.
156. M. Y. Yoon, K. M. Lee, Y. Park and S. S. Yoon, *PLoS One*, 2011, **6**, e16105.
157. E. C. Pesci, J. P. Pearson, P. C. Seed and B. H. Iglewski, *J. Bacteriol.*, 1997, **179**, 3127.
158. E. C. Pesci, J. B. Milbank, J. P. Pearson, S. McKnight, A. S. Kende, E. P. Greenberg and B. H. Iglewski, *Proc. Natl. Acad. Sci. U. S. A.*, 1999, **96**, 11229.
159. V. E. Wagner, D. Bushnell, L. Passador, A. I. Brooks and B. H. Iglewski, *J. Bacteriol.*, 2003, **185**, 2080.
160. M. Toyofuku, N. Nomura, T. Fujii, N. Takaya, H. Maseda, I. Sawada, T. Nakajima and H. Uchiyama, *J. Bacteriol.*, 2007, **189**, 4969.
161. M. Toyofuku, N. Nomura, E. Kuno, Y. Tashiro, T. Nakajima and H. Uchiyama, *J. Bacteriol.*, 2008, **190**, 7947.
162. S. S. Yoon, R. F. Hennigan, G. M. Hilliard, U. A. Ochsner, K. Parvatiyar, M. C. Kamani, H. L. Allen, T. R. DeKievit, P. R. Gardner, U. Schwab, J. J. Rowe, B. H. Iglewski, T. R. McDermott, R. P. Mason, D. J. Wozniak, R. E. Hancock, M. R. Parsek, T. L. Noah, R. C. Boucher and D. J. Hassett, *Dev. Cell*, 2002, **3**, 593.
163. D. J. Hassett, J. Cuppoletti, B. Trapnell, S. V. Lymar, J. J. Rowe, S. S. Yoon, G. M. Hilliard, K. Parvatiyar, M. C. Kamani, D. J. Wozniak, S. H. Hwang, T. R. McDermott and U. A. Ochsner, *Adv. Drug Delivery Rev.*, 2002, **54**, 1425.
164. J. H. Hammond, E. F. Dolben, T. J. Smith, S. Bhuju and D. A. Hogan, *J. Bacteriol.*, 2015, **197**, 2810.
165. M. G. Almeida, A. Serra, C. M. Silveira and J. J. Moura, *Sensors*, 2010, **10**, 11530.
166. E. Lojou, F. Cutruzzolà, M. Tegoni and P. Bianco, *Electrochim. Acta*, 2003, **48**, 1055.

167. A. S. Serra, S. R. Jorge, C. M. Silveira, J. J. Moura, E. Jubete, E. Ochoteco, G. Cabanero, H. Grande and M. G. Almeida, *Anal. Chim. Acta*, 2011, **693**, 41.
168. C. M. Silveira, P. O. Quintas, I. Moura, J. J. Moura, P. Hildebrandt, M. G. Almeida and S. Todorovic, *PLoS One*, 2015, **10**, e0129940.
169. M. Li, T. Ford, X. Li and J. D. Gu, *Environ. Sci. Technol.*, 2011, **45**, 3547.
170. C. M. Jones and S. Hallin, *ISME J.*, 2010, **4**, 633.
171. Z. Wang, X. X. Zhang, X. Lu, B. Liu, Y. Li, C. Long and A. Li, *PLoS One*, 2014, **9**, e113603.
172. R. B. Herbert Jr, H. Winbjork, M. Hellman and S. Hallin, *Water Res.*, 2014, **66**, 350.
173. D. R. Graf, C. M. Jones and S. Hallin, *PLoS One*, 2014, **9**, e114118.
174. C. M. Jones, A. Spor, F. P. Brennan, M. Breuil, D. Bru, P. Lemanceau, B. Griffiths, S. Hallin and L. Philippot, *Nat. Clim. Change*, 2014, **4**, 801.
175. C. M. Jones, B. Stres, M. Rosenquist and S. Hallin, *Mol. Biol. Evol.*, 2008, **25**, 1955.

CHAPTER 5

Structure and Function of Copper Nitrite Reductase

MASAKI NOJIRI[a]

[a]Department of Chemistry, Graduate School of Science, Osaka University, 1-1 Machikaneyama, Toyonaka, Osaka 560-0043, Japan
*E-mail: nojiri@ch.wani.osaka-u.ac.jp

5.1 Introduction

There are two forms of nitrite reductases (NiRs) that reduce nitrite (NO_2^-) to nitric oxide ($^{\bullet}NO$, abbreviated as NO). One is copper-containing NiR (CuNiR), encoded by the gene *nirK*, and the second is the haem-containing NiR, encoded by *nirS* (described in Chapter 4 of this book). These are distributed widely and are associated with the propagation of various microorganisms.[1–5] This widespread association with the global N-cycle is indicative of the importance of these enzymes in vital processes, including the natural production of greenhouse gases, such as nitrogen oxides (NO*x*).[6,7]

$$NO_3^- \rightarrow NO_2^- \rightarrow {}^{\bullet}NO \rightarrow N_2O \rightarrow N_2 \quad (5.1)$$

CuNiR reversibly catalyses the one-electron reduction of NO_2^- to NO (eqn (5.2)).

$$NO_2^- + 2H^+ + e^- \rightarrow {}^{\bullet}NO + H_2O \quad (5.2)$$

RSC Metallobiology Series No. 9
Metalloenzymes in Denitrification: Applications and Environmental Impacts
Edited by Isabel Moura, José J. G. Moura, Sofia R. Pauleta and Luisa B. Maia

Published by the Royal Society of Chemistry, www.rsc.org

The reducing electron is delivered through an inter-protein electron transfer (ET) reaction with a redox partner.[8–11] This enzyme contains two Cu-binding sites per monomeric subunit (~37 kDa) in a homotrimeric overall structure (see Figure 5.1a).[12] The two copper sites are classified as type-1 copper (T1Cu) and type-2 copper (T2Cu) and are distinguished by their spectroscopic, structural and functional properties. T1Cu is bound by two histidine (His) residues, a cysteine (Cys) and a methionine (Met) and relays an electron from the redox partner to T2Cu. It exhibits an intense characteristic colour: "blue" or "green" in the cupric (Cu^{2+}) state.[1,13] Meanwhile, T2Cu is coordinated by three His residues and serves as a site for the binding and catalytic reduction of NO_2^-. These sites are ~13 Å apart and are connected by a conserved –His–Cys– sequence segment, composed of a Cys residue ligand of T1Cu and a His residue ligand of T2Cu (see Figure 5.1b).

CuNiR was first isolated by Iwasaki *et al.* over five decades ago.[14,15] This protein is more commonly known as "blue" CuNiR. It exhibits an intense sulphur Sγ–Cys to Cu^{2+} charge-transfer absorption band at ~600 nm (ε = ~5000 M^{-1} cm^{-1}), to which is due its blue colour, as well as an unusual, small Cu hyperfine interaction observed in the low-field region in the electron paramagnetic resonance (EPR) spectrum.[16] These unique spectroscopic properties are due to the high covalency of the Cu–Sγ–Cys bond and are shared widely in all blue copper proteins containing a T1Cu site.[17–21] Moreover, T1Cu is roughly classified into two types—"axial" and "rhombic"—on the basis of

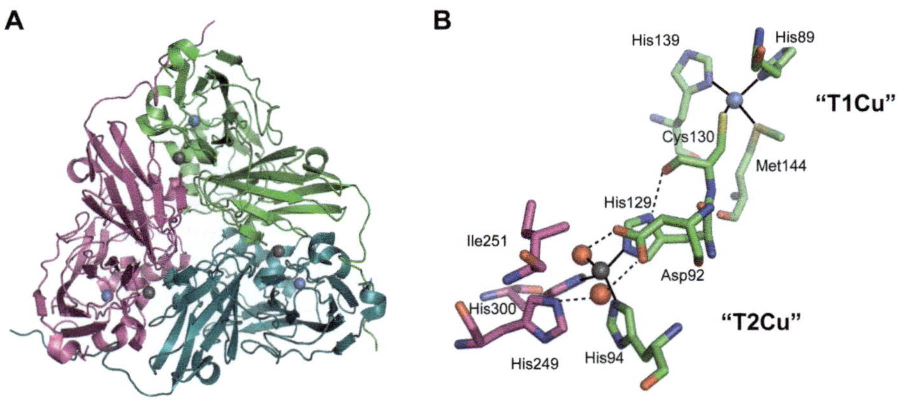

Figure 5.1 Structure of "blue" copper nitrite reductase. (a) A ribbon diagram of "blue" copper nitrite reductase isolated from *Achromobacter xylosoxidans* (Protein Data Bank identifier: 1oe1).[95] (b) A view of the two Cu-binding sites, type-1 copper (T1Cu) and type-2 copper (T2Cu) sites. The ligands and important neighbouring residues are represented as sticks. The spheres in red are water molecules. The solid lines and dotted lines indicate the coordinate bonds and H-bonds, respectively. The minor species of the alternative conformers of the Met ligand of the T1Cu site was removed from this figure for clarity. The copper atoms are represented as spheres, T1Cu in dark blue and T2Cu in grey in both panels.

the extent of rhombicity of the *g* tensor of the EPR spectrum.[21] In addition to the band at 600 nm, the absorption spectrum shows a band at 450 nm, which is weak for the axial type and stronger for the rhombic type.[13] Consequently, rhombic T1Cu centres generally are "green" rather than "blue". Several quantum-chemical calculations have revealed the character of the singly occupied molecular orbital (SOMO) at the electronic ground state of T1Cu sites.[22,23] This character depends on whether d_{x2-y2} or d_{xy} is the SOMO in the ground state.[24,25] It is essentially an anti-bonding orbital between Cu and Sγ–Cys, and the anti-bond is of pure π character for the axial type and of mixed σ–π character for the rhombic type. It has been proposed that the distortion of the T1Cu geometry involved in the contribution of the σ character to the SOMO is coupled with positional shifts of the S atoms of Met and Cys ligands (also defined by the rotation of the Met-Sδ–Cu–Sγ–Cys plane). Accordingly, it is now well-accepted that the Cu–Sδ-Met distance of the axial-type T1Cu ("blue") tends to be longer—usually ~3.1 Å—than that of the rhombic-type T1Cu ("green"), being ~2.6 Å.[23]

In the Cu^{2+} state, T2Cu has a pseudo-tetrahedral geometry with three His ligands and a water molecule in the apical position and exhibits the normal EPR spectral pattern observed in square-planer (tetragonal) Cu complexes.[16,26,27] The site is located at the interface between two subunits, generating a substrate-accessible channel connected to the protein surface. A number of other residues within the channel pocket are proposed to be important for catalytic NO_2^- reduction, particularly a histidine, an aspartate and an isoleucine, as adjudged by site-directed mutagenesis studies.[28–30] A hydrogen (H)-bonding network between Asp_{92}, His_{249} (*Achromobacter xylosoxidans* enzyme numbering), a water molecule and NO_2^- is believed to play an important role in proton uptake for catalysis, while Ile_{251} is proposed to help with the positioning of NO_2^- within the site (see Figure 5.1b). Recently, a computational study based on spectroscopic data proposed that protonation of Asp_{92} induces a significant positive shift in the reduction potential of the $T2Cu^{2+}–NO_2^-$ fragment, triggering ET in order to generate the catalytically active $T2Cu^+–NO_2^-$ intermediate.[31]

This chapter will focus on recent advances in structure–function relationships for CuNiRs and, in particular, on the geometries and reaction mechanisms at the Cu site. Owing to the limited space available, a comprehensive review is not possible. Some excellent reviews that provide an overview of CuNiR studies can be referred to for additional and more exhaustive information.[13,32,33]

5.2 The T1Cu Site

In the last decade, more than ten CuNiRs were isolated anew and characterised. Most of them belong to the "green" or "greenish-blue" (Verdigris) types, having a characteristic absorption band at 450 nm apart from the typical absorption band at ~600 nm. In Figure 5.2, a UV-visible absorption spectrum of *Geobacillus kaustophilus* CuNiR (*Gk*NiR) is presented as an example

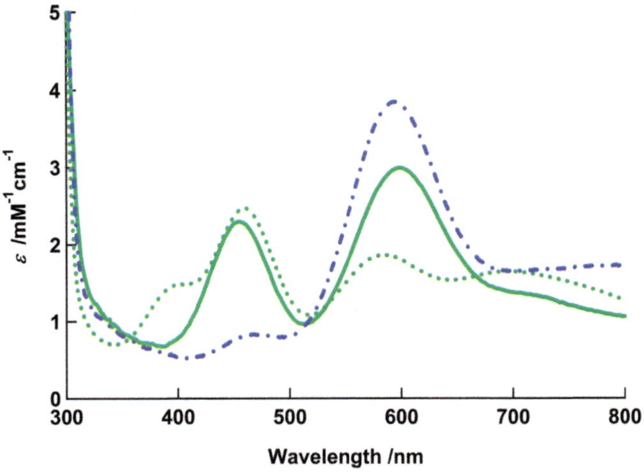

Figure 5.2 Electronic absorption spectra of "greenish-blue", "green" and "blue" copper nitrite reductases (CuNiRs). The solid line represents the spectrum of the "greenish-blue" *Geobacillus kaustophilus* CuNiR, the dotted line in green is the spectrum of the "green" *Achromobacter cycloclastes* CuNiR and the dashed–dotted line in blue is the spectrum of the "blue" *Achromobacter xylosoxidans* CuNiR.

of the "greenish-blue" type. Previously, the CuNiR isolated from *Rhodobacter sphaeroides* was referred by a similar nickname, "blue–green" NiR,[34] even though they are clearly distinguishable based on their spectroscopic and geometric parameters, with *Rhodobacter* CuNiR being closely related to the classical "green" type (*vide infra*).

In contrast to the absorption spectrum of the classical "green" type, this "greenish-blue" type does not have a (shoulder) band at ~400 nm assigned to the Sδ-Met to Cu^{2+} charge-transfer transition.[23] The loss of this band strongly suggests that the orbital for Sδ (Met) does not contribute to the SOMO for this "greenish-blue" type of T1Cu. This type of CuNiR was isolated for the first time from *Bacillus halodenitrificans*; however, its unique properties were not mentioned.[35] Nowadays, approximately half of the reported T1Cu sites of CuNiR belong to this type.[4,5,36–41]

Five crystal structures of "greenish-blue" CuNiRs have been reported, and their coordinates are available on the Protein Data Bank (PDB). In Figure 5.3a, a superposition of the structures of all T1Cu sites of this type is shown with their ligand atom positions highlighted, and the geometric parameters are listed in Table 5.1. The positional deviations between their ligand atoms are very small (≤0.1 Å), indicating that they are clustered in a single group. Interestingly, the distance between Sδ-Met and Cu atoms in all "greenish-blue" T1Cu sites is significantly shorter (~2.6 Å) comparable to those of the "green"-type T1Cu; however, the Sδ-Met and Sγ–Cys atoms are displaced by ~0.5 Å compared to those observed in the "green"-type T1Cu (see Figure 5.3b). Assuming that the equatorial *xy* plane for Cu is essentially parallel to

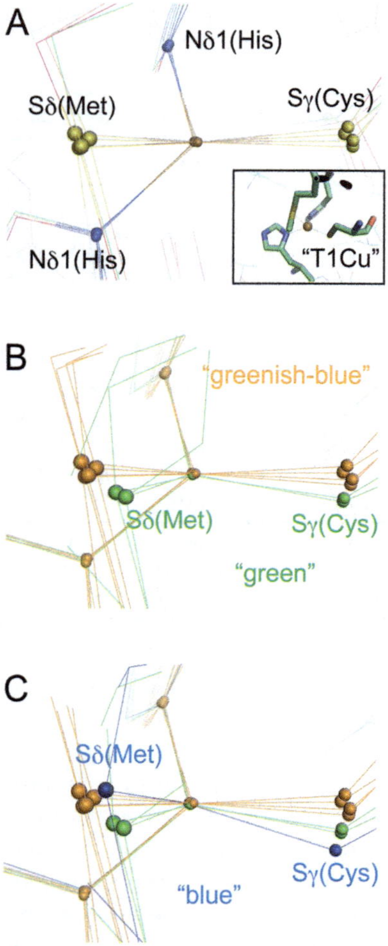

Figure 5.3 Comparison of the structures between "greenish-blue", "green" and "blue" types of copper nitrite reductases (CuNiRs). (a) Overlay of four type-1 copper (T1Cu) sites of "greenish-blue" CuNiRs (Protein Data Bank identifiers: 3wi9, 4knu, 2zoo and 2dv6 [$T1Cu_C$]), determined at a resolution of ≥2.2 Å. (b) Overlay of two T1Cu sites of "green" CuNiRs, displayed in green (Protein Data Bank identifiers: 2bw4 and 1mzy) to the structures in (a) displayed in yellow. Note that the S positions for *Rhodobacter* CuNiR T1Cu (1mzy) are very close to the "green"-type positions. (c) Overlay of the "blue"-type CuNiR (Protein Data Bank identifier: 1oe1), displayed in blue to the structures in (a), displayed in yellow. All T1Cu structures were fitted using three atoms: two His Nδ1 ligands and the Cu atoms.

the Nδ1–Cu–Nδ1 plane,[21–25] the Sδ-Met is likely to be localised at a position that is closer to the *z*-axis for Cu. The dihedral angles (θ) between the planes N–Cu–N and S–Cu–S are comparable in both types, being ~65° ("greenish-blue") and ~66° ("green"), respectively. In addition, it is notable that an interesting difference in side-chain conformation of the Met ligand exists

Table 5.1 Geometric parameters of type-1 copper sites in copper nitrite reductase enzymes (His_1 and His_2 are the first and second His ligands in the amino acid sequence).[a]

PDB identifier[b]	3wi9	4knu	2zoo	2dv0[c]	1kbw	2bw4	1snr	1mzy[d]	1oe1[d]
Type	GB	GB	GB	GB	GB	G	G	G	B
Distances (Å)									
His_1-Cu	2.07	2.10	2.13	2.14	1.99	2.04	2.08	2.11	2.02
Cys-Cu	2.13	2.23	2.22	2.13	2.10	2.23	2.22	2.19	2.20
His_2-Cu	2.03	2.10	2.06	2.10	1.90	2.03	2.07	2.16	2.03
Met-Cu	2.61	2.48	2.56	2.55	2.61	2.49	2.45	2.41	2.45
Angles (°)									
His_1-Cu-Cys	139	135	138	144	128	127	128	130	122
His_1-Cu-His_2	99	102	99	95	108	101	96	100	101
His_1-Cu-Met	82	84	81	81	82	86	90	84	88
Cys-Cu-His_2	103	107	110	104	105	108	107	107	114
Cys-Cu-Met	113	112	112	116	108	108	107	112	114
His_2-Cu-Met	117	117	113	116	128	127	130	124	116

[a]B: blue; G: green; GB: greenish-blue; PDB: protein data bank.
[b]The structures of the type-1 copper sites of *G. kaustophilus*, *N. europaea*, *P. haloplanktis*, *H. denitrificans*, *N. gonorrhoeae*, *A. cycloclstes*, *A. faecalis*, *R. sphaeroides* and *A. xylosoxidans* are taken from PDB entries 3wi9, 4knu, 2zoo, 2dv6, 1kbw, 2bw4, 1snr, 1mzy and 1oe1, respectively.
[c]Type-1 copper in the C-terminal copper nitrite reductase domain in *H. denitrificans* copper nitrite reductase.
[d]1mzy and 1oe1 have the alternative conformation of the Met ligand; therefore, the unliganded conformers are tentatively removed from this table for clarity.

between "green" and "greenish-blue" types: a "*trans* (*anti*)" conformation is observed for the "green" type and a "*gauche*" conformation is observed for the "greenish–blue" type. All of these structural features seem to support the notion that the orbital of Sδ-Met at the "greenish-blue" T1Cu site is not oriented to overlap significantly with the lobes of the ground-state d_{x2-y2} or d_{xy} orbital. This corresponds to the absence of an absorption band at ~400 nm. Moreover, the Met ligand of *Rhodobacter* CuNiR exhibits a strong absorption band at ~400 nm and a "*gauche*" conformation; however, the Sδ-Met and Sγ–Cys clearly localise to the "green"-type position (see Figure 5.3b).

A comparison of the geometric parameters of T1Cu between "greenish-blue" and "blue" CuNiR is of considerable interest. As shown in Figure 5.3c and Table 5.1, all parameters, including the side-chain conformation of the Met ligand, are quite similar. This has been previously noted in *Neisseria* CuNiR;[36] however, the resolution was not sufficiently high at that point in time, and hence it was not regarded as an important feature. The single obvious difference between these types of centre is the relative positions of the Sγ–Cys atoms, which are ~0.7 Å apart (see Figure 5.3c). If so, it is not possible to eliminate the hypothesis that the Sγ–Cys at the "greenish-blue" T1Cu site can shift without causing a large positional change in the Sδ-Met. It may be noted that this hypothesis is apparently contradictory to the "coupled-distortion" model applied widely in the interpretation of absorption spectra in blue-copper proteins.[23] Such geometrical ambiguities that have surfaced in the current high-resolution CuNiR structures should be clarified in the near future by using higher-quality structural data and theoretical studies. The present data indicate a few structural differences between "blue" and "greenish-blue" types in the second coordination sphere as follows:

(1) The Cu–Nδ1–Cγ bond angle involving solvent-exposed His ligands for the "greenish-blue" T1Cu is bent further away from linear (~155°) than those of the "blue" type (~162°; see Figure 5.4). In this context, it might be noted that a similar deviation at the T1Cu site has been observed in the 1.18 Å resolution crystal structure of the more perturbed T1Cu-containing pseudoazurin (PAz);[42]
(2) The Sγ–Cys in the "greenish-blue" T1Cu has an additional H-bond with the main-chain amide N of a neighbouring residue (see Figure 5.4).

Based on the current data of the S ligand positions and spectroscopic properties focused on the T1Cu site, it is clear that another group, in addition to the classical "blue" and "green" types and including the *Rhodobacter* "blue–green" type, exists in the CuNiR family. The number of "greenish-blue" CuNiRs already occupies about half of the majority of known CuNiRs from a variety of microorganisms and is no longer considered rare. This type has unique geometric and spectroscopic features, which may be expected to provide further insights into the relationships between the geometric and spectroscopic properties of T1Cu. In addition, a classification of CuNiRs focused on the number of amino acid residues in some loop regions has

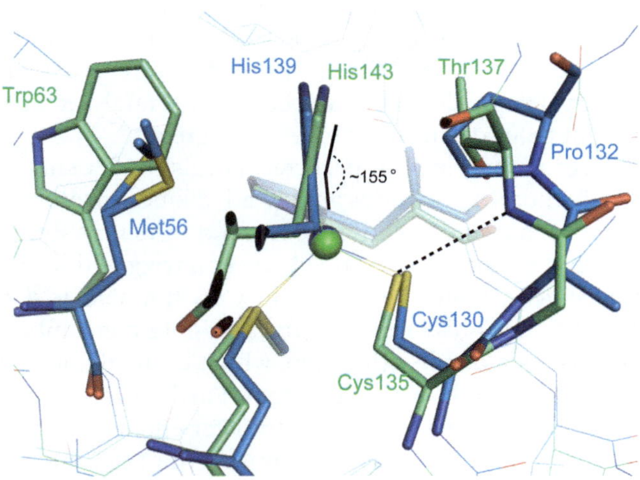

Figure 5.4 A close-up view of the superimposed type-1 copper (T1Cu) sites of a "greenish-blue" and a "blue" copper nitrite reductase in green and blue, respectively. The bond between the solvent-exposed His and Cu in the "greenish-blue" type is bent due to the interaction with Trp_{63} in the second coordination sphere. Moreover, the Sγ atom in the "greenish-blue" T1Cu sites forms an H-bond with the neighbouring amide group.

been proposed;[36] however, the definition seems to be no longer suitable, given the recently reported structural features (*e.g.*, as shown in the *Gk*NiR structure).[39] Furthermore, the relatively short Sδ-Met–Cu bond lengths in the T1Cu site (≤2.6 Å) seems to be strictly conserved in all types ("blue", "green" and "greenish-blue") of CuNiR structures, in contrast to that observed for the classical "blue" type in small blue-copper proteins and multi-copper oxidases, which are relatively long (~3.2 Å).[18–20,23,43] This apparent distance constraint is probably a functional prerequisite for the T1Cu site in CuNiRs to tune (or regulate) a reduction potential into the suitable range for inter- and intra-ETs (*vide infra*) and the following NO_2^- reduction at the T2Cu site. The loss (or weakening) of the Met ligand interaction with Cu dramatically shifts the reduction potential towards a positive potential.[34,44]

5.3 Electron Transfer

5.3.1 Intramolecular ET

Rapid and controlled intramolecular ET between T1Cu and T2Cu sites is necessary for the enzymatic function of CuNiR. Electrons must tunnel through the protein medium for about ~13 Å. A non-adiabatic process in the semi-classical Marcus theory is the usual model of choice.[45] Pioneering studies by two independent groups have demonstrated, using a pulse-radiolysis technique, that for one electron injected into CuNiR, the electron distributes between the T1Cu and T2Cu sites with appropriate forward and backward ET rate constants, according to the thermodynamic equilibrium.[46,47] After NO_2^- binds to T2Cu, this equilibrium adjusts dramatically, fully oxidising the T1Cu and reducing the T2Cu. The important physicochemical parameters regarding this intra-ET process, the reorganisation energy (λ), the activation entropy and the pH dependencies have been further analysed using a mutant and two different types of CuNiRs, "blue" and "green".[47–50]

In 2009, a large paradigm shift occurred in this field. Using the sophisticated laser-flash photolysis kinetics and detailed kinetic isotope effects data, the groups of Hasnain and Scrutton succeeded in demonstrating that this intra-ET process is a proton-coupled ET process, triggered by one proton or one water molecule at some site within the CuNiR molecule.[51] This finding indicates that the previously reported intra-ET rate constants are not true ET rate constants, as a rate-limiting step, which is an adiabatic non-ET process, is included prior to the actual ET event. Furthermore, studies on some mutants (probably) affecting the proton-transfer efficiencies have proposed that a long H-bonding network, which connects the T2Cu site to the protein surface *via* side-chain amides of two asparagine residues (Asn_{90} and Asn_{107}), a carbonyl O-Ala_{131}, two water molecules and a side-chain carboxyl group of Asp_{92}, form a channel (or route) for effective H^+ transfer (see Figure 5.5).[52–54]

Recently, a computational work has also proposed a very interesting intra-ET mechanism focused on electronic coupling (H_{DA}) and Cu–Sγ–Cys bond characters for T1Cu.[55] It has been long claimed that two possible ET pathways exist between T1Cu and T2Cu sites in CuNiR: one through the protein backbone (PATH-1) and another through the H-bond between the carbonyl O for T1Cu Cys and the imidazole Nδ1 for T2Cu His (PATH-2) (see Figure 5.6).[23] Solomon and co-workers simulated a mechanism by which the different types, π ("blue") or σ ("green"), of Cu–Sγ bond characters for T1Cu sites affect the intra-ET pathways and H_{DA}.[55] Following calculations using some model fragments, it was possible to obtain a reasonable result indicating

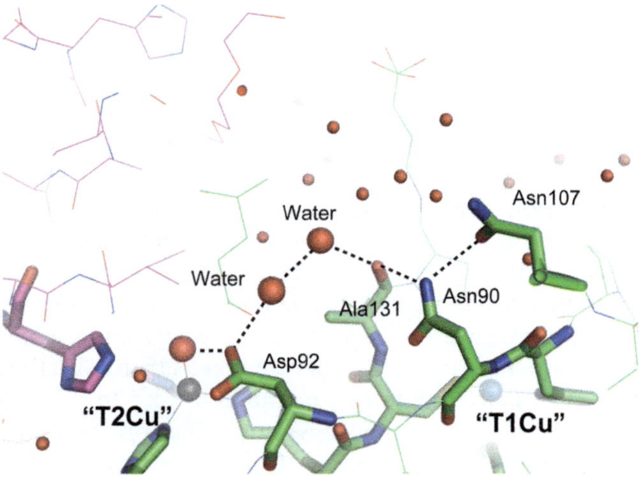

Figure 5.5 Proposed proton channel for *Achromobacter xylosoxidans*. The dotted lines indicate the proposed proton channel along with H-bonds, according to ref. 52–54. The large and small spheres in red represent the water molecules in the channel and in the cavity, respectively. Important residues are represented as sticks. Figures was prepared using Protein Data Bank identifier 2XXF. T1Cu: type-1 copper; T2Cu: type-2 copper.

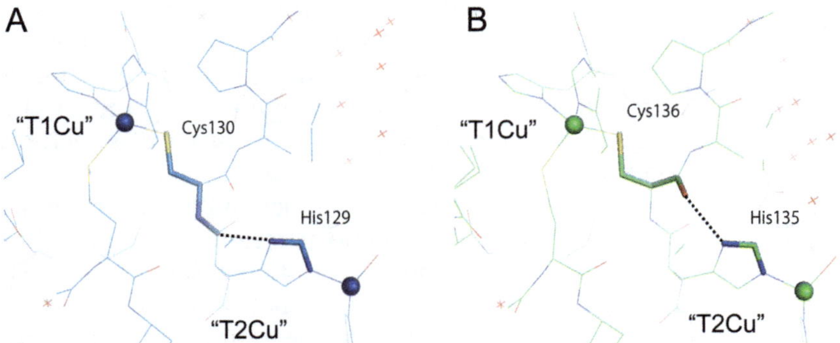

Figure 5.6 The two proposed selective intra-ET pathways. (a) "PATH 1" in the "blue" (π)-type. (b) "PATH 2" in the "green" (σ)-type. The through-space jump sites in the proposed intra-electron transfer pathways are represented by dotted lines in both panels. Pathways are proposed according to the calculations in ref. 55. T1Cu: type-1 copper; T2Cu: type-2 copper.

that the σ-type ("green") shows a larger H_{DA} than the π-type ("blue") and is more sensitive to the H-bond distance between carbonyl O-Cys and Nδ1-His (T2Cu). Based on this computational result, it has been suggested that the two possible pathways can be selectively activated depending on T1Cu geometries, in which the π-type uses "PATH 1", while the σ-type uses the "PATH 2". Moreover, it is notable that the T2Cu geometry has little effect on decreasing H_{DA} (~3 cm^{-1}) in both ET pathways because it seems to be easily influenced by various conditions, such as NO_2^- binding and pH (*vide infra*).

5.3.2 Intermolecular ET

In CuNiR, the electron for NO_2^- reduction is supplied from a physiological redox partner through the inter-protein ET reaction (inter-ET). The inter-ET reaction is a bimolecular reaction and generally obeys a simple two-step mechanism: intermolecular association and the following inter-ET reaction. Therefore, the reaction is largely affected not only by the actual ET parameters (*vide supra*), but also by features of the transient protein–protein interaction that occurs before the ET event. The redox-partner proteins of CuNiRs are usually small electron-transfer proteins, such as a cupredoxin (blue copper protein) or a cytochrome *c*, which deliver an electron from the respiratory chain complex (*e.g.*, cytochrome bc_1) to CuNiRs.[8–11,56] In a living cell, such small partner proteins repeat the delivery of electrons multiple times for effective bioenergetics and metabolism. The protein–protein interaction for individual inter-ET events consequently has an impermanent and dynamic nature; however, appropriate specificity is required in order to ensure partner recognition and directional electron flow in the biological systems.

In the last decade, two remarkable results were reported in this field. One was the characterisation of the transient ET complex of "green" CuNiR with PAz using paramagnetic nuclear magnetic resonance (NMR) spectroscopy.[57] In the paramagnetic NMR techniques, gadolinium complexes were attached specifically onto the surface of CuNiR, causing strong distance-dependent relaxation effects on the chemical shifts of PAz residues. Docking of the two proteins has been performed on the basis of NMR-derived distance restraints and the chemical shift perturbation data for PAz. Interestingly, the ensembles of the docked structures show convergence to a cluster of structures with an average root-mean-square deviation of 1.5 Å. By using the cluster, the binding interface between both proteins was analysed in detail, confirming that the interface consists of polar and non-polar residues surrounded by charged residues, consistent with a previous suggestion based on the mutation studies.[58] The average Cu–Cu distance between both T1Cu sites was estimated to be 15.5 ± 0.5 Å for the complex, and the theoretical ET pathways for all of the energy-minimised structures in the cluster were analysed. Subsequently, several possible routes were found within the 20 structures in the cluster (the 20 structures have been deposited at the PDB under entry 2P80). The two best predominant routes among them have been mentioned in that study: one involves two through-space jumps from the PAz T1Cu ligand His_{81}

to CuNiR residue Met_{84} and from Met_{84} to the CuNiR ligand His_{145}; the other involves one through-space jump from PAz His_{81} to the CuNiR residue Pro_{139}, followed by a through-bond path, Pro_{139}–Pro_{138}–Ala_{137}–Cys_{136} (T1Cu ligand in CuNiR) (see Figure 5.7).

Since the midpoint reduction potentials of both T1Cu sites are known to be ~+275 mV (PAz) and ~+213 mV (T1Cu in CuNiR) at pH 7.0 at 25 °C,[59,60] the forward inter-ET reaction from PAz to CuNiR should be a thermodynamically uphill reaction. However, second-order inter-ET rate constants from PAz to CuNiR are known to be very high (in the order of 10^5–10^6 M^{-1} s^{-1}).[61,62] Their reduction potentials are able to fluctuate concomitantly with basic physical parameters, such as temperature, ionic strengths and pH. However, the apparent vertical relationship between both reduction potentials seems to be maintained in the physiological range. Generally, such a situation should be compensated for by a strong H_{DA} in the possible ET paths *via* the intervening medium between the redox centres, large potential shifts for redox centres or something advantageous during the transient complex formation to achieve the rapid and directional inter-ET.[63,64] Further analyses for these modulating mechanisms are needed in order to deeply understand the CuNiR : PAz ET system.

The other important result was the analysis of the crystal structure of the ET complex of "blue" CuNiR with cytochrome c_{551} (cyt. c_{551}).[11] This complex structure was determined at a resolution of 1.7 Å. A model of the complex, which

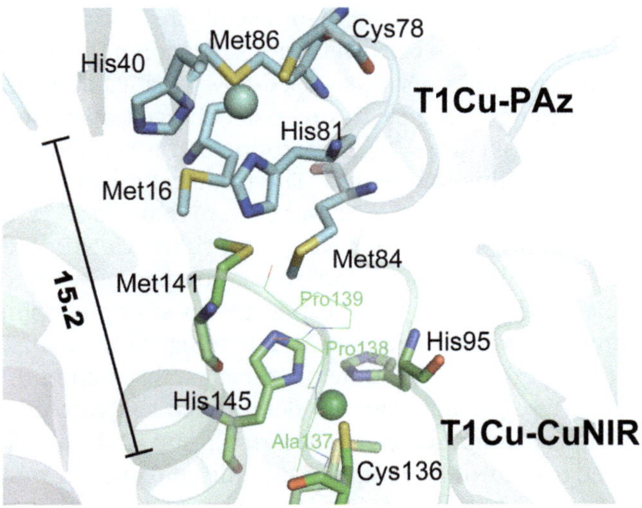

Figure 5.7 A close-up view of the interface between the type-1 copper (T1Cu) sites in the "green" copper nitrite reductase (CuNiR)–pseudoazurin (Paz) complex structure as determined by nuclear magnetic resonance. The T1Cu ligands and three Met residues intervening between both redox centres are represented as cyan sticks (PAz) or as light green ("green" CuNiR). The distance between both T1Cu atoms is 15.2 Å. The predicted electron transfer pathway between T1Cu sites is His_{81} to Met_{84} of PAz to His_{145} of CuNiR or His_{81} of PAz to Pro_{139}–Pro_{138}–Ala_{137}–Cys_{136} of CuNiR (displayed as thin sticks).

consists of one cyt. c_{551} and one trimeric CuNiR molecule within an asymmetric unit, contains 8242 protein atoms, one c-type haem group and six copper atoms (see Figure 5.8). The high-quality data allowed a detailed analysis of the interface. Within the docking interface, the total buried solvent-accessible surface area upon complex formation is estimated to be ~1080 Å^2, and the gap index (the ratio of gap volume to buried solvent-accessible surface area) is ~3.6 Å.[65] In comparison with other ET complexes, the CuNiR–cyt. c_{551} interaction was found to have a similar buried interface, with an area of ~500 Å^2 for each protein.[66,67] The interacting residues of CuNiR are localised both at the hydrophobic patch and at the "tower loop" region, near the T1Cu site. Contact between the hydrophobic patches on both proteins brings their redox centres within 10.5 Å, which is sufficiently close to allow rapid inter-ET.[68–70]

Regarding the interaction of cytochrome cd_1 NiR and its electron donors, Hajdu and collaborators have discussed several points regarding the inter-ET mechanism: one was "pseudospecificity", which is often invoked to explain the mechanism by which several electron donor proteins with very different structures, such as PAz and cytochrome *c*, can transfer electrons to the same enzyme.[71] The point of this concept is an idea that only a transient and promiscuous interaction between the surface patches of enzyme and the partner protein is required for inter-ET to occur. To elucidate this hypothesis, the crystal structure of the transient ET complex of the "blue" CuNiR from *Achromobacter xylosoxidans* and PAz from *Hyphomicrobium denitrificans*, as a possible heterogeneous electron donor suitable for pseudospecific reaction, has been analysed. The forward (PAz → CuNiR) inter-ET reaction in the present heterogeneous pair was monitored using stopped-flow kinetics at a wavelength of 444 nm, corresponding to the Sγ–Cys σ → Cu charge-transfer band for PAz ($\varepsilon = 1940\ M^{-1}\ cm^{-1}$) (see Figure 5.9).[72] Absorption of the T1Cu for "blue" CuNiR is mostly negligible at this wavelength. The resultant second-order rate constant was estimated to be $2.2 \pm 0.2 \times 10^5\ M^{-1}\ s^{-1}$ at pH 6.0 and 25 °C, *ca.* 20-times less than that of the cognate partner cyt. c_{551} for "blue" CuNiR.[11] It is interesting to note that it retains a rate constant that is comparable with that of physiological CuNiR–PAz pairs,[61,62,73] suggesting that a strong coupling between both redox centres is maintained even in a heterogeneous pair. To visualise the interaction between the two proteins, the crystal structure of the binary complex of the heterogeneous pair was determined at a resolution of 3.0 Å. Crystals of the binary complex grew in the space group $P2_13$, with $a, b, c = 153.2$ Å; they contained two CuNiR monomeric subunits and one PAz molecule in an asymmetric unit. The final *R* and *R*-free factors are 17% and 24%, respectively. In the final model, 95% of the residues are in the most favourable regions of the Ramachandran plot (Table 5.2). The coordinate and structure factor are available at the PDB under the identifier 5b1j.

During the structure determination, at first, two CuNiR monomeric subunits were found in the asymmetric unit by using molecular replacement. The two subunits independently formed the CuNiR trimers, by a crystallographic three-fold axis. Figure 5.10a shows a first ($F_O - F_C$) electron density map calculated with the phase prior to the inclusion of PAz in the model after

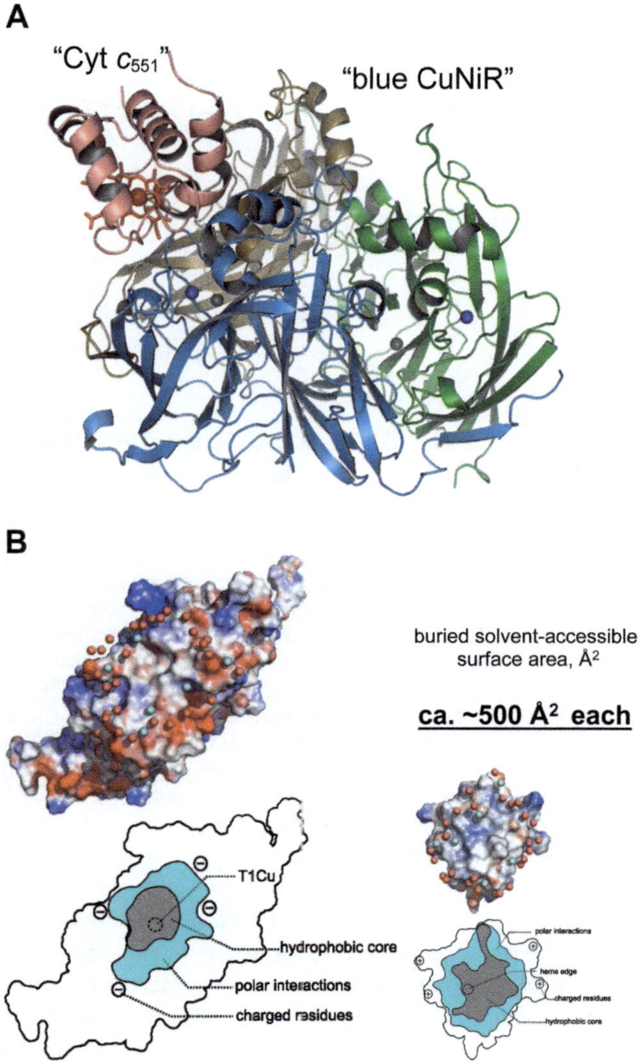

Figure 5.8 Crystal structure of the transient electron transfer complex of "blue" copper nitrite reductase (CuNiR) with cytochrome c_{551} (cyt. c_{551}). (a) Overall structure of the binary complex of "blue" CuNiR with cyt. c_{551}. Cyt. c_{551} is represented as a pink-coloured ribbon and the cyt. c_{551}-docked subunit of the "blue" CuNiR is coloured in blue. The haem group (red), type-1 copper (T1Cu) atom (dark blue) and type-2 copper (T2Cu) atom (grey) are depicted as sticks and spheres. (b) Upper image: contact surface is coloured by electrostatic potential. Eight water molecules bridging both proteins through hydrogen bonds are coloured in cyan and the other waters molecules binding to the subunit or cyt. c_{551} are coloured in red. Electrostatic potentials were calculated by the program DelPhi. The displayed potentials range from −15 (red) to +15 (blue) kTe^{-1}. Lower image: schematic representation of the interfaces of the transient complex of "blue" CuNiR with cyt. c_{551}. The entry/exit point of the electrons is surrounded by a set of hydrophobic residues, which are surrounded by a polar region. Charged residues are found on the edge of the binding site.[74]

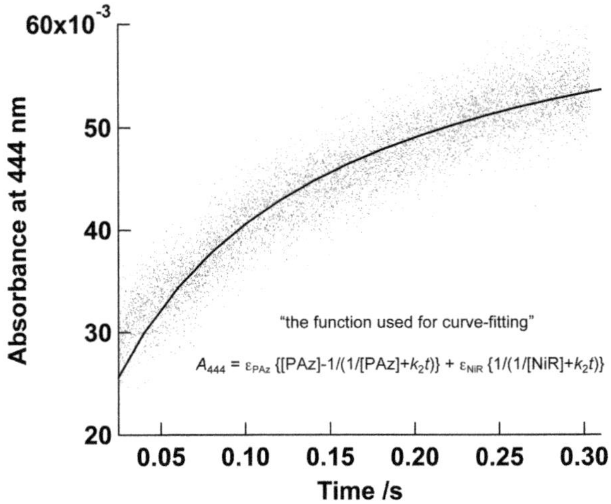

Figure 5.9 The inter-electron transfer (ET) reaction of the heterogeneous donor pseudoazurin (PAz) to "blue" copper nitrite reductase (CuNiR). The grey dots represent the observed data; the black line is a fitting curve. Stopped-flow kinetics of the inter-ET reaction were assessed in 10 mM potassium phosphate buffer (pH 6.0) at 25 °C under anaerobic conditions. In order to maintain a second-order conditions, the concentration ratios of CuNiR to PAz were always equimolar: [PAz] = [NiR]. The second-order rate constants were calculated with the modified second-order kinetics function shown in the figure, using IgorPro ver. 4.02 (WaveMetrics), in which [PAz] and [NiR] are the initial concentrations of PAz and "blue" CuNiR, respectively. At 444 nm, the extinction coefficients of the two proteins are $\varepsilon_{PAz} = 1940\ M^{-1}\ cm^{-1}$ for PAz and $\varepsilon_{NiR} = 539\ M^{-1}\ cm^{-1}$ for blue CuNiR.

molecular replacement of CuNiR alone. The map gave strong signals, sufficient to confirm the presence of a PAz molecule even at the contour level of 2.5σ. The CuNiR molecules contribute to major contacts in the crystal lattice and provide the spaces for three Paz's related with a crystallographic three-fold axis; these PAz's are bound to the docking sites on only one CuNiR trimer without steric hindrance. The main interaction in the complex structure is accomplished by direct contact between both hydrophobic patches on each T1Cu site and brings their Cu sites within ~16 Å (see Figure 5.10b). It was found that the solvent-exposed His_{81} ligand for PAz directly interacts with the carbonyl O-Ala_{86} in CuNiR within ~2.6 Å, which is proposed to be a suitable through-space jump for the inter-ET path in this heterogeneous pair complex. Very interestingly, a comparison with the previous CuNiR–cytochrome *c* complex structure reveals that the docking site is mostly identical in both cases. In other words, the present heterogeneous ET complex structure analysis clearly demonstrates that the docking site on the CuNiR molecular surface has a high site-specificity limit on the T1Cu site and wide adaptability that is not affected by the identity of the partner. To express such "pseudo-specificity", it is likely that the characteristic distribution pattern rather than

Table 5.2 Crystallographic data collection and refinement statistics.[a]

Data set	CuNiR : PAz	Cl^--bound CuNiR[b]
Crystallographic statistics		
Wavelength, Å	0.9	0.9
Resolution, Å (outer shell)	19.1–3.0 (3.05–3.00)	90.5–1.35 (1.37–1.35)
Unique reflections	23 931	86 291
Completeness, % (outer shell)	99.2 (99.1)	99.9 (99.9)
R_{merge} (outer shell)	0.093 (0.57)	0.069 (0.71)
Data redundancy (outer shell)	4.8 (4.7)	5.2 (4.1)
Average I/σ (outer shell)	18.7 (3.0)	29.1 (2.0)
Refinement statistics		
Resolution range, Å	19.1–3.0	90.5–1.35
No. of reflections ($F > 0$)	22 698	73 010
Total no. of atoms	6055	2899
Completeness of data, %	99.1	89.0
R-factor (R-free)	0.174 (0.24)	0.149 (0.173)
r.m.s. deviation		
Bond, Å	0.011	0.005
Angle, °	1.6	1.2
Ramachandran plot		
Favored, %	95.4	98.5
Allowed, %	4.0	1.5
Outlier, %	0.6	0

[a]CuNiR: cooper nitrite reductase; PAz: pseudoazurin.
[b]Preparation of the Cl^--bound form of crystal was supported in part by "High-Quality Protein Crystal Growth Experiment on JEM" promoted by Japan Aerospace Exploration Agency (JAXA).

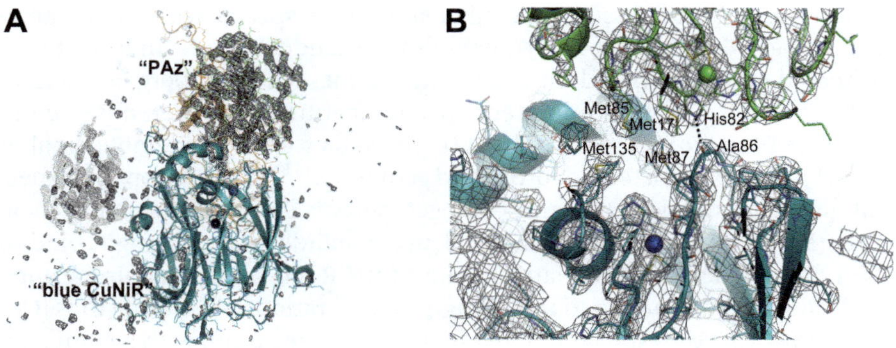

Figure 5.10 Crystal structure of the transient electron transfer complex of the "blue" copper nitrite reductase (CuNiR) with pseudoazurin (PAz). (a) First $F_O - F_C$ electron density map calculated with the phase prior to the inclusion of PAz in the model after molecular replacement. The contoured level is 2.5σ (black mesh). The final PAz and CuNiR models are represented by wires and ribbons. (b) Details of the interface showing a σ_A-weighted $2F_O - F_C$ map contoured at 1.5σ. CuNiR and PAz are shown as blue and green sticks, and both type-1 copper atoms are represented as spheres. The dotted line indicates a direct interaction between the His_{81} ligand of PAz and the carbonyl O of Ala_{86} of CuNiR.

each stereo-selectivity of polar, non-polar and charged residues surrounding the electron entry/exit sites is at least one of the key factors in the transient partner interaction, as described before (see Figure 5.9).[71,74] Furthermore, it has been speculated that direct interaction of the solvent-exposed His_{81} residue for PAz observed in both NMR and in the present X-ray structure analyses may be able to tune-up the inter-ET mechanism by affecting the electronic structure of the PAz T1Cu. Although some ambiguities remain, it might be possible to provide more useful insights into the mechanism underlying the inter-ET reaction in the near future.

5.4 The T2Cu Site and Nitrite Reduction

It is well known that in the oxidised resting state, T2Cu has a (pseudo)tetrahedral geometry with three His ligands and a water molecule in the apical position. However, the geometrical assignment of the T2Cu site often remains controversial, even though more than two decades have passed since the first report of a crystal structure of CuNiR in 1991;[12] that is, it has been unclear which direction the d_{x2-y2} ground state turns to. One of the reasons for this might be the fact that the relative positions of exogenous ligands, including water(s) and NO_2^-, lack uniformity in a large number of CuNiR structures. Previously, similar discussion has also been addressed by two independent groups.[75,76] There is a consensus on two apparent possibilities that, in the resting state, T2Cu can take two different tetra-coordinate conformations, or basically it has a square-pyramidal conformation with two exogenous ligands coordinating positions, such as a bidentate κ^2 O,O-binding mode of NO_2^-. Interestingly, EPR spectroscopic data in the resting state of CuNiRs indicate that the T2Cu site generally has a single species with the parameters $g_{//}$ = ~2.36 and $A_{//}$ = ~13 mT, exhibiting a tendency to form a penta-coordinated geometry based on the Vänngård–Piesach–Blumberg correlation plot.[13,16,31,34,77–79] However, different spectral features have been observed in the "blue" CuNiR studies, in which the $A_{//}$ values are considerable smaller (~9 mT), suggesting a tetra-coordinated geometry.[16,80,81] These apparent spectral differences suggest that the T2Cu geometry is easily influenced by measurement conditions, such as the presence of halide ions that might bind to the T2Cu. In fact, the T2Cu site of "blue" CuNiR has been structurally analysed using crystals prepared under high concentrations of chloride ions (Cl^-; ≥200 mM). These crystals diffracted up to 1.35 Å resolution with a completeness of 98%. The data statistics are listed in Table 5.2. Figure 5.11 shows a $2F_O - F_C$ difference Fourier electron density map around the T2Cu site at 1.0σ and the final model (PDB identifier: 5b1k). In addition to a dominant water molecule (or HO^-) coordinating to T2Cu, another electron density attributed to Cl^- was clearly observed at the site. In fact, Cl^- coordination to the T2Cu site has also been reported previously in a low-resolution structure.[82] The geometric parameters are shown in Table 5.3.

It is interesting to note that the NO_2^- binding affects the T2Cu EPR parameters, in particular the deviation of the $A_{//}$value (1–2 mT), and increases

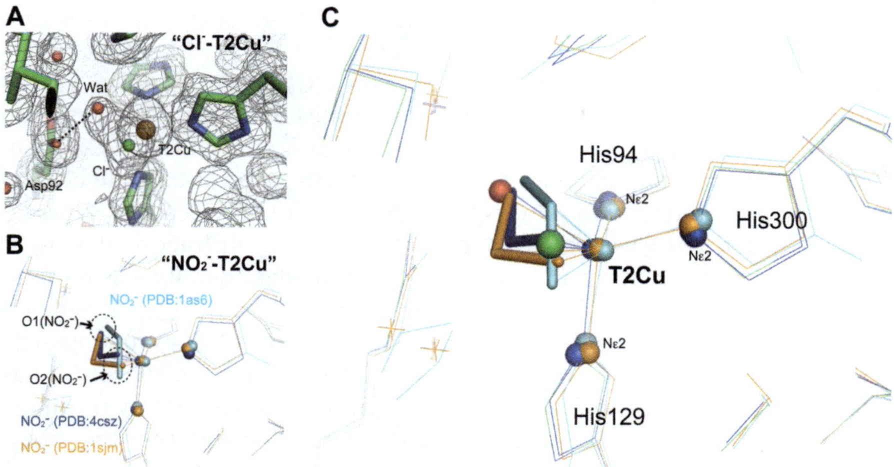

Figure 5.11 The penta-coordinated (Cl^--bound) type-2 copper (T2Cu) and NO_2^--bound form structures. (a) View of the Cl^--bound T2Cu site. The Cu, Cl^- and coordinating water molecule are represented as spheres in brown (Cu), green (Cl^-) and red (water). The $2F_O - F_C$ electron density map is overlaid on the final model structure. (b) The NO_2^--bound T2Cu site.[90] (c) Superimposition of the penta-coordinated (Cl^--bound) T2Cu and NO_2^--bound form. All of the ligand positions are overlaid at approximately same positions. PDB: Protein Data Bank.

Table 5.3 The type-2 copper geometric parameters in the "NO_2^--bound form" and "Cl^--bound form".[a,b]

	NO_2^--bound form[c]	Cl^--bound form
Distances (Å)		
His_1-Cu	2.08	2.08
His_2-Cu	2.03	2.11
His_3-Cu	2.15	2.06
O1(NO_2^-)-Cu or Wat-Cu	1.94	2.14
O2(NO_2^-)-Cu or Cl^--Cu	2.38	2.40
Angles (°)		
His_1-Cu-His_2	104	113
His_1-Cu-His_3	105	102
His_2-Cu-His_3	108	115
His_1-Cu-O1 or His_1-Cu-O1	88	100
His_2-Cu-O1 or His_2-Cu-O1	121	104
His_3-Cu-O1 or His_3-Cu-Wat	105	123
His_1-Cu-O2 or His_1-Cu-Cl^-	148	145
His_2-Cu-O2 or His_2-Cu-Cl^-	104	96
His_3-Cu-O2 or His_3-Cu-Cl^-	81	82
O1–Cu–O2 or O–Cu–Cl^-	65	52
τ^d	0.45	0.37

[a] His_1, His_2 and His_3 are the first, second and third His ligands in the amino acid sequence.
[b] $\tau = (\beta - \alpha)/60$.
[c] The NO_2^--bound form (F306C mutant of blue copper nitrite reductase [AxNiR]) structure is taken from Protein Data Bank entry 4csz.[90]
[d] The geometric index introduced Addison *et al.*[92]

spectral rhombicity, suggesting that the relative geometrical distortion occurs at the T2Cu site due to the NO_2^- binding.[34,78,80]

Nevertheless, the first crystal structure of the NO_2^--bound form of "green" CuNiR was reported in 1997,[83] and the coordinates of the centres have been widely used as a model when consider the reaction mechanism (PDB identifier: 1as6).[31] Afterwards, a new NO_2^--bound form representing the characteristic face–on interaction (k^3 O,N,O-binding mode) between NO_2^- and Cu (PDB identifier: 1sjm) was reported in 2004.[84] The X-ray diffraction data were collected using a synchrotron X-ray beam, and the data quality was remarkably higher than that of the previous structure. In this study, the side-on NO-bound form was also reported. Both the NO_2^- and NO unique coordination modes were further investigated by many spectroscopists, and it has been confirmed by theoretical and model complex studies that their unique coordination modes are caused by the steric interaction with the neighbouring Ile residue, and that the NO-bound form corresponds to the $T2Cu^+$–NO state.[85–89] At the same time, the NO_2^--bound form of a mutant (F306C; Phe_{306} is a surface-exposed residue) of the "blue" CuNiR has also been determined at 1.7 Å (PDB identifier: 4csz) (see Figure 5.11b),[90] presenting a quite similar NO_2^--coordination mode relative to that of the previous report. The NO_2^--bound form of another "green" CuNiR isolated from *Achromobacter cycloclastes* has also been reported (PDB identifier: 5akr).[91]

It is important to compare the current NO_2^--bound T2Cu structure with the resting penta-coordinated (Cl^--bound) structure. Figure 5.11c shows a superposition of both states' (NO_2^--bound and Cl^--bound forms) structures of "blue" CuNiRs. Using the penta-coordinate geometry index τ_5 introduced by Addison *et al.*,[92] the coordination geometries of both structures give values of $\tau = 0.45$ (NO_2^--bound form) and 0.37 (Cl^--bound form). This means that the geometry of the NO_2^--bound form has a slight distortion toward a trigonal bipyramidal geometry compared to the resting penta-coordinated geometry. This geometrical change is consistent with the spectral changes observed in the previous EPR spectra of the NO_2^--bound form. Moreover, this could be indicative of the d_{x2-y2} ground state, along with an equatorial plane formed by Nε2-His_{94}, Nε2-His_{300}, O1 and O2 (or Cl^-) in both structures (see Figure 5.11c). Therefore, it can be proposed that the electron from the T1Cu site is received through the axial ligand (His_{129}) during the intra-ET event. These features of the NO_2^-–T2Cu geometry seem to be also supported by a very recently analysed structure using serial femtosecond crystallography (PDB identifier: 5d4i; $\tau = 0.52$).[93]

In addition, the comparison between the NO_2^--bound form structures of both "green" and "blue" CuNiRs is of considerable interest; in particular, there is a remarkable difference between their His_{249} conformations near the T2Cu site (see Figure 5.12). This apparent discrepancy is probably due to the different crystallisation pH conditions, which are pH = 4.0 ("green") and 6.5 ("blue"). Similar conformational change of the His residue has been also observed in the *Rhodobacter* CuNiR structures under various pH and redox conditions.[76,94] Moreover, it is worth noting that the His is directly

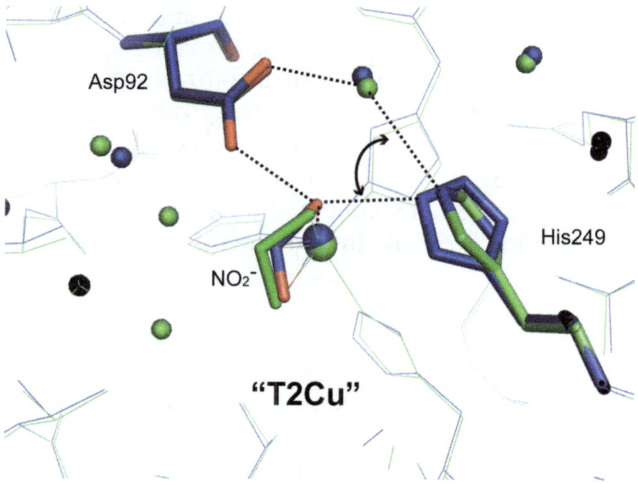

Figure 5.12 The second coordination spheres in two NO_2^--bound structures of *Alcaligenes faecalis* (pH 4.0) and *Achromobacter xylosoxidans* (pH 6.5) copper nitrite reductases in green and blue, respectively. The large conformational difference is observed at the side-chain of His_{249}. In near neutral pH, the His residue makes an H-bond with the substrate NO_2^-. T2Cu: type-2 copper.

hydrogen bond to the O1 atom of NO_2^- in the pH 6.5 structure, as the O atom adopts a (pseudo)tetrahedral-like conformation. It is likely that this conformation significantly affects the bond polarity between the N and O1 of NO_2^-, which might be expected to maximise the overlap between the d_{x2-y2} and the anti-bonding σ* orbital of the O–NO bond. Such a direct hydrogen bond between NO_2^- and the His residue will be further confirmed by a neutron diffraction study using the crystals at pH 6.5 in the near future. At present, it remains unclear which residue—either Asp_{92} or His_{249}—is a proton donor for generating the protonated NO_2^- (HONO),[31–33] but the present NO_2^--bound form analysed in the "blue" CuNiR X-ray studies might be significant in the NO_2^- reduction mechanism of CuNiR.

Acknowledgements

This research was supported by a Grant-in-Aid for the Japan Society for the Promotion of Science KAKENHI (grant numbers: 16750144, 20750137 and 23750190) and some grants from the Sumitomo Foundation, the Japan Foundation for Applied Enzymology and the Takeda Science Foundation. This study was also supported in part by "High-Quality Protein Crystal Growth Experiment on JEM" promoted by Japan Aerospace Exploration Agency (JAXA). The Russian Spacecraft "Progress" and "Soyuz" provided by the Russian Federal Space Agency were used for space transportation. Some of the space crystallisation technology was developed by the European Space

Agency (ESA) and the University of Granada. The author thanks staff for the beamlines 32XU, 41XU and 44XU at SPring-8 for kind support in the collection of X-ray data (proposal numbers 2010A1401, 2010A6531, 2010B1993, 2010B6531, 2011A6634, 2013A6841, 2013B6841, 2014A6940 and 2014B6940). Finally, the author is profoundly grateful to Drs D. Hira, H. Koteishi and Y. Fukuda, Mr R. Yoneda, Mr A. Agatahama, Ms A. Tsuda, Ms F. Shirota and Ms S. Ikebuchi, and the many other students who graduated from his laboratory, and to the emeritus Prof. S. Suzuki for kind help and useful discussions.

References

1. W. G. Zumft, *Microbiol. Mol. Biol. Rev.*, 1997, **61**, 533.
2. P. Tavares, A. S. Pereira, J. J. G. Moura and I. Moura, *J. Inorg. Biochem.*, 2006, **100**, 2087.
3. R. Bartossek, G. W. Nicol, A. Lanzen, H.-P. Klenk and C. Schleper, *Environ. Microbiol.*, 2010, **12**, 1075.
4. T. J. Lawton, K. E. Bowen, L. A. Sayavedra-Soto, D. J. Arp and A. C. Rosenzweig, *J. Biol. Chem.*, 2013, **288**, 25575.
5. D. Hira, H. Toh, C. T. Migita, H. Okubo, T. Nishiyama, M. Hattori, K. Furukawa and T. Fujii, *FEBS Lett.*, 2012, **586**, 1658.
6. D. E. Canfield, A. N. Glazer and P. G. Falkowski, *Science*, 2010, **330**, 192.
7. A. J. Thomson, G. Giannopoulos, J. Pretty, E. M. Baggs and D. J. Richardson, *Philos. Trans. R. Soc., B*, 2012, **367**, 1157.
8. M. Y. Liu, W. J. Payne and J. LeGall, *J. Bacteriol.*, 1986, **166**, 604.
9. T. Kakutani, H. Watanabe, K. Arima and T. Beppu, *J. Biochem.*, 1981, **89**, 463.
10. H. J. Wijma, G. W. Canters, S. de Vries and M. P. Verbeet, *Biochemistry*, 2004, **43**, 10467.
11. M. Nojiri, H. Koteishi, T. Nakagami, K. Kobayashi, T. Inoue, K. Yamaguchi and S. Suzuki, *Nature*, 2009, **462**, 117.
12. J. W. Godden, S. Turley, D. C. Teller, E. T. Adman, M. Y. Liu, W. J. Payne and J. LeGall, *Science*, 1991, **253**, 438.
13. S. Suzuki, K. Kataoka, K. Yamaguchi, T. Inoue and Y. Kai, *Coord. Chem. Rev.*, 2000, **190–192**, 245.
14. H. Iwasaki and T. Mori, *J. Biochem.*, 1955, **42**, 375.
15. H. Suzuki and H. Iwasaki, *J. Biochem.*, 1962, **52**, 193.
16. Z. H. L. Abraham, D. J. Lowe and B. E. Smith, *Biochem. J.*, 1993, **295**, 587.
17. E. I. Solomon, J. W. Hare and H. B. Gray, *Proc. Natl. Acad. Sci. U. S. A.*, 1976, **73**, 1389.
18. J. M. Guss and H. C. Freeman, *J. Mol. Biol.*, 1983, **169**, 521.
19. H. Nar, A. Messerschmidt, R. Huber, M. van de Kamp and G. W. Canters, *J. Mol. Biol.*, 1991, **221**, 765.
20. J. M. Guss, H. D. Bartunik and H. C. Freeman, *Acta Cryst.*, 1992, **B48**, 790.
21. E. I. Solomon and R. G. Hadt, *Coord. Chem. Rev.*, 2011, **255**, 774.
22. K. Pierloot, J. O. A. De Kerpel, U. Ryde, M. H. M. Olsson and B. O. Roos, *J. Am. Chem. Soc.*, 1998, **120**, 13156.

23. E. I. Solomon, R. K. Szilagyi, S. DeBeer George and L. Basumallick, *Chem. Rev.*, 2004, **104**, 419.
24. K. Pierloot, J. O. A. De Kerpel, U. Ryde and B. O. Roos, *J. Am. Chem. Soc.*, 1997, **119**, 218.
25. M. van Gastel, J. W. A. Coremans, H. Sommerdijk, M. C. van Hemert and E. J. J. Groenen, *J. Am. Chem. Soc.*, 2002, **124**, 2035.
26. B. G. Malmström and T. Vänngård, *J. Mol. Biol.*, 1960, **2**, 118.
27. R. Malkin, B. G. Malmström and T. Vänngård, *FEBS Lett.*, 1968, **1**, 50.
28. K. Kataoka, H. Furusawa, K. Takagi, K. Yamaguchi and S. Suzuki, *J. Biochem.*, 2000, **127**, 345.
29. M. J. Boulanger, M. Kukimoto, M. Nishiyama, S. Horinouchi and M. E. P. Murphy, *J. Biol. Chem.*, 2000, **275**, 23957.
30. M. J. Boulanger and M. E. P. Murphy, *Protein Sci.*, 2002, **12**, 248.
31. S. Ghosh, A. Dey, Y. Sun, C. P. Scholes and E. I. Solomon, *J. Am. Chem. Soc.*, 2009, **131**, 277.
32. A. C. Merkle and N. Lehnert, *Dalton Trans.*, 2012, **41**, 3355.
33. E. I. Solomon, D. E. Heppner, E. M. Johnston, J. W. Ginsbach, J. Cirera, M. Qayyum, M. T. Kieber-Emmons, C. H. Kjaergaard, R. G. Hadt and L. Tian, *Chem. Rev.*, 2014, **114**, 3659.
34. K. Olesen, A. Veselov, Y. Zhao, Y. Wang, B. Danner, C. P. Scholes and J. P. Shapleigh, *Biochemistry*, 1998, **37**, 6086.
35. G. Denariaz, W. J. Payne and J. LeGall, *Biochim. Biophys. Acta*, 1991, **1056**, 225.
36. M. J. Boulanger and M. E. P. Murphy, *J. Mol. Biol.*, 2002, **315**, 1111.
37. K. Yamaguchi, M. Kobayashi, K. Kataoka and S. Suzuki, *Biochem. Biophys. Res. Commun.*, 2003, **300**, 36.
38. A. Tsuda, R. Ishikawa, H. Koteishi, K. Tange, Y. Fukuda, K. Kobayashi, T. Inoue and M. Nojiri, *J. Biochem.*, 2013, **154**, 51.
39. Y. Fukuda, H. Koteishi, R. Yoneda, T. Tamada, H. Takami, T. Inoue and M. Nojiri, *Biochim. Biophys. Acta*, 2014, **1837**, 396.
40. K. Inatomi and L. I. Hochstein, *Curr. Microbiol.*, 1996, **32**, 72.
41. H. Ichiki, Y. Tanaka, K. Mochizuki, K. Yoshimatsu, T. Sakurai and T. Fujiwara, *J. Bacteriol.*, 2001, **183**, 4149.
42. D. Hira, M. Nojiri and S. Suzuki, *Acta Cryst.*, 2009, **D65**, 85.
43. A. Messerschmidt, A. Rossi, R. Ladenstein, R. Huber, M. Bolognesi, G. Gatti, A. Marchesini, R. Petruzzelli and A. Finazzi-Agró, *J. Mol. Biol.*, 1989, **206**, 513.
44. M. A. Hough, M. J. Ellis, S. Antonyuk, R. W. Strange, G. Sawers, R. R. Eady and S. S. Hasnain, *J. Mol. Biol.*, 2005, **350**, 300.
45. R. A. Marcus and N. Sutin, *Biochim. Biophys. Acta*, 1985, **811**, 265.
46. S. Suzuki, T. Kohzuma, Deligeer, K. Yamaguchi, N. Nakamura, S. Shidara, K. Kobayashi and S. Tagawa, *J. Am. Chem. Soc.*, 1994, **116**, 11145.
47. O. Farver, R. R. Eady, Z. H. L. Abraham and I. Pecht, *FEBS Lett.*, 1998, **436**, 239.
48. O. Farver, R. R. Eady, G. Sawers, M. Prudêncio and I. Pecht, *FEBS Lett.*, 2004, **561**, 173.

49. S. Wherland, O. Farver and I. Pecht, *ChemPhysChem*, 2005, **6**, 805.
50. K. Kobayashi, S. Tagawa, Deligeer and S. Suzuki, *J. Biochem.*, 1999, **126**, 408.
51. S. Brenner, D. J. Heyes, S. Hay, M. A. Hough, R. R. Eady, S. S. Hasnain and N. S. Scrutton, *J. Biol. Chem.*, 2009, **284**, 25973.
52. M. A. Hough, R. R. Eady and S. S. Hasnain, *Biochemistry*, 2008, **47**, 13547.
53. N. G. H. Leferink, C. Han, S. V. Antonyuk, D. J. Heyes, S. E. J. Rigby, M. A. Hough, R. R. Eady, N. S. Scrutton and S. S. Hasnain, *Biochemistry*, 2011, **50**, 4121.
54. N. G. H. Leferink, R. R. Eady, S. S. Hasnain and N. S. Scrutton, *FEBS J.*, 2012, **279**, 2174.
55. R. G. Hadt, S. I. Gorelski and E. I. Solomon, *J. Am. Chem. Soc.*, 2014, **136**, 15034.
56. J. Chen and M. Strous, *Biochim. Biophys. Acta*, 2013, **1827**, 136.
57. M. D. Vlasie, R. Fernández-Busnadiego, M. Prudêncio and M. Ubbink, *J. Mol. Biol.*, 2008, **375**, 1405.
58. M. Kukimoto, M. Nishiyama, M. Tanokura, E. T. Adman and S. Horinouchi, *J. Biol. Chem.*, 1996, **271**, 13680.
59. G. Battistuzzi, M. Bellei, M. Borsari, G. W. Canters, E. de Waal, L. J. C. Jeuken, A. Ranieri and M. Sola, *Biochemistry*, 2003, **42**, 9214.
60. H. J. Wijma, I. MacPherson, M. Alexandre, R. E. M. Diederix, G. W. Canters, M. E. O. Murphy and M. P. Verbeet, *J. Mol. Biol.*, 2006, **358**, 1081.
61. Y. Iwasaki, T. Takeuchi, E. Tamiya, I. Karube, M. Nishiyama, S. Horinouchi, T. Beppu, H. Kadoi, S. Uchiyama, S. Suzuki and M. Suzuki, *Electroanalysis*, 1992, **4**, 771.
62. Y. Astier, G. W. Canters, J. J. Davis, H. A. O. Hill, M. P. Verbeet and H. J. Wijma, *ChemPhysChem*, 2005, **6**, 1114.
63. P. Xiong, J. M. Nocek, J. Vura-Weis, J. V. Lockard, M. R. Wasielewski and B. M. Hoffman, *Science*, 2010, **330**, 1075.
64. S. Keinan, J. M. Nocek, B. M. Hoffman and D. N. Beratan, *Phys. Chem. Chem. Phys.*, 2012, **14**, 13881.
65. R. A. Laskowski, *J. Mol. Graphics*, 1995, **13**, 323.
66. P. B. Crowley and M. A. Carrondo, *Proteins*, 2004, **55**, 603.
67. J. Janin, R. P. Bahadur and P. Chakrabarti, *Q. Rev. Biophys.*, 2008, **41**, 133.
68. C. C. Moser, J. M. Keske, K. Warncke, R. S. Farid and P. L. Dutton, *Nature*, 1992, **355**, 796.
69. J. N. Onuchic, D. N. Beratan, J. R. Winkler and H. B. Gray, *Annu. Rev. Biophys. Biomol. Struct.*, 1992, **21**, 349.
70. C. C. Page, C. C. Moser, X. Chen and P. L. Dutton, *Nature*, 1999, **402**, 47.
71. P. A. Williams, V. Fulop, Y.-C. Leung, C. Chan, J. W. B. Moir, G. Howlett, S. J. Ferguson, S. E. Radford and J. Hajdu, *Nat. Struct. Biol.*, 1995, **2**, 975.
72. D. Hira, M. Nojiri, K. Yamaguchi and S. Suzuki, *J. Biochem.*, 2007, **142**, 335.
73. K. Kataoka, K. Yamaguchi, M. Kobayashi, T. Mori, N. Bokui and S. Suzuki, *J. Biol. Chem.*, 2004, **279**, 53374.

74. M. Prudêncio and M. Ubbink, *J. Mol. Recognit.*, 2004, **17**, 524.
75. N. Källrot, K. Nilsson, T. Rasmussen and U. Ryde, *Int. J. Quantum Chem.*, 2005, **102**, 520.
76. F. Jacobson, H. Guo, K. Olesen, M. Ökvist, R. Neutze and L. Sjölin, *Acta Cryst.*, 2005, **D61**, 1190.
77. J. Peisach and W. E. Blumberg, *Arch. Biochem. Biophys.*, 1974, **165**, 691.
78. M. Fittipaldi, H. J. Wijma, M. P. Verbeet, G. W. Canters, E. J. J. Groenen and M. Huber, *Biochemistry*, 2005, **44**, 15193.
79. L. Bubacco, M. van Gastel, E. J. J. Groenen, E. Vijgenboom and G. W. Canters, *J. Biol. Chem.*, 2003, **278**, 7381.
80. B. D. Howes, Z. H. L. Abraham, D. J. Lowe, T. Brüser, R. R. Eady and B. E. Smith, *Biochemistry*, 1994, **33**, 3171.
81. Z. H. L. Abraham, B. E. Smith, B. D. Howes, D. J. Lowe and R. R. Eady, *Biochem. J.*, 1997, **324**, 511.
82. F. E. Dodd, J. van Beeumen, R. R. Eady and S. S. Hasnain, *J. Mol. Biol.*, 1998, **282**, 369.
83. M. E. P. Murphy, S. Turley and E. T. Adman, *J. Biol. Chem.*, 1997, **272**, 28455.
84. E. I. Tocheva, F. I. Rosell, A. G. Mauk and M. E. P. Murphy, *Science*, 2004, **304**, 867.
85. O. M. Usov, Y. Sun, V. M. Grigoryants, J. P. Shapleigh and C. P. Scholes, *J. Am. Chem. Soc.*, 2006, **128**, 13102.
86. S. Ghosh, A. Dey, O. M. Usov, Y. Sun, V. M. Grigoryants, C. P. Scholes and E. I. Solomon, *J. Am. Chem. Soc.*, 2007, **129**, 10310.
87. K. Fujisawa, A. Tateda, Y. Miyashita, K. Okamoto, F. Paulat, V. K. K. Praneeth, A. Merkle and N. Lehnert, *J. Am. Chem. Soc.*, 2008, **130**, 1205.
88. A. C. Merkle and N. Lehnert, *Inorg. Chem.*, 2009, **48**, 11504.
89. N. Lehnert, U. Cornellssen, F. Neese, T. Ono, Y. Noguchi, K. Okamoto and K. Fujisawa, *Inorg. Chem.*, 2007, **46**, 3916.
90. N. G. H. Leferink, S. V. Antonyuk, J. A. Houwman, N. S. Scrutton, R. R. Eady and S. S. Hasnain, *Nat. Commun.*, 2014, **5**, 4395.
91. S. V. Antonyuk, R. W. Strange, G. Sawers, R. R. Eady and S. S. Hasnain, *Proc. Natl. Acad. Sci. U. S. A.*, 2005, **102**, 12041.
92. A. W. Addison, T. N. Rao, J. Reedijk, J. van Rijn and G. C. Verschoor, *J. Chem. Soc., Dalton Trans.*, 1984, 1349.
93. Y. Fukuda, K. M. Tse, T. Nakane, T. Nakatsu, M. Suzuki, M. Sugahara, S. Inoue, T. Masuda, F. Yumoto, N. Matsugaki, E. Nango, K. Tono, Y. Joti, T. Kameshima, C. Song, T. Hatsui, M. Yabashi, O. Nureki, M. E. P. Murphy, T. Inoue, S. Iwata and E. Mizohata, *Proc. Natl. Acad. Sci. U. S. A.*, 2016, **113**, 2928.
94. F. Jacobson, A. Pistorius, D. Farkas, W. de Grip, Ö. Hansson, L. Sjölin and R. Neutze, *J. Biol. Chem.*, 2007, **282**, 6347.
95. M. J. Ellis, F. E. Dodd, G. Sawers, R. R. Eady and S. S. Hasnain, *J. Mol. Biol.*, 2004, **328**, 429.

CHAPTER 6

Structure and Function of Nitric Oxide Reductases

TAKEHIKO TOSHA[a] AND YOSHITSUGU SHIRO*[a,b]

[a]RIKEN SPring-8 Center, Sayo, Hyogo 679-5148, Japan; [b]Graduate School of Life Science, University of Hyogo, Ako, Hyogo 678-1297, Japan
*E-mail: yshiro@sci.u-hyogo.ac.jp

6.1 Introduction

In living systems, there are two types of nitric oxide (NO)-generating systems: one is NO synthase (NOS) in mammal, and the other is nitrite reductase (NiR) in microbial denitrification. In the former case, the NO molecule is generated from arginine (Arg) and molecular oxygen O_2 by the oxygenation reaction catalysed by NOS, which then acts as a signal molecule on being accepted by the NO receptor, soluble guanylate cyclase. In the latter case, NO is generated by the one-electron reduction of nitrite (NO_2^-) catalysed by NiR, but due to its high cytotoxicity, it must be decomposed by NO reductases (NORs) immediately after its formation.[1–4] Thus, NO has aspects of both good and bad molecules. In this chapter, we describe the characteristic properties of NO and then the NO detoxification system, NOR, in bacterial denitrification.

RSC Metallobiology Series No. 9
Metalloenzymes in Denitrification: Applications and Environmental Impacts
Edited by Isabel Moura, José J. G. Moura, Sofia R. Pauleta and Luisa B. Maia

Published by the Royal Society of Chemistry, www.rsc.org

6.2 Properties and Reactions of NO and Metal–NO Complexes

One eminent chemical feature of NO is that the molecule has a single electron spin mainly localised at nitrogen atom,[5–7] which is indicated by electron spin resonance spectroscopy.[8] The bond length of NO is 1.15 Å and the bond order is 2.5, which are between those for O_2 (1.21 Å, double bond) and N_2 (1.10 Å, triple bond). The bond order and the radical character of NO are reasonably described by its molecular orbital that accommodates 11 valence electrons of this molecule. The three bonding orbitals (one σ2p and two π2p) are occupied, and one antibonding orbital (π*2p) has a single electron spin, making the N–O bond order be 2.5 and giving it a paramagnetic character. NO is easily one-electron oxidised to produce NO^+, which is isoelectric with CO and has a triple bond character. On the other hand, one-electron addition to NO makes NO^-, which is a species that is isoelectric with O_2.

Due to its radical character, NO is a reactive gas molecule and serves as a strong ligand of metals in biology. NO reacts with molecular oxygen quickly to produce nitrogen dioxide (NO_2). Kinetic and theoretical studies show that the reaction between NO and O_2 proceeds as follows:

$$NO + O_2 \rightarrow ONOO \text{ (or } NO{\cdot}O_2) \tag{6.1}$$

$$ONOO \text{ (or } NO{\cdot}O_2) + NO \rightarrow ONOONO \tag{6.2}$$

$$ONOONO \rightarrow 2NO_2 \tag{6.3}$$

$NO \cdot O_2$ is a weakly bound complex of NO and O_2, which is different from the nitrosyldioxyl radical (ONOO). NO_2 is a nitrosylation reagent of aromatic groups of amino acids and nucleic acids, and acidifies the aqueous solution, therefore exhibiting high toxicity. NO also reacts rapidly with O_2^-, yielding peroxynitrite:

$$NO + O_2^- \rightarrow OONO^- \tag{6.4}$$

which can damage biomolecules due to its high reactivity and is usually produced by neutrophils as an anti-bacterial reagents.

When NO binds to metal, several types of bonding are possible. The most common binding mode is the M–NO, "end-on" metal nitrosyl complex, which is seen in a number of NO-bound haem proteins such as haemoglobin–NO and myoglobin–NO. Under special conditions, other type of NO coordination to metal, an isonitrosyl, M–ON and "side-on" M(NO) species are observed.[9] The M–NO orbital interactions are similar to those found in M–CO; NO donates an electron to the metal through the σ orbital of the N atom, and the back donation from the metal dπ orbital to the π* orbital of NO is also involved.

In recognition of the covalent nature of the M–N–O interaction and the difficulty of assigning formal oxidation states to the metal and NO in M–N–O, Enemark and Feltham proposed a formalism that treated the M–N–O as a single entity.[10] In their formulism, which counts metal d electrons and the lone π electron from the NO ligand, "ferrous (Fe^{2+}) nitrosyl" and "ferric (Fe^{3+}) nitrosyl" porphyrin complexes are represented as $\{FeNO\}^7$ and $\{FeNO\}^6$, respectively.[10] The six-coordinated $\{FeNO\}^7$ species shows inverse correlation for the two vibration modes νN–O and νFe–NO. In case the back donation from the Fe to the π* orbital of NO increased, νN–O is lowered and higher νFe–NO is obtained, which is similar to Fe^{2+}–CO porphyrin complexes. For the six-coordinated $\{FeNO\}^7$ species, however, the sensitivities of the Fe–NO and N–O stretching frequencies to *cis* and *trans* effects are directly correlated.[11] Therefore, further contributions to the Fe–NO bond in addition to the π back donation must exist in these complexes.[11,12] Interestingly, the vibration frequencies of the Fe–NO unit of the $\{FeNO\}^6$ species do not seem to correlate with those of $\{FeNO\}^7$ and the carbonyl complex.[13–15]

As compared with the well-studied haem Fe–N–O system, the Fe–N–O species in non-haem Fe complexes has been less investigated. A systematic study by Lehnert and co-workers, however, demonstrated the nature of non-haem Fe^{2+} nitrosyl complexes.[16] The vibrational data in combination with DFT calculations showed that the electronic structure for non-haem ferrous nitrosyl complexes corresponds with Fe^{3+}–NO^- species.[16,17] In sharp contrast to the inverse correlation between νN–O and νFe–NO in the haem nitrosyl system, these vibrational frequencies in the non-haem iron system exhibited direct correlation (*i.e.*, lowered νN–O leads to lower νFe–NO). Since DFT analysis suggested that the NO^- ligand acts as a weak π-acceptor and strong π-donor, the π donation from the π* orbital of NO^- to the dπ orbital of non-haem iron is dominant, strengthening both νN–O and νFe–NO, which explains the observed direct correlation. It is also important to note that the magnitude of the π donation from the NO ligand is sensitive to the type of Fe ligand and its geometry.

6.3 Bacterial NOR

In 1971, the first evidence for membrane-bound NOR activity was obtained with *Alcaligenes faecalis* IAM 1015.[18] Since then, membrane-bound NOR activities were reported for *Pseudomonas stutzeri*, *Rhodobacter sphaeroides* IL106, *Halomonas halodenitrificans* and *Paracoccus denitrificans*.[2] NOR is expressed under anaerobic or O_2-limiting conditions.[19–21] Anaerobic expression of NOR is consistent with the presence of the recognition motif for the fumarate and nitrate reduction (FNR)-like regulator in the promoter region of the NOR gene, since FNR is a *trans*-acting protein of *Escherichia coli* that activates gene expression under anaerobic conditions.[22,23]

Three types of membrane-bound NORs are known.[3,4,24–26] The first isolated and most extensively studied enzyme is a cytochrome *c*-dependent NOR (cNOR) that is found only in denitrifying bacteria. cNOR consists of two subunits: NorB and NorC. The NorC subunit towards the periplasmic side has a haem *c* binding motif CXXCH, in addition to the spectroscopic characterization (*vide infra*), indicating that histidine- and methionine-coordinated haem *c*[27–32] functions as an electron acceptor from cytochrome *c* (cytochrome c_{551} or cytochrome c_{552}) or blue-copper azurin (azurin or pseudoazurin).[33–36] NorB is the larger subunit of cNOR and contains three metal centres (*i.e.*, two *b*-type haems and non-haem iron [Fe_B]). A different type of NOR, quinol-dependent NOR (qNOR), is a single-subunit enzyme and was found in *Ralstonia eutropha* H16 (Gram-negative soil bacterium),[37] *Pyrobaculum aerophilum* (archea),[38] *Synechocystis* sp. strain PCC6803 (cyanobacterium),[39] *Geobacillus stearothermophilus* (*G.s.*)[24] and some pathogens such as *Neisseria gonorrhoeae*[40] and *Neisseria meningitides*.[41] The first transmembrane (TM) α-helix and the adjacent hydrophilic region of qNOR have marked sequence similarity to NorC, and the main body of qNOR is homologous to NorB. One additional TM helix, compared with cNOR, connects NorC and NorB homologous regions in qNOR. qNOR lacks haem *c* in the hydrophilic domain and accepts electrons from quinol, rather than from a soluble protein such as cytochrome *c* or pseudoazurin. The active centre of qNOR consists of haem *b* and non-haem Fe_B, which are also intrinsically observed in cNOR. The third type of enzyme is Cu_ANOR, which has only been found in bacilli.[42–44] This type of NOR was originally designated as qCu_ANOR, since the enzyme receives electrons from both cytochrome *c* and quinol.[43,44] However, Al-Attar and de Vries recently found that cytochrome *c* but not quinol supported the NO reduction reaction, and they proposed changing the name from qCu_A-NOR to Cu_ANOR.[45] The unique feature of Cu_ANOR is that the smaller subunit of the enzyme is suggested to contain two copper atoms (Cu_A) like cytochrome oxidase, and the amino acid sequence of the smaller subunit is 65% identical to the N-terminal of the SoxB-type cytochrome oxidase (cytochrome bo_3) subunit II.[44] The larger subunit of Cu_ANOR harbours one non-haem iron and two *b*-type haems, as found in qNOR and NorB of cNOR. The subunit topologies of the three types of NOR are shown in Figure 6.1.

Discovery of the FixN-type cytochrome oxidase (FixN complex, cytochrome cbb_3)[46] and amino acid sequence analysis of the NorB subunit of cNOR[47] revealed intriguing structural similarities between these respiratory enzymes. Both NorB and the FixN-type cytochrome oxidase are predicted to have 12 TM helices like other cytochrome oxidases (SoxB- and SoxM-type enzymes) whose crystal structures are known, and six invariant His residues are assumed to bind to two haems and non-haem mononuclear metal (Cu_B in the FixN-type oxidase or Fe_B in NOR), suggesting that the molecular topologies and the positions of the metal centres in the main subunit are very similar to each other.[48–51] FixO, the haem *c* binding subunit of the FixN-type cytochrome oxidase, has similar features to the NorC subunit of cNOR, further suggesting the evolutionary link between these enzymes.[48] Therefore, the overall architecture of NOR was expected to be somewhat homologous to that of cytochrome oxidase, and NOR is classified as a member of the haem–copper oxidase superfamily. Based on these common structural features and phylogenetic analysis, Saraste and Castresana proposed that the oxygen respiratory chain developed from the anaerobic, denitrifying respiratory system.[48,49] The close relationship between these enzymes was also reported by van der Oost and others.[50]

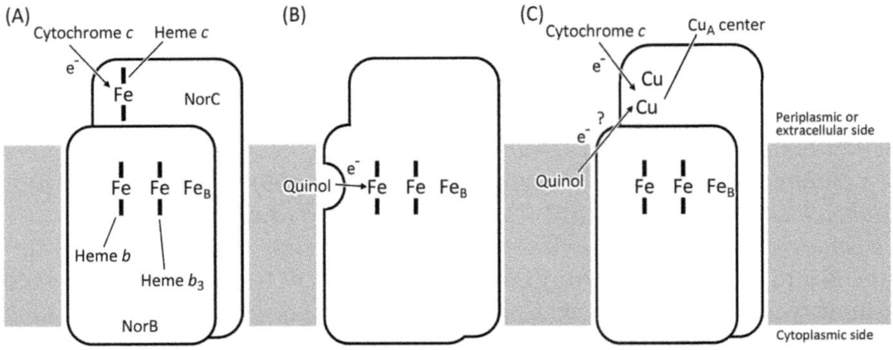

Figure 6.1 Schematic representation of three nitric oxide reductases (NORs). (a) Cytochrome *c*-dependent NOR (cNOR) consists of two subunits, NorB and NorC. The core subunit, NorB, contains haem *b* and a binuclear active centre (haem b_3 and Fe_B). The hydrophilic region of the smaller subunit, NorC, contains haem *c* and receives electrons from protein electron donors such as cytochrome *c* for the catalytic NO reduction. (b) Quinol-dependent NOR (qNOR) is a single-subunit enzyme and utilises electrons form quinol for the catalytic reaction. There is no haem *c* in the hydrophilic region of qNOR. (c) Cu_ANOR contains a di-copper centre (Cu_A centre) in the hydrophilic region. Cu_ANOR was originally suggested to receive electrons both from cytochrome *c* and quinol. A recent study, however, showed that Cu_ANOR could receive electrons only from cytochrome *c*.

6.4 Structural and Functional Knowledge Before Crystal Structures were Available

6.4.1 Metal Centres

The first isolation and purification of membrane-bound NOR was achieved for cNOR from *Ps. stutzeri.*[52] After the success of this purification, a number of purifications and characterisations of NORs from the other microorganisms (*Pa. denitrificans*, *H. halodenitrificans*, *Ps. Aeruginosa* [*Ps.a*], *Ps. nautica* and *Thermus thermophilus*) have been reported.[27,31,36,53–56] The oxidised form of cNOR from *Ps. stutzeri* has a Soret band at 412 nm (ε = 312 mM^{-1} cm^{-1}) and displays a broad absorption peak at 537 nm with a shoulder at around 558 nm.[57,58] Its reduced form has an intense Soret band at 421 nm and its α- and β-bands are found at 553 and 523 nm, respectively. The α-band at 553 nm in the reduced form of cNOR has a shoulder at around 560 nm. These spectral patterns around the α- and β-regions are consistent with the prediction that cNOR contains *b*- and *c*-type haems.[57,58]

Room temperature magnetic circular dichroism (MCD) spectra of oxidised (ferric) cNOR from *Pa. denitrificans*,[57] *Pa. halodenitrificans*[31] and *Ps. stutzeri*[30] have indicated that the enzyme has two low-spin haems and one high-spin haem. One low-spin haem is coordinated by histidine and methionine, and the other low-spin haem has a bis-histidine coordination structure. A sharp negative feature at 611 nm is assigned to a high-spin ferric haem. Low-temperature, near-infrared (IR) and visible MCD spectra show the presence of three rather than two low-spin haems, which is indicative of a temperature-dependent spin-state change.[30]

The electron paramagnetic resonance (EPR) spectra also show the presence of two low-spin haems (g_x = ~1.4, g_y = 2.25, g_z = 2.97 and g_z = 3.54 for cNOR from *Ps. stutzeri* and several other species).[31,54,57,59,60] The high-spin haem signal at g = 6 and non-haem iron signal at g = 4.3 are also observed for the enzymes, except for the preparation by Hendriks *et al.*, and these signals may represent only a few percent in terms of enzyme concentration. Most of the high-spin haem and non-haem iron is EPR silent since, as indicated by resonance Raman investigation of cNOR from *Pa. denitrificans* using ^{18}O- and ^{2}H-labeled water, these metals are antiferromagnetically coupled through a bridging ligand.[61,62] On the basis of these spectroscopic results, cNOR was expected to have haem *c* in the low-spin state with His and Met ligands, haem *b* in a low-spin state with two His ligands and a binuclear centre consisting of haem *b* (haem b_3) and non-haem iron (Fe_B). Since NOR has six conserved His residues, it was considered that haem b_3 has one His and Fe_B has three His residues as their ligands. In addition to three His residues, one of five conserved Glu residues was suggested to be a potential ligand for Fe_B on the basis of the sequence comparison between NORs and cytochrome oxidases. The haem b_3 in the five-coordinated high-spin state has a μ-oxo bridging ligand with Fe_B. On the other hand, a Mössbauer spectroscopic study on cNOR from *Ps. nautica* showed that haem b_3 iron adopts a six-coordinated low-spin state, suggesting that the g = 6 signal arises from a spin–spin coupling between low-spin haem b_3 iron and high-spin Fe_B.[36] Since the g = 6 signal has been detected in the EPR spectra for other as-purified (oxidised) cNORs, haem b_3 iron could be in a low-spin state for other cNORs. For complete understanding of the spin state of haem b_3, further spectroscopic studies are required. qNOR shows essentially the same EPR spectrum as that of cNOR, except for the lack of g_z = ~3.5,[37,38] since qNOR does not have haem *c*. Cu_ANOR also has similar EPR signals and shows additional EPR signals (g_z = 2.172, $g_{x,y}$ = 2.02, A_z = 3.54 mT), which are characteristic of the binuclear mixed-valence Cu_A centres.[44]

The reduced form of cNOR from *Pa. denitrificans* solubilised by dodecyl β-D-maltoside (DDM) has the resonance Raman haem Fe-His stretching mode at 218 cm^{-1}, which is very similar to that of cytochrome oxidase (215 cm^{-1}).[61,63,64] More recent resonance Raman study on *Pa. denitrificans* cNOR shows that the DDM-solubilised enzyme has the νFe-His mode at 207 cm^{-1}.[65] In the reduced state of qNOR from Archaeon *Pyrobaculum aerophilum*, the Fe-His stretching mode is detected at 213 cm^{-1}.[38] Since the haem Fe-His vibrational mode is only resonance enhanced in the five-coordinated high-spin ferrous haem, haem b_3 iron likely adopts the five-coordinated high-spin state in the reduced form. The five-coordinate, high-spin, histidine-ligated haem *b* exhibits the νFe-His frequency in the 200–260 cm^{-1} region, and this relatively low frequency indicates the weaker Fe-His coordination when compared with horseradish peroxidase and cytochrome *c* peroxidase, which have the ~245 cm^{-1} Fe-His stretching mode,[66] suggesting that electron donation from the proximal ligand of haem b_3 is relatively weak. The resonance Raman data for oxidised and reduced forms indicates that, upon reduction, the bridging ligand is lost and the histidine is ligated to the haem b_3.[61,62]

Contrary to the resonance Raman data on the reduced forms of *Pa. denitrificans* cNOR and *Py. aerophilum* qNOR, Moura and co-workers reported that haem b_3 iron of *Ps. nautica* cNOR is unambiguously six-coordinated low spin in the reduced sate on the basis of a Mössbauer spectroscopic analysis.[36] Currently, the Raman data on cNOR from *Ps. nautica* have not been reported. Detailed spectroscopic analysis on NORs will provide firm conclusions as to the coordination structure and spin state of haem b_3 iron.

One notable difference in the reaction catalysed by cytochrome oxidase and NOR is that whilst cytochrome oxidase reduces one molecule of O_2:

$$O_2 + 4H^+ + 4e^- \rightarrow 2H_2O \tag{6.5}$$

NOR utilises two molecules of NO to produce N_2O and H_2O:

$$2NO + 2H^+ + 2e^- \rightarrow N_2O + H_2O \tag{6.6}$$

It is therefore likely that the active site cavity volume of NOR is larger than that of cytochrome oxidase. Exposure of dithionite-reduced Cu_ANOR from *Bacillus azotoformans* to CO makes two IR bands at 1972 and 2068 cm^{-1} that are assigned to haem b_3- and Fe_B-bound CO stretching modes, respectively.[67] Photodissociation of the haem b_3–CO complex results in a perturbation of the CO band associated with Fe_B, indicating that two CO molecules can bind concomitantly at the active site of this enzyme. The CO stretching mode assignable to Fe_B–CO species has not been reported for cNOR and qNOR; however, resonance Raman and IR spectroscopic studies show that cytochrome oxidases from *T. thermophilus* and *E. coli* (cytochromes ba_3 and bo_3, respectively), both of which are capable of catalysing the reduction of NO, can accommodate two diatomic molecules in their active sites.[68]

6.4.2 Electron Transfer

In the reaction catalysed by NOR, two equivalent protons and electrons are consumed:

$$2NO + 2H^+ + 2e^- \rightarrow N_2O + H_2O\ (E_0'\,[\text{pH } 7.0] = +1.18\ \text{V};\ \Delta G_0' = -73.2\ \text{kcal mol}^{-1}) \tag{6.7}$$

The cytochrome bc_1 complex and cytochrome *c* can supply electrons to the active site in cNOR.[33,34] However, myxothiazol, a cytochrome bc_1 complex inhibitor, cannot completely block NO reduction in *Rhodobacter capsulatus*.[34] In addition, the significant rate of NO reduction by mutant strains lacking the cytochrome bc_1 complex or cytochrome c_2 implies that there is an electron transport system that bypasses the cytochrome bc_1 complex and cytochrome c_2. Indeed, experiments using mutant strains of *Pa. denitrificans* lacking cytochrome *c*, pseudoazurin or both proteins indicate that pseudoazurin can substitute for cytochrome *c* in electron transfer to cNOR.[35,69] Electrostatic interactions are important between cytochrome *c* and cNOR since the activity decreases rapidly with increasing ionic strength.[70] Cytochrome *c* or other electron donors supply electrons to the cNOR active site through haem *c* in the NorC subunit and haem *b* in the NorB subunit, which is analogous to the electron transfer pathway in cytochrome oxidases.[71,72] This electron transfer pathway is supported by the electron redistribution analysis by flow-flash experiments using partially reduced CO-bound cNOR.[73]

The proposed electron flow is also consistent with the fact that the reduction potentials for haems *c* and *b* in cNOR from *Ps. stutzeri* are 280 and 322 mV, respectively. Detailed electrochemical measurements done for cNOR from *Pa. denitrificans* with potentiometry coupled to UV/visible absorption spectroscopy revealed the haem *c*, haem *b* and Fe_B have reasonable redox potentials of 310, 345 and 320 mV, respectively. However, the redox potential for haem b_3 is unexpectedly lower than the other three metal centres (60 mV).[57] The authors of this study suggested that the unusually low redox potential of haem b_3 can cause the avoidance of the formation of a stable haem $b_3(Fe^{2+})$–NO complex during turnover, which may be an inhibited state of the enzyme. The isolated soluble domain of the NorC subunit has a reduction potential of 183 mV, suggesting that the complexation with NorB lowers the reduction potential by 180 mV.[74] The redox potential measurements in the presence of substrate analogue CO showed that the substrate binding could raise the redox potential for haem b_3 to equivalent potentials for the other redox centres and regulate the enzymatic reaction.[75] The analysis of the redox potential for cNOR from *Ps. nautica* by using cyclic voltammetry indicated that haem b_3 iron and Fe_B have substantially lower redox potentials than those of haem *c* and *b*; the redox potentials are −369, −162, 43 and 208 mV for Fe_B, haem b_3, *b* and *c*, respectively.[76] The presence of substrate NO induces no shift of the redox potential of haem b_3 in cNOR from *Ps. nautica*, while oxygen molecules, which can be catalytically reduced by cNOR, change the redox potential.[76–78]

While cytochrome *c* cannot support the NO reduction reaction of qNOR, NO is reduced in the presence of quinol (2-methyl-1,4-naphtoquinol).[37] Since no haem *c* is contained in qNOR, the electrons from quinol are first supplied to haem *b*, and then the binuclear centre accepts the electrons. The quinol binding site of qNOR is not defined, but a crystal structure and simulation study on cytochrome bo_3, a member of the cytochrome oxidase family that accepts electrons from quinol, suggested that the enzyme binds quinol around the haem *b*. Thus, the quinol binding site of qNOR might also be close to haem *b*, and electrons are transferred from haem *b* to the active site.

The NOR from *B. azotoformans*, Cu_ANOR, receives electrons from protein electron donors for the catalytic NO reduction. Similarly to cytochrome oxidase, the Cu_A centre receives electrons at first, and then the binuclear centre receives electrons through haem *b*. Suharti *et al.* have reported that horse heart cytochrome *c* cannot support the Cu_ANOR reaction, but they found that only cytochrome c_{551} out of three *c*-type cytochromes from *B. azotoformans* could stimulate the NO reduction reaction.[44]

6.4.3 Proton Transfer

In 1985, the first data showing that the protons required for NOR-dependent NO reduction are consumed on the periplasmic side were obtained from whole-cell measurements. Membrane potential measurement of *R. capsulatus* using endogenous carotenoid pigments shows that the isoascorbate-DAD (2,3,5,6-tetramethyl-*p*-phenylenediamine)-driven NOR reaction is not electrogenic, supporting the idea that electrons and protons are supplied from the same side of the membrane.[79] Other data showing that cNOR is not electrogenic come from experiments with liposome-reconstituted cNOR.[73,80] The absence of respiratory control for steady-state NO and O_2 reduction by proteoliposome-containing cNOR suggests that both NO and O_2 reduction by cNOR are electroneutral. Drachev *et al.* have developed apparatus that can directly measure the membrane potential produced by membrane proteins.[81] Electrometric measurements based on a similar system are used for NO and O_2 reduction by cNOR-reconstituted liposomes, and the results indicate that no electrochemical gradient builds up during the catalytic reaction.[73,80] The concept regarding proton uptake from the periplasmic side is thus firmly established by these electrometric and flow-flash experiments for cNOR; however, there is little experimental information on the direction of the proton uptake for qNOR and Cu_ANOR. Very recent work by Al-Attar and de Vries suggested that Cu_ANOR from *B. azotoformans* utilises cytoplasmic protons for catalytic NO reduction and creates an electrochemical gradient across the membrane.[45] Adelroth and co-workers tried to further understand the proton transfer mechanism by using flow-flash techniques,[80,82–86] and they found that proton uptake is coupled with the oxidation of haems *b* and *c* during O_2 reduction by cNOR.[86] They also indicated that the proton-coupled haem oxidation process has pH dependency with $pK_a = 6.6$. Consequently, a residue with $pK_a = 6.6$ might be involved in proton transfer to the active site.

A potential proton entry site and shuttling pathway are proposed from mutagenesis analyses and the homology model of the NorB subunit based on *E. coli* cytochrome bo_3 oxidase.[84] Highly conserved Glu_{122} and Glu_{125} (*Pa. denitrificans* cNOR numbering) are presumed to be proton entry sites, and Glu_{198} and Glu_{267}, which are commonly found in NOR but not in cytochrome oxidases, provide protons to the active site. One should note, however, that the homology model is not complete since the NorC subunit is not included, and Glu_{122} and Glu_{125} belong to a loop region that is difficult to model due to its flexibility.

6.5 Functional Characterisation Based on the Crystal Structures of NORs

Shiro and co-workers successfully determined the crystal structures of two distinct types of bacterial NORs: cNOR from the Gram-negative *Ps.a.* and qNOR from the Gram-positive *G.s.* in 2010 and 2012, respectively.[87,88] The structural data for these two types of NOR enzymes have provided us with structural bases for the catalytic generation of N_2O and shed light on the molecular evolution of this respiratory enzyme.[89–95] In this section, we describe the structural properties of cNOR and qNOR and discuss the possible mechanism for the NO reduction reaction and for the functional conversion during molecular evolution.

6.5.1 Overall Structures

In cNOR, the NorB subunit contains 12 TM helices and the NorC subunit has one TM helix and a subsequent α-helical hydrophilic domain, typical of a cytochrome *c* fold, in the periplasmic side (Figure 6.2). qNOR contains 14 TM helices in the membrane-spanning region and an α-helical hydrophilic domain located on the extracellular side. The topologies of the helices of the TM region in the two NORs are basically similar. One extra TM helix in qNOR (TM helix II) connects the C-terminal end of the hydrophilic domain, which is equivalent to the NorC subunit of cNOR, with the N-terminal of TM helix III, which corresponds to the N-terminal of the NorB subunit of cNOR.

The arrangement of TM helices shows strong similarity between NORs and cytochrome oxidases. For example, the interaction of TM helices between NorB and NorC in cNOR is very similar to the interaction between the N and O subunits observed in the cbb_3 type of cytochrome oxidase from *Ps. stutzeri.*[96] The cytochrome cbb_3 oxidase also contains the hydrophilic region with a haem *c* in the O subunit, although another large hydrophilic subunit (P subunit) containing multiple haem *c* molecules is associated with the other subunits. Since cytochrome cbb_3 oxidase has a low activity for the reduction of NO to N_2O, it is thought to be evolutionarily the closest to NOR. The cytochrome cbb_3 oxidase is found in many pathogenic proteobacteria that colonise microaerobic host tissues and diazotrophs that can simultaneously perform aerobic respiration and nitrogen fixation.

NOR has five conserved Glu residues,[3,48,50,69,97,98] all of which are known to be functionally important in the catalytic NO reduction reaction: Glu_{135}, Glu_{138}, Glu_{211}, Glu_{215} and Glu_{280} in *Ps.a.* cNOR and Glu_{429}, Glu_{432}, Glu_{512}, Glu_{516} and Glu_{581} in *G.s.* qNOR (Figure 6.2). The crystal structures unambiguously showed that the spatial positions of these Glu residues are the same in both cNOR and qNOR. The structures showed that three of the five conserved Glu residues are located in close proximity to the non-haem metal of the binuclear centre, while the remaining two are positioned near the water cluster, which is located around the Ca^{2+} binding site (*vide infra*).

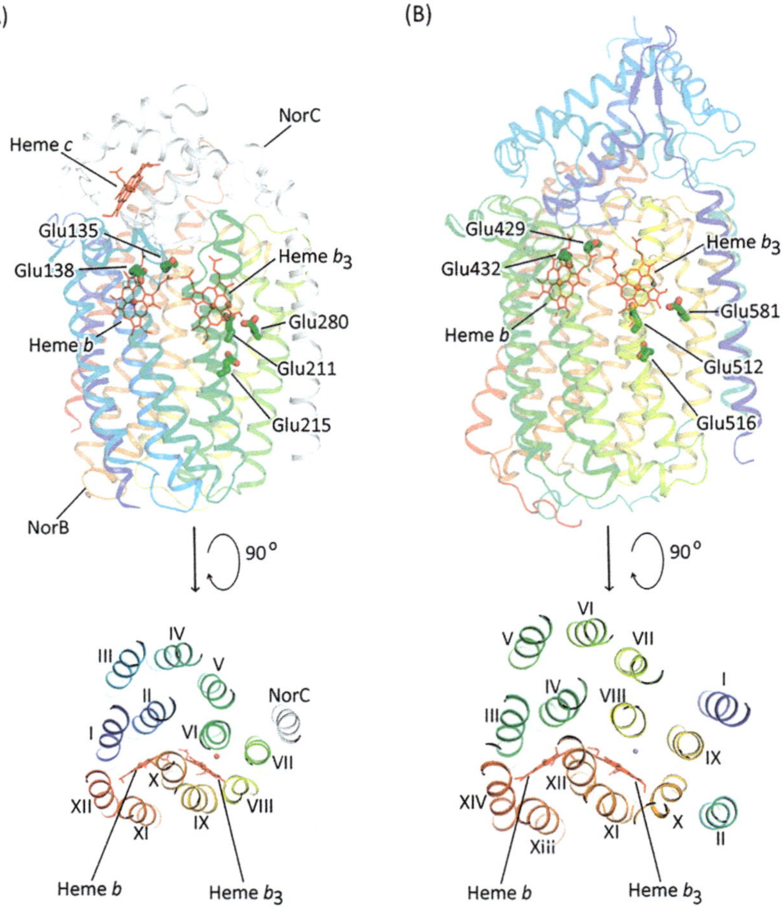

Figure 6.2 Overall structures of nitric oxide reductases (NORs). (a) The crystal structure of *Pseudomonas aeruginosa*. Cytochrome *c*-dependent NOR (cNOR) is shown as a ribbon in various colours (NorB) and white (NorC). Haem molecules are shown as red sticks, and five conserved Glu residues are shown as bold sticks. The upper structure is viewed parallel to the membrane and the lower panel is from the cytoplasmic side. (b) The crystal structure of *Geobacillus stearothermophilus*. Quinol-dependent NOR (qNOR) is shown as a ribbon. The qNOR structure shown is in the same orientation as in the *Ps. aeruginosa* cNOR.

6.5.2 Electron Transfer Pathway

The protein folds and metal species of the hydrophilic region in the periplasmic or extracellular side differ significantly within the haem copper oxidase superfamily. In contrast to the cytochrome *c* fold in cNOR and cytochrome *cbb*$_3$ oxidase, the cupredoxin fold is found in the hydrophilic region in cytochrome *ba*$_3$ and *aa*$_3$ oxidases, and two copper ions (Cu_A site) in this region play an important role in the electron transfer.[99–103] Cytochrome *bo*$_3$ oxidase also has a cupredoxin fold, but lacks a Cu_A metal centre in the hydrophilic region, because the physiological electron donor is not a protein electron donor but a quinol compound (ubiquinol) located inside the membrane.[104]

The electrons utilised in NO reduction by cNOR are donated from electron donors (cytochrome c_{551}, cytochrome c_{552}, azurin or pseudoazurin) to haem *c* in the hydrophilic (globular) domain of the NorC subunit. The haem *c* has His_{65} and Met_{112} as axial iron ligands. The electrons are then transferred to the catalytic binuclear centre (haem b_3 and Fe_B) *via* haem *b*. The iron-to-iron distance between haems *c* and *b* in cNOR is 20.3 Å and the distance between haems *b* and b_3 in cNOR is 14.1 Å. These distances are comparable to those between the Cu_A site and haem *a* (~20 Å) and between haems *a* and a_3 (~14 Å) in the aa_3 cytochrome oxidase of *Pa. denitrificans.*[99,100]

The qNOR hydrophilic domain exhibits substantial structural homology with the cytochrome *c* domain of the NorC subunit in cNOR, while there is no marked sequence similarity between them. Several bulky hydrophobic residues (Tyr_{84}, Tyr_{98}, Tyr_{170}, Tyr_{171}, Phe_{175}, Leu_{183} and Phe_{214}) in the *G.s.* qNOR compensate for the loss of the hydrophobic haem macrocycle to retain the cytochrome *c* fold (Figure 6.3). Menaquinol is a physiological electron donor to the binuclear centre through haem *b* for NO reduction by qNOR. In the co-crystal structure of *G.s.* qNOR with a quinol analogue, 2-heptyl hydroxyquinoline *N*-oxide (HQNO), HQNO was identified at the enzyme surface 10 Å away from the haem *b* iron and close to the head group region of the extracellular side of the membrane. The structure suggests that Asp_{746}, His_{328} and Glu_{332} are responsible for the binding of the quinol analogue to *G.s.* qNOR.

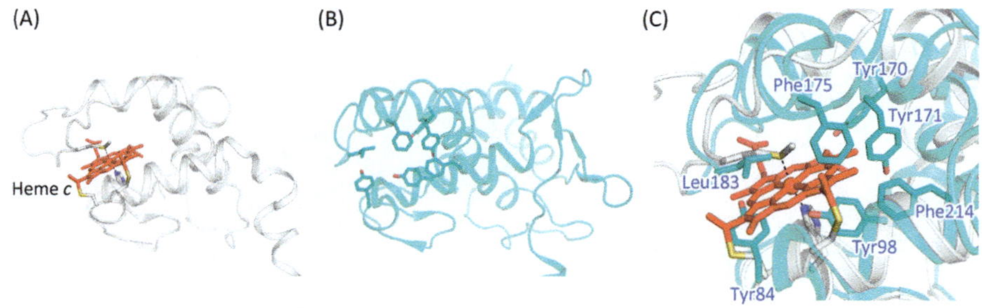

Figure 6.3 Structures of the cytochrome *c* domain of NorC in cytochrome *c*-dependent nitric oxide reductase (cNOR) and the corresponding domain of quinol-dependent nitric oxide reductase (qNOR). (a) Cytochrome *c* domain of NorC subunit (*Pseudomonas aeruginosa* cNOR) is shown as a white ribbon. Haem *c* is shown as a red stick. (b) The corresponding hydrophilic domain in *Geobacillus stearothermophilus*. qNOR is shown as a cyan ribbon with the same orientation as in NorC. (c) Haem *c* binding site of NorC (white ribbon) and the corresponding region of qNOR (cyan ribbon) are shown in the superposed structures. Haem *c* in NorC is shown as a red stick. The bulky hydrophobic residues in the hydrophilic domain of qNOR, which are located at the position that is equivalent to the haem *c* binding site in cNOR, are shown as cyan sticks.

Since these three residues are highly conserved in qNORs, but not in cNORs, these residues generally recognise quinol in qNOR as an electron donor.

Concerning the electron transfer pathway from haem *b* to haem b_3, the edge-to-edge distance in cNOR (4.3 Å) does not fall in the range of van der Waals contacts, but is shorter than those of cytochrome oxidases (4.7–5.7 Å) (Figure 6.4). Only cytochrome cbb_3 oxidase shows van der Waals contacts between the haem *b* and b_3 methyl groups (3.6 Å).[96] The tertiary structures of cytochrome oxidases show that a well-conserved Phe is present between the haem edges. However, this well-conserved Phe in cytochrome oxidases is replaced by Gly in cNOR. Instead, a Phe_{352} residue from another amino acid sequence position of cNOR seems to complement the absence of well-conserved Phe. The haem *b*, which adopts a six-coordinated low-spin iron by the coordination of two His residues (His_{60} and His_{349} in *Ps.a.* cNOR; His_{355} and His_{653} in *G.s.* qNOR), mediates electron transfer from the electron donor (haem *c* in cNOR; menaquinol in qNOR) to the non-haem metal and haem b_3 binuclear centre.

A Ca^{2+} is located between the propionates of the two haems *b* and b_3 in both NORs (Figure 6.4). The Ca^{2+} ligands are the 7-propionate of haem b_3, the

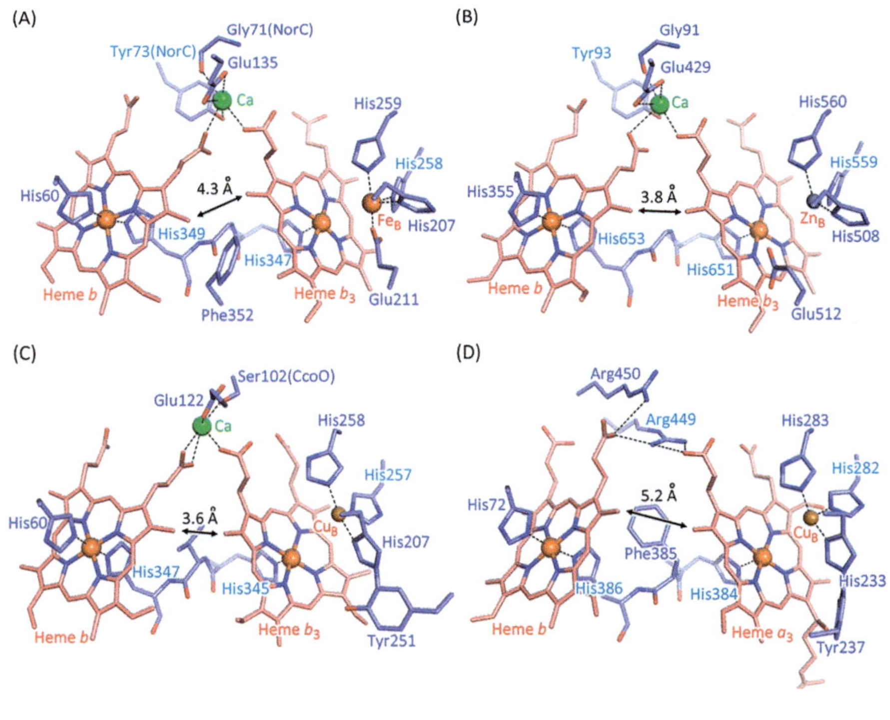

Figure 6.4 Arrangement of metal centres of nitric oxide reductases (NORs) and cytochrome oxidases. (a) *Pseudomonas aeruginosa* cytochrome *c*-dependent NOR (Protein Data Bank identifier: 3O0R). (b) *Geobacillus stearothermophilus* quinol-dependent NOR (Protein Data Bank identifier: 3AYF). (c) Cytochrome cbb_3 oxidase (Protein Data Bank identifier: 3MK7). (d) Cytochrome ba_3 oxidase (Protein Data Bank identifier: 1XME).

6-propionate of haem *b*, Glu carboxylate (Glu_{135} in NorB of *Ps.a.* cNOR; Glu_{429} in *G.s.* qNOR), Gly carbonyl oxygen (Gly_{71} in NorC of *Ps.a.* cNOR; Gly_{91} in *G.s.* qNOR), Tyr phenolate (Tyr_{73} in NorC of *Ps.a.* cNOR; Tyr_{93} in *G.s.* qNOR) and one water molecule. The presence of Ca^{2+} bridging two haem propionates was also observed in cytochrome cbb_3 oxidase, a microaerobic respiratory enzyme, with a similar coordination structure, but the equivalent position in the other cytochrome oxidases involved in the aerobic respiratory chain is occupied by positively charged side chains of two conserved Arg residues (*e.g.*, Arg_{449} and Arg_{450} in cytochrome ba_3 oxidase).[99–104] Although the specific role of Ca^{2+} in the catalytic reaction remains unclear, Ca^{2+} is essential for the NOR-catalysed reaction, which is evident from the loss of Ca^{2+} and NO reduction activity as a result of mutations of the Ca^{2+} ligands.[55,85,88] One of the possible roles of Ca^{2+} is the maintenance of the electron transfer pathway from haem *b* to haem b_3 and/or the modulation of the redox potentials of the two haems for achieving effective electron transfer.

6.5.3 Active Site Structures

The structures of the binuclear centres are quite similar in cNOR and qNOR (Figure 6.5); one His imidazole is the axial ligand for the haem b_3 (His_{347} in *Ps.a.* cNOR; His_{651} in *G.s.* qNOR), three His residues (His_{207}, His_{258} and His_{259} in *Ps.a.* cNOR; His_{508}, His_{559} and His_{560} in *G.s.* qNOR) are ligands for the non-haem metal and the haem b_3 iron and non-haem metal are bridged by one oxygen atom (μ-oxo-bridge), as was predicted by vibrational spectroscopic studies.

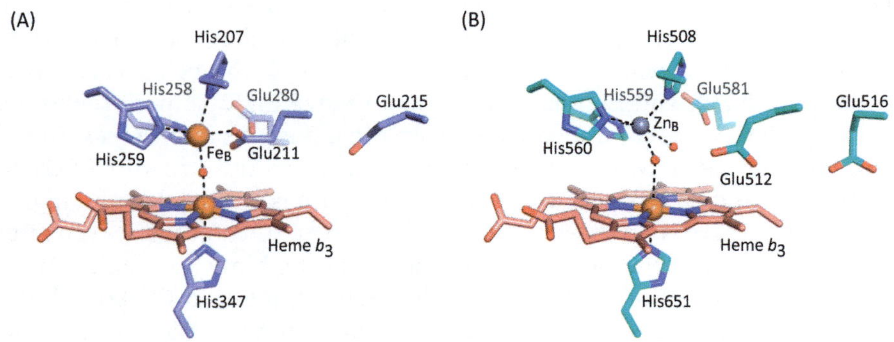

Figure 6.5 Binuclear centres of nitric oxide reductases (NORs). (a) *Pseudomonas aeruginosa* cytochrome *c*-dependent NOR (cNOR) in the oxidised state. Three of five conserved Glu residues (Glu_{211}, Glu_{215} and Glu_{280}) are located in close proximity to the binuclear centre. His_{347} coordinates to haem b_3 iron. Three His residues (His_{207}, His_{258} and His_{259}) and one Glu residue (Glu_{211}) serve as ligands for Fe_B. An oxo ligand (red ball) bridges haem b_3 iron and Fe_B. (b) *Geobacillus stearothermophilus* quinol-dependent NOR in the oxidised state. Contrary to cNOR, Fe_B is replaced with a Zn iron (Zn_B), and a Glu residue (Glu_{512}) is not a ligand for Zn_B in the crystal structure. Instead of Glu coordination, one extra water molecule coordinates to Zn_B (red ball).

However, some structural differences are observed in the coordination structure of the non-haem metal, presumably due to differences in the non-haem metal between *Ps.a.* cNOR and *G.s.* qNOR. In *Ps.a.* cNOR, the non-haem metal is an iron (Fe_B), to which the carboxylate of Glu_{211} coordinates, as was expected. On the other hand, the metal in this site is a zinc (Zn_B) in *G.s.* qNOR, in which the equivalent Glu residue in *G.s.* qNOR (Glu_{512}) is no longer a ligand for Zn_B, and the resulting vacant position of the Zn_B coordination sphere becomes occupied by one extra water molecule. Since Zn is an inactive metal for the redox chemistry, the isolated Zn_B-substituted *G.s.* qNOR was inactive for the NO reduction. Therefore, it was considered that Fe_B might have been accidentally substituted by Zn_B during the expression and purification process of *G.s.* qNOR.

Although in the crystal structure the carboxylate group of Glu_{512} is located a considerable distance from Zn_B (5 Å), a molecular dynamics (MD) simulation suggests that this side chain can move toward the Zn_B coordination sphere in *G.s.* qNOR. Such structural flexibility around the Glu residue at the TM helix is possibly due to the presence of the conserved Gly residue adjacent to a Glu residue (Gly_{212} in *Ps.a.* cNOR; Gly_{513} in *G.s.* qNOR). This conserved Glu in NOR (Glu_{211} in *Ps.a.* cNOR; Glu_{512} in *G.s.* qNOR) has been considered to play an important role in facilitating N–O bond cleavage during the production of N_2O and H_2O, possibly as the terminal proton donor, together with two other conserved Glu residues near the binuclear centre (Glu_{215} and Glu_{280} in *Ps.a.* cNOR; Glu_{516} and Glu_{581} in *G.s.* qNOR).

6.5.4 Molecular Mechanism of NO Reduction by NOR

For the reduction of NO by bacterial NOR, two NO molecules must be accommodated to the haem b_3 and Fe_B binuclear centre, in sharp contrast to cytochrome oxidase, in which one O_2 molecule binds to the binuclear centre that consists of haem and non-haem Cu (Cu_B). Interestingly, some cytochrome oxidases that can accommodate two diatomic molecules (CO/NO) exhibited substantial NOR activity.[68,105] Inspection of the active site structure of cNOR (Figure 6.5) indicates that the coordination sphere is so crowded and highly packed that there is no space to accommodate two NO molecules, even after the dissociation of the bridging μ-oxo ligand at the binuclear centre. However, the structures of cNOR in the fully reduced, reduced CO-bound and acetaldoxime-bound forms[93] show that two of diatomic molecules (or four atoms) could be accommodated to the binuclear centre of NOR (Figure 6.6).

In terms of the coordination structure of NO in the active site of the short-lived intermediate, three different mechanisms were proposed (Figure 6.7).[13,106] The first one is a *trans* mechanism, where Fe of haem b_3 and Fe_B bind one NO molecule, respectively, to form an iron–nitrosyl dimer intermediate. The close proximity of the two nitrogen atoms promotes N–N bond formation *via* an electrophilic attack or *via* a radical coupling process. Rapid-freezing EPR spectroscopic studies on *Ps.a.* cNOR detected the haem b_3–NO and Fe_B–NO species, which supports the *trans* mechanism.[54] The vibrational spectroscopic characterisation of CO-bound Cu_ANOR is also consistent with

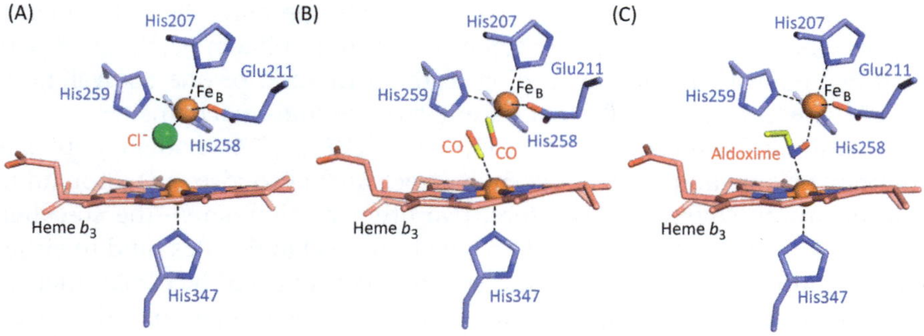

Figure 6.6 Structures of the binuclear centre of cytochrome *c*-dependent nitric oxide reductase in the reduced and ligand-bound states. (a) The reduced state (Protein Data Bank identifier: 3WFB). A bridging oxo ligand observed in the oxidised state (see Figure 6.5a) is dissociated from the binuclear centre. (b) CO-bound state (Protein Data Bank identifier: 3WFC). Two CO molecules (yellow sticks) coordinate to haem b_3 iron and Fe_B. The binding of CO shows no remarkable structural change in the binuclear centre. (c) Acetoaldoxime-bound state (Protein Data Bank identifier: 3WFD). Acetoaldoxime bridges haem b_3 iron and Fe_B. Acetaldoxime binding does not induce structural change as observed in the CO-bound state.

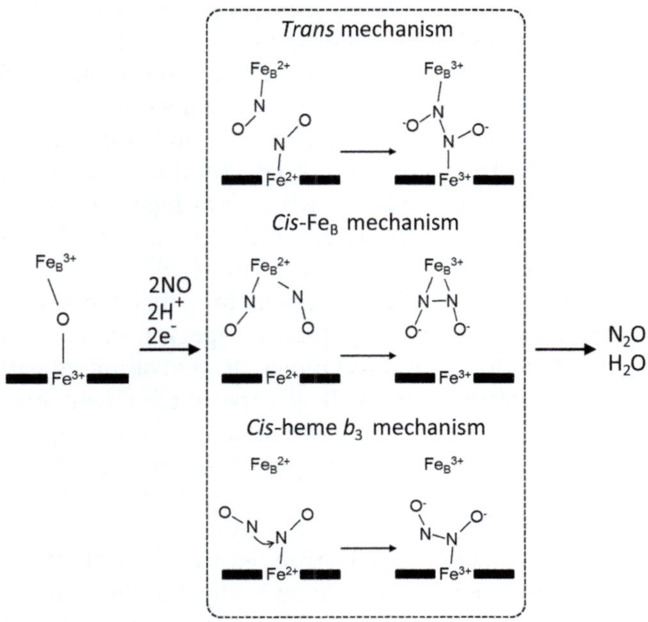

Figure 6.7 Three possible mechanisms in terms of coordination structures of the binuclear centre with nitric oxide (NO) in NO reductase.

this *trans* mechanism.[13,67] A NOR model complex prepared by Collman and co-workers can bind NO to both haem Fe^{2+} and non-haem $Fe_B{}^{2+}$, and reacts with two equivalents of NO, leading to the formation of one equivalent of N_2O.[107] These results are also consistent with the *trans* mechanism.

In contrast, however, another haem Fe^{2+}–NO/$Fe_B{}^{2+}$–NO model complex is thermally stable and does not seem to react further to give N_2O upon addition of proton source.[108] Therefore, two other mechanisms—the so-called *cis* mechanisms—were proposed, in which two NO molecules bind to either haem b_3 iron or Fe_B. Thomson and co-workers proposed the *cis*-Fe_B mechanism because a rather stable $\{FeNO\}^7$ species at haem b_3, a potential "dead end" product, should not be formed during the turnover in this model.[109–111] The considerably low redox potential of haem b_3 relative to the other redox sites haem *c*, haem *b* and Fe_B could suppress the formation of stable ferrous haem b_3–NO species, which supports the *cis*-Fe_B mechanism.[57] However, spectroscopic study on *Ps. nautica* cNOR showed that, despite the low redox potential of haem b_3, reduction of all four redox centres, including haem b_3, was required for catalytic NO reduction.[36] Spectroscopic analysis also demonstrated that ferrous haem b_3 adopted a six-coordinate structure, suggesting that an unknown sixth ligand for haem b_3 iron would result in the low affinity to NO, thereby facilitating the formation of a ferrous–dinitrosyl complex at Fe_B.[36] Since a subsequent electrochemical study indicated that the presence of NO induced no shift of the redox potential of haem b_3, it is highly plausible that the binding of NO to haem b_3 iron is not involved in the catalytic reaction.[76] Taken together with a steady-state kinetic study in which the consecutive binding of two NO substrates in the catalytic NO reduction was observed, Moura and co-workers revised the *cis*-Fe_B mechanism.[78] In the revised *cis*-Fe_B mechanism, first NO molecule coordinates to ferrous Fe_B, because an unknown ligand (possibly hydroxide ligand) occupies the haem b_3 coordination site. Then, a second NO molecule coordinates to ferrous Fe_B or directly attacks the NO ligand of Fe_B, producing an Fe_B–hyponitrite intermediate. In this case, haem b_3 with a sixth ligand could stabilise the Fe_B–hyponitrite intermediate and/or assist the cleavage of the N–O bond of hyponitrite species to form N_2O and a water molecule.

In the alternative *cis* mechanism—a *cis*-haem b_3 mechanism—the first NO molecule binds to haem b_3 to form a $\{FeNO\}^7$ species, which is then electrophilically attacked by the second NO molecule. As compared with the other two proposed mechanisms, there is little experimental evidence for the *cis*-haem b_3 mechanism. However, the mechanism looks analogous to, but not the same as, that which has been proposed for fungal NOR (cytochrome P450nor). In the case of P450nor, the key step is the formation of the short-lived intermediate $\{FeNO\}^8$ by the two-electron (H^-) reduction of the ferric haem–NO complex, while the one-electron reduced form $\{FeNO\}^7$ of P450nor never reacts with another NO.[112] It is noteworthy that theoretical calculation favours the *cis*-haem b_3 mechanism.[113,114] In this proposed mechanism, the first NO coordinates with haem b_3, and an O atom of the coordinated NO interacts with Fe_B. The calculation also indicated that the *trans* mechanism was energetically unfavourable.[114] Structural characterisation of the reaction intermediate(s) is required in order to draw a complete picture of the NO reduction mechanism in NOR.

Figure 6.8 displays one of the possible mechanisms on the basis of the spectroscopic and crystallographic studies on *Ps.a.* cNOR.

i. In the resting state (**1**), the haem iron and Fe_B in the ferric form are bridged by the O ligand, consistent with the crystal structure. The μ-oxo-bridged coordination structure is in good agreement with the structure proposed based on the resonance Raman study.
ii. A strong ESR signal was observed at $g = 6$ after single turnover in cNOR, suggesting that the haem b_3 iron is in a ferric and five-coordinated state. It is likely that the non-bridged structure of the binuclear centre, rather than the μ-oxo-bridged structure, is involved in the catalytic turnover of cNOR. This state is designated as the fully oxidised state (**2**).
iii. The order of the reduction of the two irons and the NO binding is still unclear. However, as a result, two NO molecules bound to the binuclear centre in the fully reduced state (*e.g.*, in the *trans* mechanism, which we proposed based on data obtained from a freeze-quenched EPR study) are shared by ferrous haem b_3 and Fe_B (**3**). Most recently, Lu and co-workers supported this *trans* coordination of two NO molecules using the model system of NOR, which was constructed in modified myoglobin by mutagenesis techniques.[115]

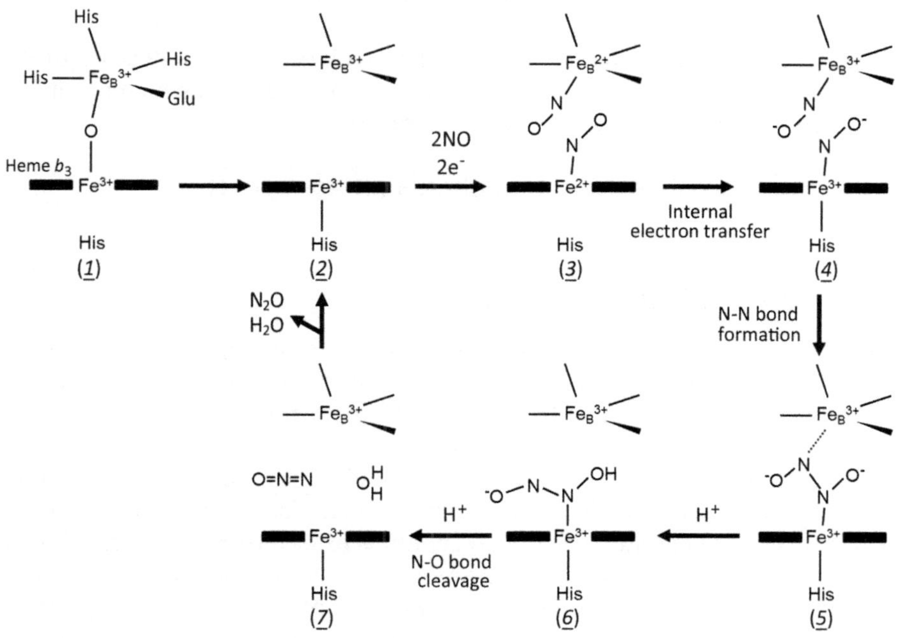

Figure 6.8 Proposed reaction mechanism of the nitric oxide (NO) reductase-catalysed NO reduction reaction.

iv. After the internal electron transfers from the ferrous irons to the bound NO molecules (**4**), the close proximity of two NO^- molecules on the haem b_3 and Fe_B could promote N–N bond formation, producing a hyponitrite intermediate as a transient species. Richardson and co-workers proposed that such a hyponitrite could bridge the two irons of haem b_3 and Fe_B in the reaction intermediate (**5**). In this hypothetical structure, the Fe–Fe distance would be expected to be at least 4.4 Å. In other words, the irons must shift by 0.4 Å from their original (fully oxidised state) positions to generate sufficient space for the accommodation of two NO molecules. Such an elongated Fe–Fe distance was observed in the *G.s.* qNOR structure (haem b_3 Fe–Zn_B distance: 4.6 Å)[88] and ligand-bound forms of *Ps.a.* cNOR (haem b_3 Fe–Fe_B distance: 4.4 Å),[93] suggesting that the positional shift of irons would be possible during the reduction of NO by NOR. The subsequent protonation of the oxygen atom in the hyponitrite intermediate could produce a protonated hyponitrite species that could reside on the haem b_3 coordination site (**6**).

v. Protons, which are transferred from bulk water through the delivery pathway (*vide infra*), could facilitate the N–O bond cleavage of the hyponitrite to produce N_2O and H_2O (**7**).

The chemical reaction of the NO reduction by bacterial NOR is referred to as a "disproportionation" reaction:

$$2(NO^-) \rightarrow \text{hyponitrite} + 2H^+ \rightarrow N_2O + H_2O \tag{6.8}$$

On the other hand, a theoretical study based on the *Ps.a.* cNOR structure[114] proposed that the *cis*-haem b_3 mechanism is most possible, which is consistent with the so-called "donor–acceptor" reaction:

$$(NO)^{2-} + NO \rightarrow \text{hyponitrite} + 2H^+ \rightarrow N_2O + H_2O \tag{6.9}$$

6.5.5 Proton Transfer Pathway

The NO reduction reaction requires protons for the cleavage of the N–O bond in the transient hyponitrite species to produce N_2O and H_2O. The catalytic protons would be transferred from bulk water to the buried active site through a specific pathway consisting of a water channel and/or a hydrogen-bonding network. In the *Ps.a.* cNOR structure, although neither an obvious channel nor a hydrogen-bonding network from the cytoplasmic side to the binuclear centre has been identified, two channels extend from the water cluster located around the haem b and b_3 propionates (Ca^{2+} ligands) to the periplasmic side of the enzyme through the NorB/NorC interface (Figure 6.9). Asp_{198} (NorB) and Glu_{57} (NorC) on the protein surface are probably key residues that comprise the proton entry site, because both acidic residues are highly conserved in cNOR. It was proposed that the hydrophilic channels from the

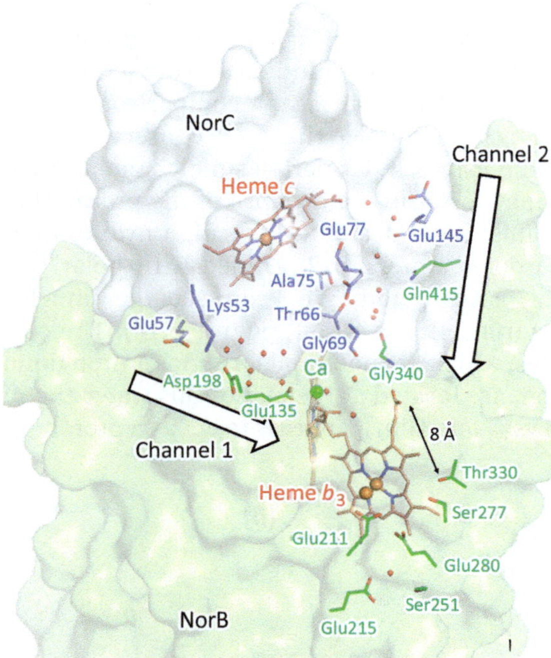

Figure 6.9 Possible proton pathway for catalytic nitric oxide (NO) reduction in *Pseudomonas aeruginosa* cytochrome *c*-dependent NO reductase (cNOR). There are hydrophilic channels from the protein surface to the water cluster located around the haem propionate groups of haem b_3 in the NorB/NorC interface. Although there is an 8.0 Å gap between the haem propionate group and Thr_{330}, a hydrogen-bonding network connects Thr_{330} and Glu_{211} which is a potential terminal proton donor for the catalytic NO reduction. The hydrophilic channels and hydrogen-bonding network could function as proton transfer pathways in cNOR.

periplasmic side identified in the cNOR structure could serve as pathways for the transfer of protons to the catalytic site for the NO reduction reaction. The proposal based on the reported structure is completely consistent with extensive evidence obtained from electrochemical, biochemical and flow-flash kinetic data, all of which unambiguously indicate that electrons and protons are supplied from the same periplasmic side of the molecule and that the cNOR reaction is non-electrogenic.[73,79,80] The mutation studies and MD simulation also supported this suggestion.[55,116,117] However, in our cNOR structure, the water cluster around the Ca^{2+} and haem propionate groups is not connected to the Fe_B ligand Glu_{211}, which is a possible terminal proton donor in the NO reduction reaction (Figure 6.9). It is possible that some conformational changes could be induced during the catalytic turnover to create a proton transfer pathway from the periplasmic side to the essential Glu_{211} and Glu_{280} residues at the binuclear centre through the water cluster near the Ca^{2+} binding site.

In sharp contrast to cNOR, although a water cluster is also present around the Ca^{2+} in the qNOR, there is no plausible proton transfer pathway such as an obvious water channel and/or a hydrogen-bond network from the extracellular (periplasmic) surface. A comparison of the structures of *G.s.* qNOR and *Ps.a.* cNOR revealed that a key conserved Asp residue (Asp_{198} in *P.a.* cNOR) at the extracellular/periplasmic channel in cNOR is absent in qNOR, which may result in the collapse of the proton transfer pathway from the extracellular side in qNOR. Instead, a hydrophilic channel that contains many ordered waters and is lined with polar residues (Glu_{516}, Gln_{545}, Glu_{591}, Glu_{594}, Asn_{622} and Tyr_{660}) connects the binuclear centre with the cytoplasmic surface in *G.s.* qNOR (Figure 6.10). The conserved Glu_{512} and Glu_{581} residues, which are potential terminal proton donors for the NO reduction reaction, are located at the end of this water channel, and a salt bridge between Lys_{597} and Glu_{281} appears to act as a proton entry site at the

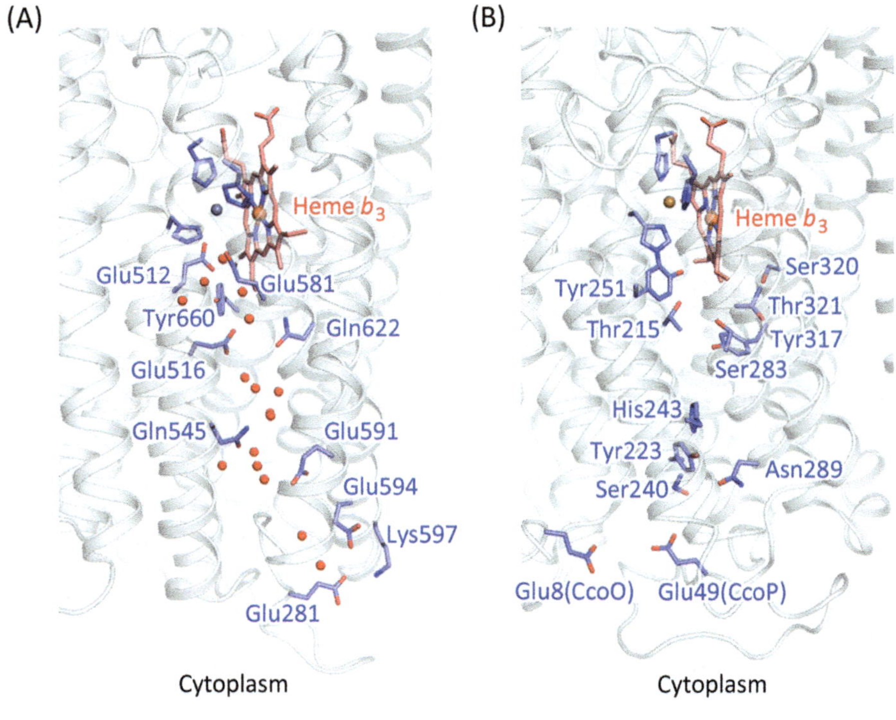

Figure 6.10 Possible proton pathway for catalytic nitric oxide (NO) reduction in *Geobacillus stearothermophilus* (*G.s.*) quinol-dependent NO reductase (qNOR). (a) Water-containing hydrophilic channel from the cytoplasm to the binuclear active centre in *G.s.* qNOR. Structure-guided mutagenesis and molecular dynamics simulation suggest that the water channel functions as a proton transfer pathway for the catalytic NO reduction. (b) Proton transfer pathway called the K-pathway in cytochrome *cbb*$_3$ oxidase. The structure shown is the same orientation as in *G.s.* qNOR.

cytoplasmic side. A MD simulation using the *G.s.* qNOR structure supported the hypothesis that protons enter the channel through a Lys_{597}–Glu_{281} salt bridge, and are then transferred through the chain of water molecules in the water channel,[88] before being donated to the binuclear centre through the carboxylate groups of the Glu_{512} and Glu_{581} residues. Furthermore, the mutation of Lys_{597} abolished NO reduction activity. Therefore, in the case of qNOR, the catalytic protons appear to be supplied from the cytoplasmic side to the binuclear centre. This conclusion for qNOR is inconsistent with the previous conclusions obtained from cNOR studies. Further studies will be required in order to confirm the characteristic property of the proton transfer in qNOR.

6.6 Molecular Evolution of Proton Pumps in Respiratory Enzymes

In both cNOR and qNOR, there is no continuous channel and/or hydrogen-bonding network between the cytoplasmic and extracellular/periplasmic sides, and thus protons do not permeate through the TM region of NORs. These observations are consistent with the fact that NORs have no proton pumping ability. However, it is notable that the direction of the catalytic proton transfer is entirely opposite between cNOR and qNOR, although these two types of NOR enzymes share a number of structural and functional properties.

It is also interesting that the location of the water channel in qNOR overlaps with the K-pathway of cytochrome oxidase, which functions as a proton transfer pathway for the catalytic reduction of O_2 and as a proton pumping pathway. A close look at the K-pathway region of the microaerobic cytochrome *cbb*$_3$ oxidase—a close relative of NOR in a phylogenetic analysis—shows that several functionally important residues in the K-pathway of cytochrome *cbb*$_3$ oxidase are superimposable on the residues that line the water channel of *G.s.* qNOR (Figure 6.10). Therefore, the water channel in qNOR could be a prototype of the proton pumping pathway that is found in respiratory enzymes, providing a clue as to how the respiratory enzymes acquired their proton pumping ability. It is not straightforward to create proton pumping ability in NOR by simple mutagenesis, because nature has designed elegant proton pumping machinery such as an energy-driven gate for pumping protons in cytochrome oxidases. For example, in mitochondrial cytochrome *aa*$_3$ oxidase, Yoshikawa and Tsukihara and their co-workers proposed that the conformational change coupled with the redox change and O_2 reduction could be responsible for the unidirectional delivery of protons.[102,118–121] However, further functional studies based on the structures of the respiratory enzymes promise to pave the way to elucidating the evolutionary events in the transition from anaerobic to aerobic respiration, and can permit the function of the respiratory enzymes to be controlled by protein engineering.

References

1. D. J. Richardson and N. J. Watmough, *Curr. Opin. Chem. Biol.*, 1999, **3**, 207.
2. W. G. Zumft, *Microbiol. Mol. Biol. Rev.*, 1997, **61**, 533.
3. W. G. Zumft, *J. Inorg. Biochem.*, 2005, **99**, 194.
4. W. G. Zumft, in *The smallest biomolecules: ciatomics and their interactions with haem proteins*, ed. A. Gosh, Elsevier, Amsterdam, 2008, p. 327.
5. D.-H. Lee, B. Mondal and K. D. Karlin, in *Activation of Small Molecules*, ed. W. B. Tolman, Wiley, Weinheinm, 2006, p. 43.
6. J. A. McCleverty, *Chem. Rev.*, 2004, **104**, 403.
7. J. S. Stamler, D. J. Singel and J. Loscalzo, *Science*, 1992, **258**, 1898.
8. M. W. Wong, R. H. Nobes, W. J. Bouma and L. Radom, *J. Chem. Phys.*, 1989, **91**, 2971.
9. P. Coppens, I. Novozhilova and A. Kovalevsky, *Chem. Rev.*, 2002, **102**, 861.
10. J. H. Enemark and R. D. Feltham, *Coord. Chem. Rev.*, 1974, **13**, 339.
11. D. P. Linder and K. R. Rodgers, *Inorg. Chem.*, 2005, **44**, 1367.
12. D. P. Linder, K. R. Rodgers, J. Banister, G. R. Wyllie, M. K. Ellison and W. R. Scheidt, *J. Am. Chem. Soc.*, 2004, **126**, 14136.
13. P. Moenne-Loccoz, *Nat. Prod. Rep.*, 2007, **24**, 610.
14. S. Hu and J. R. Kincaid, *J. Am. Chem. Soc.*, 1991, **113**, 9760.
15. E. Obayashi, K. Tsukamoto, S. Adachi, S. Takahashi, M. Nomura, T. Iizuka, H. Shoun and Y. Shiro, *J. Am. Chem. Soc.*, 1997, **119**, 7807.
16. T. C. Berto, M. B. Hoffman, Y. Murata, K. B. Landenberger, E. E. Alp, J. Zhao and N. Lehnert, *J. Am. Chem. Soc.*, 2011, **133**, 16714.
17. C. A. Brown, M. A. Pavlosky, T. E. Westre, Y. Zhang, B. Hedman, K. O. Hodgson and E. I. Solomon, *J. Am. Chem. Soc.*, 1995, **117**, 715.
18. T. Matsubara and H. Iwasaki, *J. Biochem.*, 1971, **69**, 859.
19. A. Remde and R. Conrad, *FEMS Microbiol. Lett.*, 1991, **80**, 329.
20. H. Korner, *Arch. Microbiol.*, 1993, **159**, 410.
21. A. M. Jones and T. C. Hollocher, *Biochim. Biophys. Acta*, 1993, **1144**, 359.
22. R. W. Ye, D. Haas, J. O. Ka, V. Krishnapillai, A. Zimmermann, C. Baird and J. M. Tiedje, *J. Bacteriol.*, 1995, **177**, 3606.
23. H. M. Hassan and H. C. Sun, *Proc. Natl. Acad. Sci. U. S. A.*, 1992, **89**, 3217.
24. J. Hendriks, A. Oubrie, J. Castresana, A. Urbani, S. Gemeinhardt and M. Saraste, *Biochim. Biophys. Acta*, 2000, **1459**, 266.
25. S. J. Field, F. H. Thorndycroft, A. D. Matorin, D. J. Richardson and N. J. Watmough, *Methods Enzymol.*, 2008, **437**, 79.
26. S. de Vries and I. Schroder, *Biochem. Soc. Trans.*, 2002, **30**, 662.
27. G. J. Carr and S. J. Ferguson, *Biochem. J.*, 1990, **269**, 423.
28. A. P. de Boer, J. van der Oost, W. N. Reijnders, H. V. Westerhoff, A. H. Stouthamer and R. J. van Spanning, *Eur. J. Biochem.*, 1996, **242**, 592.
29. H. Arai, Y. Igarashi and T. Kodama, *Biochim. Biophys. Acta*, 1995, **1261**, 279.
30. M. R. Cheesman, W. G. Zumft and A. J. Thomson, *Biochemistry*, 1998, **37**, 3994.

31. N. Sakurai and T. Sakurai, *Biochemistry*, 1997, **36**, 13809.
32. P. Girsch and S. de Vries, *Biochim. Biophys. Acta*, 1997, **1318**, 202.
33. T. Satoh, *Arch. Microbiol.*, 1984, **139**, 179.
34. L. C. Bell, D. J. Richardson and S. J. Ferguson, *J. Gen. Microbiol.*, 1992, **138**, 437.
35. J. W. B. Moir and S. J. Ferguson, *Microbiology*, 1994, **140**, 389.
36. C. G. Timoteo, A. S. Pereira, C. E. Martins, S. G. Naik, A. G. Duarte, J. J. Moura, P. Tavares, B. H. Huynh and I. Moura, *Biochemistry*, 2011, **50**, 4251.
37. R. Cramm, A. Pohlmann and B. Friedrich, *FEBS Lett.*, 1999, **460**, 6.
38. S. de Vries, M. J. Strampraad, S. Lu, P. Moenne-Loccoz and I. Schroder, *J. Biol. Chem.*, 2003, **278**, 35861.
39. A. Busch, B. Friedrich and R. Cramm, *Appl. Environ. Microbiol.*, 2002, **68**, 668.
40. T. C. Householder, E. M. Fozo, J. A. Cardinale and V. L. Clark, *Infect. Immun.*, 2000, **68**, 5241.
41. M. F. Anjum, T. M. Stevanin, R. C. Read and J. W. Moir, *J. Bacteriol.*, 2002, **184**, 2987.
42. Suharti and S. de Vries, *Biochem. Soc. Trans.*, 2005, **33**, 130.
43. Suharti, H. A. Heering and S. de Vries, *Biochemistry*, 2004, **43**, 13487.
44. Suharti, M. J. Strampraad, I. Schroder and S. de Vries, *Biochemistry*, 2001, **40**, 2632.
45. S. Al-Attar and S. de Vries, *FEBS Lett.*, 2015, **589**, 2050.
46. J. A. Garcia-Horsman, E. Berry, J. P. Shapleigh, J. O. Alben and R. B. Gennis, *Biochemistry*, 1994, **33**, 3113.
47. W. G. Zumft, C. Braun and H. Cuypers, *Eur. J. Biochem.*, 1994, **219**, 481.
48. M. Saraste and J. Castresana, *FEBS Lett.*, 1994, **341**, 1.
49. M. Saraste, *Antonie van Leeuwenhoek*, 1994, **65**, 285.
50. J. van der Oost, A. P. de Boer, J. W. de Gier, W. G. Zumft, A. H. Stouthamer and R. J. van Spanning, *FEMS Microbiol. Lett.*, 1994, **121**, 1.
51. J. Hendriks, U. Gohlke and M. Saraste, *J. Bioenerg. Biomembr.*, 1998, **30**, 15.
52. W. G. Zumft, D. J. Gotzmann, K. Frunzke and A. Viebrock, in *Inorganic Nitrogen Metabolism*, ed. W. R. Ullrich, P. J. Aparicio, P. J. Syrett and F. Castillo, Springer Verlag, Berlin, 1987, p. 61.
53. J. Hoglen and T. C. Hollocher, *J. Biol. Chem.*, 1989, **264**, 7556.
54. H. Kumita, K. Matsuura, T. Hino, S. Takahashi, H. Hori, Y. Fukumori, I. Morishima and Y. Shiro, *J. Biol. Chem.*, 2004, **279**, 55247.
55. L. A. Schurig-Briccio, P. Venkatakrishnan, J. Hemp, C. Bricio, J. Berenguer and R. B. Gennis, *Proc. Natl. Acad. Sci. U. S. A.*, 2013, **110**, 12613.
56. K. Conrath, A. S. Pereira, C. E. Martins, C. G. Timoteo, P. Tavares, S. Spinelli, J. Kinne, C. Flaudrops, C. Cambillau, S. Muyldermans, I. Moura, J. J. Moura, M. Tegoni and A. Desmyter, *Protein Sci.*, 2009, **18**, 619.
57. K. L. Gronberg, M. D. Roldan, L. Prior, G. Butland, M. R. Cheesman, D. J. Richardson, S. Spiro, A. J. Thomson and N. J. Watmough, *Biochemistry*, 1999, **38**, 13780.

58. S. J. Field, L. Prior, M. D. Roldan, M. R. Cheesman, A. J. Thomson, S. Spiro, J. N. Butt, N. J. Watmough and D. J. Richardson, *J. Biol. Chem.*, 2002, **277**, 20146.
59. J. Hendriks, A. Warne, U. Gohlke, T. Haltia, C. Ludovici, M. Lubben and M. Saraste, *Biochemistry*, 1998, **37**, 13102.
60. D. H. Kastrau, B. Heiss, P. M. Kroneck and W. G. Zumft, *Eur. J. Biochem.*, 1994, **222**, 293.
61. P. Moenne-Loccoz and S. de Vries, *J. Am. Chem. Soc.*, 1998, **120**, 5147.
62. P. Moenne-Loccoz, O.-M. H. Richter, H. W. Huang, I. M. Wasser, R. A. Ghiladi, K. D. Karlin and S. de Vries, *J. Am. Chem. Soc.*, 2000, **122**, 9344.
63. T. Ogura, K. Honnami, T. Oshima, S. Yoshikawa and T. Kitagawa, *J. Am. Chem. Soc.*, 1983, **105**, 7781.
64. P. V. Argade, Y. C. Ching and D. L. Rousseau, *Science*, 1984, **225**, 329.
65. E. Pinakoulaki and C. Varotsis, *Biochemistry*, 2003, **42**, 14856.
66. J. Teraoka and T. Kitagawa, *J. Biol. Chem.*, 1981, **256**, 3969.
67. S. Lu, Suharti, S. de Vries and P. Moenne-Loccoz, *J. Am. Chem. Soc.*, 2004, **126**, 15332.
68. T. Hayashi, M. T. Lin, K. Ganesan, Y. Chen, J. A. Fee, R. B. Gennis and P. Moenne-Loccoz, *Biochemistry*, 2009, **48**, 883.
69. F. H. Thorndycroft, G. Butland, D. J. Richardson and N. J. Watmough, *Biochem. J.*, 2007, **401**, 111.
70. M. Itoh, S. Mizukami, K. Matsuura and T. Satoh, *FEBS Lett.*, 1989, **244**, 81.
71. J. R. Winkler, B. G. Malmstrom and H. B. Gray, *Biophys. Chem.*, 1995, **54**, 199.
72. J. J. Regan, B. E. Ramirez, J. R. Winkler, H. B. Gray and B. G. Malmstrom, *J. Bioenerg. Biomembr.*, 1998, **30**, 35.
73. J. H. Hendriks, A. Jasaitis, M. Saraste and M. I. Verkhovsky, *Biochemistry*, 2002, **41**, 2331.
74. A. Oubrie, S. Gemeinhardt, S. Field, S. Marritt, A. J. Thomson, M. Saraste and D. J. Richardson, *Biochemistry*, 2002, **41**, 10858.
75. S. J. Field, M. D. Roldan, S. J. Marritt, J. N. Butt, D. J. Richardson and N. J. Watmough, *Biochim. Biophys. Acta*, 2011, **1807**, 451.
76. C. M. Cordas, A. G. Duarte, J. J. Moura and I. Moura, *Biochim. Biophys. Acta*, 2013, **1827**, 233.
77. C. M. Cordas, A. S. Pereira, C. E. Martins, C. G. Timoteo, I. Moura, J. J. G. Moura and P. Tavares, *ChemBioChem*, 2006, **7**, 1878.
78. A. G. Duarte, C. M. Cordas, J. J. G. Moura and I. Moura, *Biochim. Biophys. Acta*, 2014, **1837**, 375.
79. J. P. Shapleigh and W. J. Payne, *J. Bacteriol.*, 1985, **163**, 837.
80. J. Reimann, U. Flock, H. Lepp, A. Honigmann and P. Adelroth, *Biochim. Biophys. Acta*, 2007, **1767**, 362.
81. L. A. Drachev, A. A. Jasaitis, A. D. Kaulen, A. A. Kondrashin, E. A. Liberman, I. B. Nemecek, S. A. Ostroumov, A. Semenov and V. P. Skulachev, *Nature*, 1974, **249**, 321.
82. P. Lachmann, Y. Huang, J. Reimann, U. Flock and P. Adelroth, *J. Biol. Chem.*, 2010, **285**, 25531.

83. U. Flock, P. Lachmann, J. Reimann, N. J. Watmough and P. Adelroth, *J. Inorg. Biochem.*, 2009, **103**, 845.
84. U. Flock, J. Reimann and P. Adelroth, *Biochem. Soc. Trans.*, 2006, **34**, 188.
85. U. Flock, F. H. Thorndycroft, A. D. Matorin, D. J. Richardson, N. J. Watmough and P. Adelroth, *J. Biol. Chem.*, 2008, **283**, 3839.
86. U. Flock, N. J. Watmough and P. Adelroth, *Biochemistry*, 2005, **44**, 10711.
87. T. Hino, Y. Matsumoto, S. Nagano, H. Sugimoto, Y. Fukumori, T. Murata, S. Iwata and Y. Shiro, *Science*, 2010, **330**, 1666.
88. Y. Matsumoto, T. Tosha, A. V. Pisliakov, T. Hino, H. Sugimoto, S. Nagano, Y. Sugita and Y. Shiro, *Nat. Struct. Mol. Biol.*, 2012, **19**, 238.
89. L. Salomonsson, J. Reimann, T. Tosha, N. Krause, N. Gonska, Y. Shiro and P. Adelroth, *Biochim. Biophys. Acta*, 2012, **1817**, 1914.
90. Y. Shiro, *Biochim. Biophys. Acta*, 2012, **1817**, 1907.
91. Y. Shiro, H. Sugimoto, T. Tosha, S. Nagano and T. Hino, *Philos. Trans. R. Soc., B*, 2012, **367**, 1195.
92. T. Tosha and Y. Shiro, *IUBMB Life*, 2013, **65**, 217.
93. N. Sato, S. Ishii, H. Sugimoto, T. Hino, Y. Fukumori, Y. Sako, Y. Shiro and T. Tosha, *Proteins*, 2014, **82**, 1258.
94. E. Terasaka, N. Okada, N. Sato, Y. Sako, Y. Shiro and T. Tosha, *Biochim. Biophys. Acta*, 2014, **1837**, 1019.
95. P. Moenne-Loccoz and J. A. Fee, *Science*, 2010, **330**, 1632.
96. S. Buschmann, E. Warkentin, H. Xie, J. D. Langer, U. Ermler and H. Michel, *Science*, 2010, **329**, 327.
97. N. J. Watmough, G. Butland, M. R. Cheesman, J. W. Moir, D. J. Richardson and S. Spiro, *Biochim. Biophys. Acta*, 1999, **1411**, 456.
98. G. Butland, S. Spiro, N. J. Watmough and D. J. Richardson, *J. Bacteriol.*, 2001, **183**, 189.
99. S. Iwata, C. Ostermeier, B. Ludwig and H. Michel, *Nature*, 1995, **376**, 660.
100. C. Ostermeier, S. Iwata, B. Ludwig and H. Michel, *Nat. Struct. Biol.*, 1995, **2**, 842.
101. T. Tsukihara, H. Aoyama, E. Yamashita, T. Tomizaki, H. Yamaguchi, K. Shinzawa-Itoh, R. Nakashima, R. Yaono and S. Yoshikawa, *Science*, 1996, **272**, 1136.
102. S. Yoshikawa, K. Shinzawa-Itoh, R. Nakashima, R. Yaono, E. Yamashita, N. Inoue, M. Yao, M. J. Fei, C. P. Libeu, T. Mizushima, H. Yamaguchi, T. Tomizaki and T. Tsukihara, *Science*, 1998, **280**, 1723.
103. T. Soulimane, G. Buse, G. P. Bourenkov, H. D. Bartunik, R. Huber and M. E. Than, *EMBO J.*, 2000, **19**, 1766.
104. J. Abramson, S. Riistama, G. Larsson, A. Jasaitis, M. Svensson-Ek, L. Laakkonen, A. Puustinen, S. Iwata and M. Wikstrom, *Nat. Struct. Biol.*, 2000, **7**, 910.
105. T. Hayashi, I. J. Lin, Y. Chen, J. A. Fee and P. Moenne-Loccoz, *J. Am. Chem. Soc.*, 2007, **129**, 14952.
106. N. J. Watmough, S. J. Field, R. J. Hughes and D. J. Richardson, *Biochem. Soc. Trans.*, 2009, **37**, 392.

107. J. P. Collman, Y. Yang, A. Dey, R. A. Decreau, S. Ghosh, T. Ohta and E. I. Solomon, *Proc. Natl. Acad. Sci. U. S. A.*, 2008, **105**, 15660.
108. I. M. Wasser, H. W. Huang, P. Moenne-Loccoz and K. D. Karlin, *J. Am. Chem. Soc.*, 2005, **127**, 3310.
109. C. S. Butler, H. E. Seward, C. Greenwood and A. J. Thomson, *Biochemistry*, 1997, **36**, 16259.
110. N. J. Watmough, M. R. Cheesman, C. S. Butler, R. H. Little, C. Greenwood and A. J. Thomson, *J. Bioenerg. Biomembr.*, 1998, **30**, 55.
111. K. L. Gronberg, N. J. Watmough, A. J. Thomson, D. J. Richardson and S. J. Field, *J. Biol. Chem.*, 2004, **279**, 17120.
112. Y. Shiro, M. Fujii, T. Iizuka, S. Adachi, K. Tsukamoto, K. Nakahara and H. Shoun, *J. Biol. Chem.*, 1995, **270**, 1617.
113. L. M. Blomberg, M. R. Blomberg and P. E. Siegbahn, *Biochim. Biophys. Acta*, 2006, **1757**, 31.
114. M. R. Blomberg and P. E. Siegbahn, *Biochemistry*, 2012, **51**, 5173.
115. H. Matsumura, T. Hayashi, S. Chakraborty, Y. Lu and P. Moenne-Loccoz, *J. Am. Chem. Soc.*, 2014, **136**, 2420.
116. J. ter Beek, N. Krause, J. Reimann, P. Lachmann and P. Adelroth, *J. Biol. Chem.*, 2013, **288**, 30626.
117. A. V. Pisliakov, T. Hino, Y. Shiro and Y. Sugita, *PLoS Comput. Biol.*, 2012, **8**, e1002674.
118. T. Tsukihara, K. Shimokata, Y. Katayama, H. Shimada, K. Muramoto, H. Aoyama, M. Mochizuki, K. Shinzawa-Itoh, E. Yamashita, M. Yao, Y. Ishimura and S. Yoshikawa, *Proc. Natl. Acad. Sci. U. S. A.*, 2003, **100**, 15304.
119. S. Yoshikawa, K. Muramoto, K. Shinzawa-Itoh, H. Aoyama, T. Tsukihara, K. Shimokata, Y. Katayama and H. Shimada, *Biochim. Biophys. Acta*, 2006, **1757**, 1110.
120. K. Muramoto, K. Ohta, K. Shinzawa-Itoh, K. Kanda, M. Taniguchi, H. Nabekura, E. Yamashita, T. Tsukihara and S. Yoshikawa, *Proc. Natl. Acad. Sci. U. S. A.*, 2010, **107**, 7740.
121. M. Kubo, S. Nakashima, S. Yamaguchi, T. Ogura, M. Mochizuki, J. Kang, M. Tateno, K. Shinzawa-Itoh, K. Kato and S. Yoshikawa, *J. Biol. Chem.*, 2013, **288**, 30259.

CHAPTER 7

Insights into Nitrous Oxide Reductase

SOFIA R. PAULETA*[a], CÍNTIA CARREIRA[a] AND ISABEL MOURA[a]

[a]UCIBIO, REQUIMTE, Dept. Química, FCT-UNL, Campus da Caparica, 2829-516 Caparica, Portugal
*E-mail: srp@fct.unl.pt

7.1 Introduction

Nitrous oxide reductase (N_2OR) is the metalloenzyme that performs the last step of bacterial denitrification, the two-electron reduction of nitrous oxide to dinitrogen and water. This reaction completes the nitrogen cycle and prevents the release of nitrous oxide, a potent greenhouse gas and ozone-depleting agent, into the atmosphere.

The reduction of nitrous oxide is thermodynamically favourable, but a catalyst is required to overcome the large activation barrier of this reaction in the gas phase ($\Delta G^{o\prime} = -339.5$ kJ mol^{-1},[1,2] with an activation energy of around 250 kJ mol^{-1}),[3] as it consists of a spin-forbidden process.[4] Nevertheless, nature has designed a metal centre to overcome this high activation barrier, which is the "CuZ centre", the catalytic centre of N_2OR. Besides this metal centre, this enzyme has another multicopper centre, the binuclear CuA centre, which is the electron transferring centre.

RSC Metallobiology Series No. 9
Metalloenzymes in Denitrification: Applications and Environmental Impacts
Edited by Isabel Moura, José J. G. Moura, Sofia R. Pauleta and Luisa B. Maia

Published by the Royal Society of Chemistry, www.rsc.org

The "CuZ centre" is a tetranuclear copper sulphide cluster that has been isolated in two different forms: CuZ as (4Cu2S) or CuZ* as (4Cu1S). Defining the tetranuclear copper centre structure and its interaction with nitrous oxide is fundamental to understand the catalytic properties of this enzyme. In recent years, new insights into the structure of the catalytic centre and of its catalytic properties have arisen through the solution of the crystal structure of the "CuZ centre" and by the analysis of the spectroscopic and kinetic properties of CuZ(4Cu2S) and CuZ*(4Cu1S) by different research groups.

7.2 Biogenesis of N_2OR

The gene that encodes N_2OR, *nosZ*, has been identified in the genomes of several organisms, mainly in α-, β- and γ-proteobacteria that can perform the complete denitrification pathway under limiting oxygen tensions as a less efficient alternative to aerobic respiration.[5] Nevertheless, some bacteria and archaea genomes that present the *nosZ* gene lack the *nirS* or *nirK* genes that encode nitrite reductase, and thus in fact cannot denitrify.[1,5]

Genome analysis of the non-denitrifying microorganisms indicates that the presence of *nosZ* is, in some cases, associated with the *nrfA* gene (which encodes the dissimilatory nitrite reductase that is responsible for the last step of the dissimilatory metabolic pathway of nitrite ammonification), but in others, their genomes do not encode any known nitrite reductases.[6–9] Therefore, the characterisation of these organisms and understanding their role in the global consumption of nitrous oxide is imperative in order to better comprehend the ecological controls on N_2O emissions and greenhouse effects.

In fact, the global analysis of the locus where *nosZ* is located and the primary sequence of N_2OR show that there are two main groups of these organisms. One group harbours the typical *nosZ* gene, encoding the N_2OR that is better characterised, while the other group encodes an "atypical" *nosZ* gene. This "atypical" *nosZ* gene is included in a gene cluster containing genes that encode *c*- and *b*-type cytochromes, membrane iron–sulphur proteins and the known accessory N_2OR genes *nosDLFY* (*vide infra*) in different arrangements.[7] The proteins encoded in these genes might constitute an electron transport system from menaquinol to the periplasmic N_2OR, with a similar role to NosR (*vide infra*).[10]

The only "atypical" N_2OR isolated to date is the one from *Wolinella* (*Wo.*) *succinogenes*, revealing the presence of a *c*-type haem.[11–15] In fact, such N_2ORs, with an additional C-terminal domain with the canonical *c*-type haem binding motif –CXXCH–, have been identified in a few organisms from the *Campylobacter*, *Sulfurimonas, Wolinella* and *Denitrovibrio* genera, with the first three being host-associated organisms from the ε-proteobacteria group. The primary sequence of these "atypical" N_2ORs reveals the presence of a signal peptide that is Sec-dependent (Sec relates to a secretory pathway), as would be expected if we take into account the requirement of the periplasmic cytochrome *c* maturation systems for its biosynthesis.[16] Interestingly, in the other "atypical" N_2ORs without the additional C-terminal domain, its primary sequence still presents a similar signal peptide in most cases, which has not received a clear explanation to date. Other "atypical" N_2ORs have been identified in *Geobacillus thermodenitrificans*,[17] *Anaeromyxobacter dehalogenans*[7] and other *Bacillus* species.[18–20] In some Gram-positive bacteria, N_2OR has been identified as being a transmembrane enzyme, but the reason for the presence of multiple *nosZ* gene copies in these genomes is still unknown (see Chapter 15 of this book for more details).[21]

In most proteobacteria, the typical *nosZ* gene is part of a locus that encodes other enzymes/proteins that are proposed to assist in the activity of N_2OR or to be involved in the assembly of its copper centres, especially the "CuZ centre". In addition, these typical N_2ORs are transported to the periplasm in a folded state, most probably already as a dimer in the apo-form,[22,23] through the twin-arginine translocation (Tat) system,[24] as indicated by the presence of a 35–40 signal peptide containing the consensus motif –RRXF/L–.[25,26] While the "CuZ centre" is quite unique and only found in N_2OR, the CuA centre is common to cytochrome *c* oxidase.[27] Thus, the CuA centre has been proposed to be assembled post-translocationally of the folded N_2OR polypeptide through the Tat system,[23,24,28] probably using a molecular system that is common to cytochrome *c* oxidase.[29]

These auxiliary genes are usually organised in the locus as *nosRZDFYL* and can be preceded or followed by the *nosX* gene, mainly in α-proteobacteria. The location of *nosR* is also variable, being either at the end of the gene cluster or after *nosZ*.[2] In some cases, there is more than one copy of these accessory genes, which sometimes encode truncated non-functional proteins due to the presence of premature stop codons. The number of transcriptional units that is transcribed from the *nos* gene cluster is also variable between organisms, ranging from one single hexacistronic transcriptional unit as in the case of *Pseudomonas* (*Ps.*) *aeruginosa*[30] to three as in *Ps. stutzeri* (formerly known as *Ps. perfectomarina*),[31–33] *Paracoccus* (*Pa.*) *denitrificans*[34,35] and *Sinorhizobium meliloti*,[36] despite comprising different genes.

The gene *nosR* is conserved in all organisms that encode *nosZ*, with the exception of ε-proteobacteria and Gram-positive bacteria, and has been proposed to be involved in the transcription of *nosZ* and also of the *nosDFYLtatE* operon in *Ps. stutzeri*.[31,33,37,38] NosR was first isolated from *Ps. stutzeri* and shown to be a transmembrane flavoprotein containing iron–sulphur centres.[38] In fact, the flavin cofactor is covalently bound through the conserved Thr_{163} residue to the periplasmic domain, while the cytoplasmic domain presents a polyferredoxin signature that is proposed to bind two [4Fe–4S] centres. Although these characteristics are not of a transcription regulator (as it clearly lacks a DNA binding domain), its role in the transcription of *nosZ* might be exerted through the interaction with the transcription regulator DnrD.[38]

The other functions of NosR have been associated with the *in vivo* high activity of N_2OR and are also connected with the insertion of CuA and "CuZ" centres into the enzyme.[38,39] These different aspects of the biosynthesis and activity of N_2OR might in fact be related to the different features of the accessory NosR. A series of mutant strains lacking some NosR domains or with point mutations in some of the conserved motifs of this protein was crucial to elucidating its function.[38] Thus, on the one hand, the periplasmic flavin-containing domain was shown to be essential for whole-cell nitrous oxide reduction as a putative electron/proton donor (or electron sink) from the quinone pool,[38] while the other—the intermembrane –CX_3CP– motif—seems to be essential to maintaining N_2OR with a "CuZ centre" as CuZ(4Cu2S).[38] These motifs might be implicated in –SH redox chemistry or reversible metal binding, which might be required to orient NosR transmembrane helices properly.[1] Nevertheless, the mutant strain with point mutations in some of the –CX_3CP– motif, and from which N_2OR was isolated with a "CuZ centre" as CuZ*(4Cu1S), presented a higher whole-cell nitrous oxide reduction than a NosR mutant strain in the cytoplasmic iron–sulphur domain, from which N_2OR with a "CuZ centre" as CuZ(4Cu2S) was isolated.[38] The hypothesis that NosR might be involved not just in the maturation of N_2OR, but also in its activity, was strengthened by the identification of a supramolecular complex between NosR and N_2OR in *Ps. aeruginosa*.[40]

The conserved group of genes *nosDFYL*, which is common to typical and "atypical" *nosZ* gene clusters, has been proposed to be associated with "CuZ centre" assembly in the periplasmic space.[41–43] NosDFY is proposed to form an ABC transporter with NosD, located in the periplasm, and being transported by the Sec system similarly to the "atypical" NosZ, while NosF and NosY are located in the cytoplasm and in the cytoplasmic membrane, respectively.[41] The role of NosDFY has been associated with sulphur transport and its insertion into the "CuZ centre", since copper insertion does not seems to depend on these genes.[35] Thus, NosF, which has ATPase activity, might couple ATP/GTP hydrolysis with the energy-dependent transfer of sulphur across NosY, the cytoplasmic membrane pore.[33,44] This membrane pore is composed of NosY, a 30 kDa membrane protein with five span regions, which contacts NosD, a 45 kDa periplasmic protein of as-yet unknown function.[35] The other component of this transporter might be NosL; although it is not essential for "CuZ centre" assembly, its role in N_2OR biogenesis cannot be ruled out since, in most organisms, the *nosL* gene is co-transcribed with *nosDFY*, and in *Ps. aeruginosa*, it is part of the supramolecular complexes of N_2OR and other denitrification enzymes.[40] NosL is a small lipoprotein of around 20 kDa that is proposed to be an outer membrane protein facing the periplasmic space.[35] NosL, proposed to be a chaperone, binds specifically to Cu^{1+}, and its structure was determined in the apo-form by NMR as consisting of two homologous domains with an unusual β-β-α-β topology,[45–47] and thus might be involved in copper delivery to apo-N_2OR. NosL was recently reported to strongly interact with NosR and also NorC (a subunit of *c*-type nitric oxide reductase) in *Ps. aeruginosa*,[40] strengthening the notion that it might be involved in the maturation process of N_2OR.

The other component of the *nosZ* gene cluster that has only been detected in α- and β-proteobacteria is *nosX*, which encodes a 34 kDa periplasmic flavoprotein belonging to the ApbE protein family (known to be involved in iron–sulphur biosynthesis).[48] NosX might have a role in maintaining N_2OR in a functional and active state *in vivo*, similarly to NosR.[49] In fact, in a *Pa. denitrificans nosXnirX* double knockout strain (*nirX* is a homologue of *nosX*), no nitrous oxide reduction was detected by the whole cells, and N_2OR was isolated in this background with a "CuZ centre" as CuZ*(4Cu1S).[50]

Therefore, NosR and/or NosX (or a NosX homologue) seem to be important in maintaining N_2OR in an active state *in vivo* and are essential for whole-cell nitrous oxide reduction. Thus, it can be speculated that their function would be to assist N_2OR in attaining the reduced state of the "CuZ centre" and/or to be involved in a sulphur shift activation mechanism of N_2OR (*vide infra*).

7.3 The Structure of N_2OR

7.3.1 The Overall Structure

As mentioned, N_2OR contains two multicopper sites, a binuclear CuA electron-transferring centre and a tetranuclear copper sulphide catalytic centre, named the "CuZ centre". In fact, although the structure of the CuA centre was already anticipated as its spectroscopic properties were similar to those of the CuA centre from cytochrome *c* oxidase,[27] the structure of the "CuZ centre" was only revealed after the determination of the first X-ray structure of this enzyme.[51]

The first X-ray structure of a N_2OR was determined for the one isolated from *Marinobacter* (*Ma.*) *hydrocarbonoclasticus* (formerly known as *Ps. nautica*) and solved at 2.4 Å by the group of Cambillau and collaborators in the year 2000.[51] This structure revealed that the enzyme is a homodimer, with each monomer having two domains, each binding one of the copper centres (see Figure 7.1). The N-terminal domain has a seven-bladed β-propeller fold, with the "CuZ centre" inserted in the middle, while the C-terminal has a cupredoxin fold composed of nine β-strands that coordinates the CuA centre (see Figure 7.1).

The two centres in the same monomer are 40 Å apart, but they are only 10–11 Å apart when considering CuA and "CuZ" centres from two different monomers (see Figure 7.1). Thus, N_2OR is a functional homodimer since the electron transfer between the CuA and "CuZ" centres has to occur between the centres located in different monomers, as the distance of 40 Å does not enable efficient intramolecular electron transfer.[52]

7.3.2 The Copper Centres—Structure and Coordination

The CuA centre is very similar to the copper centre present in cytochrome *c* oxidase.[27,53] It is a binuclear copper centre with two bridging S^{γ} atoms of two cysteine side chains (Cys_{561} and Cys_{565}) binding the two copper atoms. The two other ligands of CuA_1 are one $N^{\varepsilon 2}$ of a histidine (His_{526}) and one S^{δ} of a methionine (Met_{572}), while CuA_2 is coordinated by one $N^{\varepsilon 2}$ of a histidine (His_{569}) and the carbonyl atom of a tryptophan (Trp_{563}) (numbering of the residues is according to the primary sequence of *Pa. denitrificans* N_2OR) (see

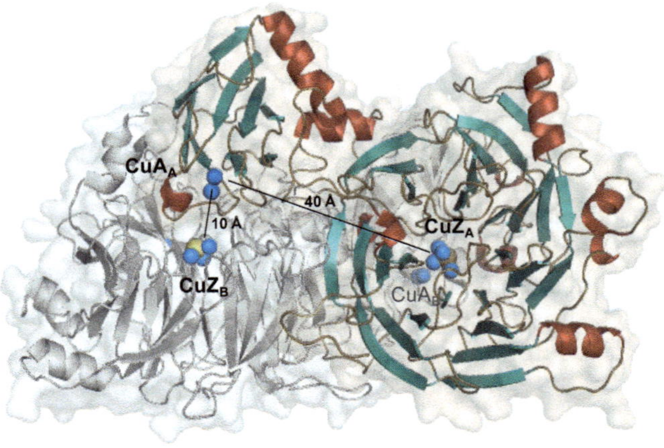

Figure 7.1 Structure of the *Marinobacter hydrocarbonoclasticus* nitrous oxide reductase functional homodimer. The backbone is represented with the identified secondary structure as a cartoon, with one dimer coloured by the secondary structure and the other in light grey. The protein surface is shown coloured according to the subunit, with one monomer coloured light yellow and the other light grey. The distances between the CuA and "CuZ" centres of the two monomers are represented. The CuA and "CuZ" centres are represented by spheres, in which the copper atoms are coloured in blue and the sulphur atoms are in yellow. Figure prepared with Pymol using Protein Data Bank identifier 1QNI.

Figure 7.2a). The CuA centre is located in a loop near the C-terminus between two β-strands.

The first X-ray structure revealed that the "CuZ centre" was unique in biology, as it is a tetranuclear copper site and an oxygen was identified as being the central atom coordinating the four copper atoms.[51] The reanalysis of this structure[54] was performed shortly after taking into account the fact that other spectroscopic data, together with elementary analysis, indicated the presence of an inorganic sulphide.[55] Thus, the "CuZ centre" was defined as a tetranuclear μ_4-sulphide-bridged copper centre, with the four copper atoms being coordinated by seven conserved histidine side chains either by their $N^{\varepsilon 2}$ (His_{80}, His_{128}, His_{270}, His_{325} and His_{376}) or $N^{\delta 1}$ (His_{79} and His_{437}) (numbering of the residues is according to the primary sequence of *Pa. denitrificans* N_2OR) in a distorted tetrahedron geometry (see Figure 7.2a). As can be observed from its structure, all of the copper atoms are coordinated by two histidine residues, with the exception of Cu_{IV}, which led to the proposal of this being the substrate binding site (the unoccupied coordination site, as in other metalloenzymes).[56]

Since the year 2000, other structures of N_2OR have been solved at higher resolution for enzymes isolated from *Pa. denitrificans* (1.6 Å)[54,57] and *Achromobacter* (*Ac.*) *cycloclastes* (1.86 Å),[58] which confirmed the presence of the bridging sulphide and copper centre geometries, but differ in the atom that is found in between the Cu_I and Cu_{IV} atoms of the "CuZ centre".

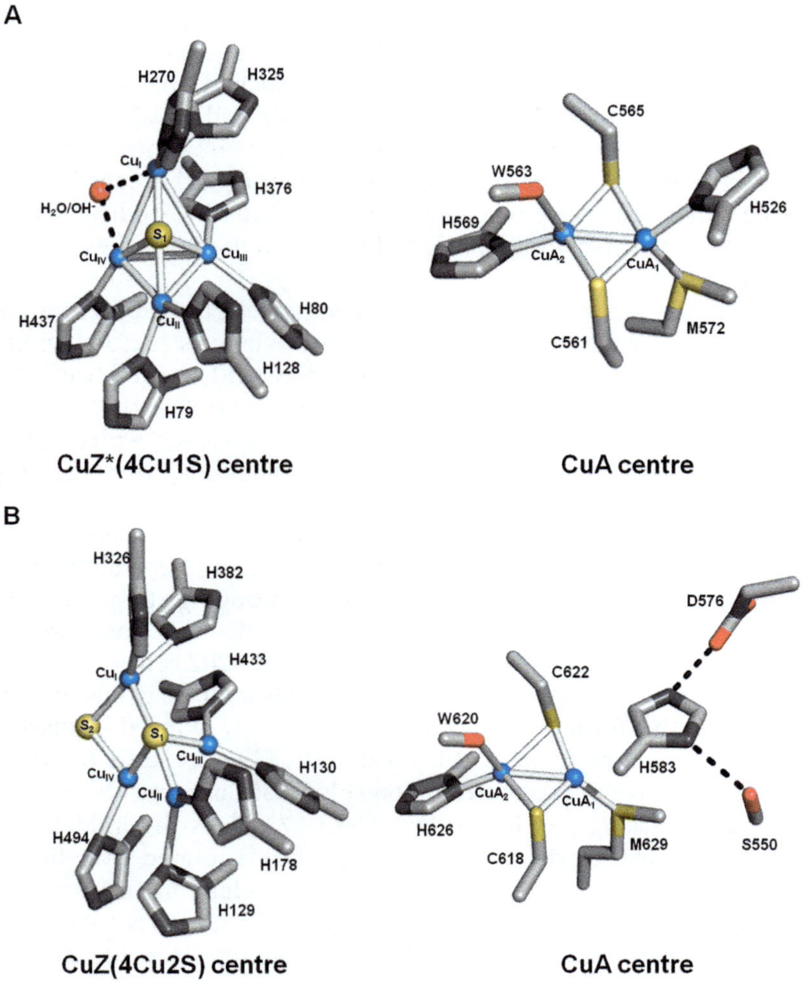

Figure 7.2 Structure of nitrous oxide reductase (N_2OR) copper centres. (a) Representation of the CuA and "CuZ" centres of the *Paracoccus denitrificans* N_2OR with its ligands. In the CuA centre, the copper atoms are numbered 1–2. In the "CuZ centre", the copper atoms are numbered between I and IV. The "CuZ centre" is as CuZ*(4Cu1S) in the [$1Cu^{2+}:3Cu^{1+}$] redox state and it contains one sulphur atom named S_1 and a water/hydroxyl atom at the edge of Cu_I–Cu_{IV}. (b) Representation of the CuA and "CuZ" centres of the *Pseudomonas stutzeri* N_2OR with its ligands. In the CuA centre, the copper atoms are numbered 1–2. In the "CuZ centre", the copper atoms are numbered between I and IV. The "CuZ centre" is as CuZ(4Cu2S) in the [$2Cu^{2+}:2Cu^{1+}$] redox state and contains two sulphur atoms named S_1 and S_2. In the CuA centre, His_{583} is not bound to the CuA_1 atom. Figures were prepared with Pymol using Protein Data Bank identifier 1FWX in (a) and Protein Data Bank identifier 3SBP in (b).

An oxygen atom from either a water molecule or a hydroxyl ion was modelled binding to Cu_{IV} in the cases of *Ma. hydrocarbonoclasticus* and *Pa. denitrificans* N_2ORs,[54] while in the case of the enzyme isolated from *Ac. cycloclastes*, two oxygen atoms, attributed to a water molecule and a hydroxyl ion, were identified bridging the Cu_I and Cu_{IV} atoms of the "CuZ centre".[58] Nevertheless, it is important to point out that the exact nature of this edge ligand was only elucidated from spectroscopic and computational studies (*vide infra* Section 7.4), as the crystallographic studies only highlighted the presence of an oxygen atom.

Note that these structures differ in the redox state of the CuA centre, which was reduced in the first two and oxidised in the latter structure, while the "CuZ centre" was in the same form as CuZ*(4Cu1S) in the $[1Cu^{2+}:3Cu^{1+}]$ redox state. Moreover, the structure of the "CuZ centre" in the presence of the inhibitor iodide was obtained for *Ac. cycloclastes* N_2OR,[58] which indicated that this atom was bound between the Cu_I and Cu_{IV} atoms, strengthening the proposal of this site being the substrate binding site.

More recently, a structure of N_2OR isolated from *Ps. stutzeri* was determined at 1.7 and 2.1 Å,[59] in which the "CuZ centre" was in a different form and redox state. In these structures, a different bridging atom between the Cu_I and Cu_{IV} atoms of the "CuZ centre" was identified and was modelled to be a sulphur atom (see Figure 7.2b).[59] This form of "CuZ centre" was named CuZ(4Cu2S), while the previous form was CuZ*(4Cu1S). As will be described in Section 7.4, this form of "CuZ centre" can be easily reduced/oxidised, and in this X-ray structure, it was in the $[2Cu^{2+}:2Cu^{1+}]$ redox state. Besides this difference, a change in the coordination sphere of CuA_1 was observed, with the CuA_1 atom losing one ligand (see Figure 7.2b), as the imidazole ring of His_{583} (equivalent to His_{526} in *Pa. denitrificans*) is rotated away from this copper atom, which might have implications for electron entry, as will be discussed in Chapter 11 of this book. The crystals of *Ps. stutzeri* N_2OR in the fully oxidised state were pressurised with the substrate nitrous oxide, and it was observed that the imidazole ring of His_{583} rotated back, coordinating CuA_1, and the substrate molecule was not coordinated directly to the "CuZ centre", but was located between the CuA and "CuZ" centres (*vide infra* Section 7.5).[59]

Recently, the structure of *Shewanella* (*Sh.*) *denitrificans* N_2OR was solved in the apo-form in order to understand the role of the calcium ion that is conserved in all of the known N_2OR structures and is present in each monomer.[22] Previously, it was postulated that this calcium ion could have a role in maintaining the functional N_2OR dimer.[57] However, these recent studies revealed that this is not the case; in fact, in the absence of calcium ions, N_2OR is a dimer, but its two domains stand apart, which might be crucial for the maturation machinery to be able to assemble the CuA and "CuZ" centres.[22] Moreover, in the presence of calcium ions, the arrangement of the residues that coordinate the CuA and "CuZ" centres in the apo-N_2OR is similar to the one found in the holo-N_2OR. This is an indication that the geometry of the copper sites is imposed by the polypeptide, as occurs in most cupredoxins,[60,61] and is known as the entatic or rack-induced state hypothesis by Vallee and Williams,[62,63] which is proposed to minimise the reorganisation energy of the electron transfer process and enhance the electron transfer rates. One other feature that was observed in the reconstituted CuA centre of *Sh. denitrificans* apo-N_2OR was that the imidazole ring of the equivalent His_{526} in *Pa. denitrificans*, which coordinates CuA_1 atom, is flipped away, as is observed in *Ps. stutzeri* N_2OR.[22]

7.4 Spectroscopic Properties

In 1972, a copper-containing protein was purified from the cell extracts of *Alcaligenes (Al.) faecalis*,[64] which was later shown to have a similar visible spectrum to an enzyme from *Ps. stutzeri* that was for the first time reported to have N_2OR activity *in vitro* using reduced methyl viologen as an electron donor.[65] Over the years, N_2OR has been isolated from different organisms, such as *Rhodopseudomonas sphaeroides* f. sp. *denitrificans*,[66] *Pa. denitrificans*,[67] *Ac. cycloclastes*,[68] *Ps. aeruginosa*,[69] *Flexibacter canadensis*,[70] *Pa. pantotrophus*,[71] *Thiobacillus denitrificans*,[72] *Ac. xyloxidans*,[73] *Ma. hydrocarbonoclasticus*[74] and *Hyphomicrobium denitrificans*,[75] with different forms of the enzyme being reported.

These isolated forms of N_2OR differ in terms of the oxidation states of the CuA and "CuZ" centres. Indeed, N_2OR can be isolated with the CuA centre in either the oxidised or the reduced state, with the "CuZ centre" being either mainly CuZ(4Cu2S) or CuZ*(4Cu1S) (see Table 7.1). In order to isolate N_2OR with the "CuZ centre" mainly as CuZ(4Cu2S), the purification procedure has to be performed under anoxic conditions,[76] or under oxic conditions but in a fast procedure without freezing the intermediate fractions,[77] as

Table 7.1 Different forms of nitrous oxide reductase that can be isolated or prepared *in vitro*, indicating their old denomination, the redox state of each copper centre and the form of the "CuZ centre".

Nitrous oxide reductase name (old denomination)	CuA centre oxidation state	"CuZ centre" oxidation state	"CuZ centre" form	Isolation/ preparation
Purple, fully oxidised	$[Cu^{+1.5}:Cu^{1.5+}]$	$[2Cu^{2+}:2Cu^{1+}]$	CuZ(4Cu2S)	Anoxic
Purple, CuA_{red}	$[Cu^{+1}:Cu^{1+}]$	$[2Cu^{2+}:2Cu^{1+}]$	CuZ(4Cu2S)	Anoxic
Blue, fully red	$[Cu^{+1}:Cu^{1+}]$	$[1Cu^{2+}:3Cu^{1+}]$	CuZ(4Cu2S)	Anoxic
Pink/purple	$[Cu^{+1.5}:Cu^{1.5+}]$	$[1Cu^{2+}:3Cu^{1+}]$	CuZ*(4Cu1S)	Oxic, anoxic low pH growth
Blue, CuA_{red}	$[Cu^{+1}:Cu^{1+}]$	$[1Cu^{2+}:3Cu^{1+}]$	CuZ*(4Cu1S)	Oxic, anoxic low pH growth
Fully red	$[Cu^{+1}:Cu^{1+}]$	$[4Cu^{1+}]$	CuZ*(4Cu1S)	Very slow reduction

oxygen was shown to modify the "CuZ centre" as CuZ(4Cu2S), turning it into CuZ*(4Cu1S).[76] The enzyme with the "CuZ centre" mainly as CuZ*(4Cu1S) has been isolated under anoxic conditions from cells grown at a constant moderately acidic pH (unpublished data from the authors, Carreira *et al.*), and more usually when the purification procedure is performed under oxic conditions.[74,76,78] This form has gained physiological relevance since it was isolated from mutant strains in the accessory proteins, such as from the *Pa. denitrificans nosXnirX* double knockout mutant,[50] and also from *Ps. stutzeri* mutant strains that only express part of the *nosR* gene.[38]

In the early literature, the enzyme in its different oxidation states was categorised by its colour in solution as purple, pink and blue, which corresponded to different oxidation states of the CuA centre and to different forms and oxidation states of the "CuZ centre". In fact, this denomination is not accurate and generated some confusion, and thus recently the denomination of CuZ(4Cu2S) and CuZ*(4Cu1S) with the indication of the redox state of the "CuZ" and CuA centres has been adopted (see Table 7.1).

Nonetheless, N_2OR cannot be isolated with the "CuZ centre" in a single form, either as CuZ(4Cu2S) or CuZ*(4Cu1S), and is usually purified with its "CuZ centre" as a mixture of the two,[76,77,79] even when the enzyme is isolated under anoxic conditions. N_2OR samples have been reported to have a "CuZ centre" with a percentage of CuZ(4Cu2S) ranging from 5% to 80%.

The presence of a mixture of CuZ/CuZ* centres in N_2OR samples made the analysis of the spectroscopic and kinetic properties of the "CuZ centre"—and, as a consequence, of N_2OR—more difficult. These properties, as well as the ones of the CuA centre, will be discussed in the next section, treating each centre separately for simplicity, although the spectroscopic and kinetic properties of the enzyme are the combination of those of the two centres and depend on the ratio of CuZ : CuZ*.

Since the "CuZ centre" is a tetranuclear copper centre and the CuA centre is a binuclear copper cluster, several oxidation states could be envisioned for N_2OR. Table 7.1 lists the different forms of the enzyme that have been isolated or prepared *in vitro*, indicating the oxidation states of both copper centres together with the type of "CuZ centre" and the old denomination of these forms of the enzyme.

7.4.1 The CuA Centre

The CuA centre is a binuclear copper centre that is located, as mentioned in Section 7.3, in the C-terminal domain of N_2OR. The residues that coordinate this copper centre were identified earlier on through the analysis of the first primary sequences of this enzyme and their comparison with those of cytochrome *c* oxidase.[34,80,81] Later, these residues were confirmed by site-directed mutagenesis[82] and in the X-ray crystallographic structures of N_2OR.

The spectroscopic properties of the N_2OR CuA centre are similar to those of the same centre from cytochrome *c* oxidase, with maximum absorption bands at 480 nm ($\approx$4000 M^{-1} cm^{-1}), 540 nm ($\approx$4000 M^{-1} cm^{-1}) and 780 nm

(≈3000 M^{-1} cm^{-1}) (see Figure 7.3a[iii] and Table 7.2).[53,83,84] In fact, this similarity was established by comparison of the room temperature and low-temperature MCD spectrum of N_2OR and cytochrome *c* oxidase.[43,85] In addition, its characteristic X-band electron paramagnetic resonance (EPR) spectrum in the oxidised form, with an axial signal with $g_{\parallel} = 2.18$ and $g_{\perp} \approx 2.02$ with a seven-line hyperfine coupling ($A_{\parallel}$ = 3.8 mT) (see Figure 7.3b[iii]) also pointed out that CuA was a binuclear centre.[42,78,82] Further analysis of the EPR spectrum at different frequencies led to the proposal that there is one unpaired electron (S = 1/2) shared between two copper nuclei (I = 3/2) that are considered

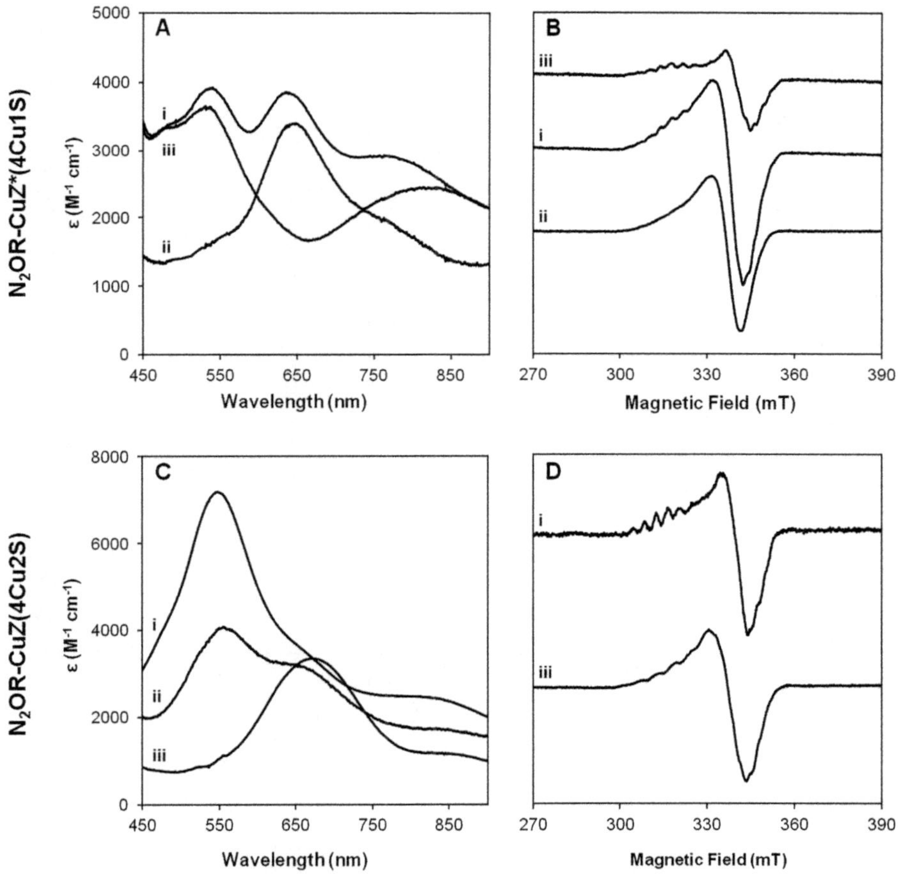

Figure 7.3 Visible and electron paramagnetic resonance (EPR) spectra of nitrous oxide reductase (N_2OR), mainly with the "CuZ centre" as CuZ*(4Cu1S) (a and b) or as CuZ(4Cu2S) (c and d). (a) and (b) show the visible and EPR spectra, respectively, of N_2OR-CuZ*(4Cu1S) in the fully oxidised form (i) and dithionite-reduced form (ii) and the spectrum of the CuA centre (iii). (c) and (d) show the visible and EPR spectra, respectively, of N_2OR-CuZ(4Cu2S) in the fully oxidised form (i), ascorbate-reduced form (ii) and dithionite-reduced form (iii). The extinction coefficients were determined using the concentration of the N_2OR monomer.

Table 7.2 Summary of "CuZ centre" properties. The different oxidation states of the four copper atoms in the "CuZ centre" are presented for each of the CuZ centre forms, together with their spectroscopic and kinetic properties.[a]

Oxidation state	Name[b]	Cu_I–Cu_{IV} edge	Spin state	Turnover number	Visible absorption[c]	EPR	Reference
$[4Cu^{2+}S]^{6+}$	N.O.	—	$S = 0, 1, 2$	—	—	—	
$[3Cu^{2+}:1Cu^{1+}:S]^{5+}$	N.O.	—	$S = 1/2, 3/2$	—	—	—	
$[2Cu^{2+}:2Cu^{1+}:S]^{4+}$	CuZ_{ox}	Bridging S	$S = 0$	None	545 mn (≈5000 M^{-1} cm^{-1})	Silent	76, 78, 93, 99
$[1Cu^{2+}:3Cu^{1+}:S]^{3+}$	CuZ_{red}	Bridging SH^-	$S = 1/2$	0.6 h^{-1}	670–685 nm (≈4500 M^{-1} cm^{-1})	$g_\parallel = 2.150$, $g_\perp = 2.035$, $A_\parallel = 5.6$ mT[d]	67, 75, 77–79, 93, 94, 99
	CuZ*	OH^- bridging Cu_I–Cu_{IV}		None	640 nm (≈4000 $M^{-1}cm^{-1}$)	$g_\parallel = 2.160$, $g_\perp = 2.040$, $A_\parallel = 6.1$ mT/$A_\parallel = 2.4$ mT[e]	55, 74, 76, 85, 93–95
	CuZ^0[f]	H_2O bound to Cu_I/Cu_{IV}		>200 s^{-1}	680 nm (≈4000 M^{-1} cm^{-1})	$g_\parallel = 2.160$, $g_\perp = 2.037$	90
$[4Cu^{1+}:S]^{2+}$	CuZ*	Empty/H_2O	$S = 0$	>200 s^{-1}	No bands	Silent	78, 79, 98–100

[a] N.O.: not observed yet.
[b] CuZ is CuZ(4Cu2S) and CuZ* is CuZ*(4Cu1S).
[c] The molar extinction coefficient is given by the concentration of the nitrous oxide reductase monomer.
[d] Consider three identical hyperfine coupling constants.
[e] With a 5 : 2 ratio.
[f] CuZ^0 is an intermediate of the CuZ*(4Cu1S) catalytic cycle.

equivalent,[83,86,87] which explains the intense and well-resolved hyperfine coupling lines. Moreover, the small hyperfine coupling constant indicates that there is high covalent delocalisation of the unpaired spin onto the cysteine thiolate ligands.[88] Thus, the CuA centre can be represented as $[1Cu^{1.5+}:1Cu^{1.5+}]$ in the oxidised mixed-valence state.

The visible spectrum of the CuA centre in the mixed-valence state is characterised by Scys → Cu charge transfer bands and intervalence $\psi \rightarrow \psi^*$ charge transfer transitions associated with the mixed-valence nature of the CuA centre,[53,83,86,89] which disappear upon reduction, since in this redox state of $[1Cu^{+1}:1Cu^{+1}]$, both copper atoms of the CuA centre have a $3d^{10}$ configuration, and so the CuA centre is diamagnetic ($S = 0$) and colourless, not contributing to either the visible or the EPR spectra of N_2OR.[59,74,76]

The reduction potential of the CuA centre has been determined by potentiometric titration to be +260 mV at pH 7.5 for *Pa. pantotrophus* N_2OR,[76] and a similar value was obtained for other N_2ORs isolated from *Ma. hydrocarbonoclasticus*[74,90] and *Ps. stutzeri*.[78]

Although the CuA centre was also an object of intense study using other spectroscopic techniques, such as X-ray absorption spectroscopy and resonance Raman, which also contributed to elucidating its electronic features,[88,91] this centre still needs further exploration. In fact, the recent X-ray structure of *Ps. stutzeri* N_2OR showed that one of the copper atoms of the CuA centre, CuA_1, lost the coordinating histidine ligand (*vide supra*, Section 7.3), but this had only a small impact on the spectroscopic properties of the CuA centre, contrary to what would be expected given that removal of a strongly bound equatorial copper ligand should perturb the spin distribution of this centre.[92]

7.4.2 The "CuZ Centre"

The nature of this centre was only unravelled after the X-ray structure of a N_2OR was solved (Section 7.3),[51,54] showing that it is a tetranuclear copper centre coordinated only by imidazole rings from conserved histidine residues. In fact, mostly due to the usually low Cu:protein ratios that were obtained for the first enzyme preparations,[67,78] this centre was considered to be a binuclear centre in the earlier spectroscopic studies.[85] Nevertheless, the presence of a sulphur moiety was always postulated and required to explain the different spectroscopic data that were gathered, especially the visible absorption band at around 660 nm that did not disappear upon reduction with dithionite and was attributed to a ligand-to-metal charge transfer transition.[85] Later, this sulphur was identified as an inorganic sulphur atom,[55] and the structural data revealed that it was in fact a bridging sulphur atom between the four copper atoms. The presence of a second sulphur atom was only established after the structure determination of *Ps. stutzeri* N_2OR.[59] This fact may be explained by considering that the "CuZ centre" might be lost during the purification and thus full CuZ occupancy is not attained (lower copper content than expected), the sulphur determination is not very accurate and there is heterogeneity of N_2OR samples (mixture of CuZ/CuZ*).

As mentioned, the "CuZ centre" in N_2OR can exist as either CuZ(4Cu2S) or CuZ*(4Cu1S), which differ in their redox, spectroscopic and catalytic properties, and this will be discussed in the next sections.

7.4.2.1 *CuZ*(4Cu1S)*

The visible spectrum of N_2OR isolated under oxic conditions is characterised by a strong absorption band at 640 nm (≈4000 M^{-1} cm^{-1}) (see Figure 7.3a[ii]) after reduction of the CuA centre with ascorbate or dithionite.[55,74,76,85,93] The EPR spectra of this sample have a broad axial signal (with $g_{\parallel}$ = 2.16 and $g_{\perp}$ ≈ 2.04), with a not well-resolved odd number of hyperfine split lines in the parallel region (see Figure 7.3b[ii]).[74,85,94] Moreover, earlier studies clearly indicated that this form of "CuZ centre" could not be easily reduced nor oxidised.[74,76] These features were attributed to the presence of CuZ*(4Cu1S).

Considering that CuZ*(4Cu1S) could not be further reduced and that it presents EPR and visible spectra, it was considered that at least one of the copper atoms had to be oxidised. The determination of the oxidation state of the four copper atoms in CuZ* was accomplished by Cu K-edge X-ray absorption spectroscopy, which can distinguish between [$3Cu^{2+}$: $1Cu^{1+}$] and [$1Cu^{2+}$: $3Cu^{1+}$] redox states, with the later model better fitting the experimental data.[94,95] Thus, CuZ*(4Cu1S) was proposed to have only one oxidised copper atom, with the other three being reduced. In addition, complementary studies using variable-temperature and variable-field MCD determined the spin state of the CuZ*(4Cu1S) ground state to be $S_{total} = 1/2$.[94,95]

The axial signal with $g = 2.16 > g \approx 2.04 > 2.0$ indicates that this spin resides in a Cu d_{x2-y2} orbital.[96] Therefore, in order to determine the spin distribution of that single spin over the "CuZ centre", the Q-band and X-band EPR spectra were analysed.[94] In the first analysis of these EPR spectra, it was considered that in fact there were two oxidised copper atoms with a spin density ratio of 5 : 2 that corresponded to the ratio of the two hyperfine coupling constants ($A_{\parallel}$ = 6.1 mT for the more intense and $A_{\parallel}$ = 2.4 mT for the other; pH 7.0) (see Table 7.2).[94,95] The identification of these two copper atoms of the "CuZ centre" was possible by DFT calculations using the CuZ*(4Cu1S) structure determined for the same organism (*Ma. hydrocarbonoclasticus* N_2OR solved at 2.4 Å; in these EPR samples, the "CuZ centre" was close to 100% CuZ*[4Cu1S]) and considering different molecules in the edge of Cu_I–Cu_{IV}.[94,95] In these studies, it was clear that the dominant oxidised copper atom of CuZ*(4Cu1S) is Cu_I (with higher spin density),[94,95,97] a fact that was attributed to its higher coordination number (four) when compared with the other copper atoms. Two models were postulated: in one, the spin density would be distributed between Cu_I and Cu_{II}, with a water ligand bridging Cu_I–Cu_{IV}; and in the other, the spin density would be distributed between Cu_I and Cu_{IV}, with a hydroxide bridging Cu_I and Cu_{IV}.[94,95] This last hypothesis was later supported with the interpretation of the X-band and Q-band EPR spectra together with analysis of the resonance Raman spectra at different pHs.[98] DFT calculations taking into account this experimental data using the *Ma. hydrocarbonoclasticus*

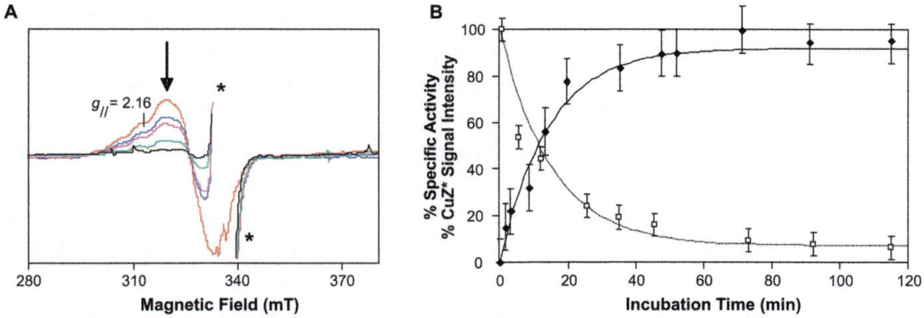

Figure 7.4 Activation of CuZ*(4Cu1S). The reductive activation of CuZ*(4Cu1S) during prolonged incubation with reduced methyl viologen was followed by the decrease in the electron paramagnetic resonance (EPR) signal of CuZ*(4Cu1S) in the $[1Cu^{2+}:3Cu^{1+}]$ oxidation state (a), with a concomitant increase in specific activity of nitrous oxide reductase (N_2OR) (b, black diamonds). The intensity of the EPR signal (open squares) and specific activity of *Marinobacter hydrocarbonoclasticus* N_2OR (black diamonds) were fitted with a rate constant of $k = 0.07$ min^{-1}.[103] The maximum specific activity of this sample was 190 µmol N_2O min^{-1} mg^{-1}, which corresponds to a CuZ*(4Cu1S) percentage close to 90%. *Methyl viologen radical.

N_2OR "CuZ centre" structure provided a description of the ground-state wave function of the CuZ*(4Cu1S), which is partially delocalised, with Cu_I being the dominantly oxidised copper (26%), while Cu_{IV} was the second copper that has a significantly spin density (13%), with the bridging sulphur atom having 29%.[95,98,99] There is also some spin delocalisation over the other two copper atoms—Cu_{II} (9%) and Cu_{III} (4%)[99]—with hyperfine coupling not resolved in the EPR spectra, which contributes to the broadening of this spectrum (see Figure 7.3b[ii] compared to Figure 7.3d[iii], *vide infra*).[97] In this model, a lysine residue (Lys_{397} according to the *Ma. hydrocarbonoclasticus* primary sequence) close to the "CuZ centre" is hydrogen bonded to the hydroxide bridging ligand at lower pH (*vide infra*, Section 7.5.1).[98]

Although several oxidation states could be foreseen for this form of "CuZ centre" (see Table 7.2), CuZ*(4Cu1S) cannot be easily reduced nor oxidised, but can be reduced to $[4Cu^{1+}]$ after prolonged incubation with reduced methyl or benzyl viologen.[79,100,101] Such a reduction can be followed by the loss of the EPR signal (see Figure 7.4a)[100] due to the formation of the diamagnetic $S = 0$ species (as expected for a $3d^{10}$ system), which is also optically silent. The reduction potential of this reaction has not yet been established, as it is an irreversible process under the experimental conditions used.[90]

7.4.2.2 CuZ(4Cu2S)

N_2OR can be isolated with the "CuZ centre" mainly as CuZ(4Cu2S) when its purification procedure is anoxic. In this form, the visible spectrum of N_2OR in the fully oxidised state is characterised by a strong absorption at 550 nm, with a shoulder on each side (480 nm and 635 nm) and another absorption band at around 800 nm (see Figure 7.3c[i]). Upon reduction of the CuA centre, the absorption bands corresponding to CuZ(4Cu2S) can be identified as a strong absorption at 545 nm (≈5000 M^{-1} cm^{-1}) with a weak, lower-energy shoulder at 635 nm (see Figure 7.3c[ii]),[76,93,99] which have a S → Cu charge transfer character. These transitions could arise either from the μ_4S^{2-} or μ_2S ligand, between Cu_I and Cu_{IV}, but DFT calculations combined with analysis of the vibration modes in the resonance Raman spectrum predict that these transitions come mainly from the μ_4S^{2-} bridging ligand.[99]

This form of "CuZ centre" is in the [$2Cu^{2+}$:$2Cu^{1+}$] redox state and can be reversibly reduced to [$1Cu^{2+}$:$3Cu^{1+}$] with a midpoint potential of 60 mV at pH 7.5.[76] This reduction can be followed by visible spectroscopy through the appearance of a 670–685 nm (≈4500 M^{-1} cm^{-1}) absorption band[76,78,99] that characterises CuZ(4Cu2S) in the [$1Cu^{2+}$:$3Cu^{1+}$] redox state (see Figure 7.3c[iii]).

In terms of spin state, the reduced form of CuZ(4Cu2S) is S = 1/2, as is CuZ*(4Cu1S), while the oxidised [$2Cu^{2+}$:$2Cu^{1+}$] form was shown to be S = 0 by MCD[93] and does not contribute to the EPR spectrum of N_2OR (see Figure 7.3d[ii]), which has the characteristic seven hyperfine coupling lines on the g parallel region from the CuA centre.[99,102] Nevertheless, since there are no reports to date of N_2OR with a homogeneous form of "CuZ centre", there is always a contribution from CuZ*(4Cu1S), S = 1/2, to the EPR spectrum of the oxidised N_2OR that is isolated under anoxic conditions.[77,79,99] On the other hand, CuZ(4Cu2S) cannot be further reduced to the [$4Cu^{1+}$] redox state, an advantage that has been used to study the spectroscopic properties of CuZ(4Cu2S) without the contribution of CuZ*(4Cu1S).[79,99] Using these experimental conditions, it was possible to elucidate the electronic features of CuZ(4Cu2S) in the [$1Cu^{2+}$:$3Cu^{1+}$] redox state.[99] The EPR spectrum of the CuZ(4Cu2S) [$1Cu^{2+}$:$3Cu^{1+}$] redox state has an axial signal with $g_\parallel > g_\perp > 2.0$ and five evenly spaced hyperfine lines in the g parallel region (see Figure 7.3d[iii]).[99] Similarly to what was determined for CuZ*(4Cu1S), the axial nature of the spectrum indicates that the unpaired electron spin is dominantly in a d_{x2-y2} orbital on each Cu atom, even if delocalised over several of the copper nuclei of the "CuZ centre". By simulating the X-band, X-band second derivative and Q-band EPR spectra, $g_\parallel = 2.15$ and $g_\perp = 2.04$ can be estimated, with three identical hyperfine coupling constants of $A_\parallel = 5.6$ mT.[99] These values are similar to those of CuZ*(4Cu1S),[98] and also to those of the *Pa. pantotrophus* "CuZ centre" mainly as CuZ(4Cu2S),[102] but the spin density is now distributed mainly over three of the copper atoms, Cu_I (17%), Cu_{II} (11%) and Cu_{IV} (10%), with the remaining spin density being over μ_4-sulphide (34%), μ_2-SH^- (16%) and Cu_{III} (6%).[99]

The protonation state of the μ_2-sulphur ligand was studied in the two oxidation states, [$2Cu^{2+}$:$2Cu^{1+}$] and [$1Cu^{2+}$:$3Cu^{1+}$]. In the oxidised state, CuZ(4Cu2S) was shown to have a μ_2-sulphide ligand since the two vibrations at 350 and 405 cm^{-1}, assigned to Cu–S stretch mode, have ^{34}S isotope shifts of −5.6 and −5.8 cm^{-1}, respectively, and do not present any deuterium shift between pD6 and pD10.[99] Therefore, the only pK_a of the edge ligand that is consistent with the data is one that is lower than 5.5.

In the reduced state, the Cu_I–Cu_{IV} edge of CuZ(4Cu2S) is a SH^- ligand, as the two vibration modes observed in the resonance Raman spectrum of this state at 426 cm^{-1} and 461 cm^{-1} have a H/D isotope shift, and the DFT calculations using this model also well explains the spin density distribution used to simulate the EPR spectrum. The pK_a of this bridging thiolate ligand was estimated to be 11–12 by computational methods, as the resonance Raman lines do not change up to a pH of 10.0.[99]

It is important to mention that the electron spin delocalisation observed in both the CuA and "CuZ" centres contributes to a low site reorganisation energy during the redox processes, which leads to a rapid and efficient electron transfer for the CuA centre and stabilisation of the oxidised form of the "CuZ centre" after substrate reduction, which is required for a rapid turnover.[92,95]

7.5 Kinetic Properties of N_2OR

The specific activity of N_2OR has been measured for the enzyme isolated under oxic and anoxic conditions and with the "CuZ centre" in different redox states. The *in vitro* catalytic activity has been mainly measured using reduced methyl or benzyl viologen as an electron donor and monitoring the amount of nitrous oxide consumed over time by gas chromatography,[65,78,103] or the absorption decay of the reduced artificial electron donor.[78,104,105]

N_2OR with its "CuZ centre" mainly in one of the forms described above—CuZ*(4Cu1S) or CuZ(4Cu2S)—does not present a high catalytic activity after isolation, which is not consistent with the high rate of nitrous oxide reduction that is observed by whole cells growing under denitrifying conditions (48–72 μmol N_2O min^{-1} mg^{-1}, value determined by measuring the consumption of nitrous oxide and an immunochemical estimation of the N_2OR concentration in the cell).[78] In fact, the specific activity of N_2OR with the "CuZ centre" as either CuZ or CuZ*, in different oxidation states, estimated using reduced methyl/benzyl viologen as an electron donor under anaerobic conditions, is smaller, being between 1 and 10 μmol N_2O min^{-1} mg^{-1},[76,77] than the one expected *in vivo* (48–72 μmol N_2O min^{-1} mg^{-1}).[78] Therefore, it can be postulated that the enzyme is isolated in an unready state and it is foreseen that it requires an activation step. The catalytic activity of each form of "CuZ centre" and its activation will be discussed, as well as the substrate binding site and the catalytic mechanism.

The physiological electron donor has been established for some N_2ORs isolated from *Pa. pantotrophus*,[71,106] *Ma. hydrocarbonoclasticus*,[104] *Ac. cycloclastes*[107] and *Wo. succinogenes*[12–14] as being a small electron shuttle protein, either a *c*-type cytochrome or a cupredoxin—pseudoazurin or azurin. The specific activity attained using these reduced electron donors is still small, even if N_2OR has been activated (*vide infra*); thus, it can be speculated that this enzyme requires other accessory factors (either NosR or a NosX homologue) to attain maximum activity.

The pH effect on the activity of N_2OR has been studied mainly for the enzyme that is not activated and using an artificial electron donor. The optimum pH activity was determined to be between pH 8.0 and 9.5 for several N_2ORs.[71,75,78,104] In the case of the *Ma. hydrocarbonoclasticus* N_2OR, the pH profile of the activity has a pK_a of 6.6 when methyl viologen is used as an electron donor, with high activity being maintained at high pH values.[104] By contrast, using the physiological electron donor cytochrome c_{552}, the pH profile of the activity of the *Ma. hydrocarbonoclasticus* N_2OR is different, with the activity decreasing at higher pHs, with a pK_a of 8.3.[104] However, in the case of the *Ac. cycloclastes* N_2OR, a different profile is observed for the effect of pH on the activity, with two maxima being observed, which depend on the pH at which the enzyme is activated.[108] These results raised the question of whether the activation of the enzyme was not just due to a reduction of CuZ*(4Cu1S) to the [$4Cu^{1+}$] redox state, but involved a protonation/deprotonation of a ligand or the transfer of two hydrides (*vide infra*).[108,109]

7.5.1 CuZ and CuZ*

In order to determine the catalytic activity of the *Ma. hydrocarbonoclasticus* N_2OR isolated under oxic conditions, in our laboratory, the assay was performed by adding a N_2O-saturated solution to a cuvette containing reduced methyl viologen and N_2OR.[74] During such assays, it was observed by serendipity that, depending on the period that the enzyme was incubated with reduced methyl viologen, different specific activities were attained, which increased with the incubation period.

In fact, it was later shown that this period corresponded to an activation step, during which CuZ*(4Cu1S) is being reduced from the [$1Cu^{2+}$:$3Cu^{1+}$] to the fully reduced [$4Cu^{1+}$] state.[100] This activation was observed by the increase in specific activity that was concomitant with the disappearance of the EPR signal of CuZ*(4Cu1S) (with a rate constant of $k = 0.07$ min^{-1} for both) as the $S = 1/2$ species was decreasing, due to the formation of a $S = 0$ species—the fully reduced [$4Cu^{1+}$] state of CuZ*(4CuS1) (see Figure 7.4b).[100] This indicates that the catalytically active form of CuZ*(4Cu1S) is its fully reduced form. Later, this activation was also observed for the *Ac. cycloclastes* N_2OR through visible and EPR spectroscopies and nitrogen labelling experiments.[101,108]

The activation mechanism is still a poorly understood process, since it is a very slow reduction ($k = 0.07$ min^{-1}) that requires a very potent reducing agent, such as reduced methyl viologen ($E_m \approx -441$ mV).[110] Nevertheless, as Chan *et al.* point out,[101] activation with either dithionite ($E_m \approx -470$ mV),[111] which has a similar reducing power as methyl viologen, or benzyl viologen ($E_m \approx -360$ mV)[110] do not activate the enzyme to the same extent: a specific activity of 124 U mg^{-1} was obtained after incubation with reduced methyl viologen compared with 62 U mg^{-1} and 34 U mg^{-1} for reduced benzyl viologen and dithionite, respectively.[101] This may indicate that it is not just the reducing power, but the accessibility of the reducing agent to the "CuZ centre" (as intramolecular electron transfer does not seem to be thermodynamically favourable under these conditions) that is important in this process. Therefore, the different charge of dithionite and methyl viologen,[112,113] as well as the contamination of dithionite solution with bisulphite, may explain these results.[111] These authors have later reported that methyl viologen in fact binds to the "CuZ centre", and although it can reduce the "CuZ centre", its binding inhibits its catalytic activity.[107] However, these results have not been reproduced by other research groups.

This activation mechanism is associated with protonation/deprotonation events, since it was observed that the rate of reduction of CuZ*(4Cu1S) decreases with an increase in pH, with a pK_a of 9.0.[98] At pH 7.0, the rate of reduction was determined to be 0.18 min^{-1}, and it decreases to 0.034 min^{-1} at pH 10.5, a fact that was attributed to the tuning of the reduction potential of the "CuZ centre", which would be more positive when a highly conserved lysine residue near Cu_I–Cu_{IV} edge is protonated and hydrogen bonded to the bridging hydroxide, accelerating the reduction rate of CuZ*(4Cu1S) (Lys_{397} according to the *Ma. hydrocarbonoclasticus* primary sequence). This pH effect in the activation mechanism might also explain the pH profile of the catalytic activity of the *Ac. cycloclastes* N_2OR.[108]

While the catalytically relevant form of CuZ*(4Cu1S) is its fully reduced state, CuZ(4Cu2S) cannot be further reduced to that redox state,[59,79,99] and therefore enzyme preparations with a high content in CuZ(4Cu2S) were always reported to have a lower specific activity than those containing a high CuZ*(4Cu1S) content,[74,77] and the specific activity correlates well with the percentage of this form of "CuZ centre" present in the N_2OR samples.[79] This fact was an indication that CuZ(4Cu2S) may not be catalytically competent. In order to clarify whether CuZ(4Cu2S) could react with nitrous oxide and to determine its specific rate constant of nitrous oxide reduction, single-turnover experiments were performed, which demonstrated that the oxidised form of CuZ(4Cu2S)—[$2Cu^{2+}$: $2Cu^{1+}$]—does not react with the substrate,[79,106] but the reduced form—[$1Cu^{2+}$: $3Cu^{1+}$]—does, though at a very low rate of 0.6 h^{-1},[79] which cannot be considered relevant in the catalytic cycle of N_2OR that has a turnover number of 321 s^{-1}.[104]

Therefore, these results clearly demonstrate that only fully reduced CuZ*-(4Cu1S), for which the lower limit of the rate of reaction with nitrous oxide was determined to be 200 s^{-1},[79] can be part of the rapid turnover of this enzyme under steady-state conditions and is consistent with the expected N_2OR-specific activity *in vivo* of 48–72 μmol N_2O min^{-1} mg^{-1}.

These single-turnover experiments of either CuZ(4Cu2S) [$1Cu^{2+}$: $3Cu^{1+}$] or fully reduced CuZ*(4Cu1S) [$4Cu^{1+}$] with nitrous oxide showed that the reaction is a two-electron process, since the final visible, MCD and resonance Raman spectra show the presence of an oxidised CuA centre and CuZ(4Cu2S) [$2Cu^{2+}$: $2Cu^{1+}$] or CuZ*(4Cu1S) [$1Cu^{2+}$: $3Cu^{1+}$].[79,90] In fact, these experiments enabled the observation for the first time of one intermediate species of the N_2OR catalytic cycle (see Figure 7.5), named CuZ^0.[90]

An increase in N_2OR catalytic activity has also been observed when the enzyme, isolated under either anoxic or oxic conditions, is dialysed against

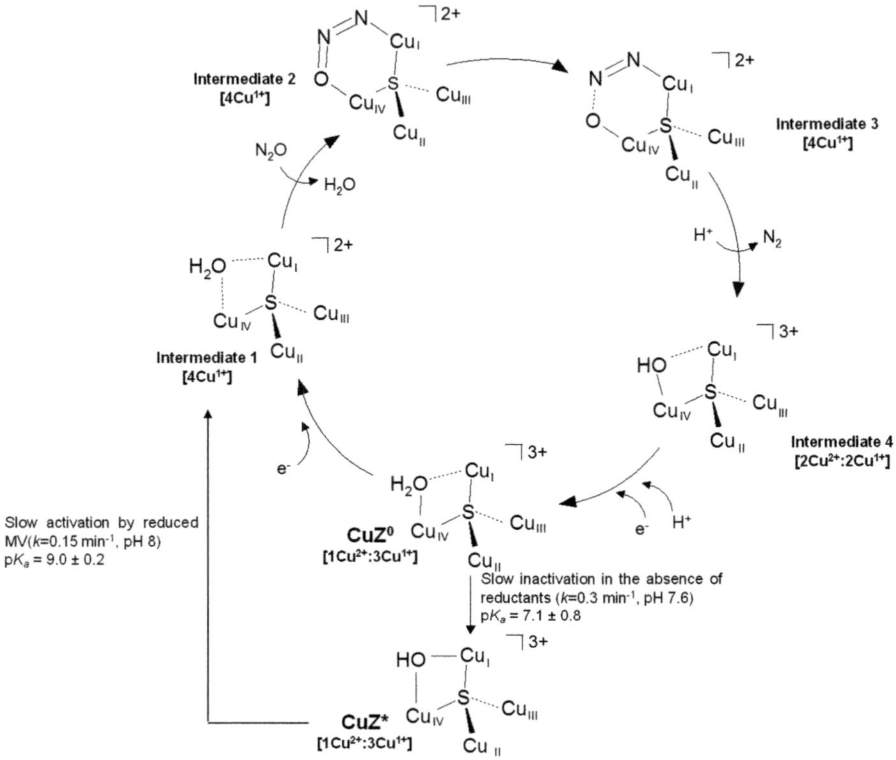

Figure 7.5 Schematic representation of the proposed catalytic cycle of nitrous oxide reductase with the "CuZ centre" as CuZ*(4Cu1S). The different intermediate species were proposed based on DFT calculations and on experimental observation and spectroscopic characterisation of CuZ^0.

a high-pH buffer.[69,71,78] The increase in activity seems to be higher for the enzyme isolated under anoxic conditions in the cases of *Ps. stutzeri*, *Ps. aeruginosa* and *Pa. pantotrophus* N_2ORs.[69,71,78] However, the molecular mechanism for this activation remains elusive, as no spectroscopic or structural studies have been reported for the enzyme itself after prolonged incubation at high pH values. Moreover, in the case of CuZ*(4Cu1S), only small changes have been reported due to pH (for pH 10).[98] Thus, further characterisation of the "CuZ centre" as CuZ(4Cu2S) after prolonged incubation at higher pHs (>10) is required in order to understand this mechanism and what would be its relevance for the *in vivo* activity and catalytic cycle of N_2OR.

7.5.2 CuZ^0

Single-turnover experiments of fully reduced CuZ*(4Cu1S) with nitrous oxide were monitored by visible and EPR spectroscopy.[90] These experiments showed a rapid (within 1 s) oxidation of the CuA centre through the increase of the characteristic absorption bands at 480, 540 and 800 nm and also the development of an absorption band at 680 nm, which had not been previously observed.[90] This novel absorption feature of N_2OR slowly decayed (with a $k = 0.3\ min^{-1}$), which was concomitant with the appearance of the characteristic CuZ*(4Cu1S) absorption band at 640 nm.[79,90]

The specific activity was measured during this period and showed that the intermediate species, characterised by the 680 nm ($\approx 4000\ M^{-1}\ cm^{-1}$) absorption band, has a high turnover number, similarly to the fully reduced CuZ*(4Cu1S).[90] Thus, this intermediate species, named CuZ^0, is part of the catalytic cycle of N_2OR (see Figure 7.5).

Although full spectroscopic and structural characterisation of this intermediate species in underway, its spectroscopic properties indicate that it is a form of "CuZ centre" in the [$1Cu^{2+}:3Cu^{1+}$] redox state, which can be easily reduced to [$4Cu^{1+}$] during catalysis. In fact, CuZ^0 has an axial EPR signal with a $g_{\parallel}$ of 2.16 and a $g_{\perp}$ of 2.037.[90]

The inactivation process (decay of CuZ^0 to CuZ*[4Cu1S]) is pH dependent, with a pK_a of 7.1 as estimated by electrochemical experiments.[90] This slow conversion of CuZ^0 to CuZ* can be attributed to a protonation/deprotonation process that may be followed by a structural rearrangement in the "CuZ centre" (see Figure 7.5). Such a process has been associated with the deprotonation of a water ligand bound to either Cu_I or Cu_{IV} in CuZ^0,[2,90] which would decrease the reduction potential of the "CuZ centre", with it being more difficult for the CuA centre to reduce it during catalysis. In fact, such a pK_a has been observed in the steady-state kinetics of N_2OR (pK_a of 8.3) when its physiological redox partner was used as an electron donor.[90,104] Moreover, an isotope effect on the catalytic activity of N_2OR is consistent with a solvent-exchangeable proton in the rate-limiting step of the catalytic cycle,[109] which supports the hypothesis of CuZ^0 having a water molecule bound either to Cu_I or Cu_{IV}.

Table 7.2 summarises the possible oxidation states of the "CuZ centre", which have been observed and which ligands are found at the Cu_I–Cu_{IV} edge, as well as their spin states, spectroscopic properties and turnover numbers.

7.5.3 Substrate Binding Site and Catalytic Cycle

N_2OR has been isolated with the "CuZ centre" mainly as CuZ(4Cu2S) or CuZ*(4Cu1S). Although it was recently shown that the only form of N_2OR that can attain high turnover numbers is that with CuZ*(4Cu1S) in its fully reduced [$4Cu^{1+}$] state, the awareness that this form occurs when CuZ(4Cu2S) is exposed to oxygen has led some research groups to regard this form as a degraded form of the enzyme that never occurs *in vivo* during catalysis. Another point that has been raised is the high reducing power that is required to activate N_2OR, which makes it difficult for it to take place in the periplasmic space of the Gram-negative bacteria where this enzyme is usually located. Even so, one might conceive that the fully reduced form of the enzyme could be attained after an activation that includes a μ_2-sulphur shift, similarly to what has been observed in other enzymes,[114] in an as-yet unknown process. This mechanism might involve the auxiliary proteins NosR and/or the NosX homologue, known to be essential to attain maximum nitrous oxide reduction in whole cells and thus forming with N_2OR a supramolecular complex like the one observed in *Ps. aeruginosa*.[40] During this activation process, the "CuZ centre" would also be reduced to its fully reduced state, enabling binding of nitrous oxide. Thus, the CuZ(4Cu2S) form of the "CuZ centre" would be regarded as a protective form of the enzyme, occurring when substrate or electrons are not available and catalysis cannot take place. The observation that there was a lag phase in the stoichiometric reduction of N_2OR, with CuA reduced and CuZ(4Cu2S) in the $[2Cu^{2+}:2Cu^{1+}]$ oxidation state, by reduced cytochrome *c*[106,109] might be evidence for this conformational change that needs to occur in CuZ(4Cu2S) *in vivo*. Formation of such a complex would also increase the reduction potential of the "CuZ centre" as CuZ(4Cu2S) from +60 mV to +150 mV.

One other question that has been raised regards the identification of the nitrous oxide binding site in N_2OR and, in particular, in the "CuZ centre". Since the first X-ray crystallographic studies, the substrate nitrous oxide has been proposed to bind between Cu_I and Cu_{IV},[54,58] where a water-derived molecule was identified.[56] In fact, DFT calculations using these "CuZ centre" coordinates (a CuZ*[4Cu1S] form) indicated that the lowest-energy structure for the binding of nitrous oxide is in a bent μ-1,3-N_2O bridging mode between Cu_I and Cu_{IV}, with a 139° N–N–O angle (see Figure 7.6), for the fully reduced $[4Cu^{1+}]$ form (-117 kJ mol^{-1}), while an end-on coordinating mode is preferred when CuZ*(4Cu1S) is in the $[1Cu^{2+}:3Cu^{1+}]$ redox state, in both cases with the terminal nitrogen atom of N_2O coordinating to Cu_I (-75 kJ mol^{-1}).[100,101]

The binding of nitrous oxide in a bent μ-1,3-N_2O bridging mode to the fully reduced CuZ*(4Cu1S) centre is efficient due to strong back-bonding from the Cu_I and Cu_{IV} atoms of the "CuZ centre", which makes the N_2O ligand an efficient electron acceptor.[100,115] In fact, the CuZ* → N_2O charge transfer and strong interaction of the bent μ-1,3-N_2O ligand with those two redox-active copper atoms provide a low activation energy pathway for the N–O cleavage process.[100,115] This charge transfer is reflected in the negative charge (-0.53 a.u.) of the bound N_2O ligand and in the elongation of the N–N and N–O bonds.[100,115] The N–O bond is thus weakened, facilitating its cleavage through the simultaneous transfer of two electrons from the reduced "CuZ centre", which can be seen as an efficient two-electron reducing agent. The loss of the two electrons by the reduction of nitrous oxide can be efficiently delocalised over the bridging μ_4-sulphide of the "CuZ centre", contributing to the low reorganisation energy of this process and to the stabilisation of the "CuZ centre" after substrate reduction, allowing rapid re-reduction of the catalytic centre during the catalytic cycle.[100,115]

The binding of nitrous oxide in a bent μ-1,3 bridging mode is also proposed to be stabilised by the hydrogen bonding between a neighbouring conserved lysine residue near the Cu_I–Cu_{IV} edge and a nitrous oxide O atom (Lys_{397} according to the *Ma. hydrocarbonoclasticus* primary sequence).[57,115]

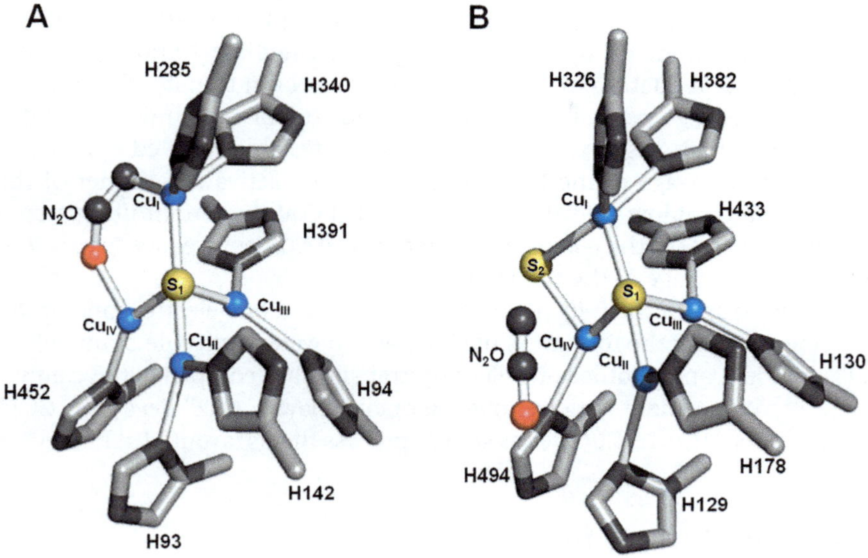

Figure 7.6 Nitrous oxide binding at the active site of nitrous oxide reductase (N_2OR). (a) Proposed binding structure of N_2O in a μ-1,3 bridging mode at the edge of Cu_I–Cu_{IV} based on structural, electronic and theoretical calculations for the "CuZ centre" as CuZ*(4Cu1S) in the [$4Cu^{1+}$] redox state,[103] and modelled on the *Achromobacter cycloclastes* N_2OR structure. (b) Substrate binding mode observed in a N_2O-pressurised crystal of the *Pseudomonas stutzeri* N_2OR with the "CuZ centre" as CuZ(4Cu2S) in the [$2Cu^{2+}$: $2Cu^{1+}$] redox state. Figures were prepared with Pymol using Protein Data Bank identifier 1FWX in (a) and Protein Data Bank identifier 3SBR in (b).

As mentioned before, CuZ(4Cu2S) in the [$1Cu^{2+}$: $3Cu^{1+}$] redox state is not competent in turnover, but reacts with nitrous oxide in a two-electron process that appears to involve one electron from the reduced "CuZ centre" and another from the CuA centre. This observation is consistent with the crystallographic studies of nitrous oxide-pressurised crystals of the *Ps. stutzeri* N_2OR isolated under anoxic conditions,[59] which showed the presence of a nitrous oxide molecule localised between the CuA and "CuZ" centres (see Figure 7.6). In fact, the observation of a nitrous oxide molecule trapped in the pressurised crystals is consistent with the slow rate of reduction observed in the single-turnover experiments (0.6 h^{-1}),[79] as otherwise other crystallographic methods would have needed to be employed in order to observe such an intermediate state.

The analysis of the intramolecular electron transfer routes for the reduction of the substrate, as well as the conformational changes that occur in the coordination sphere of the CuA centre, will be discussed in Chapter 11 of this book, and thus will not be presented here.

The proposed catalytic mechanism of the "CuZ centre" showing the binding of nitrous oxide to the fully reduced CuZ*(4Cu1S) centre is presented in

Figure 7.5. The fully reduced state (intermediate 1) binds nitrous oxide in a μ-1,3 bridging mode on the Cu_I–Cu_{IV} edge (intermediate 2 and 3), which is followed by its reduction in a two-electron/one-proton transfer that induces N–O bond cleavage, releasing N_2. The hydroxide remains bound to Cu_I (intermediate 4), which has a $[2Cu^{2+}:2Cu^{1+}]$ redox state. This coupled e^-/H^+ transfer with the cleavage of the N–O bond lowers the activation barrier of this step.[98,115] Thus, at low pH values, it is proposed that the rate-limiting step of the catalytic cycle is the electron transfer *via* the CuA centre (see Figure 7.5), and not the cleavage of the N–O bond.[115]

In order to attain the fully reduced state, there is a one-electron/one-proton reducing step with the formation of CuZ^0, with a water molecule bound either to Cu_I or Cu_{IV}, and another one-electron transfer that completes the catalytic cycle.[98,115] When there is not enough reducing power, CuZ^0 decays to CuZ*(4Cu1S) in the $[1Cu^{2+}:3Cu^{1+}]$ redox state, a process that is favoured at higher pH.

7.6 Concluding Remarks

During these last 45 years since a "typical" N_2OR was first isolated, many advances regarding its spectroscopic data, catalytic properties and structure have been made. These studies revealed that the enzyme can be isolated with its catalytic centre, the "CuZ centre", in two forms—CuZ(4Cu2S) or CuZ*(4Cu1S)—depending on the genetic background and purification procedure (anoxic or oxic) used. To date, it is clear that CuZ*(4Cu1S) can be activated to a form that is catalytically relevant and can explain the high rate at which the bacterial cells producing N_2OR reduce nitrous oxide under denitrifying conditions. Nevertheless, there is still a debate as to how this form is attained *in vivo*. It is possible to envisage that the isolated CuZ(4Cu2S) form of the "CuZ centre" is in fact a protected form that occurs in the absence of substrate or when not enough reducing power is available to complete the catalytic cycle. The activation mechanism would thus involve a sulphur shift *in vivo*, assisted by accessory proteins, for which the encoding genes are, in most cases, co-transcribed with *nosZ*, NosR and/or the NosX homologue, and this would also be responsible for maintaining the "CuZ centre" in a fully reduced state (which has been shown to have the lowest activation energy for binding nitrous oxide). These proteins are known to be essential for whole cells to be able to reduce nitrous oxide, and in fact, it was recently shown that they form a supramolecular complex together with the other denitrification enzymes.[40]

Besides these questions, the identification and characterisation of the other intermediate species in the catalytic cycle still need to be completed.

Acknowledgements

The authors thank Edward Solomon, Esther Pierce, Simone Dell'Acqua, Christian Cambillau and Oliver Einsle for the fruitful discussions and collaborative work that led to many advances and new points of view regarding

N_2OR. The authors would also like to thank Fundação para a Ciência e Tecnologia (FCT) for the financial support provided (PTDC/BIA-PRO/098882/2008 to SRP, PTDC/QUI-BIQ/116481/2010 and PTDC/BBB-BQB/0129/2014 to IM and SFRH/BD/87898/2012 scholarship to CC). This work was supported by the Unidade de Ciências Biomoleculares Aplicadas-UCIBIO, which is financed by national funds from FCT/MEC (UID/Multi/04378/2013) and co-financed by the ERDF under the PT2020 Partnership Agreement (POCI-01-0145-FEDER-007728). SRP is an IF fellow supported by FCT.

References

1. W. Zumft and H. Korner, in *Biology of the Nitrogen Cycle*, ed. H. Bothe, S. J. Ferguson and W. E. Newton, Elsevier, 2007, ch. 5, pp. 67–81.
2. S. R. Pauleta, S. Dell'Acqua and I. Moura, *Coord. Chem. Rev.*, 2013, **257**, 332.
3. K. Jones, in *Comprehensive Inorganic Chemistry*, ed. J. C. Bailar, H. J. Emelöus, R. Nyholm and A. F. Trottman-Dickenson, Pergamon Press, Oxford, 1975, vol. 2, ch. 19, pp. 147–388.
4. G. A. Vaughan, P. B. Rupert and G. L. Hillhouse, *J. Am. Chem. Soc.*, 1987, **109**, 5538.
5. W. G. Zumft, *Microbiol. Mol. Biol. Rev.*, 1997, **61**, 533.
6. L. H. Orellana, R. L. Rodriguez, S. Higgins, J. C. Chee-Sanford, R. A. Sanford, K. M. Ritalahti, F. E. Loffler and K. T. Konstantinidis, *MBio*, 2014, **5**, e01193.
7. R. A. Sanford, D. D. Wagner, Q. Wu, J. C. Chee-Sanford, S. H. Thomas, C. Cruz-Garcia, G. Rodriguez, A. Massol-Deya, K. K. Krishnani, K. M. Ritalahti, S. Nissen, K. T. Konstantinidis and F. E. Loffler, *Proc. Natl. Acad. Sci. U. S. A.*, 2012, **109**, 19709.
8. I. Verbaendert, S. Hoefman, P. Boeckx, N. Boon and P. De Vos, *FEMS Microbiol. Ecol.*, 2014, **89**, 162.
9. K. Chourey, S. Nissen, T. Vishnivetskaya, M. Shah, S. Pfiffner, R. L. Hettich and F. E. Loffler, *Proteomics*, 2013, **13**, 2921.
10. J. Simon, O. Einsle, P. M. Kroneck and W. G. Zumft, *FEBS Lett.*, 2004, **569**, 7.
11. M. Luckmann, D. Mania, M. Kern, L. R. Bakken, A. Frostegard and J. Simon, *Microbiology*, 2014, **160**, 1749.
12. S. Teraguchi and T. C. Hollocher, *J. Biol. Chem.*, 1989, **264**, 1972.
13. C. Zhang, A. M. Jones and T. C. Hollocher, *Biochem. Biophys. Res. Commun.*, 1992, **187**, 135.
14. C. S. Zhang and T. C. Hollocher, *Biochim. Biophys. Acta*, 1993, **1142**, 253.
15. C. S. Zhang, T. C. Hollocher, A. F. Kolodziej and W. H. Orme-Johnson, *J. Biol. Chem.*, 1991, **266**, 2199.
16. J. Simon and L. Hederstedt, *FEBS J.*, 2011, **278**, 4179.
17. X. Liu, C. Gao, A. Zhang, P. Jin, L. Wang and L. Feng, *FEMS Microbiol. Lett.*, 2008, **289**, 46.
18. C. M. Jones, A. Welsh, I. N. Throback, P. Dorsch, L. R. Bakken and S. Hallin, *FEMS Microbiol. Ecol.*, 2011, **76**, 541.

19. D. Mania, K. Heylen, R. J. van Spanning and A. Frostegard, *Environ. Microbiol.*, 2014, **16**, 3196.
20. D. Mania, K. Heylen, R. J. van Spanning and A. Frostegard, *Environ. Microbiol.*, 2016, DOI: 10.1111/1462-2920.13124.
21. K. Heylen and J. Keltjens, *Front. Microbiol.*, 2012, **3**, 371.
22. L. K. Schneider and O. Einsle, *Biochemistry*, 2016, **55**, 1433.
23. A. Dreusch, D. M. Burgisser, C. W. Heizmann and W. G. Zumft, *Biochim. Biophys. Acta*, 1997, **1319**, 311.
24. M. P. Heikkila, U. Honisch, P. Wunsch and W. G. Zumft, *J. Bacteriol.*, 2001, **183**, 1663.
25. T. Palmer, F. Sargent and B. C. Berks, *Trends Microbiol.*, 2005, **13**, 175.
26. B. C. Berks, T. Palmer and F. Sargent, *Adv. Microb. Physiol.*, 2003, **47**, 187.
27. H. Beinert, *Eur. J. Biochem.*, 1997, **245**, 521.
28. N. Minagawa and W. G. Zumft, *BioMetals*, 1988, **1**, 117.
29. B. P. Dash, M. Alles, F. A. Bundschuh, O. M. Richter and B. Ludwig, *Biochim. Biophys. Acta*, 2015, **1847**, 202.
30. H. Arai, M. Mizutani and Y. Igarashi, *Microbiology*, 2003, **149**, 29.
31. H. Cuypers, A. Viebrock-Sambale and W. G. Zumft, *J. Bacteriol.*, 1992, **174**, 5332.
32. H. Cuypers, J. Berghofer and W. G. Zumft, *Biochim. Biophys. Acta*, 1995, **1264**, 183.
33. U. Honisch and W. G. Zumft, *J. Bacteriol.*, 2003, **185**, 1895.
34. F. U. Hoeren, B. C. Berks, S. J. Ferguson and J. E. McCarthy, *Eur. J. Biochem.*, 1993, **218**, 49.
35. W. G. Zumft, *J. Mol. Microbiol. Biotechnol.*, 2005, **10**, 154.
36. P. Holloway, W. McCormick, R. J. Watson and Y. K. Chan, *J. Bacteriol.*, 1996, **178**, 1505.
37. K. U. Vollack and W. G. Zumft, *J. Bacteriol.*, 2001, **183**, 2516.
38. P. Wunsch and W. G. Zumft, *J. Bacteriol.*, 2005, **187**, 1992.
39. L. Velasco, S. Mesa, C. A. Xu, M. J. Delgado and E. J. Bedmar, *Antonie Van Leeuwenhoek*, 2004, **85**, 229.
40. J. M. Borrero-de Acuna, M. Rohde, J. Wissing, L. Jansch, M. Schobert, G. Molinari, K. N. Timmis, M. Jahn and D. Jahn, *J. Bacteriol.*, 2016, **198**, 1401.
41. P. Wunsch, M. Herb, H. Wieland, U. M. Schiek and W. G. Zumft, *J. Bacteriol.*, 2003, **185**, 887.
42. J. Riester, W. G. Zumft and P. M. Kroneck, *Eur. J. Biochem.*, 1989, **178**, 751.
43. D. M. Dooley, M. A. McGuirl, A. C. Rosenzweig, J. A. Landin, R. A. Scott, W. G. Zumft, F. Devlin and P. J. Stephens, *Inorg. Chem.*, 1991, **30**, 3006.
44. W. G. Zumft, A. Viebrock-Sambale and C. Braun, *Eur. J. Biochem.*, 1990, **192**, 591.
45. L. M. Taubner, M. A. McGuirl, D. M. Dooley and V. Copie, *Biochemistry*, 2006, **45**, 12240.
46. M. A. McGuirl, J. A. Bollinger, N. Cosper, R. A. Scott and D. M. Dooley, *J. Biol. Inorg. Chem.*, 2001, **6**, 189.

47. L. M. Taubner, M. A. McGuirl, D. M. Dooley and V. Copie, *J. Biomol. NMR*, 2004, **29**, 211.
48. J. M. Boyd, J. A. Endrizzi, T. L. Hamilton, M. R. Christopherson, D. W. Mulder, D. M. Downs and J. W. Peters, *J. Bacteriol.*, 2011, **193**, 887.
49. N. F. Saunders, J. J. Hornberg, W. N. Reijnders, H. V. Westerhoff, S. de Vries and R. J. van Spanning, *J. Bacteriol.*, 2000, **182**, 5211.
50. P. Wunsch, H. Korner, F. Neese, R. J. van Spanning, P. M. Kroneck and W. G. Zumft, *FEBS Lett.*, 2005, **579**, 4605.
51. K. Brown, M. Tegoni, M. Prudencio, A. S. Pereira, S. Besson, J. J. Moura, I. Moura and C. Cambillau, *Nat. Struct. Biol.*, 2000, **7**, 191.
52. J. R. Winkler, *Curr. Opin. Chem. Biol.*, 2000, **4**, 192.
53. J. A. Farrar, F. Neese, P. Lappalainen, P. M. H. Kroneck, M. Saraste, W. G. Zumft and A. J. Thomson, *J. Am. Chem. Soc.*, 1996, **118**, 11501.
54. K. Brown, K. Djinovic-Carugo, T. Haltia, I. Cabrito, M. Saraste, J. J. Moura, I. Moura, M. Tegoni and C. Cambillau, *J. Biol. Chem.*, 2000, **275**, 41133.
55. T. Rasmussen, B. C. Berks, J. Sanders-Loehr, D. M. Dooley, W. G. Zumft and A. J. Thomson, *Biochemistry*, 2000, **39**, 12753.
56. I. Moura, S. R. Pauleta and J. J. Moura, *J. Biol. Inorg. Chem.*, 2008, **13**, 1185.
57. T. Haltia, K. Brown, M. Tegoni, C. Cambillau, M. Saraste, K. Mattila and K. Djinovic-Carugo, *Biochem. J.*, 2003, **369**, 77.
58. K. Paraskevopoulos, S. V. Antonyuk, R. G. Sawers, R. R. Eady and S. S. Hasnain, *J. Mol. Biol.*, 2006, **362**, 55.
59. A. Pomowski, W. G. Zumft, P. M. Kroneck and O. Einsle, *Nature*, 2011, **477**, 234.
60. C. S. Nobrega, I. H. Saraiva, C. Carreira, B. Devreese, M. Matzapetakis and S. R. Pauleta, *Biochim. Biophys. Acta*, 2016, **1857**, 169.
61. S. Najmudin, S. R. Pauleta, I. Moura and M. J. Romao, *Acta Crystallogr., Sect. F: Struct. Biol. Cryst. Commun.*, 2010, **66**, 627.
62. B. L. Vallee and R. J. Williams, *Proc. Natl. Acad. Sci. U. S. A.*, 1968, **59**, 498.
63. L. A. Abriata, A. J. Vila and M. Dal Peraro, *J. Biol. Inorg. Chem.*, 2014, **19**, 565.
64. T. Matsubara and H. Iwasaki, *J. Biochem.*, 1972, **71**, 747.
65. W. G. Zumft and T. Matsubara, *FEBS Lett.*, 1982, **148**, 107.
66. W. P. Michalski, D. H. Hein and D. J. D. Nicholas, *Biochim. Biophys. Acta*, 1986, **872**, 50.
67. S. W. Snyder and T. C. Hollocher, *J. Biol. Chem.*, 1987, **262**, 6515.
68. C. L. Hulse and B. A. Averill, *Biochem. Biophys. Res. Commun.*, 1990, **166**, 729.
69. C. K. SooHoo and T. C. Hollocher, *J. Biol. Chem.*, 1991, **266**, 2203.
70. A. M. Jones, T. C. Hollocher and R. Knowles, *FEMS Microbiol. Lett.*, 1992, **92**, 205.
71. B. C. Berks, D. Baratta, J. Richardson and S. J. Ferguson, *Eur. J. Biochem.*, 1993, **212**, 467.
72. U. H. Hole, K. U. Vollack, W. G. Zumft, E. Eisenmann, R. A. Siddiqui, B. Friedrich and P. M. Kroneck, *Arch. Microbiol.*, 1996, **165**, 55.

73. S. Ferretti, J. G. Grossmann, S. S. Hasnain, R. R. Eady and B. E. Smith, *Eur. J. Biochem.*, 1999, **259**, 651.
74. M. Prudencio, A. S. Pereira, P. Tavares, S. Besson, I. Cabrito, K. Brown, B. Samyn, B. Devreese, J. Van Beeumen, F. Rusnak, G. Fauque, J. J. Moura, M. Tegoni, C. Cambillau and I. Moura, *Biochemistry*, 2000, **39**, 3899.
75. K. Yamaguchi, A. Kawamura, H. Ogawa and S. Suzuki, *J. Biochem.*, 2003, **134**, 853.
76. T. Rasmussen, B. C. Berks, J. N. Butt and A. J. Thomson, *Biochem. J.*, 2002, **364**, 807.
77. S. Dell'Acqua, S. R. Pauleta, J. J. Moura and I. Moura, *Philos. Trans. R. Soc. London, B*, 2012, **367**, 1204.
78. C. L. Coyle, W. G. Zumft, P. M. Kroneck, H. Korner and W. Jakob, *Eur. J. Biochem.*, 1985, **153**, 459.
79. E. M. Johnston, S. Dell'Acqua, S. Ramos, S. R. Pauleta, I. Moura and E. I. Solomon, *J. Am. Chem. Soc.*, 2014, **136**, 614.
80. A. Viebrock and W. G. Zumft, *J. Bacteriol.*, 1988, **170**, 4658.
81. W. G. Zumft, A. Dreusch, S. Lochelt, H. Cuypers, B. Friedrich and B. Schneider, *Eur. J. Biochem.*, 1992, **208**, 31.
82. J. M. Charnock, A. Dreusch, H. Korner, F. Neese, J. Nelson, A. Kannt, H. Michel, C. D. Garner, P. M. Kroneck and W. G. Zumft, *Eur. J. Biochem.*, 2000, **267**, 1368.
83. P. M. Kroneck, W. A. Antholine, J. Riester and W. G. Zumft, *FEBS Lett.*, 1988, **242**, 70.
84. P. M. Kroneck, W. A. Antholine, J. Riester and W. G. Zumft, *FEBS Lett.*, 1989, **248**, 212.
85. J. A. Farrar, A. J. Thomson, M. R. Cheesman, D. M. Dooley and W. G. Zumft, *FEBS Lett.*, 1991, **294**, 11.
86. W. E. Antholine, D. H. Kastrau, G. C. Steffens, G. Buse, W. G. Zumft and P. M. Kroneck, *Eur. J. Biochem.*, 1992, **209**, 875.
87. P. M. Kroneck, W. E. Antholine, D. H. Kastrau, G. Buse, G. C. Steffens and W. G. Zumft, *FEBS Lett.*, 1990, **268**, 274.
88. K. R. Williams, D. R. Gamelin, L. B. LaCroix, R. P. Houser, W. B. Tolman, T. C. Mulder, S. de Vries, B. Hedman, K. O. Hodgson and E. I. Solomon, *J. Am. Chem. Soc.*, 1997, **119**, 613.
89. D. R. Gamelin, D. W. Randall, M. T. Hay, R. P. Houser, T. C. Mulder, G. W. Canters, S. de Vries, W. B. Tolman, Y. Lu and E. I. Solomon, *J. Am. Chem. Soc.*, 1998, **120**, 5246.
90. S. Dell'Acqua, S. R. Pauleta, P. M. Paes de Sousa, E. Monzani, L. Casella, J. J. Moura and I. Moura, *J. Biol. Inorg. Chem.*, 2010, **15**, 967.
91. M. L. Alvarez, J. Ai, W. Zumft, J. Sanders-Loehr and D. M. Dooley, *J. Am. Chem. Soc.*, 2001, **123**, 576.
92. E. I. Solomon, D. E. Heppner, E. M. Johnston, J. W. Ginsbach, J. Cirera, M. Qayyum, M. T. Kieber-Emmons, C. H. Kjaergaard, R. G. Hadt and L. Tian, *Chem. Rev.*, 2014, **114**, 3659.
93. J. A. Farrar, W. G. Zumft and A. J. Thomson, *Proc. Natl. Acad. Sci. U. S. A.*, 1998, **95**, 9891.

94. P. Chen, S. DeBeer George, I. Cabrito, W. E. Antholine, J. J. Moura, I. Moura, B. Hedman, K. O. Hodgson and E. I. Solomon, *J. Am. Chem. Soc.*, 2002, **124**, 744.
95. P. Chen, S. I. Gorelsky, S. Ghosh and E. I. Solomon, *Angew. Chem., Int. Ed. Engl.*, 2004, **43**, 4132.
96. E. I. Solomon, *Comments Inorg. Chem.*, 1984, **3**, 227.
97. P. Chen, I. Cabrito, J. J. Moura, I. Moura and E. I. Solomon, *J. Am. Chem. Soc.*, 2002, **124**, 10497.
98. S. Ghosh, S. I. Gorelsky, S. D. George, J. M. Chan, I. Cabrito, D. M. Dooley, J. J. Moura, I. Moura and E. I. Solomon, *J. Am. Chem. Soc.*, 2007, **129**, 3955.
99. E. M. Johnston, S. Dell'Acqua, S. R. Pauleta, I. Moura and E. I. Solomon, *Chem. Sci.*, 2015, **6**, 5670.
100. S. Ghosh, S. I. Gorelsky, P. Chen, I. Cabrito, J. J. Moura, I. Moura and E. I. Solomon, *J. Am. Chem. Soc.*, 2003, **125**, 15708.
101. J. M. Chan, J. A. Bollinger, C. L. Grewell and D. M. Dooley, *J. Am. Chem. Soc.*, 2004, **126**, 3030.
102. V. S. Oganesyan, T. Rasmussen, S. Fairhurst and A. J. Thomson, *Dalton Trans.*, 2004, 996.
103. K. Frunzke and W. G. Zumft, *J. Chromatogr. A*, 1984, **299**, 477.
104. S. Dell'acqua, S. R. Pauleta, E. Monzani, A. S. Pereira, L. Casella, J. J. Moura and I. Moura, *Biochemistry*, 2008, **47**, 10852.
105. J. K. Kristjansson and T. C. Hollocher, *J. Biol. Chem.*, 1980, **255**, 704.
106. T. Rasmussen, T. Brittain, B. C. Berks, N. J. Watmough and A. J. Thomson, *Dalton Trans.*, 2005, **21**, 3501.
107. K. Fujita, M. Hirasawa-Fujita, D. E. Brown, Y. Obara, F. Ijima, T. Kohzuma and D. M. Dooley, *J. Inorg. Biochem.*, 2012, **115**, 163.
108. K. Fujita and D. M. Dooley, *Inorg. Chem.*, 2007, **46**, 613.
109. K. Fujita, J. M. Chan, J. A. Bollinger, M. L. Alvarez and D. M. Dooley, *J. Inorg. Biochem.*, 2007, **101**, 1836.
110. G. D. Watts and A. Burns, *Biochem. J.*, 1975, **152**, 33.
111. S. G. Mayhew, *Eur. J. Biochem.*, 1978, **85**, 535.
112. G. Cheddar, T. E. Meyer, M. A. Cusanovich, C. D. Stout and G. Tollin, *Biochemistry*, 1989, **28**, 6318.
113. Y. Furukawa, K. Ishimori and I. Morishima, *Biochemistry*, 2000, **39**, 10996.
114. N. M. Cerqueira, P. A. Fernandes, P. J. Gonzalez, J. J. Moura and M. J. Ramos, *Inorg. Chem.*, 2013, **52**, 10766.
115. S. I. Gorelsky, S. Ghosh and E. I. Solomon, *J. Am. Chem. Soc.*, 2006, **128**, 278.

CHAPTER 8

Model Compounds for Molybdenum Nitrate Reductases

AMIT MAJUMDAR[a] AND SABYASACHI SARKAR*[b]

[a]Department of Inorganic Chemistry, Indian Association for the Cultivation of Science, Jadavpur, Kolkata-700032, India; [b]Nanoscience and Synthetic Leaf Laboratory, Indian Institute of Engineering Science and Technology, Shibpur, Howrah-711103, India
*E-mail: abya@iitk.ac.in

8.1 Introduction

8.1.1 Nitrate Reduction and Its Implication in Biology

Denitrification is an important, four-step reductive process that is related to the nitrogen cycle in which certain anaerobic bacteria uses nitrogen oxide species, such as nitrate (NO_3^-), nitrite (NO_2^-), nitric oxide (NO) and nitrous oxide (N_2O), as terminal electron acceptors instead of O_2. Most of the denitrifying organisms prefer to use oxygen as the terminal electron acceptor. The denitrification process, however, provides an alternative anaerobic survival mechanism for such organisms. Nitrogen and inorganic nitrogen compounds, such as NO_3^-, NO_2^-, NO, N_2O and ammonia (NH_3), are the key components of the global nitrogen cycle. Nitrate reduction to ammonia is a two-step process that involves the reduction of NO_3^- to NO_2^-, which in turn

RSC Metallobiology Series No. 9
Metalloenzymes in Denitrification: Applications and Environmental Impacts
Edited by Isabel Moura, José J. G. Moura, Sofia R. Pauleta and Luisa B. Maia

Published by the Royal Society of Chemistry, www.rsc.org

is reduced to ammonium (NH_4^+). Reduction of nitrate to nitrite (NO_3^-/NO_2^-; $E_0' = +433$ mV)[1] generates a large amount of energy, which in turn is utilised by the organism to sustain its life. In assimilatory nitrate reduction process, nitrate is first reduced to nitrite, which in turn is converted into ammonia *via* assimilatory nitrite reductases. Nitrate is also reduced to nitrite in dissimilatory nitrate reduction processes and the nitrite can be further metabolised to nitrogen or ammonia by denitrification or dissimilatory nitrate reduction to ammonia, respectively. A detailed description of the denitrification process may be found in Chapter 1.

8.1.2 Classification and Active Site Structures of Nitrate Reductase

The nitrate reductases are classified into four different groups, namely eukaryotic nitrate reductases (eukNR), periplasmic nitrate reductases (Nap), respiratory nitrate reductases (Nar) and assimilatory nitrate reductases (Nas).[2] According to the Hille classification of mononuclear molybdenum enzymes,[3] Nar, Nap and prokaryotic Nas belong to the DMSO reductase (DMSOR) family, while eukNR can be regarded as a member of the sulphite oxidase (SO) family.

The active site of the Nap, monomeric NapA (oxidised form), from *Desulfovibrio desulfuricans* was the first structurally characterised (1.9 Å resolution)[4] nitrate reductase. The molybdenum atom is six coordinate and resides in a distorted trigonal prismatic geometry (Figure 8.1a).[4] Two dithiolene sulphur atoms of pyranopterin dithiolate present as the guanine dinucleotide (Mo–S = 2.2–2.4 Å) provides four coordination. The side chain of Cys_{140} (Mo–S = 2.5 Å) provides the fifth coordination, while a terminal oxo group (Mo–O ≈ 2.2 Å)

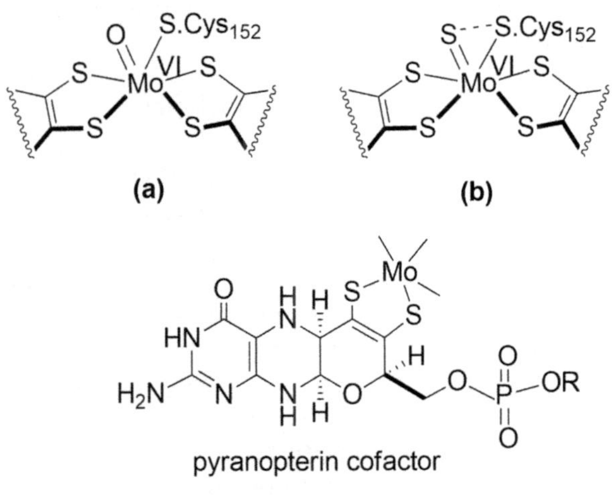

Figure 8.1 Originally proposed (a)[10] and revised (b)[14] active site structure of NapA. Also shown is the pyranopterin cofactor.

completes the six-fold coordination environment around the molybdenum centre. The crystal structure showed that the Mo atom deviates by ~0.7 Å from the mean plane defined by the four dithiolene sulphur atoms. In subsequent structural characterisation of NapAB, from the sixth ligand to the Mo(VI) centre was proposed as a coordinated water or hydroxo ligand instead of a terminal oxo. Such descriptions of coordinated oxo/hydroxo ligands were consistent with the spectroscopically characterised Mo(V) state, where the coordination of a water or hydroxyl group to the Mo centre is generally accepted. Thus, the oxidised active site of NapA can be represented as $[Mo^{VI}(OH_x)(S_{Cys})(S_2Pd)_2]$ ($x = 1, 2$). The reduced desoxo site $[Mo^{IV}(S_{Cys})(S_2Pd)_2]^{1-}$ was proposed as mediating the reduction of nitrate to nitrite.[4]

The NapA structure from *D. desulfuricans* was revisited,[5] and a careful examination of the anomalous scattering parameters of oxygen and sulphur suggested that the terminal ligand to the Mo(VI) centre is sulphide instead of an oxo group.[5] It was suggested that a partial persulphide bond is formed between the terminal sulphide and the cysteine sulphur. Structural characterisation at 1.9 Å resolution of a second dissimilatory nitrate reductase (Nar) from the facultative anaerobe *Escherichia coli* has been reported and, based on the findings, the reduced active site has been formulated as containing a $[Mo^{IV}(O_2CAsp)(S_2Pd)_2]$ site with an unsymmetrically coordinated carboxylate group (Mo–O = 1 to ~1.9 Å and Mo–O = 2 to ~2.4 Å).[6] A detailed description of the active structures of the nitrate reductases may be found in Chapter 3.

8.1.3 Mechanism of Nitrate Reduction

The possibility of nitrate reduction by an atom transfer reaction was first conceived by Williams in 1973,[7] while the possibility of the minimal reaction (eqn (8.1)) as a possible enzymatic process was proposed by Wentworth.[8] Subsequently, this reaction was incorporated into mechanistic proposals for nitrate reductase activity.[9,10]

$$Mo^{IV}O + NO_3^- = Mo^{VI}O_2 + NO_2^- \quad (8.1)$$

An initial catalytic mechanism for nitrate reduction was proposed based on the crystal structure of NapA from *D. desulfuricans*[4] and is shown in Figure 8.2.[11] The penta-coordinate Mo(IV) centre of the reduced active site binds NO_3^- *via* one oxygen atom to give a six-coordinate centre. The bound NO_3^- is then reduced to NO_2^- followed by its release from the Mo centre. The oxygen atom transfer (OAT) reaction is a key step in all of the mononuclear Mo- and W-containing oxido-reductase class of enzymes. Formally, the OAT process involves the transfer of a neutral oxygen atom (O), which is equivalent to the coupled transfer of 2 e^- and one O^{2-} in opposite directions. These types of reactions are defined by Anbar and Taube to be simple inner-sphere reactions[12] and are discussed in detail elsewhere.[13,14] A slightly different mechanism, however, was proposed for *E. coli* NapA[15] and is also shown in Figure 8.2.

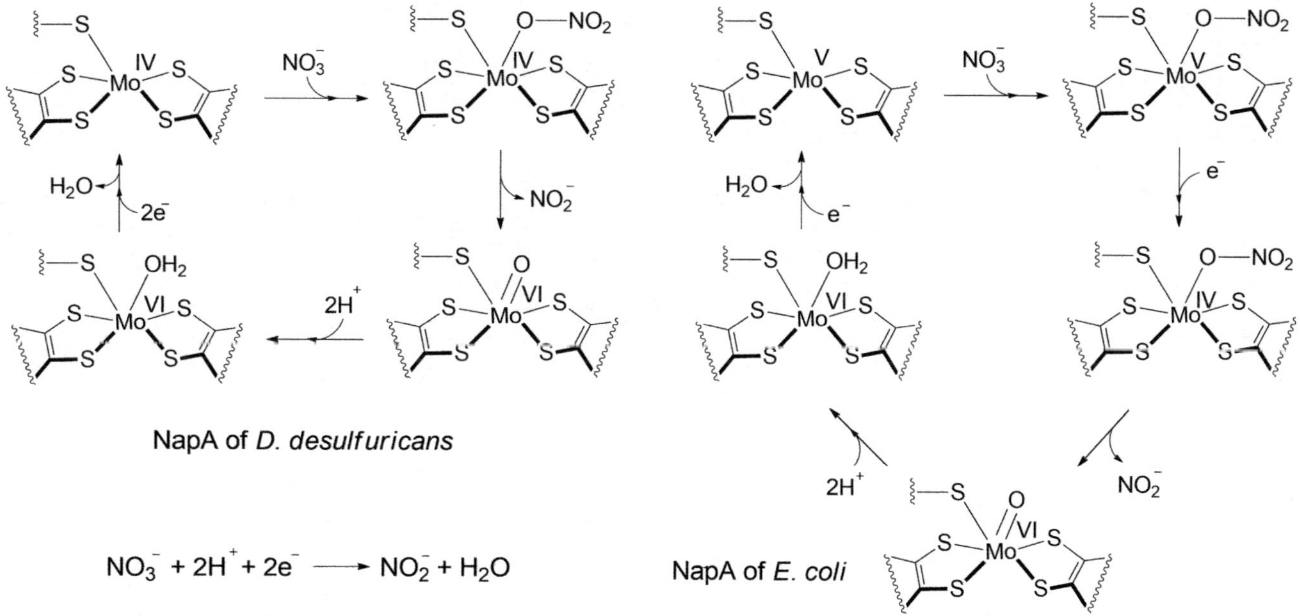

Figure 8.2 Proposed mechanism for the catalytic reduction of nitrate by NapA of *Desulfovibrio desulfuricans* and *Escherichia coli*. Adapted from ref. 11 with permission from The Royal Society of Chemistry.

The revised crystal structures of NapA from *D. desulfuricans* and *Cupriavidus necator* suggested the presence of a terminal sulphide group instead of a terminal oxo group and also that the reduced state of the active site contains a persulphide bridge between the sulphide and the cysteine sulphur atom. This revelation suggests the involvement of both Mo and ligand-based redox chemistry instead of the originally proposed, solely molybdenum-based redox chemistry in the catalytic cycle of the enzyme.[16] Based on these new observations and subsequent studies, the mechanism of nitrate reduction has been revised and a detailed discussion on the mechanistic aspects may be found in Chapter 3.

8.2 Model Chemistry

8.2.1 Introduction

Bioinorganic modelling chemistry in general mainly follows two approaches. These approaches are: (i) biomimetic or structural modelling, where the primary target is to construct the coordination environment around the concerned metal atom as closely as possible to that which is present in the active site structure; and (ii) bioinspired or functional modelling, where the goal is to achieve the enzymatic function under ambient conditions, irrespective of the coordination environment around the metal atom.[17,18] A more ambitious approach would be the structural–functional modelling approach,[18] in which both structural and functional features may be optimally realised; this is an ultimate but quite infrequently achieved goal in bioinorganic modelling chemistry.

Bioinorganic modelling chemistry of nitrate reductase in particular and that of mononuclear Mo and W enzyme active sites[18–21] uses either dithiolene- or non-dithiolene-type ligand systems, as are shown in Figure 8.3. A large number of model systems has been reported[18–21] based on the ligands shown in Figure 8.3. Only a few of those could react with the respective biological substrates. Many of the model systems could react only with proxy substrates while others meet the criteria of structural models. Extended dithiolene systems such as pterin and related ligands have been used in modelling studies and a detailed discussion of such systems has appeared recently.[22]

A major challenge in the modelling chemistry of Mo- and W-mediated OAT reactions is to avoid the propensity of the Mo(VI) dioxo and Mo(IV) monooxo species to form comproportionated Mo(V) μ-oxo dimers with core formulation $Mo^{V}_2O_3$.[13,14] This problem was solved by the use of sterically encumbered non-dithiolene ligands (Figure 8.3), and many such reports are available.[23–30] Other synthetic problems in molybdenum and tungsten bioinorganic chemistry involve the formation of tris chelate compounds with +IV or +V oxidation states of the central metal atom, the difficulty in stabilising penta-coordinated Mo/W desoxo complexes and the difficulty in preparing monodithiolene complexes. Such problems and the strategies that have been developed in order to overcome those are discussed elsewhere.[31]

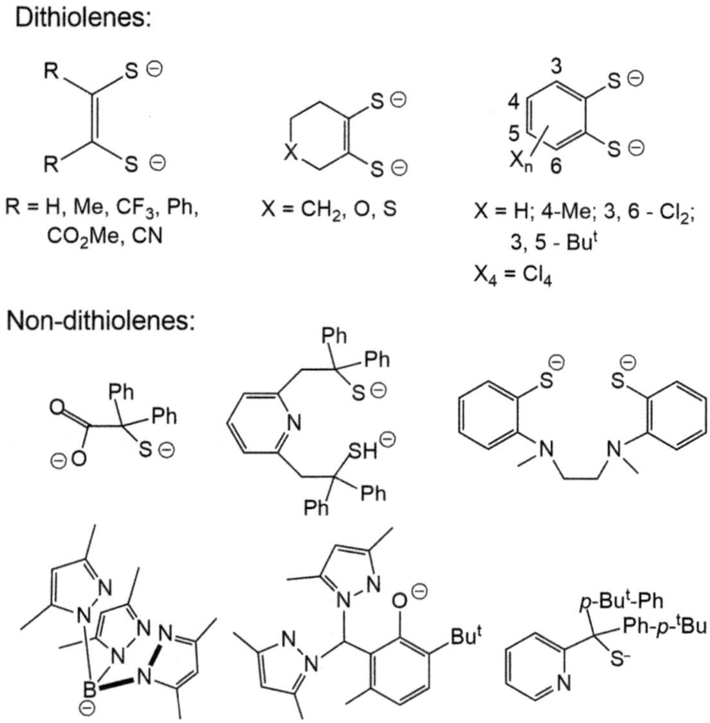

Figure 8.3 Different dithiolene- and non-dithiolene-type ligands used in modelling studies of mononuclear Mo and W enzymes. Adapted from ref. 31 with permission from The Royal Society of Chemistry.

8.2.2 Early Studies

An appreciable number of model reactions for nitrate reductase activity were reported almost 40 years ago. Several model systems for nitrate reductase activity are reported, but all of those reactions involve Mo(V)/Mo(VI) couples and produce NO or nitrogen(V) oxide as the reduced product.[32–35] Guymon and Spence showed that monomeric Mo(V) species could reduce NO_3^- to NO in aqueous media (tartrate buffer, pH 2.2–3.5) while the dimeric species remained inactive.[32] The reaction was found to be first order with respect to nitrate and half order with respect to Mo(V), and the average rate constant was reported to be 2.14 (±0.072) × 10^{-3} mol^{-1} s^{-1}. The reduction of NO_3^- by monomeric oxo Mo(V) complexes has also been investigated in non-aqueous solvents. In 1975, Garner and co-workers studied the reduction of nitrate using $MoOCl_3(Ph_3PO)_2$ in dichloromethane. The complex $MoOCl_3(Ph_3PO)_2$ reacts rapidly with NO_3^-, producing NO_2 in a one-electron step. A mechanism that involves preliminary dissociation of the Ph_3PO ligand *trans* to oxo, followed by coordination of NO_3^-, rearrangement to a *cis*-nitrato complex and electron transfer was proposed and is shown in Figure 8.4a.

Both the first and the third reaction could be observed and the rate constants for both of the steps were determined. The rearrangement was claimed

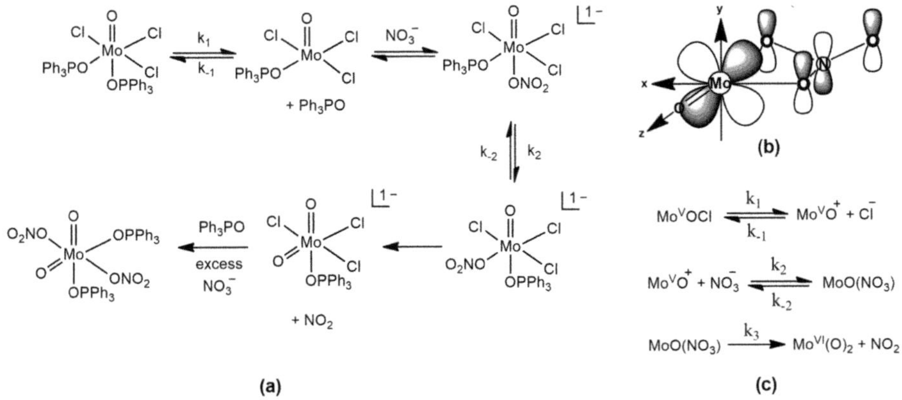

Figure 8.4 Mechanism of nitrate reduction by $MoOCl_3(Ph_3PO)_2$ (a); proposed necessary overlap between the Mo(d_{xy}) and oxygen (π^*) orbital of NO_3^- (b); nitrate reduction by [Mo^VOCl]-type complexes (c). Adapted from ref. 37 with permission from The Royal Society of Chemistry.

to be necessary in order to obtain proper overlap between the Mo(d_{xy}) and oxygen (π^*) orbital of NO_3^- (Figure 8.4b). The product of the reaction, NO_2, undergoes disproportionation in water to generate NO_3^- and NO^- and thus mimics the action of the enzyme nitrate reductase. Although the parent compound also reacts with NO_2, the reaction is sluggish and produces NOCl. The complex $[MoOCl_3(Ph_3PO)_2]^{1-}$ could also reduce NO_2^- to NO by an analogous mechanism.[36,37] However, the rearrangement reaction proceeds 38-times more slowly than in the case of NO_3^-. Such a sluggish reaction with NO_2^- was attributed to the binding of NO_2^- with a Mo centre *via* the nitrogen instead of the oxygen atom. Such a situation may inhibit the necessary rearrangement to a *cis* oxygen-bound species and thus slow down the reaction. Based on these results, Garner and co-workers suggested[36,37] an enzymatic mechanism wherein NO_3^- binds at a hydrophobic Mo(v) centre, followed by electron transfer and loss of NO_2. The NO_2 then undergoes a disproportionation reaction to generate NO_3^- and NO_2^- in an aqueous region of the cell. The rearrangement step was compared with the means by which the enzyme discriminates between NO_3^- and an inhibitor of the enzyme, NO_2^-.

A variety of monomeric oxo Mo(v) complexes such as $MoOClL^1$ (L^1 = α,α′-bipyridyl, *o*-phenanthroline), $MoOClL^2$ (L^2 = 8-hydroxyquinoline, 8-mercaptoquinoline), $MoOCl(CH_3OH)L^3$ (L^3 = *o*-[salicylidenimino]phenol) and $MoOClL^4$ (L^4 = *o*-bis[salicylidenimino]benzene) were reported to exhibit similar reactions in DMF solution (Figure 8.4c).[38–40] NO_2 was the product for most of the compounds, except $MoOClL^4$. The first step in such reactions was generally, although not exclusively, found to be rate controlling. A *cis*-nitrato complex allowing proper orbital overlap (Figure 8.4b) was again proposed to be necessary for electron transfer. The rates of the reaction depended on the nature of the ligand and the geometry (namely *cis* or *trans* for Cl^-), and the differences observed in experimental rate laws for different complexes were accounted for by the differences in the relative values of the rate constants of the reaction. Polymer-anchored Mo(v) complexes[41] were also reported to react with NO_3^- in a manner similar to that described previously by Spence and co-workers.[38]

Several other complexes such as MoOCl(mee) (mee = *N,N′*-dimethyl-*N,N′*-bis[2-mercaptoethyl] ethylenediamine) and $[MoO(PhS)_4]^-$ were found to react very rapidly with NO_3^-, but no NO_2^- could be detected upon addition of water.[39] Mass spectroscopic analysis of the gas above the reaction solution revealed the presence of NO and N_2O. The presence of these gases indicated that the one-electron-reduced product NO_2 reacted rapidly with the initial complexes. This possibility was confirmed by bubbling NO_2 into a solution of these complexes, which resulted in rapid oxidation of the initial complexes. No such reaction could be observed with the other Mo(v) complexes that were synthesised: $(NH_4)_2MoOCl_5$, $(NH_4)MoOCl_4(CH_3CN)$, *cis*-MoOCl(phen), *cis*-MoOCl(bpy), *trans*-MoOCl(bpy), *cis*-$MoOCl(ox)_2$, *cis*-$MoOCl(tox)_2$, *cis*-$MoOCl(CH_3OH)$(sap), *cis*-$MoOCl(acac)_2$, $MoO(ox)_2(CH_3Ph_2P)$ and $MoO(tox)_2(CH_3Ph_2P)$.[39] Although there is no evident reason as to why certain complexes react with NO_2 while others remain non-reactive, in some cases, a reasoning may be suggested. The complex $MoOCl(sal_2phen)$ (sal_2phen = disalicylaldehyde *o*-phenylenediamine) does not react with NO_3^- because the planarity of the ligand does not allow it to form either the *cis*-nitrato intermediate complex or the *cis*-dioxo Mo(vi) product. The pentacoordinated complex $MoO(S_2C_6H_3CH_3)_2^-$ may not react with NO_3^- because the chelating ligand does not allow the rearrangement to the *cis*-nitrato complex. Alternatively, a doubly charged nitrato complex may not be thermodynamically favourable in DMF.

Ketchum and co-workers showed later on that in strongly acidic conditions (1 M *p*-toluenesulphonic acid), a Mo(iii) complex, $Mo(H_2O)_6^{3+}$, could mediate the reduction of NO_3^- to NO_2.[42] The reaction (eqn (8.2)) was studied spectrophotometrically under anaerobic conditions and the second-order rate constant of the reaction was found to be 2.92 (±0.27) × 10^{-2} mol^{-1} s^{-1}, which is 10-times faster than that reported by Guymon and Spence.[32] Nitrite trapping reagents were found to facilitate the completion of the reaction.

$$2Mo(H_2O)_6^{3+} + 2NO_3^- \rightarrow Mo_2O_4(H_2O)_6^{2+} + 2NO_2^- + 4H^+ + 4H_2O \qquad (8.2)$$

During the early-stage development of nitrate reductase activity, not all of the biochemical information related to these enzymes was available. The development of new synthetic strategies and modifications of the existing ones in order to achieve the model nitrate reductase activity have always been helped by newer results from biochemical studies. Clearly, the enzymatic reduction of NO_3^- in a two-electron step will be much more attractive than the one-electron/disproportionation reaction of Mo(v) complexes. Based on this view and the finding that the active site of nitrate reductase is a desoxo Mo(iv) moiety, the reduction of nitrate to nitrite by desoxo-Mo(iv) complexes was investigated.

8.2.3 Model Chemistry Using Non-Dithiolene Ligands

Abiological reduction of nitrate has been accomplished previously with Mo(III) and Mo(V) complexes and were discussed before in Section 8.2.2. However, in a number of those systems, the initial reaction product was never NO_2^-, but rather was NO_2. Reduction of nitrate by a polymer-anchored tripeptide molybdenum complex to afford nitrite and a Mo(VI) species was first claimed in 1980, although supporting details for such a claim were not available.[41] Nitrate reduction by a Mo(IV) complex with a non-dithiolene coordination was first successfully demonstrated by Holm and co-workers.[43–45] It was shown that the complex $Mo^{IV}O(L\text{-}NS_2)(DMF)$ ($L\text{-}NS_2$ = 2,6-bis(2,2-diphenyl-2-mercaptoethyl)pyridine(2-)) can be oxidised to $Mo^{IV}O_2(L\text{-}NS_2)$ by excess nitrate in DMF, but the process was found to be complicated due to a rapid bleaching reaction between nitrite and the Mo(VI) complex (Figure 8.5).[45] The presence of sulphamic acid (1.5 eq./Mo) as a nitrite scavenger[46,47] was necessary in order to trap the produced nitrite (eqn (8.3)), which otherwise bleaches the Mo(VI) chromophore.

$$NO_2^- + H_2NSO_3H \rightarrow N_2 + HSO_4^- + H_2O \qquad (8.3)$$

A quantitative and well-behaved reduction of NO_3^- to NO_2^- (in the presence of excess NO_3^-) was achieved later on by employing the complex $Mo^{IV}O(L\text{-}NS_2)$(DMF) in the presence of a small excess (1.5 eq./Mo) of sulphamic acid as a nitrite scavenger.[43] This quantitative nitrate reduction process was characterised by saturation kinetics in which nitrate reversibly forms a substrate–Mo(IV) complex that ultimately gives rise to the products nitrite and the Mo(VI) complex following a first-order pathway with $k_1 = 1.49\ (\pm 0.05) \times 10^{-3}\ s^{-1}$.

8.2.4 Model Chemistry Using Dithiolene Ligands

Penta-coordinated bis(dithiolene) Mo^{IV}/W^{IV} complexes were targeted as possible analogue systems for NapA.[4] Pentacoordinated desoxo Mo(IV)/W(IV) bis(dithiolene) complexes $[M^{IV}(QC_6H_2\text{-}2,4,6\text{-}Pr^i_3)(S_2C_2Me_2)_2]^{1-}$ (M = Mo, W;

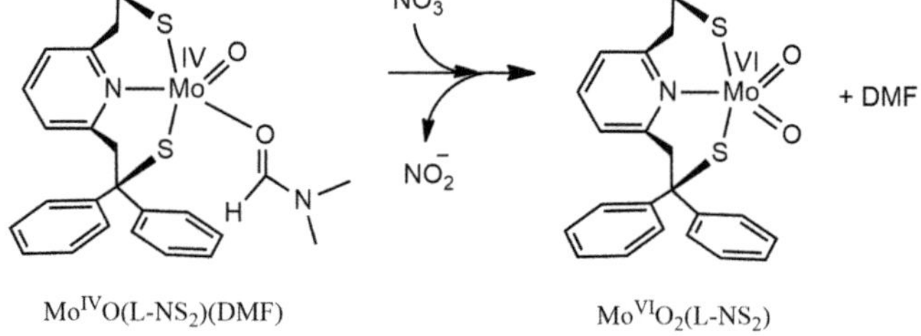

Figure 8.5 Nitrate reduction by a non-dithiolene complex, $Mo^{IV}O(L\text{-}NS_2)$(DMF). Adapted with permission from ref. 43. Copyright (1989) American Chemical Society.

Q = O, S) were synthesised[48–50] and their viability as possible model systems for nitrate reductase activity (Figure 8.2, NapA) was investigated.[50] The complexes $[Mo^{IV}(QC_6H_2\text{-}2,4,6\text{-}Pr^i_3)(S_2C_2Me_2)_2]^{1-}$ (Q = O, S), however, were inadmissible for the examination of nitrate reduction.[50] The phenolate complex (Mo^{IV}, Q = O) showed an extremely sluggish reaction in acetonitrile solution, even in the presence of excess nitrate. Reaction systems containing the thiolate complex (Mo^{IV}, Q = S; Figure 8.6a) and excess nitrate in both acetonitrile and DMF solutions underwent decomposition to form the thermodynamically stable tris chelate compound $[Mo^{IV}(S_2C_2Me_2)_3]^{1-}$ (identified spectrophotometrically), and this decomposition rate was found to be much faster than that of the required OAT reaction with nitrate. Nevertheless, an intrinsic ability of this complex to support OAT was demonstrated by the clean reduction of $(CH_2)_4SO$ to $(CH_2)_4S$. The tungsten compounds could reduce nitrate to nitrite (Figure 8.6a) following second-order kinetics (Q = O, 6.1×10^{-3} M^{-1} s^{-1}; Q = S, 1.7×10^{-1} M^{-1} s^{-1}) and were proposed to involve associative transition states (Figure 8.6b).[50]

The oxidised tungsten species $[W^{VI}O(QC_6H_2\text{-}2,4,6\text{-}Pr^i_3)(S_2C_2Me_2)_2]^{1-}$ (Q = O, S), originally synthesised by an OAT reaction with Ph_3AsO, were characterised by single-crystal X-ray structure determination (Figure 8.7c for Q = S). Desoxo pentacoordinated Mo(IV) bis(dithiolene) complexes are generally quite unstable and the synthesis and characterisation of such Mo(IV) complexes were made possible *via* the use of a sterically demanding phenolate/thiolate group, $QC_6H_2\text{-}2,4,6\text{-}Pr^i_3$ (Q = O, S). However, it is not difficult to guess that the presence of such a bulky axial ligand (Figure 8.7b) is supposed to hinder the approach of NO_3^- towards the Mo(IV) centre as well. Thus, an excellent

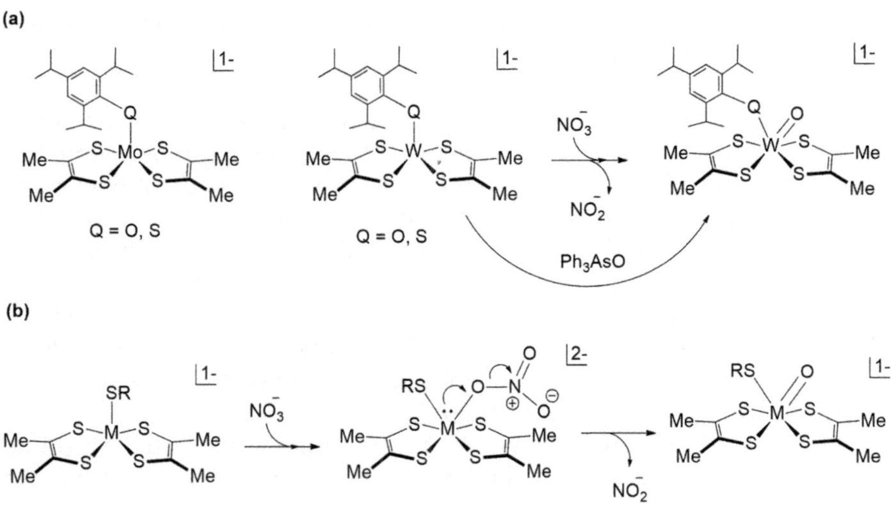

Figure 8.6 Bis-dithiolene Mo/W complexes as structural (Mo) and functional models (W) for dissimilatory nitrate reductase activity (a). Also shown is the proposed associative reaction pathway for nitrate reduction (b). Adapted with permission from ref. 50. Copyright (2005) American Chemical Society.

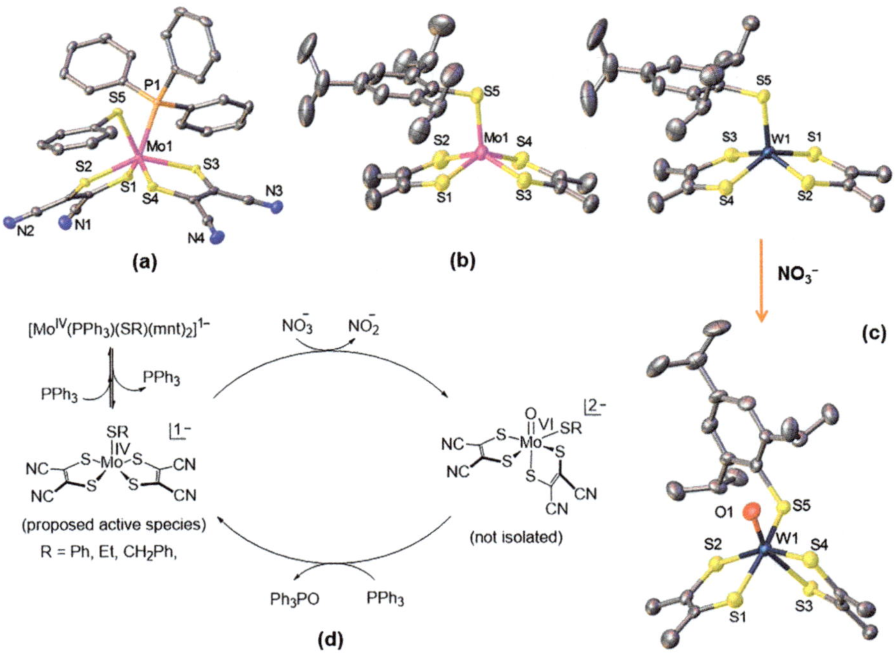

Figure 8.7 Molecular structures for $[Mo^{IV}(SPh)(PPh_3)(mnt)_2]^{1-}$ (a) as representatives of the best functional models, $[Mo^{IV}(SR)(PPh_3)(mnt)_2]^{1-}$ and the catalytic cycle involving $[Mo^{IV}(SR)(PPh_3)(mnt)_2]^{1-}$ as the catalyst and NO_3^- and PPh_3 as the oxidising and reducing substrates (d). Also shown are the molecular structures of the best structural model (b) and its W(IV) analogue, along with the nitrate-oxidised W(VI) compound (c). Adapted with permission from ref. 52. Copyright (2008) American Chemical Society.

structural model (Figure 8.7b) for the Nap active site could be synthesised, but at the expense of its functional activity. In an attempt to overcome this problem, complex $[Mo^{IV}(SPh)(S_2C_2Me_2)_2]^{1-}$ was targeted as the model active site. Unfortunately, the irradiation of the hexacoordinated complex $[Mo^{IV}(CO)(SPh)(S_2C_2Me_2)_2]^{1-}$ in order to remove CO resulted in decomposition of the parent complex to form $[Mo^{V}(S_2C_2Me_2)_3]^{1-}$.[48] This result indicated that such pentacoordinate Mo(IV) bis(dithiolene) complexes are highly unstable and hence the problem of direct handling of such unstable molecules could be avoided by the synthesis of hexacoordinated Mo(IV) where the sixth ligand (L) may easily dissociate from the Mo(IV) centre.

Based on the aforementioned idea, the compounds $[Mo^{IV}(SR)(PPh_3)(mnt)_2]^{1-}$ (mnt is maleonitriledithiolate/1,2-dicyano ethylene dithiolate, R = Ph, Et, CH_2Ph) were synthesised[51,52] as "protected" model systems for the reduced active site of NapA.[4] These compounds undergo the rapid reversible dissociation of PPh_3 and release the active pentacoordinated species $[Mo^{IV}(SR)(mnt)_2]^{1-}$ in solution. The presence of excess PPh_3 was actually necessary to stabilise the compounds in their hexacoordinated form in solution. These compounds were shown to reduce NO_3^- to NO_2^- (Figure 8.8a)

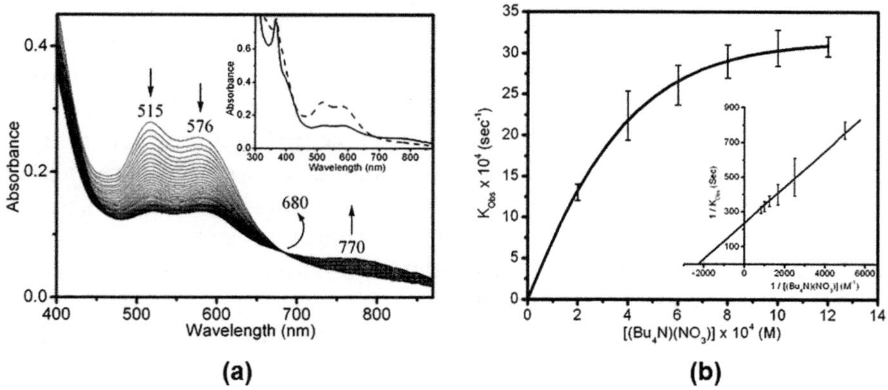

Figure 8.8 (a) Spectral changes in the reaction of 1×10^{-4} M of $[Mo(mnt)_2(PPh_3)(SPh)](Et_4N)]$ with 5×10^{-4} M of $[Bu_4N][NO_3]$ in CH_2Cl_2 at 25 °C. Total time: 4 min; scan rate: 2 s $scan^{-1}$. Inset: initial (dotted line) and final spectra (solid line). (b) Dependence of the rate of reaction of $[Mo(mnt)_2(PPh_3)(SPh)](Et_4N)]$ with 2–12 equiv. of $[Bu_4N][NO_3]$ in CH_2Cl_2 at 25 °C on $[Bu_4N][NO_3]$. Inset: the corresponding double reciprocal plot. Adapted with permission from ref. 51. Copyright (2006) American Chemical Society.

following the Michaelis–Menten saturation kinetics (Figure 8.8b) with $V_{max} = 3.2 \times 10^{-2}\ s^{-1}$ (R = Et),[52] $4.2 \times 10^{-3}\ s^{-1}$ (R = Ph)[51] and $7.1 \times 10^{-3}\ s^{-1}$ (R = CH_2Ph).[52] Nitrite generated in the system containing the complexes $[Mo^{IV}(SR)(PPh_3)(mnt)_2]^{1-}$ (R = Ph, Et, CH_2Ph) and $[Bu_4N][NO_3]$ was measured following a colorimetric method (λ_{max} = 550 nm) using the Griess reagent, which is a 1 : 1 mixture of 0.1% N/1-napthylethylenediamine dihydrochloride in water and 1% sulphanilamide in 5% phosphoric acid. Nitrite generated in these model nitrate reduction systems corresponded to ~90% of the theoretical conversion for a complete reaction.

The oxidised species $[Mo^{VI}(O)(SR)(mnt)_2]^{1-}$ produced during the reduction of nitrate was reduced back to $[Mo^{IV}(SR)(mnt)_2]^{1-}$ by the dissociated PPh_3. Utilising these observations, a catalytic cycle involving NO_3^- and PPh_3 as the oxidising and reducing substrates, respectively, was established (Figure 8.7c), with turnover numbers of 10 (R = Et), 50 (R = Ph) and 40 (R = CH_2Ph) $mmol^{-1}\ s^{-1}$.[52] The putative oxidised species $[Mo^{VI}(O)(SR)(mnt)_2]^{1-}$, however, could not be isolated, due to its participation in the successive reactions with the NO_2^- produced in the reaction. A routine workup procedure of the resultant solution yielded the well-known orange–brown species $[Mo^{IV}(NO)_2(mnt)_2]^{2-}$.[53,54] In order to rationalise the clean reduction of NO_3^- by the complex $[Mo^{IV}(SPh)(mnt)_2]^{1-}$ (Figure 8.8a), generated in solution by the reversible dissociation of coordinated PPh_3 from the parent complex $[Mo^{IV}(SPh)(PPh_3)(mnt)_2]^{1-}$ (Figure 8.7a) compared with the inability of the complex $[Mo^{IV}(SC_6H_2\text{-}2,4,6\text{-}Pr^i_3)(S_2C_2Me_2)_2]^{1-}$ (Figure 8.7b) to reduce nitrate, two more complexes, $[Mo^{IV}(SC_6H_4\text{-}p\text{-}Me)(PPh_3)(mnt)_2]^{1-}$ and $[Mo^{IV}(SC_6H_4\text{-}o\text{-}COOH)(PPh_3)(mnt)_2]^{1-}$, were synthesised that could reduce nitrate only incompletely.[55] This differing reactivity towards nitrate reduction was attributed to the increasing substitution at the thiophenolate ring, which hinders the approach of NO_3^- toward the Mo(IV) centre, and this hindrance reaches its maximum in the case of the complex $[Mo^{IV}(SC_6H_2\text{-}2,4,6\text{-}Pr^i_3)(S_2C_2Me_2)_2]^{1-}$ (Figure 8.7b), thus rendering it unable to react with nitrate.[55] Moreover, substitution of the essential thiolate coordination rather than replacing the dissociable PPh_3 in $[Mo^{IV}(SPh)(PPh_3)(mnt)_2]^{1-}$ (Figure 8.7a) by halides and thiocyanate yielded the isostructural complexes $[Mo^{IV}(PPh_3)(X)(mnt)_2]^{1-}$ (X = Cl, Br, I, NCS), which remained inactive towards nitrate reduction. This is somewhat similar to the dead end-type inhibition often encountered in native systems, although in NapA they act as competitive inhibitors.

The reduced active site of a second class of dissimilatory nitrate reductases contains the unit $[Mo^{IV}(O_2CAsp)(S_2pd)_2]$ with an unsymmetrically coordinated carboxylate group.[6] Although there are not many reports of analogue systems, the complex $[W^{IV}(O_2CPh)(S_2C_2Me_2)_2]^{1-}$ may be regarded as a tungsten analogue that mimics the immediate coordination environment of the active site.[56] Recently, several bis(ene-1,2-dithiolato) oxocarboxylato–Mo(VI) complexes have been synthesised and characterised by spectroscopic and kinetic studies in relation to the active site of type II DMSOR.[57] However, X-ray structural characterisation data for such compounds are still awaited.

8.3 Problems, Strategies and Future Scope

Simple Mo(III) and Mo(V) complexes were shown to be capable of reducing NO_3^-,[32–40] but in most of these cases, the immediate reduction product was NO_2 instead of NO_2^-. Moreover, realistic modelling chemistry of NapA[4] required the use of dithiolene-type ligands, which are considered to be small molecule mimics of the pterindithiolate ligand. This reasoning promoted the use of dithiolene ligands in the nitrate reductase modelling chemistry, as well as in the modelling chemistry of mononuclear molybdenum and tungsten enzymes in general.[18–20,31] The main challenge appeared to be the stabilisation of pentacoordinated bis(dithiolene) Mo(IV)–monothiolate complexes, which was ultimately achieved by the judicious choice of sterically demanding thiolate molecules as the axial ligand. The resulting molybdenum complex, although it could be isolated and fully characterised, did not react with nitrate.[50] The steric bulk at the axial position of the pentacoordinated Mo(IV) bis(dithiolene) complex helped to realise the best structural model for nitrate reductase, while the same steric bulk prevented the approach of the substrate (NO_3^-) towards the Mo(IV) centre. Such a situation was beautifully tackled by the synthesis of "protected" hexacoordinated Mo(IV) complexes, which release the active pentacoodinated Mo(IV) complex in solution as a result of the reversible dissociation of the coordinated triphenylphosphine. The active complexes thus released could mediate a clean reduction of NO_3^- to NO_2^- and the reaction followed the Michaelis–Menten saturation kinetics.[51,52] However, the oxidised Mo(VI) complex could not be isolated and thus kept the challenge alive. Moreover, re-examination of the Nap structure suggests the sixth molybdenum ligand to be a terminal sulphide instead of a terminal oxo group, and a redox interplay between molybdenum and sulphur has been invoked, whereby the (partial) formation of a disulphide bond may influence the two-electron reduction of Mo(VI) (see Chapter 3 of this book). Such a redox interaction has long been proposed by Stiefel in Mo–S systems.[58] The introduction of such an interpretation in native system demands the involvement of both molybdenum and ligand-based redox chemistry in the catalytic cycle of nitrate reductase, instead of the previously accepted molybdenum-based redox chemistry. Currently, there is no nitrate reductase model system that demonstrates such chemistry, and this situation necessitates the synthesis of analogues that may be tested for such a reactivity pattern.

References

1. R. K. Thauer, K. Jungermann and K. Decker, *Bacteriol. Rev.*, 1977, **41**, 100.
2. J. F. Stolz and P. Basu, *ChemBioChem*, 2002, **3**, 198.
3. R. Hille, *Chem. Rev.*, 1996, **96**, 2757.
4. J. M. Dias, M. E. Than, A. Humm, R. Huber, G. P. Bourenkov, H. D. Bartunik, S. Bursakov, J. Calvete, J. Caldeira, C. Carneiro, J. J. G. Moura, I. Moura and M. J. Romão, *Structure*, 1999, **7**, 65.
5. S. Najmudin, P. González, J. Trincão, C. Coelho, A. Mukhopadhyay, N. F. S. A. Cerqueira, C. Romão, I. Moura, J. G. Moura, C. Brondino and M. Romão, *JBIC, J. Biol. Inorg. Chem.*, 2008, **13**, 737.
6. M. G. Bertero, R. A. Rothery, M. Palak, C. Hou, D. Lim, F. Blasco, J. H. Weiner and N. C. J. Strynadka, *Nat. Struct. Mol. Biol.*, 2003, **10**, 681.
7. R. J. P. Williams, *Biochem. Soc. Trans.*, 1973, **1**, 1.
8. R. A. D. Wentworth, *Coord. Chem. Rev.*, 1976, **18**, 1.
9. E. J. Hewitt, B. A. Notton and C. D. Garner, *Biochem. Soc. Tarns.*, 1979, **7**, 629.
10. E. J. Hewitt and B. A. Notton, in *Molybdenum and Molybdenum Containing Enzymes*, ed. M. P. Coughlan, Pergamon Press, New York, 1980.
11. C. Sparacino-Watkins, J. F. Stolz and P. Basu, *Chem. Soc. Rev.*, 2014, **43**, 676.
12. M. Anbar and H. Taube, *J. Am. Chem. Soc.*, 1958, **80**, 1073.
13. R. H. Holm, *Chem. Rev.*, 1987, **87**, 1401.
14. R. H. Holm, *Coord. Chem. Rev.*, 1990, **100**, 183.
15. B. J. N. Jepson, S. Mohan, T. A. Clarke, A. J. Gates, J. A. Cole, C. S. Butler, J. N. Butt, A. M. Hemmings and D. J. Richardson, *J. Biol. Chem.*, 2007, **282**, 6425.
16. P. J. Gonzalez, M. G. Rivas, C. S. Mota, C. D. Brondino, I. Moura and J. J. G. Moura, *Coord. Chem. Rev.*, 2013, **257**, 315.
17. S. Groysman and R. H. Holm, *Biochemistry*, 2009, **48**, 2310.
18. A. Majumdar and S. Sarkar, *Coord. Chem. Rev.*, 2011, **255**, 1039.
19. J. H. Enemark, J. J. A. Cooney, J.-J. Wang and R. H. Holm, *Chem. Rev.*, 2004, **104**, 1175.
20. H. Sugimoto and H. Tsukube, *Chem. Soc. Rev.*, 2008, **37**, 2609.
21. C. Schulzke, *Eur. J. Inorg. Chem.*, 2011, 1189.
22. P. Basu and S. J. N. Burgmayer, *Coord. Chem. Rev.*, 2011, **255**, 1016.
23. J. M. Berg and R. H. Holm, *J. Am. Chem. Soc.*, 1984, **106**, 3035.
24. J. M. Berg and R. H. Holm, *J. Am. Chem. Soc.*, 1985, **107**, 917.
25. R. H. Holm and J. M. Berg, *Acc. Chem. Res.*, 1986, **19**, 363.
26. C. J. Doonan, D. A. Slizys and C. G. Young, *J. Am. Chem. Soc.*, 1999, **121**, 6430.

27. S. A. Roberts, C. G. Young, C. A. Kipke, W. E. Cleland, K. Yamanouchi, M. D. Carducci and J. H. Enemark, *Inorg. Chem.*, 1990, **29**, 3650.
28. P. D. Smith, A. J. Millar, C. G. Young, A. Ghosh and P. Basu, *J. Am. Chem. Soc.*, 2000, **122**, 9298.
29. P. Basu, B. W. Kail and C. G. Young, *Inorg. Chem.*, 2010, **49**, 4895.
30. V. N. Nemykin, S. R. Davie, S. Mondal, N. Rubie, M. L. Kirk, A. Somogyi and P. Basu, *J. Am. Chem. Soc.*, 2002, **124**, 756.
31. A. Majumdar, *Dalton Trans.*, 2014, **43**, 8990.
32. E. P. Guymon and J. T. Spence, *J. Phys. Chem.*, 1966, **70**, 1964.
33. J. T. Spence, in *Metal Ions in Biological Systems*, ed. H. Sigel, Mercel Dekker, New York, 1976, vol. 5, p. 279.
34. C. D. Garner, M. R. Hyde, F. E. Mabbs and V. I. Routledge, *Nature*, 1974, **252**, 579.
35. C. D. Garner, M. R. Hyde and F. E. Mabbs, *Nature*, 1975, **253**, 623.
36. C. D. Garner, M. R. Hyde, F. E. Mabbs and V. I. Routledge, *J. Chem. Soc., Dalton Trans.*, 1975, 1175.
37. C. D. Garner, M. R. Hyde, F. E. Mabbs and V. I. Routledge, *J. Chem. Soc., Dalton Trans.*, 1975, 1180.
38. R. D. Taylor, P. G. Todd, N. D. Chasteen and J. T. Spence, *Inorg. Chem.*, 1979, **18**, 44.
39. J. T. Spence, in *Molybdenum and Molybdenum Containing Enzymes*, ed. M. P. Coughlan, Pergamon Press, Oxford, 1980, p. 100.
40. C. D. Garner, M. R. Hyde and F. E. Mabbs, *Inorg. Chem.*, 1976, **15**, 2327.
41. J. Topich, *Inorg. Chim. Acta*, 1980, **46**, L97.
42. P. A. Ketchum, R. C. Taylor and D. C. Young, *Nature*, 1976, **259**, 202.
43. J. A. Craig and R. H. Holm, *J. Am. Chem. Soc.*, 1989, **111**, 2111.
44. J. A. Craig, E. W. Harlan, B. S. Snyder, M. A. Whitener and R. H. Holm, *Inorg. Chem.*, 1989, **28**, 2082.
45. J. M. Berg and R. H. Holm, *J. Am. Chem. Soc.*, 1985, **107**, 925.
46. W. N. Carson, *Anal. Chem.*, 1951, **23**, 1016.
47. R. C. Brasted, *Anal. Chem.*, 1952, **24**, 1111.
48. B. S. Lim, J. P. Donahue and R. H. Holm, *Inorg. Chem.*, 2000, **39**, 263.
49. J. Jiang and R. H. Holm, *Inorg. Chem.*, 2004, **43**, 1302.
50. J. Jiang and R. H. Holm, *Inorg. Chem.*, 2005, **44**, 1068.
51. A. Majumdar, K. Pal and S. Sarkar, *J. Am. Chem. Soc.*, 2006, **128**, 4196.
52. A. Majumdar, K. Pal and S. Sarkar, *Inorg. Chem.*, 2008, **47**, 3393.
53. N. G. Connelly, J. Locke, J. A. McCleverty, D. A. Phipps and B. Ratcliff, *Inorg. Chem.*, 1970, **9**, 278.
54. J. A. McCleverty, in *Progress in Inorganic Chemistry*, ed. F. A. Cotton, Interscience Publishers, A division of John Wiley and Sons, United States of America, 1968, vol. 10, p. 49.
55. A. Majumdar, K. Pal and S. Sarkar, *Dalton Trans.*, 2009, 1927.
56. K.-M. Sung and R. H. Holm, *Inorg. Chem.*, 2001, **40**, 4518.
57. H. Sugimoto, M. Sato, L. J. Giles, K. Asano, T. Suzuki, M. L. Kirk and S. Itoh, *Dalton Trans.*, 2013, **42**, 15927.
58. E. I. Stiefel, in *Molybdenum and Molybdenum Containing Enzymes*, ed. M. P. Coughlan, Pergamon Press, New York, 1980, p. 43.

CHAPTER 9

Model Compounds for Nitric Oxide Reductase

ABHISHEK DEY*[a] AND SOMDATTA GHOSH DEY*[a]

[a]Department, of Inorganic Chemistry, Indian Association for the Cultivation of Science, 2A & 2 B, Raja S. C. Mullick Road, Jadavpur, Kolkata, 700032, India
*E-mail: icad@iacs.res.in, icsgd@iacs.res.in

9.1 Introduction

Nitric oxide (NO) is a key intermediate in microbial denitrification that involves reduction of NO_2^- to N_2O *via* NO.[1–4] The denitrification process is a form of anaerobic respiration that can lead to ATP synthesis in some organisms.[5] The NO produced during the process in also used as a primary defence mechanism against invading pathogens.[2] In higher organisms, the role of NO is diverse and includes cell signalling, respiratory control, dilation of blood vessels and defence against pathogens, amongst others.[6,7] The product of NO reduction, N_2O, is a greenhouse gas and has been shown to be a strong industrial oxidant.[8–10] NO also acts as a precursor to reactive species like peroxynitrite, which is involved in apoptotic cell death.[11,12] The important role of NO and N_2O in biology has, logically, led to substantial research in its chemistry and biochemistry.

NO is a gas that is sparingly soluble in water under ambient conditions and, like O_2, it is paramagnetic. NO is known to interact very strongly with several transition metals and has a strong affinity for iron in its formal 2+

RSC Metallobiology Series No. 9
Metalloenzymes in Denitrification: Applications and Environmental Impacts
Edited by Isabel Moura, José J. G. Moura, Sofia R. Pauleta and Luisa B. Maia

Published by the Royal Society of Chemistry, www.rsc.org

state.[13–16] The geometric and electronic structures of NO-bound transition metal complexes are intriguing and have been a matter of investigation for more than five decades.[17–21] In particular, NO is a known non-innocent ligand due to its capacity to oxidise and reduce the metal centre it binds.[22,23] As a result, Enemark–Faltham nomenclature had to be developed to describe these transition metal nitrosyl complexes.[24]

Three subclasses of metalloenzymes, all of which use iron-based active sites, are known to catalyse the reduction of NO to N_2O in nature. These are the cytochrome *c*- or quinol-dependent haem bacterial NO reductases (NORs), the flavin- or NADPH-dependent non-haem di-iron NORs and the cytochrome P450-dependent fungal NO synthases (P450nor).[25–27] The reaction mechanism of flavin-dependent NOR has been extensively investigated using advanced spectroscopic techniques and has been reviewed in the recent literature.[28–31] Similarly, though limited, several groups have been investigating the reaction mechanism of P450nor.[28–30,32,33] The cytochrome *c*- or quinol-dependent NORs belong to the haem-Cu oxidase family, which also includes cytochrome *c* oxidase (CcO) as its member. The similarities in the structure and function of CcO and NOR have often been traced to their common ancestry.[34,35] Both of these membrane-bound enzymes involve an electron transfer chain that shuttles electrons from cytochromes/quinols to the active site involved in the reduction of NO or O_2. However, evidence of NO reduction resulting in the creation of a proton gradient that is coupled to ATP synthesis is scanty.[5] The active site of NOR contains a histidine-bound haem a_3 cofactor and a three-histidine and one-glutamate-bound non-haem cofactor (Fe_B) in the distal side (Figure 9.1, left).[36] This is quite similar to the organisation of the active site of CcO, where the distal Fe is replaced by Cu (Figure 9.1, right). The other significant difference between the two active sites is the substitution of the tyrosine residue in CcO with two glutamates in NOR. CcO and NOR possess a "yin–yang" relationship with each other.[37] The role of NO in regulating the function of CcO was established a while ago. NO is known to reversibly inhibit the function of CcO.[38–41] Similarly, O_2 is known

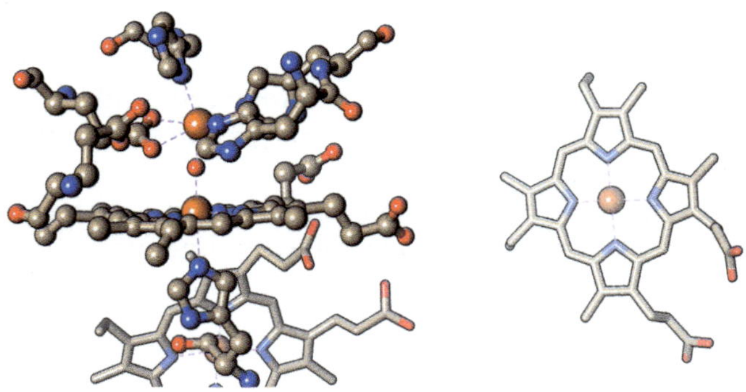

Figure 9.1 The active site of cytochrome *c* nitric oxide reductase (top, ball stick) and the active site of bovine CcO (bottom). Colour scheme: Fe, orange; Cu, brown; N, blue; O, red; and C, beige.

to inhibit the NO reduction at NOR active sites. Sluggish NO reduction is reported for some ba_3-type CcOs and, recently, O_2 reduction by NOR has been reported.[42–47]

The use of synthetic model complexes that mimic the structure and function of natural enzymes provides a distinct advantage when investigating finer details of electronic structure and reaction mechanism over conventional site-directed mutagenesis. The synthesis of smaller analogues is relatively easy and often more economical. The ability to achieve any desirable chemical modifications allowing premeditated changes in the primary ligand environment and the second sphere is difficult to achieve in a natural system, as one is often limited by the stability of the resultant mutant protein. Synthetic systems allow defined spatial control of the perturbation that is introduced, not affected by complications associated with folding of the mutant protein. However, without exception, the most elegant synthetic systems have not yet been able to match the catalytic efficiency of the natural enzymes. That does not deter the synthetic pursuit of structural and functional models, which have yielded unprecedented insight into the electronic structures of these metalloenzyme active sites. In particular, the synthetic structural and functional models of iron–sulphur clusters in electron transfer proteins, copper nitrite reductases, non-haem iron proteins, oxygen-binding haem proteins and molybdenum/tungsten proteins have enhanced the understanding of these systems substantially.[48–56]

Developments in synthetic modelling have been complemented by the use of biosynthetic models. In a biosynthetic model, the ligand framework is provided by a peptide. Peptides with known folding behaviours have been used to generate models of multi-haem proteins, Cu_A analogues, iron–sulphur clusters and O_2-activating enzymes, amongst others.[57–61] While this approach is limited by the availability of functional groups, the resultant constructs are water soluble, allowing investigations under physiological conditions. Of course, synthetic peptides beyond a certain number of amino acids may not fold into the quaternary structures adopted by the native enzyme, which impedes the modelling of larger proteins. Alternatively, unnatural active sites have been created inside known proteins by a series of site-directed mutagenesis processes. This approach offers more creative space contingent on the stability of the parent protein.

This chapter will focus on the developments made in the understanding of the structure–function correlations and reaction mechanisms of haem-dependent bacterial NORs using synthetic complexes and biosynthetic models.

9.2 Synthetic Models of Haem/Non-Haem NOR

9.2.1 Design and Synthesis

The synthetic strategies for the preparation of the synthetic haem NOR model complexes that have been developed has been summarised into four categories: (i) face selection and insertion of the proximal imidazole ligand

(Scheme 9.1); (ii) synthesis of the glutamic acid mimic (Scheme 9.2); (iii) attachment of distal imidazole ligands (Scheme 9.3); and (iv) introduction of metals (Scheme 9.4).[62–64]

9.2.1.1 *Face Selection and Insertion of the Proximal Imidazole Ligand*

Nitrobenzaldehyde couples with pyrrole to generate a tetranitroporphyrin, which, when reduced, forms a mixture of four atropisomers of meso-tetra(*ortho*-aminophenyl)porphyrin.[65] This is derivatised by the addition of one equivalent of trityl bromide.[66] The reaction mixture is heated and stirred over alumina, which causes slow isomerisation until one isomer is selectively adsorbed on the alumina, containing three amine groups on one side of the porphyrin ring and the bulky trityl group on the opposite side. This desired isomer is washed away from the alumina, followed by the protection of the three amine groups with trifluoroacetyl groups. Subsequently, the trityl group is removed and an imidazole-substituted acid chloride introduced as the proximal imidazole ligand.[49,67] This imidazole-substituted acid chloride can be derivatised as required. For ^{19}F nuclear magnetic resonance studies, a CF_3 group can be introduced or a terminal alkyne group can be appended to covalently

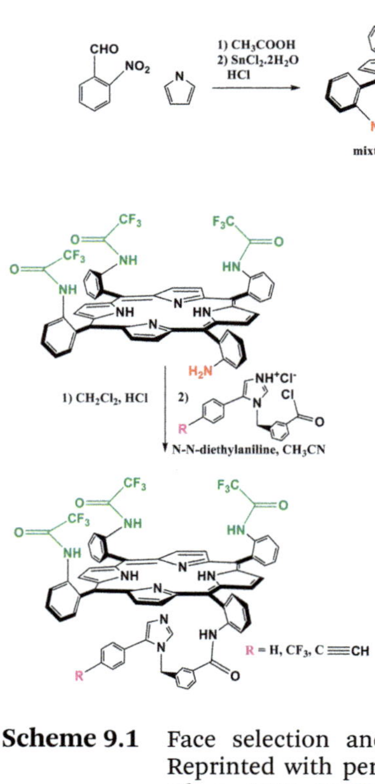

Scheme 9.1 Face selection and insertion of the proximal imidazole ligand. Reprinted with permission from J. P. Collman and S. Ghosh, *Inorg. Chem.*, 2010, **49**, 5798. Copyright © 2010, American Chemical Society.

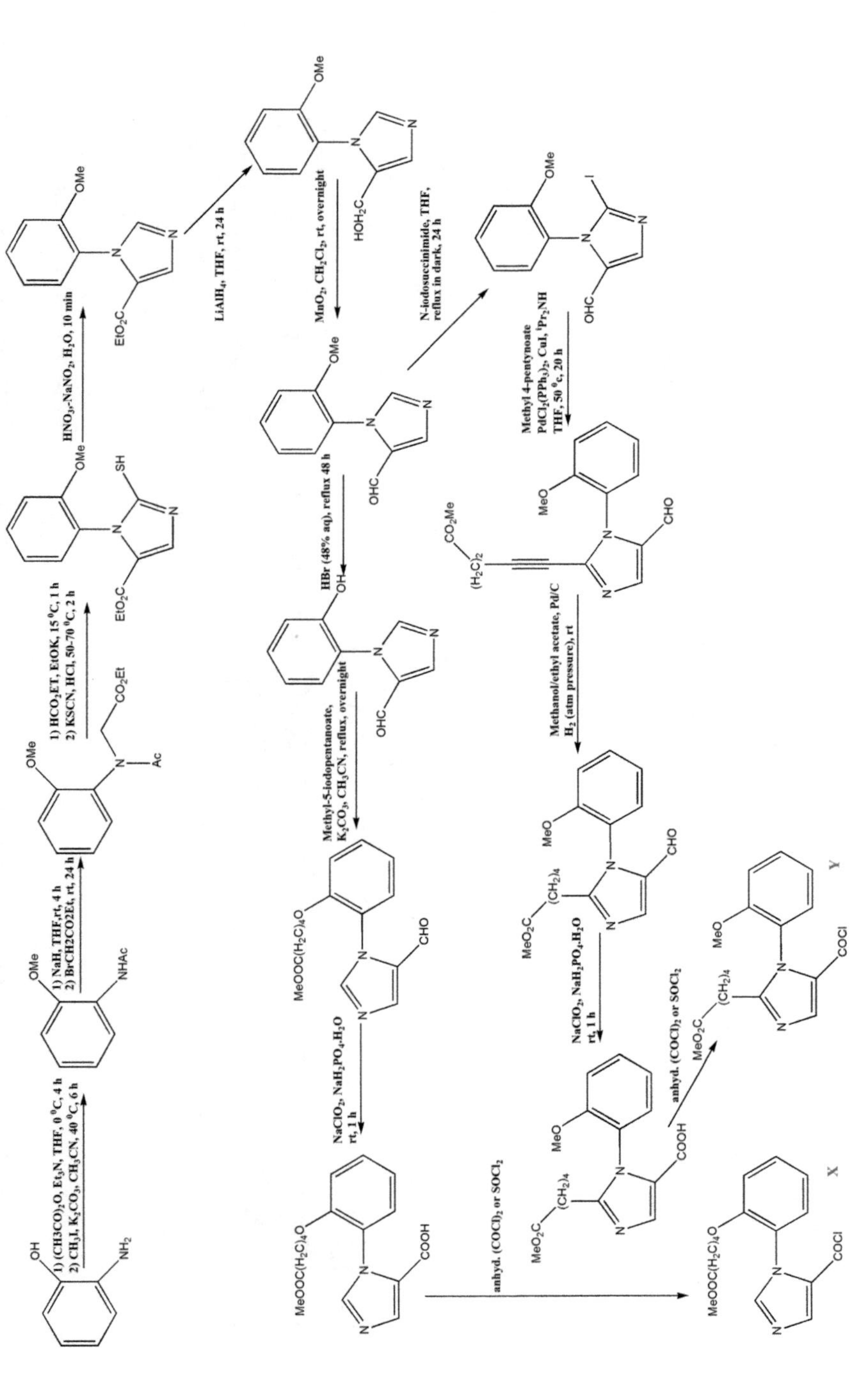

Scheme 9.2 Synthesis of imidazole containing an aliphatic carboxylic acid.

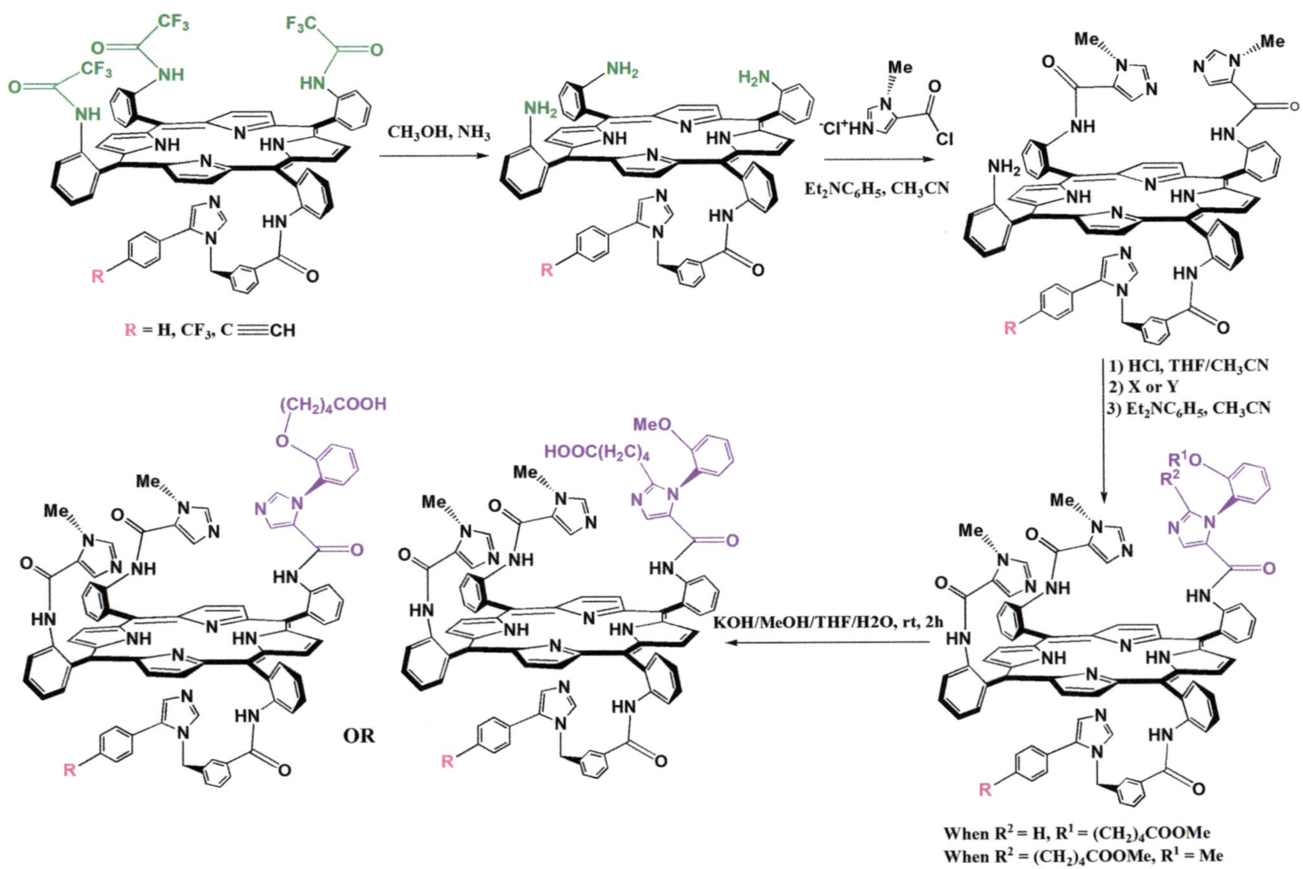

Scheme 9.3 Attachment of distal imidazole ligands. Reprinted with permission from J. P. Collman and S. Ghosh, *Inorg. Chem.*, 2010, **49**, 5798. Copyright © 2010, American Chemical Society.

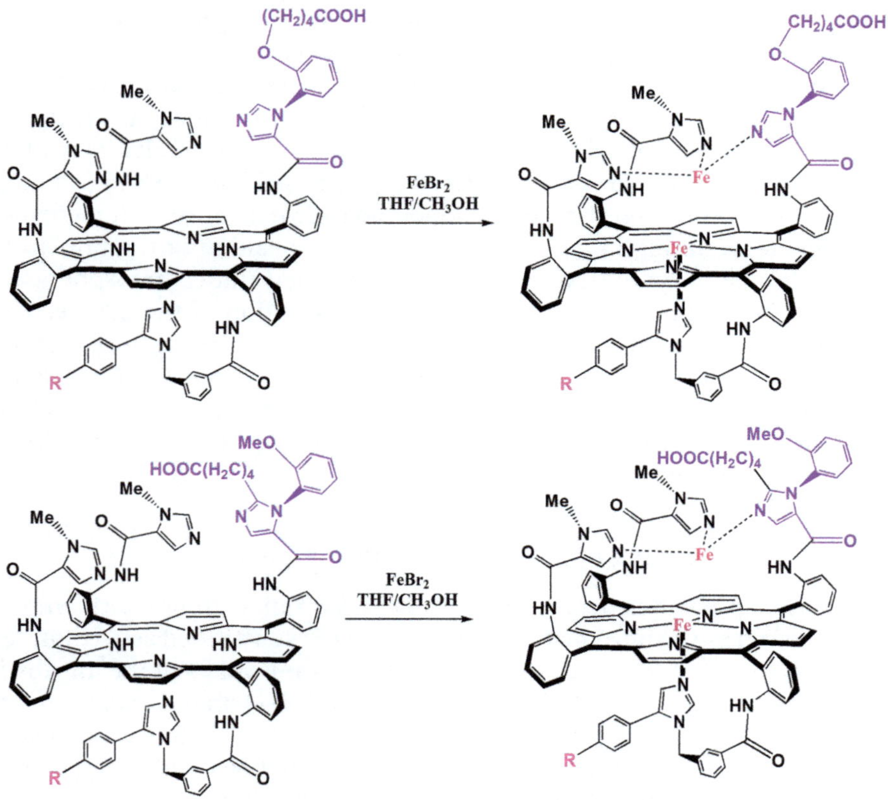

Scheme 9.4 Introduction of metals. Reprinted with permission from J. P. Collman and S. Ghosh, *Inorg. Chem.*, 2010, **49**, 5798. Copyright © 2010, American Chemical Society.

attach to electrodes using "click chemistry" in order to study the electro-catalytic two-electron reduction of NO to N_2O (Scheme 9.1).[68]

9.2.1.2 *Synthesis of the Glutamic Acid Mimic*

The amino and hydroxy groups of 2-aminophenol are first protected as the acetamide and methyl ether, respectively. This, when treated with NaH followed by ethyl bromoacetate, results in the formation of the N-aryl glycine ester. Formylation with HCO_2Et–EtOK in benzene followed by cycloaddition with HSCN forms a mercaptoimidazole ester intermediate, which, when oxidised by HNO_3–$NaNO_2$, forms the imidazole-ester derivative. The imidazole-ester is reduced to the corresponding alcohol by $LiAlH_4$ in THF, which can then be oxidised by excess MnO_2 in CH_2Cl_2 to generate the aldehyde functionality. On refluxing the compound in 48% aqueous HBr, the phenol derivative can be acquired from the methyl ether derivative. O-alkylated-imidazole aldehyde can be formed on refluxing the phenol derivative and methyl 5-iodopentanoate in the presence of excess K_2CO_3 in dry acetonitrile. Subsequent oxidation of the O-alkylated-imidazole aldehyde with $NaClO_2$/NaH_2PO_4 leads to the formation of the pure O-alkylated imidazole acid.

The methyl phenyl ether derivative of the imidazole aldehyde on refluxing with 2 equivalents of N-iodosuccinimide in dry THF results in the formation of an iodoimidazole derivative. A Pd-catalysed Sonogashira coupling reaction has been used to react the iodoimidazole derivative with methyl 4-pentynoate, resulting in derivatisation at the C2 position of the imidazole ring. Subsequent hydrogenation of the alkyne moiety under atmospheric H_2 pressure in the presence of Pd/C led to hydrogenation of the alkyne. The aldehyde group can be oxidised by $NaClO_2$/NaH_2PO_4, resulting in the formation of the C-alkylated imidazole acid. Both the O-alkylated and C-alkylated imidazole acids can be readily converted to their corresponding acyl chlorides by reaction with oxalyl chloride or thionyl chloride under anhydrous conditions (Scheme 9.2).

9.2.1.3 Attachment of Distal Imidazole Ligands

The next step involves deprotection of the three amine groups with ammonia, followed by the introduction of two imidazole-substituted acid chlorides (Scheme 9.3).[63,67,69] This leads to *cis* and *trans* regio-isomers of the unsubstituted amine ligand. The desired *cis*-isomer can be separated using a rotating chromatron. The remaining free amine group is then derivatised using the O-alkylated or C-alkylated imidazole-substituted acid chloride (X or Y, Scheme 9.2). The methyl ester groups are then saponified, generating the free bases (Scheme 9.3).

9.2.1.4 Introduction of Metals

The free bases possess all of the key features of the NOR active site: porphyrin, proximal imdazole, two distal imidazoles and a third imidazole containing an aliphatic carboxylic acid. The remaining step is illustrated in Scheme 9.4, where Fe is introduced using a ferrous salt, which binds to both the porphyrin ring as well as to the distal imidazole ligands, leading to the formation of the reduced active catalyst.

9.2.2 NO Reactivity of Synthetic Haem/Non-Haem NOR Models

A synthetic model of NOR was synthesised that incorporates the key features of the active site, including a haem Fe centre with a proximal imidazole ligand and a non-haem Fe_B centre with a tris-imidazole coordination on the distal side of haem (Figure 9.2).[70] The glutamate residue present in the NOR active site is lacking in this synthetic model. The reaction of NO with the haemFe^{II} analogue of the model (in the absence of the Fe_B centre) results in

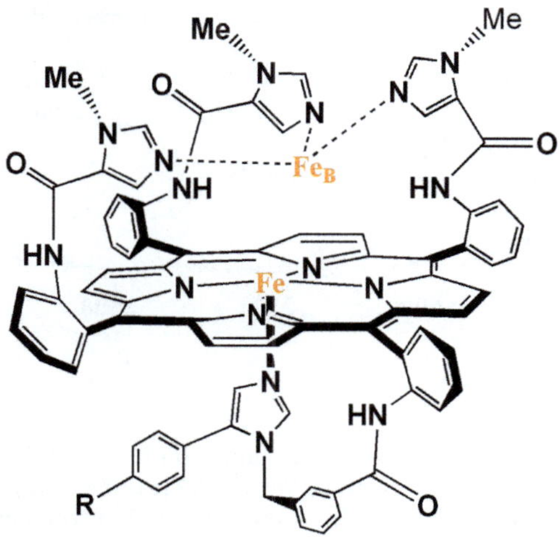

Figure 9.2 Synthetic model of nitric oxide reductase.

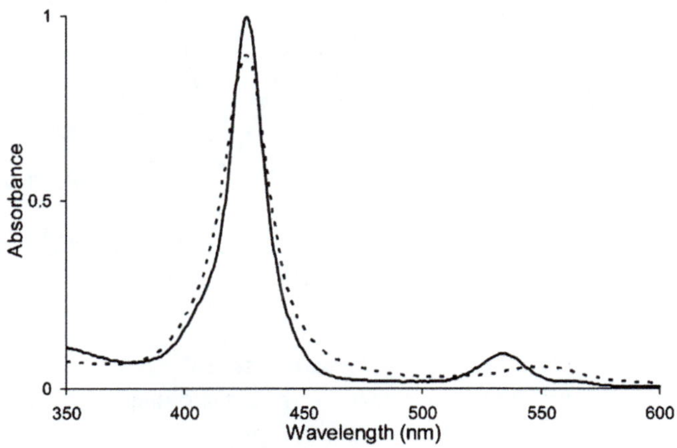

Figure 9.3 Absorption spectra of haemFe^{II} (solid line) and haemFe^{II}–NO (dotted line). Reproduced from ref. 37 with permission from the Royal Society of Chemistry.

the formation of a mono-nitrosyl haemFe^{II}–NO adduct. It is characterised by a decrease in the Soret band intensity and a red shift of the Q band in the absorption spectrum (Figure 9.3) and the formation of a low-spin, $S = 1/2$ electron paramagnetic resonance (EPR) signal (Figure 9.4, solid line), typical of six coordinate ferrous–NO complexes. The mono-nitrosyl haemFe^{II}–NO adduct has been further characterised by vibrational spectroscopy. Fourier transform infrared (FTIR) spectrum shows a new band at 1630 cm^{-1}, which shifts by ~30 cm^{-1} to 1600 cm^{-1} on using isotopically enriched ^{15}NO (Figure 9.5a). Similarly, the resonance Raman data of the haemFe^{II}–NO adduct show a shift of the Fe–N stretching frequency from 581 cm^{-1} to 545 cm^{-1} with ^{15}NO

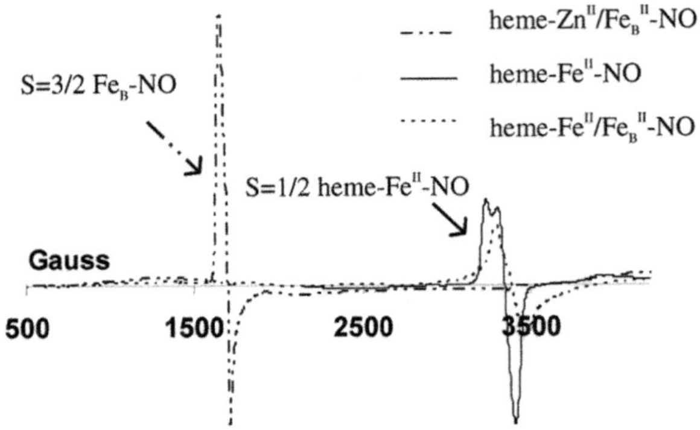

Figure 9.4 Electron paramagnetic resonance data of the haemFe^{II}–nitric oxide (NO) (solid line), haemZn^{II}/Fe$_B$^{II}–NO (dotted line) and haemFe^{II}/Fe$_B$^{II}–NO (dotted-to-dashed line) complexes. Reproduced from ref. 37 with permission from the Royal Society of Chemistry.

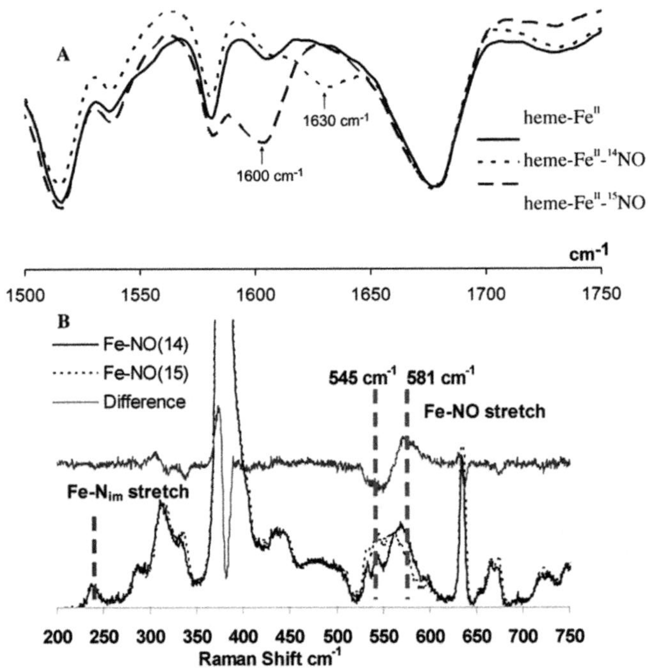

Figure 9.5 (a) Infrared spectra of haemFe^{II} (solid line), haemFe^{II}–^{14}NO (dotted line) and haemFe^{II}–^{15}NO (dashed line) derivatives and (b) resonance Raman spectra of haemFe–^{14}NO (solid line) and haemFe–^{15}NO (dotted line). Reproduced from ref. 37 with permission from the Royal Society of Chemistry. NO: nitric oxide.

(Figure 9.5b, excited at 425 nm), confirming the formation of a haemFe^{II}–NO complex. The presence of a Fe–$N_{Imidazole}$ stretch observed at 238 cm^{-1} provides further evidence for the presence of a six-coordinate iron nitrosyl complex (Figure 9.5b).

Exposing NO to the haemZn^{II}/Fe_B^{II} complex leads to the formation of the mono-nitrosyl Fe_B^{II}–NO adduct. Zn^{II} is loaded into the haem site to make it EPR silent. There is a slight blue shift of the Soret (1 nm) and Q (0.5 nm) bands of the nitrosylated adduct relative to the haemZn^{II}/Fe_B^{II} absorption spectrum. This negligible perturbation of the Soret band likely indicates that NO binding to the Fe_B centre does not bridge with the haemZn centre. The EPR spectrum shows a characteristic $g = 3.92$, $S = 3/2$ signal of a non-haem Fe–NO species (Figure 9.4). The FTIR spectrum of the mono-nitrosyl Fe_B^{II}–NO adduct shows a new band at 1810 cm^{-1}, absent in the haem-Zn^{II}/Fe_B^{II} complex, which gets shifted to 1774 cm^{-1} upon ^{15}NO substitution, implying the formation of a nitrosyl adduct (Figure 9.6). The Fe_B–NO complex could not be characterised by resonance Raman, as the spectrum was obscured by porphyrin vibrations.

The fully reduced NOR model (*i.e.*, haemFe^{II}/Fe_B^{II}) when exposed to NO results in a blue shift of the Soret band and a red shift of the Q band in the absorption spectrum (Figure 9.7). The EPR spectrum shows a low-spin Fe^{III} $S = 1/2$ signal (Figure 9.4). Spin integration of the data indicates the presence of only one Fe^{III} species. The FTIR spectrum of the NO adduct of the haemFe^{II}/Fe_B^{II} complex shows a new band at 1924 cm^{-1} relative to the starting haemFe^{II}/Fe_B^{II} complex, which shifts by ~37 cm^{-1} to 1887 cm^{-1} upon ^{15}NO substitution, consistent with the formation of haemFe^{III}–NO (Figure 9.8a). The resonance Raman data show a Fe–NO vibration at 589 cm^{-1} that shifts to 580 cm^{-1} on ^{15}NO substitution (Figure 9.8b), further indicating the formation of a haemFe^{III}–NO species.

Thus, the FTIR and resonance Raman data confirm the formation of a haemFe^{III}–NO species that is diamagnetic and hence EPR silent. The EPR data

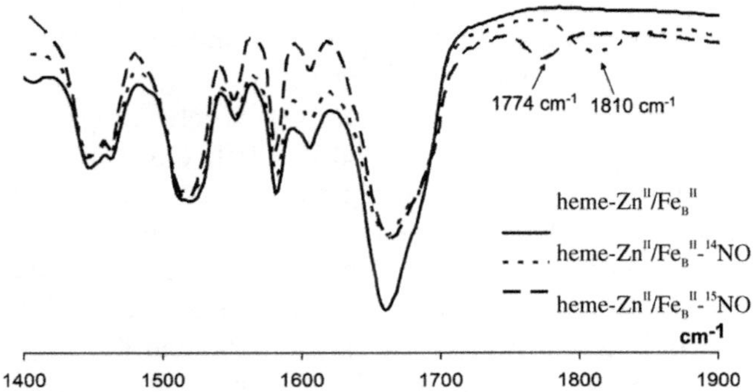

Figure 9.6 Infrared spectra of haemZn^{II}/Fe_B^{II} (solid line), haemZn^{II}/Fe_B^{II}–^{14}NO (dotted line) and haemZn^{II}/Fe_B^{II}–^{15}NO (dashed line) derivatives. Reproduced from ref. 37 with permission from the Royal Society of Chemistry. NO: nitric oxide.

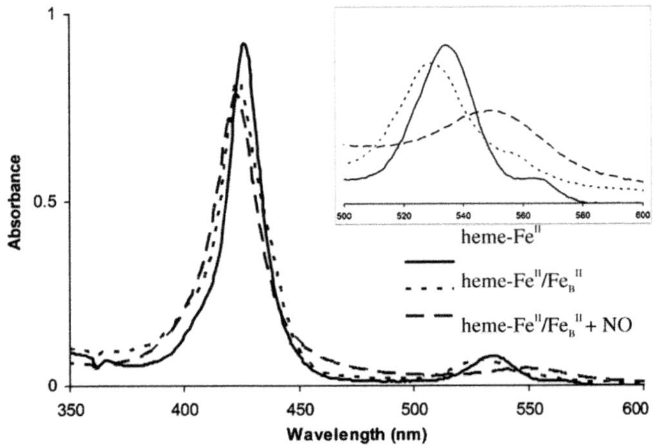

Figure 9.7 Absorption spectra of haemFe^{II} (solid line), haemFe^{II}/Fe$_B^{II}$ (dotted line) and haemFe^{II}/Fe$_B^{II}$–nitric oxide (NO) (dashed line). Inset: expanded Q band region. Reproduced from ref. 37 with permission from the Royal Society of Chemistry.

showing an $S = 1/2$ signal imply the formation of a Fe$_B^{III}$–OH species. Thus, NO addition to haemFe^{II}/Fe$_B^{II}$ generates haemFe^{III}–NO/Fe$_B^{III}$–OH (Scheme 9.5d). This has been confirmed by the formation of the same product on adding NO to the fully oxidised haemFe^{III}/Fe$_B^{III}$ complex, followed by the addition of 1 equivalent of sodium methoxide (analogous to OH). The species generated in the latter case had spectroscopic features similar to those obtained upon reacting the fully reduced haemFe^{II}/Fe$_B^{II}$ model with NO. When NO is exposed to the mixed valence state complex (*i.e.*, haemFe^{III}/Fe$_B^{II}$), it generates a mixture of haemFe^{III}–NO/Fe$_B^{II}$–NO and haemFe^{II}–NO/Fe^{III} (Scheme 9.5c). While the mixed-valence complex has the haemFe oxidised and the Fe$_B$ reduced, in the presence of NO, an equilibrium between the haemFe^{III}/Fe$_B^{II}$ and the haemFe^{II}/Fe$_B^{III}$ is attained. The haemFe^{III}/Fe$_B^{II}$ complex binds two molecules of NO to generate the haemFe^{III}–NO/Fe$_B^{II}$–NO species, while the haemFe^{II}/Fe$_B^{III}$ species can only bind one molecule of NO (high-spin non-haem Fe^{III} does not bind NO), generating the haemFe^{II}–NO/Fe^{III} species.

The protein nitrous oxide reductase (N_2OR), which catalyses the two-electron reduction of N_2O to N_2 in bacterial denitrification, has been used to detect the generation of N_2O in the reaction of NO with haemFe^{II}/Fe$_B^{II}$. It has been shown that N_2O is produced quantitatively in this reaction, where two molecules of NO are used to generate one molecule of N_2O. N_2O is not detected when NO is reacted with the mixed-valence haemFe^{III}/Fe$_B^{II}$ species, indicating that only the fully reduced bis-ferrous haemFe^{II}/Fe$_B^{II}$ form is the active form of the synthetic NOR model, which can reduce two molecules of NO to N_2O under ambient conditions. It must be mentioned that apart from the functional model described here, an elegant synthetic model of haem copper oxidase has also been demonstrated to be capable of reducing NO to N_2O under ambient conditions.[71,72]

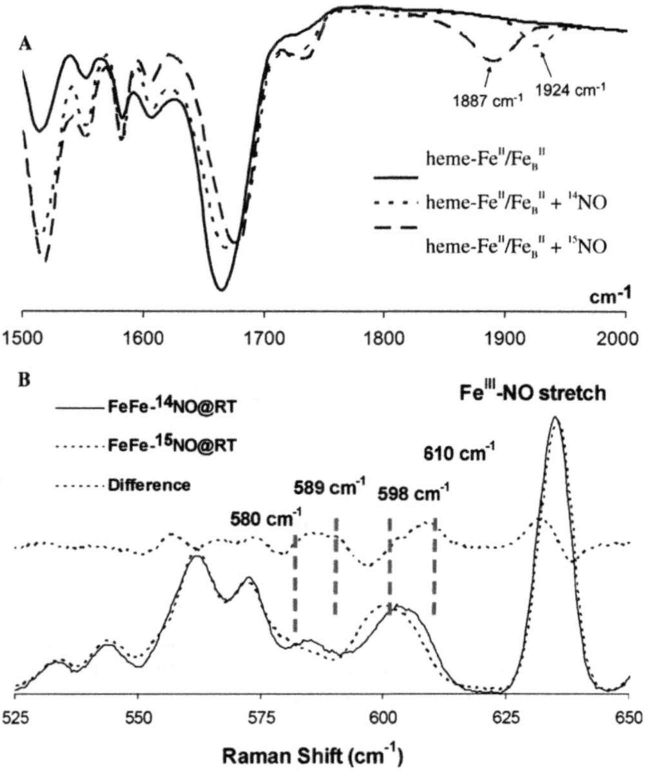

Figure 9.8 (a) Infrared spectra of haemFe^{II}/Fe$_B$^{II} (solid line), haemFe^{II}/Fe$_B$^{II}–^{14}NO (dotted line) and haemFe^{II}/Fe$_B$^{II}–^{15}NO (dashed line) derivatives and (b) resonance Raman spectra of haemFe^{II}/Fe$_B$^{II}–^{14}NO (solid line) and haemFe^{II}/Fe$_B$^{II}–^{15}NO (dotted line). Reproduced from ref. 37 with permission from the Royal Society of Chemistry. NO: nitric oxide.

9.2.3 Mechanism of NO Reduction by a Synthetic Haem/Non-Haem NOR Model

NORs reduce two molecules of NO to generate one molecule of N_2O. There are two proposed mechanisms of NO reduction.[2,73] The "*trans*" mechanism involves binding of two molecules of NO, one each, to the two reduced Fe centres of NOR (haemFe^{II} and Fe$_B$^{II}), followed by N–N bond formation and further rearrangement, leading to the formation of N_2O and a bis-ferric resting active site.[74,75] Alternatively, the "*cis*" mechanism proposes binding of both the NO molecules to one of the Fe centres, most likely the Fe$_B$ centre.[32,76] Additionally, it has also been suggested that two molecules of NO bind consecutively to the haemFe^{III} site, forming a hypo-nitrite-like intermediate, which decays to produce N_2O, H_2O and haemFe^{III}.[76–78]

NO has been exposed to the fully reduced functional NOR model (*i.e.*, haemFe^{II}/Fe$_B$^{II}) (Figure 9.2) in dichloromethane solution and, subsequently, reaction intermediates have been trapped at different temperatures with

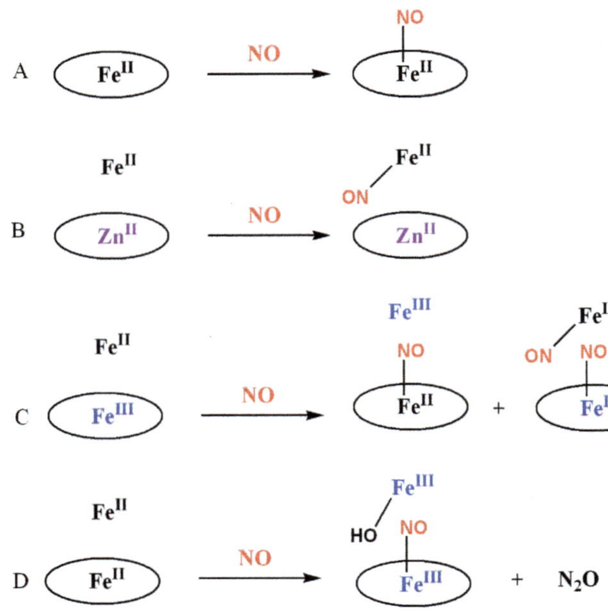

Scheme 9.5 Reactions of nitric oxide (NO) with (a) haemFe, (b) haemZn^{II}/Fe$_B$, (c) haemFe^{III}/Fe$_B$^{II} and (d) haemFe^{II}/Fe$_B$^{II}. Reproduced from ref. 37 with permission from the Royal Society of Chemistry.

the aim of elucidating the reaction mechanism of NO reduction by the synthetic functional NOR model.[74] Intermediate A is trapped at −80 °C, which has an EPR spectrum with $g = 3.92$ (Figure 9.9, red), identical to that of the haemZn^{II}/Fe$_B$^{II}–NO complex (Figure 9.9, pink), implying the formation of an $S = 3/2$ mononitrosyl Fe$_B$–NO species. The υ_4 band in the resonance Raman spectrum is very sensitive to the charge and the spin state of the bound haemFe species. The rR data of intermediate A in the high-energy region (Figure 9.10a, red) shows a υ_4 band at 1356.5 cm^{-1}, characteristic of haemFe^{II}, indicating that NO does not bind at the haemFe^{II} centre in this intermediate. On warming the reaction to −40 °C, intermediate B is generated, which is EPR silent (Figure 9.9, blue). The υ_4 band of intermediate B is observed at 1366 cm^{-1} (Figure 9.10a, blue), implying the formation of a haemFe^{II}–NO species. Another vibration observed at 587 cm^{-1}, which shifts to 559 cm^{-1} on ^{15}NO substitution, is assigned as the Fe–NO stretching frequency of the haemFe^{II}–NO species (Figure 9.10b). Warming the reaction mixture to room temperature produces a haemFe^{III}–NO/Fe$_B$^{III}–OH species (Figures 9.9 and 9.10).

Thus, intermediate A has been identified as a haemFe^{II}/Fe$_B$–NO species, based on its $S = 3/2$ EPR signal and the υ_4 band of the haemFe^{II} centre in the resonance Raman spectrum. Intermediate B has a haemFe^{II}–NO species, as indicated by its characteristic resonance Raman bands. Intermediate B can be described as either a mononitrosyl haemFe^{II}–NO species, the NO of which is derived from Fe$_B$ at elevated temperatures, or it could be a bis-nitrosyl species. When NO

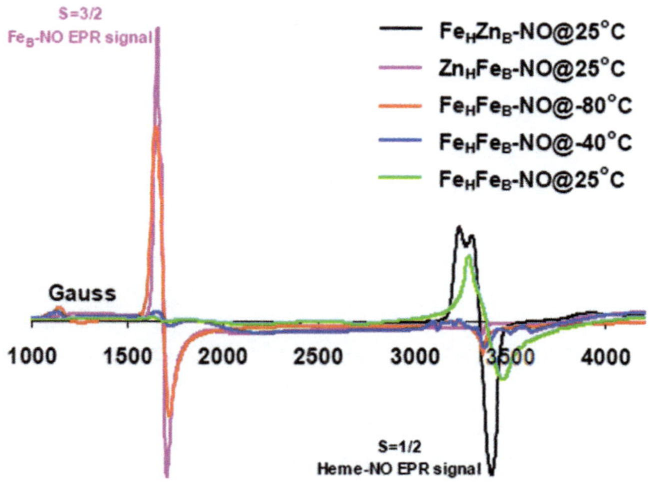

Figure 9.9 Electron paramagnetic resonance (EPR) data of the intermediates and final product in the reaction of nitric oxide (NO) with the bis-ferrous model. All samples are 1 mM in CH_2Cl_2. Data were obtained at 4 K and 10 mW microwave power. Reproduced from ref. 37 with permission from the Royal Society of Chemistry.

is added to haemFe^{II}/Fe$_B{}^{II}$ in DMF, a bis-nitrosyl species is formed, with both an S = 3/2 EPR signal (associated with haemFe^{II}/Fe$_B{}^{II}$–NO) and an S = 1/2 EPR signal (associated with haemFe^{II}–NO/Fe$_B{}^{II}$) and a resonance Raman band at 579 cm^{-1} (associated with haemFe^{II}–NO/Fe$_B{}^{II}$). In this reaction, N_2O could not be detected. Additionally, it has been observed that the haemZn^{II}/Fe$_B{}^{II}$–NO species shows no NO dissociation at room temperature. These observations suggest that intermediate B is a bis-nitrosyl species. The two NOs are close enough that they can interact and result in an EPR-silent diamagnetic species.

This synthetic functional model of NOR supports a "*trans*" mechanism of NO reduction by NORs (Scheme 9.6). In the first step, NO binds to the Fe$_B$ centre, generating a mononitrosyl high-spin Fe$_B{}^{II}$–NO species (intermediate A), followed by binding of a second NO molecule to haemFe, forming haem-Fe^{II}–NO/Fe$_B{}^{II}$–NO (intermediate B), which then generates the oxidised bis-ferric end product and N_2O, thus directly supporting the "*trans*" mechanism of NO reduction to N_2O by NORs. In these reactions, NO may be formed *via* the generation of an iron nitroxyl species (HNO). HNO generated from such a reaction can dimerise to generate N_2O. However, the reduction potential of a haemFe^{II}–NO species is much lower (−0.9 V *vs.* SCE) than the reduction potential of Fe$_B$ (−0.2 V *vs.* SCE) in non-aqueous solvents, which essentially precludes the formation of haemFe nitroxyl species under these conditions (*i.e.*, in the absence of stronger reducing agents).

The same mechanism, originally proposed by these synthetic models, have now been established to be operative in both the enzymatic active site as well as in biosynthetic models of NOR.[79,77] This aptly justifies the use of synthetic models in investigating the reactivity of metalloenzyme active sites.

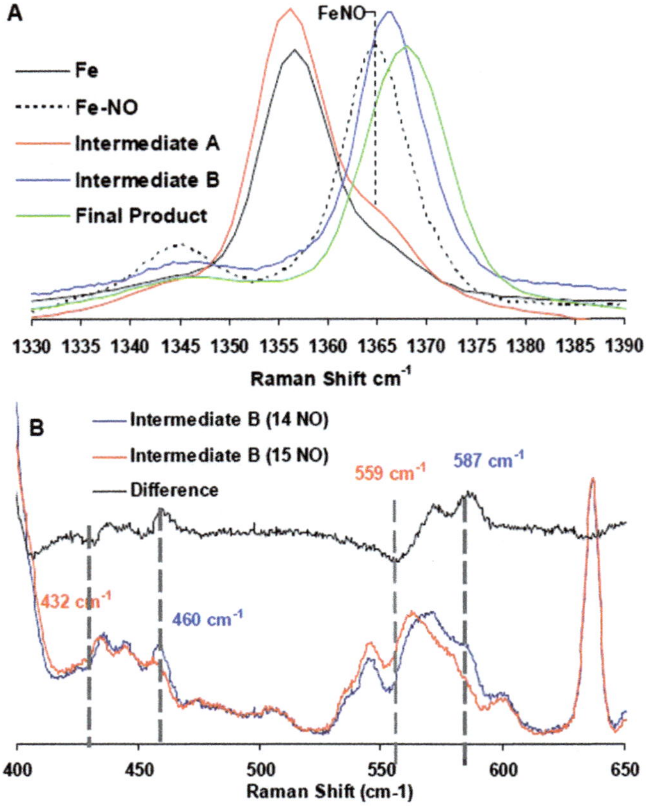

Figure 9.10 Resonance Raman data of the starting bis-ferrous complex, intermediates and end product (a) high-energy region (includes haemFe^{II}–nitric oxide [NO] data for reference) and (b) low-energy region. Reproduced from ref. 37 with permission from the Royal Society of Chemistry.

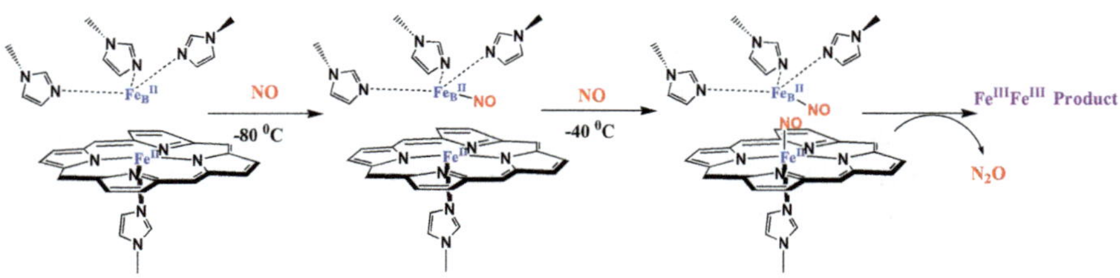

Scheme 9.6 Proposed mechanism of N_2O formation from nitric oxide (NO) by NO reductase. Reproduced from ref. 37 with permission from the Royal Society of Chemistry.

9.3 Engineered Myoglobin-Based NOR Model

9.3.1 Synthesis

A structural and functional model of NOR has been designed and synthesized that incorporates the three histidines and one glutamate present in the active site of NOR in the distal pocket of myoglobin (Mb) (Figure 9.11).[80] The Fe_B centre is designed in sperm whale Mb.[81] An overlay of the crystal structures of structurally analogous CcO and Mb identifies His_{64} as one of the distal histidine ligands of the Fe_B site of NOR. Leu_{29} and Phe_{43} are mutated to histidines to create the non-haem Fe_B metal-binding site. Val_{68} is mutated to glutamate in the Mb active site based on its proximity and angle to the haem site and the three histidine ligands of Fe_B. The designed deoxy Fe_BMb protein (haemFe is reduced) when loaded with Fe^{II} shows a slight shift of the Soret band from 433 to 434 nm, and the 557 nm band in the Q band region splits into two bands at 550 and 572 nm. However, no such spectral changes are observed when Fe^{II} is added to wild-type Mb or Fe_BMb in the absence of the glutamate residue (a Fe_B site with only three histidine residues). This clearly indicates that the histidine residues as well as the glutamate residue are essential for Fe binding in the distal pocket of NOR. A high-resolution crystal structure of the Fe_BMb (1.72 Å) reveals that the Fe_B site is five coordinate, bound to three histidines (2.18 Å to His_{29}, 2.12 Å to His_{43} and 2.20 Å to His_{64}), one to glutamate (2.19 Å to Glu_{68}) and a water molecule (2.11 Å). There is a weak interaction between Fe_B and the other O atom of Glu_{68} at 3.28 Å. The Fe_B is 4.8 Å away from the haem iron. The *in silico*-designed protein model overlaps well with the obtained crystal structure (Figure 9.11). Deoxy

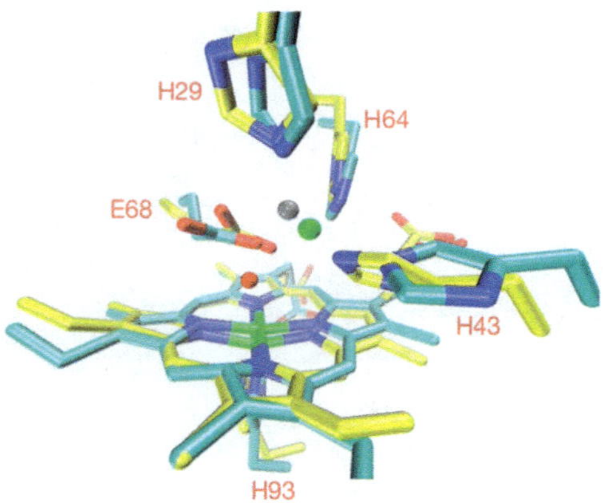

Figure 9.11 Overlay of Fe_BMb model (yellow) with the crystal structure of Fe^{II}–Fe_BMb (1.72 Å resolution, cyan). Fe^{II} of the Fe_B site is represented by a green sphere; Zn^{II} (grey sphere) is used to model the Fe_B site. A water molecule is represented by a red sphere. Reprinted with permission from N. Yeung, Y.-W. Lin, Y.-G. Gao, X. Zhao, B. S. Russell, L. Lei, K. D. Miner, H. Robinson and Y. Lu, *Nature*, 2009, **462**, 1079. Copyright © 2009, Nature Publishing Group.

Fe_BMb, in the absence of Fe_B (reduced haemFe^{II} site), is EPR silent. When the redox partner, oxidised blue copper azurin, is added to this, a high-spin $g = \sim 6$ haemFe^{III} EPR signal is observed, suggesting that azurin can oxidise the haemFe^{II} to haemFe^{III}. The EPR intensity of the high-spin haemFe^{III} signal decreases in the presence of the occupied Fe_B site, implicating that the Fe_B^{II} is spin-coupled to the haemFe^{III}, as observed in native NOR. This is further supported by the fact that there is a dramatic 112 mV (from -158 ± 4 mV to -46 ± 2 mV *vs.* NHE) increase of the reduction potential of the haemFe in the presence of Fe_B. Reduced Fe_BMb when reacted with excess NO produced ~10% N_2O, as confirmed by absorption spectroscopy and GC/MS. Thus, the designed Fe_BMb is a structural and functional model of NOR.

Subsequently, this model has been fine tuned to incorporate a second glutamate residue by mutating Ile_{107} in the second coordination sphere of the Fe_B site (Fe_BMb2, Figure 9.12). The 1.62 Å resolution crystal structure shows that there is a hydrogen bonding interaction between the Glu_{107} residue and the water-derived ligand coordinated to Fe_B. There is a significant conformational rearrangement of the Glu_{68} residue in the presence of the hydrogen bonding interaction between Glu_{107} and the water ligand. Similar to the Fe_BMb variant, the reduction potential of Fe_BMb2 increases from -134 ± 3 mV to -64 ± 3 mV *vs.* NHE in the presence of Fe_B, though the effect is less drastic compared to Fe_BMb. The Fe_B site can bind different metal ions like Cu and Zn, thereby providing an opportunity to compare the role of these metal ions in the same protein scaffold. In contrast to the Fe_BMb variant,

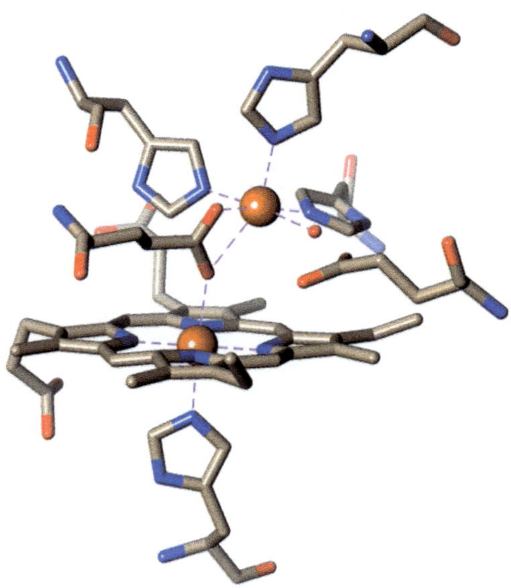

Figure 9.12 Crystal structure of Fe^{II}–I107E Fe_BMb (Protein Data Bank identifier: 3M39). Reprinted with permission from N. Yeung, Y.-W. Lin, Y.-G. Gao, X. Zhao, B. S. Russell, L. Lei, K. D. Miner, H. Robinson and Y. Lu, *Nature*, 2009, **462**, 1079. Copyright © 2009, Nature Publishing Group.

this Fe_BMb2 variant generates ~24% N_2O, indicating that the second glutamate residue plays an important role in NO reduction, possibly facilitating proton uptake during NO reduction.[82,83] A five-coordinate haem–NO species has been detected in NORs.[29,84,85] A six-coordinate haemFe–NO species is observed in Fe^{II}-bound Fe_BMb lacking the second Glu (E107).[79] However, a five-coordinate haemFe–NO species is formed when Fe^{II}, Cu^I or Zn^{II} is bound to the Fe_B site of Fe_BMb2, further implicating the role of Glu_{107} in NOR activity. This is clearly evidenced by the EPR data of the Zn_B adduct, where the characteristic signal for five-coordinate haemFe–NO can be observed (Figure 9.13). This signal cannot be seen when Fe_B or Cu_B are present, as both of these redox-active centres react with NO.

The haem site of NOR has a high affinity for binding NO (K_{eq} for NO binding to ferrous haems is ~10^{10}–10^{12} M^{-1}), and haem nitrosyl species have dominating spectroscopic features that mask the features of the Fe_B nitrosyl complexes. Because NORs are integral membrane proteins, replacing the haem cofactor(s) to selectively probe the effect of NO binding to the Fe_B site is non-trivial. However, in the biosynthetic model, the haem can be replaced by the isostructural Zn/protoporphyrin IX (ZnPP), thus allowing the spectroscopic characterisation of NO binding to the Fe_B site (Figure 9.14). The absorption spectrum of ZnPP-reconstituted Fe_BMb (*i.e.*, $ZnPPFe_BMb$) has bands at 427, 553 and 595 nm,

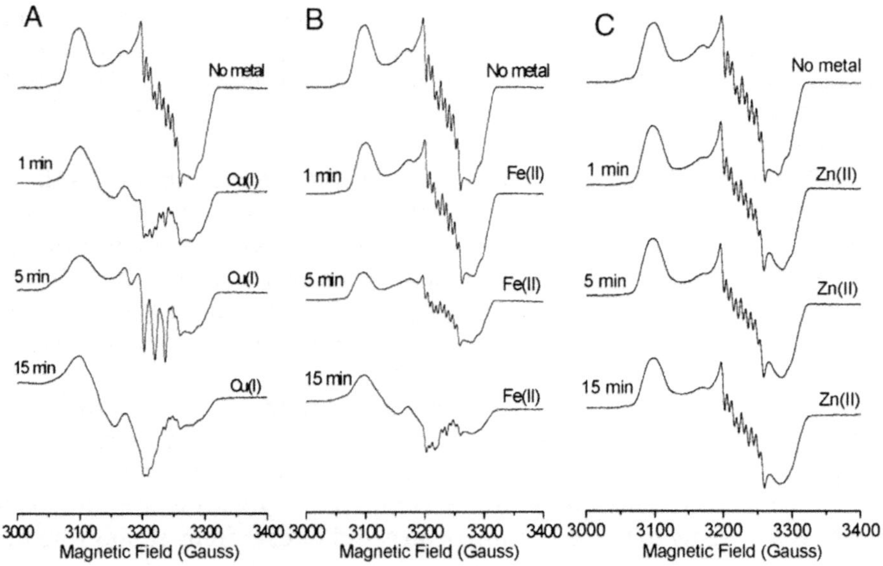

Figure 9.13 Electron paramagnetic resonance spectra of deoxy I107E Fe_BMb (0.5 mM) with no metal bound in the presence of nitric oxide (NO) after 5 minutes (top line), with 2 eq. Fe^{II} (a), Cu^I (b) or Zn^{II} (c) incubated with excess NO (~200 eq.) for 1, 5 and 15 minutes. Reproduced with permission from Y.-W. Lin, N. Yeung, Y.-G. Gao, K. D. Miner, L. Lei, H. Robinson and Y. Lu, *J. Am. Chem. Soc.*, 2010, **132**, 9970. Copyright © 2010 American Chemical Society.

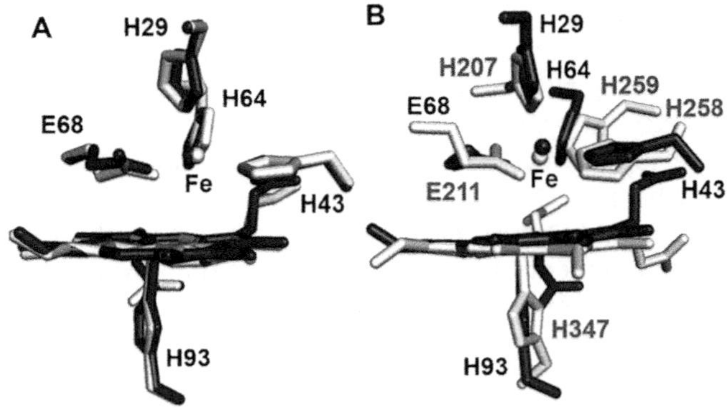

Figure 9.14 Replacement of haem in Fe_BMb (A) with Zn/protoporphyrin IX (ZnPP), thus yielding $ZnPPFe_BMb$ (B). Figures were generated in PyMol using Protein Data Bank codes 3K9Z and 4MXL, respectively. Reprinted with permission from S. Chakraborty, J. Reed, M. Ross, M. J. Nilges, I. D. Petrik, S. Ghosh, S. Hammes-Schiffer, J. T. Sage, Y. Zhang, C. E. Schulz and Y. Lu, *Angew. Chemie, Int. Ed.*, 2014, **53**, 2417. WILEY-VCH Verlag GmbH & Co. KGaA, Weinheim.

which are distinctly different from free ZnPP and Fe_BMb, indicating the incorporation of ZnPP in haem-extracted Fe_BMb. The crystal structure of $ZnPPFe_BMb$ at 1.5 Å resolution also shows the incorporation of ZnPP. The overlay of the crystal structures of Fe^{II}-loaded $ZnPPFe_BMb$, Fe_BMb and cytochrome *c*-dependent NOR (cNOR) show that these structures are very similar to each other (Figure 9.15), indicating that the Fe_B site of Fe^{II}–$ZnPPFe_BMb$ is a structural analogue of the non-haem site of cNOR and Fe_BMb.

Loading the Fe_B site with 1.0 equivalent of $FeCl_2$ results in a red shift of the Soret band from 427 nm to 429 nm (Figure 9.16). Interestingly, there is a three-fold increase in the Fe^{II} binding affinity of Fe_B when haem is replaced with ZnPP in Fe_BMb. Adding 5–20 equivalents of NO to Fe^{II}–$ZnPPFe_BMb$ results in two well-resolved doublets at $g = 4.36$, 3.58 and 4.13, 3.73 (Figure 9.17), leading to the formation of a $\{FeNO\}^7$ ferrous nitrosyl complex with an $S = 3/2$ ground state.[86]

Mossbauer and QM/MM calculations describe the $\{FeNO\}^7$ complex in Fe^{II}–$ZnPPFe_BMb$–NO as a high-spin ferrous ($S = 2$) species antiferromagnetically coupled with the NO radical ($S = 1/2$) [Fe^{2+}–$NO^•$]. This radical nature of the NO moiety can facilitate the radical coupling with a second haem-bound NO, leading to the formation of an N–N bond, thereby supporting the proposed *trans* mechanism of NO reduction by NORs.

9.3.2 Mechanism

The mechanism of N–N bond formation in the wild-type enzyme has been a matter of debate. Three popular pathways are considered for the two-proton, two-electron reduction of NO to N_2O. These are: the *trans* pathway,

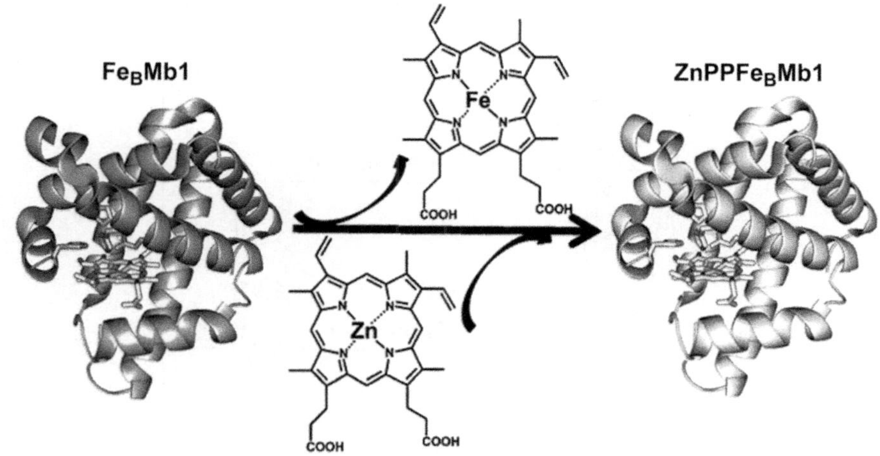

Figure 9.15 Overlay of Fe^{II}–$ZnPPFe_BMb$ (Protein Data Bank identifier: 4MXK, dark grey) with Fe^{II}–Fe_BMb (Protein Data Bank identifier: 3K9Z, light grey) and cytochrome *c*-dependent nitric oxide reductase (Protein Data Bank identifier: 3O0R, light grey). Figures were generated using PyMol. Reprinted with permission from S. Chakraborty, J. Reed, M. Ross, M. J. Nilges, I. D. Petrik, S. Ghosh, S. Hammes-Schiffer, J. T. Sage, Y. Zhang, C. E. Schulz and Y. Lu, *Angew. Chemie, Int. Ed.*, 2014, **53**, 2417. WILEY–VCH Verlag GmbH & Co. KGaA, Weinheim.

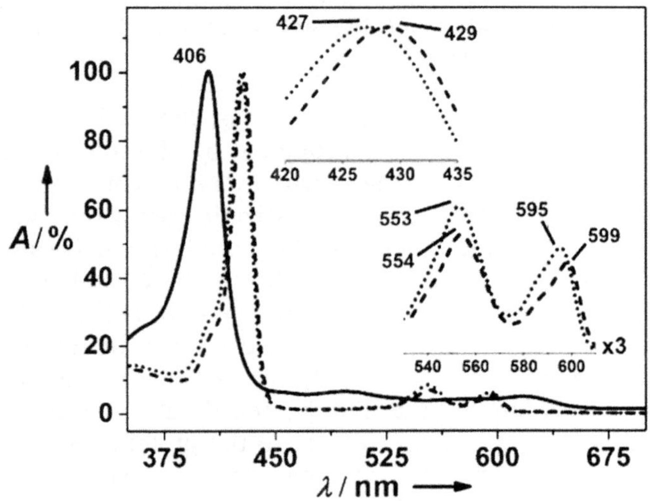

Figure 9.16 UV/visible spectra of Fe_BMb (solid line), $ZnPPFe_BMb$ (dotted line) and $ZnPPFe_BMb$ in the presence of 1.0 eq. Fe^{II} (dashed line) in 50 mm bis-tris buffer pH 7.3. Peak positions in the Soret and visible regions of $ZnPPFe_BMb$ and Fe^{II}–$ZnPPFe_BMb$ are shown as insets. Reprinted with permission from S. Chakraborty, J. Reed, M. Ross, M. J. Nilges, I. D. Petrik, S. Ghosh, S. Hammes-Schiffer, J. T. Sage, Y. Zhang, C. E. Schulz and Y. Lu, *Angew. Chemie, Int. Ed.*, 2014, **53**, 2417. WILEY–VCH Verlag GmbH & Co.KGaA, Weinheim.

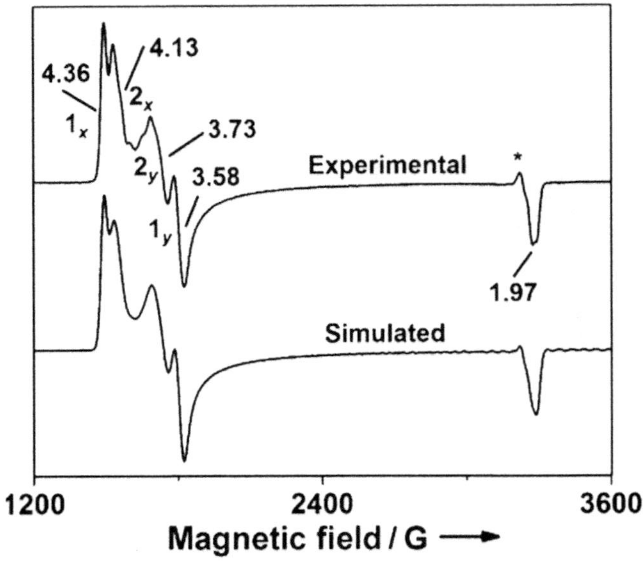

Figure 9.17 X-band electron paramagnetic resonance spectrum of a sample containing 0.7 mm $ZnPPFe_BMb$ in the presence of 1.0 eq. $FeCl_2$ and 20 eq. of nitric oxide in 50 mm bis-tris buffer pH 7.3, 5 K and the simulated spectrum. *A radical-type peak (<1%). Reprinted with permission from S. Chakraborty, J. Reed, M. Ross, M. J. Nilges, I. D. Petrik, S. Ghosh, S. Hammes-Schiffer, J. T. Sage, Y. Zhang, C. E. Schulz and Y. Lu, *Angew. Chemie, Int. Ed.*, 2014, **53**, 2417. WILEY-VCH Verlag GmbH & Co. KGaA, Weinheim.

where two NOs bind the two ferrous centres; the *cis* haem b_3 pathway, where the N–N bond formation is promoted by binding to a haem dinitrosyl adduct; and the *cis* Fe_B pathway, where a non-haem dinitrosyl adduct is invoked (Scheme 9.7). Mechanistic investigation using the wild-type NOR has been difficult due to the presence of several haem active sites in addition to the catalytic site. A five-coordinate haemFe^{II}–NO adduct was initially observed in ba_3-type CcO that has reasonable NOR activity.[29,85] However, the same group reported a histidine-bound six-coordinate haemFe^{II}–NO adduct in a bb_3-type CcO.[84] Considering the fact that the latter had lower NOR activity, it was implicit that the five-coordinate haemFe^{II}–NO, where the histidine holding the haem cofactor to the protein dissociates, could be key to the NOR activity. However, investigations using functional synthetic model complexes identified a six-coordinate haemFe^{II}–NO complex as the likely intermediate.[70,74] The same investigation, using a Zn_B analogue and Zn-porphyrin-Fe_B^{II} analogue, identified both Fe_B^{II}–NO, a haemFe^{II}–NO adduct and a haemFe^{II}–NO/Fe_B^{II}–NO dinitrosyl adduct as intermediates in the reaction. The study also indicated that the binding of NO to haem was preceded by the formation of a non-haem Fe_B^{II}–NO species. A few years later, using an almost identical approach, the formation of both haemFe^{II}–NO

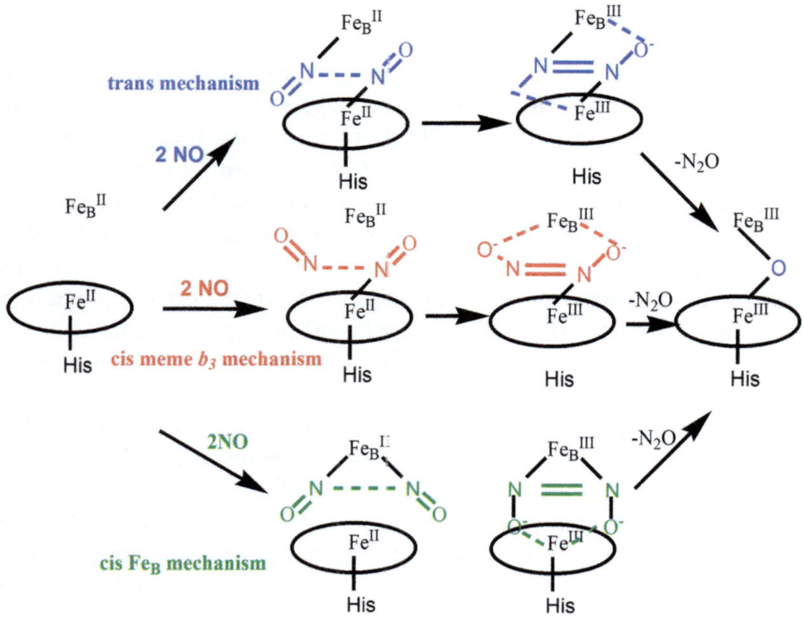

Scheme 9.7 Proposed mechanistic pathways for nitric oxide reductase.

and $Fe_B{}^{II}$–NO was demonstrated in the functional biosynthetic models.[79,77] These investigations suggested that the mechanism likely proceeded *via* the formation of a *trans* NO adduct, against the proposal from QM/MM calculations.[76,78,87–89] The likely formation of a *trans* bis-nitrosyl complex was also suggested by the binding of two CO molecules to the two Fe centres in the active site.[90] Finally, the putative *trans* hyponitrite intermediate was observed, which validated the *trans* mechanism proposal from synthetic and biosynthetic models.[29]

A key component in the mechanism of NO reduction is the role played by the glutamates. At least two glutamates were found to be essential for the activity for the enzyme and its biosynthetic models.[82,91–93] These glutamates were proposed to be essential to providing protons to the active site during NO reduction.[92,94,95] The first synthetic model for NOR did not include the glutamate residues, but was found to be functional. However, the biosynthetic model with only one glutamate, although initially claimed to be functional, was later found to be non-functional unless another glutamate (*i.e.*, two glutamates in total) was introduced in the vicinity of the active site.[80,77] These glutamates were also proposed to alter the reduction potential of the distal Fe_B centre, altering the overall energetics of NO reduction.[96] While synthetic models were synthesised containing these glutamates and their potential for tuning the Fe_B potential was demonstrated, the NOR reactivities of these models are yet to be demonstrated.[62]

9.4 NOR *versus* CcO

NOR and CcO are both members of the haem copper oxidase family of enzymes with possibly similar evolutionary origins. CcO is the terminal enzyme in the respiratory chain of higher organisms, located in their mitochondrial membrane that reduces O_2 to H_2O as the source of energy. A proton gradient across the bilayer membrane is generated during O_2 reduction in CcO, which provides the driving force for ATP synthesis. Like NOR, the CcO active site has a haem ligated to the protein by a proximal histidine ligand. In contrast to the presence of the Fe_B site in the distal pocket of NOR, there is a distal Cu_B site coordinated by three histidine ligands. In addition to the bimetallic site, there is a conserved, post-translationally modified tyrosine ligand covalently attached to one of the histidines coordinated to Cu_B. While NORs can reduce O_2, some CcOs are known to reduce NO. NO is even known to reversibly inhibit CcO, playing the role of respirator regulator. Although O_2 is not known to regulate NOR activity, it is known to be reduced by the haem/non-haem site. This is primarily due to the presence of similar reactive elements in the active sites of these enzymes. Haem is known to bind both NO and O_2 with high affinity. Similarly, both Fe_B and Cu_B are known to bind both O_2 and NO in their reduced states. Although $Fe_B{}^{II}$ can be anticipated to bind NO with much higher affinity, it is not clear if this factor alone can be deemed responsible for the difference in reactivity of CcO and NOR. The presence of tyrosine in CcO distinguishes the local environment of CcO from that of NOR, which has glutamate. The availability of synthetic functional models of both CcO and NOR makes it feasible to investigate the "yin–yang" relationship between these two enzymes in details.

9.4.1 O_2 Reduction by a Synthetic Haem/Non-Haem NOR Model

The reaction of O_2 with the fully reduced functional NOR model (*i.e.*, haem-$Fe^{II}/Fe_B{}^{II}$) was studied using heterogeneous electrocatalysis. Alkyl thiols with desired chain length are covalently attached onto electrode surfaces by forming self-assembled monolayer (SAM) on gold (Au) surfaces or interdigitated array (IDA) electrodes. This is followed by the coupling of the catalyst to the alkyl thiol by a Cu^I catalysed "click" reaction.[97] The "click" reaction couples an alkyne-terminated porphyrin catalyst to azide-terminated alkyl thiols, resulting in the 1,5 cycloadditive triazole formation (Figure 9.18).[98]

O_2 on reduction by $4e^-/4H^+$ generates H_2O, $2e^-/2H^+$ produces H_2O_2 and $1e^-$ produces $O_2{}^-$, with the latter two species being termed partially reduced oxygen species (PROS).[99] Rotating ring disc electrodes (RRDEs) or IDA electrodes have been utilised to detect and quantify the amount of PROS leaked during the electrocatalytic reduction of O_2. In the RRDE technique, a ring (Pt or Au) is held at a potential where it specifically oxidises/reduces the product/by-product of the electrocatalytic reaction occurring on the Au disc electrode

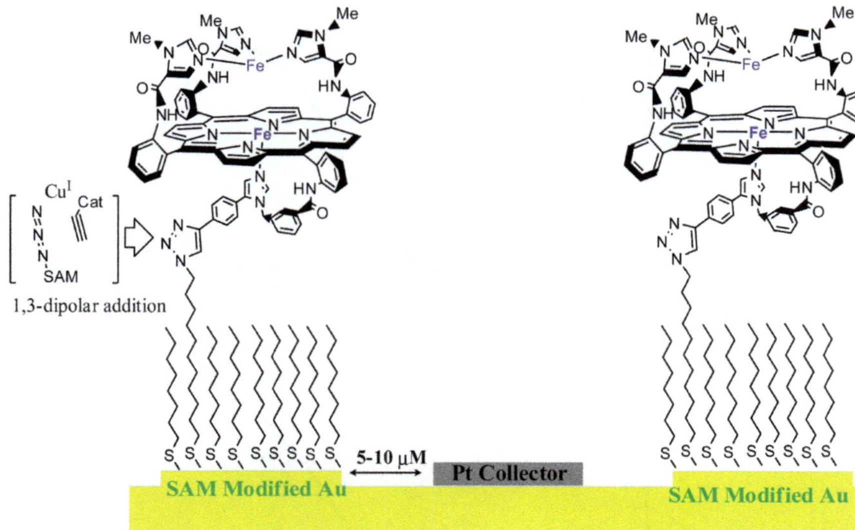

Figure 9.18 An alkyne-terminated catalyst clicked to an azide-terminated thiol by using Cu^{I}-catalysed 1,3-dipolar addition (*i.e.*, "click" reaction) on interdigitated arrays (representative example). Reproduced from ref. 37 with permission from the Royal Society of Chemistry. SAM: self-assembled monolayer.

it encircles. IDAs have two sets of alternate arrays of electrodes, one of which has the catalyst immobilised on it, while the other set has electrodeposited Pt metal, which acts as the collector. The Pt electrode is held at a positive potential (900 mV *vs.* NHE), where it oxidises both O_2^- and H_2O_2, which is produced in the adjacent electrodes during O_2 reduction, back to O_2. Thus, while the catalyst-functionalised Au working electrode detects the O_2 reduction current (*i.e.*, electrons transferred to O_2), the adjacent Pt electrode synchronously detects any H_2O_2 or O_2^- produced due to incomplete O_2 reduction.

The reduction of O_2 by the functional NOR model was studied by IDA electrodes. The cyclic voltammogram of the haemFe^{II} catalyst under fast electron transfer (~500 s^{-1}, using 11-azido-undecanethiol with octanethiol as a diluent) shows ~89% O_2 reduction and ~11% PROS formation (Figure 9.19a). However, for the haemFe^{II}/Fe$_B^{II}$ catalyst, PROS formation was reduced to ~3% (Figure 9.19b). Under slow electron transfer (~1 s^{-1}, using 18-azido-octadecanethiol with hexadecanethiol as a diluent), the PROS generation for the haemFe^{II} catalyst dramatically increases to ~20% (Figure 9.19c), indicating a greater degree of hydrolysis of the haemFe^{III}–O_2^- intermediate of the O_2 reaction under conditions of slow electron transfer. In the presence of Fe$_B$, the PROS generation decreases to ~6% (Figure 9.19d) under the same conditions.[100] Thus, the presence of additional redox-active centres reduces the PROS generation, as is also observed in functional models of CcO.[98]

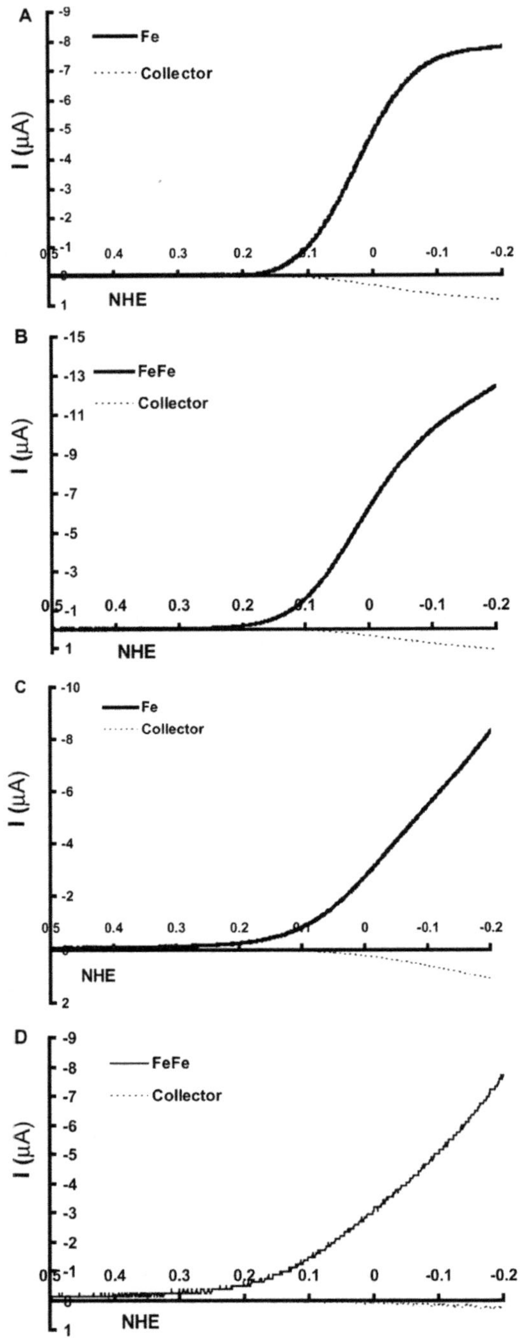

Figure 9.19 Electrocatalytic O_2 reduction current (in black) *vs.* Pt collector current (dashed black) for fast (a and b) and slow (c and d) self-assembled monolayers. Reproduced from ref. 37 with permission from the Royal Society of Chemistry.

9.4.2 Mechanism of O_2 Reaction by a Synthetic Haem/Non-Haem NOR Model

The mechanism of O_2 reduction by the functional NOR model has been investigated by trapping reaction intermediates at low temperatures in organic solvents and characterising them by resonance Raman spectroscopy.[100] The resonance Raman spectrum of the haemFe^{II}/Fe$_B{}^{II}$ catalyst has a characteristic ν_4 band at 1357.5 cm^{-1}, indicating the presence of a Fe^{II} species (Figure 9.20a). The shift of the ν_4 band to 1368 cm^{-1} and the presence of the ν_2 band at 1568 cm^{-1} on exposure to O_2 at −80 °C indicates the formation of a six-coordinate low-spin haemFe^{III} species.[101] The lower-frequency region of the spectrum shows two O_2 isotope-sensitive bands. One band is observed at 822 cm^{-1}, which shifts by 52 cm^{-1}, and the other is at 583 cm^{-1}, which shifts by 32 cm^{-1} on O^{18} isotopic substitution (Figure 9.20b). These vibrational data indicate the presence of a haemFe^{III}–peroxide species. Since two electrons are required for reducing oxygen to peroxide, the second electron (one electron is derived from haemFe^{II}) is likely to be provided by the Fe$_B$ site. Though no Fe$_B$–O stretch could be detected, it has been suggested that a bridging η_2 end-on peroxy species is generated, based on the high O–O stretching vibration and the low-spin, strong ligand field environment of the haem moiety. The low-frequency region of the spectrum shows two O_2 isotope-sensitive bands at 756 and 808 cm^{-1}, which shift by 25 and 32 cm^{-1}, respectively (Figure 9.20c). The Fe–O vibrations for Fe^{IV}=O porphyrin species have been observed between at 745 and 850 cm^{-1}, and those for non-haem ferryl species are observed between 800 and 840 cm^{-1}, depending on the spin state and different *trans* ligands. The presence these two O_2 isotope-sensitive bands at 756 and 808 cm^{-1} indicates that there are two distinct Fe^{IV}–oxo ferryl species in this intermediate, which would result in the homolytic O–O bond cleavage of the peroxy intermediate. The reaction has been further supported by the oxo atom transfer reactivity of the ferryl species. Triphenyl phosphine oxide has been detected by gas chromatography when the haemFe^{II}/Fe$_B{}^{II}$ complex was reacted with O_2 in the presence of triphenyl phosphine. On further increasing the temperature, the *trans* bis-ferryl intermediate decays to generate a low-spin haemFe^{III} end product. The proposed reaction mechanism of the O_2 reactivity with the NOR model is illustrated in Scheme 9.8. This reaction is catalytic as the product generated after 2 equivalents of oxo-atom transfer is the fully reduced, reactive bis-ferrous form. Three turnovers have been observed.

9.4.3 NO Reactivity of a Synthetic Functional Model of CcO

A model complex has been synthesised that incorporates a haem with a covalently attached imidazole on the proximal side of the porphyrin ring (mimicking the haem a$_3$ site), a tris-imidazole distal binding pocket for Cu (mimicking Cu$_B$) and a covalently attached phenol ligand to a distal imidazole ligand (mimicking the Tyr$_{244}$ residue), thereby incorporating all of the

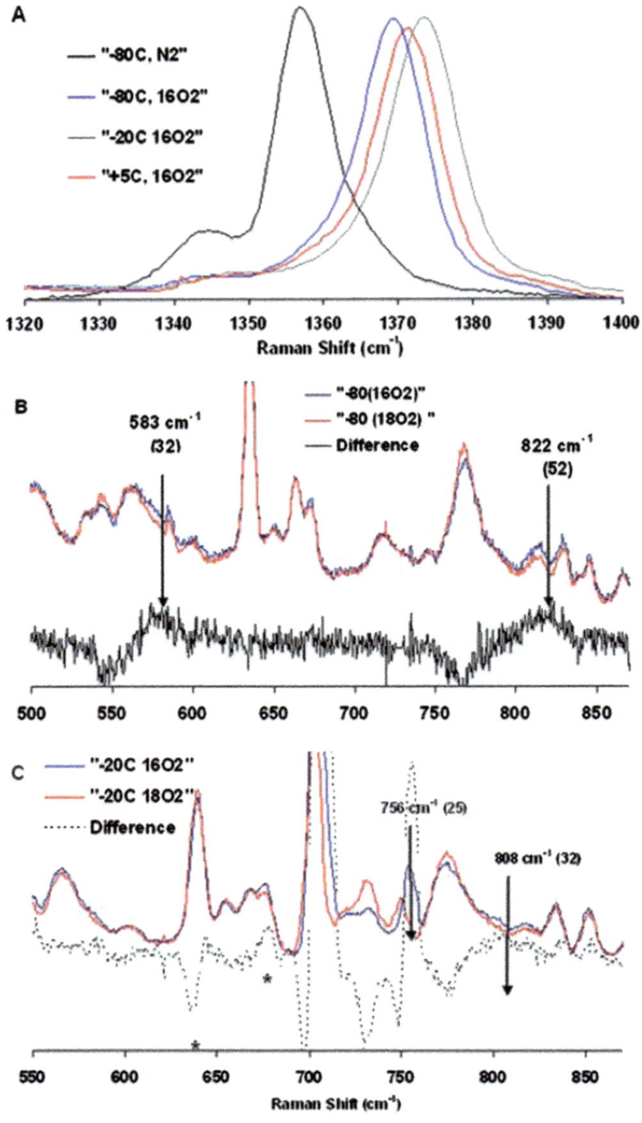

Figure 9.20 Resonance Raman data with 424 nm excitation obtained for a sample quenched after the addition of oxygen at −80 °C: (a) high-energy region; (b) low-energy region. (c) Low-energy region data obtained for a sample quenched after warming the −80 °C sample to −20 °C. Reproduced from ref. 37 with permission from the Royal Society of Chemistry.

key structural features of the CcO active site (Figure 9.21).[68] These model complexes reduce O_2 under physiological conditions with turnover numbers >10^4. The PROS generated during the electrocatalytic reduction of O_2 by the CcO model has been detected and quantified by RRDE. Under physiologically relevant slow electron transfer (from the electrode, using slow SAM),

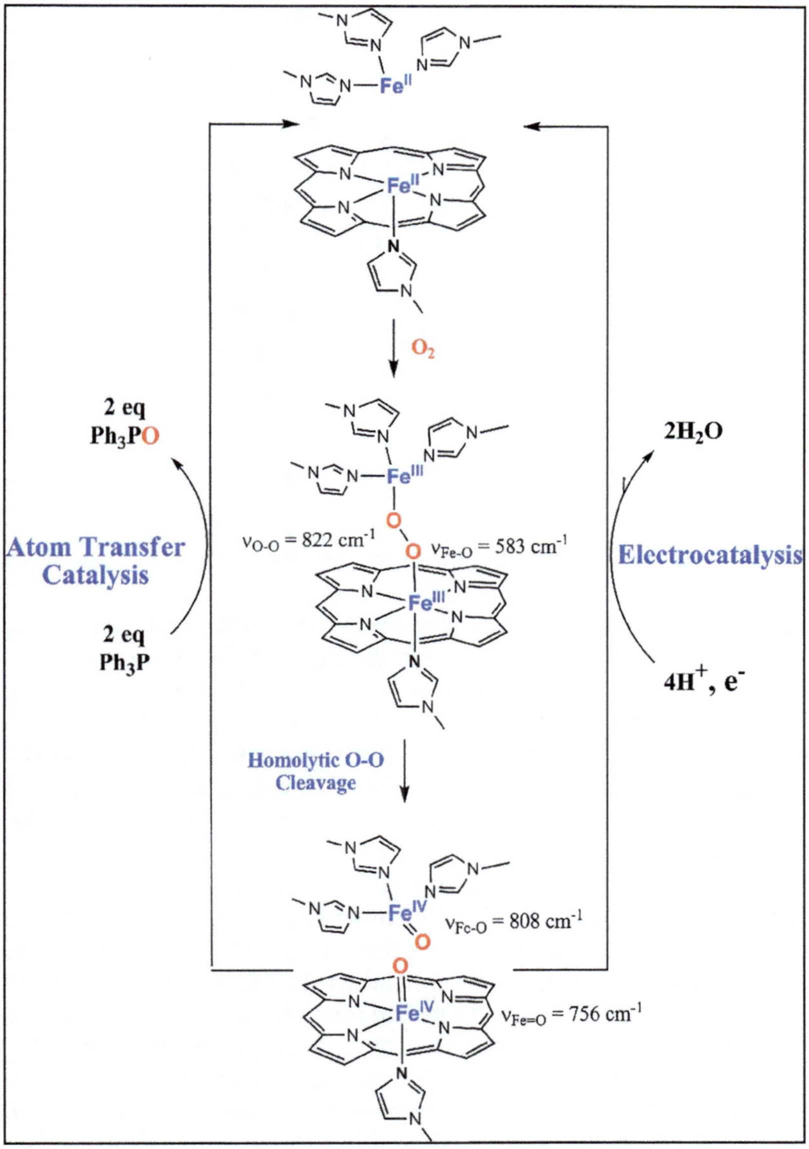

Scheme 9.8 Proposed mechanism for the reaction of O_2 with the haemFe^{II}/Fe$_B^{II}$ functional nitric oxide reductase model. Reproduced from ref. 37 with permission from the Royal Society of Chemistry.

the haemFe^{II} catalyst (in the absence of Cu$_B$ and phenol) degrades rapidly, likely because of excessive PROS produced during O_2 reduction.[98] The haem-Fe^{II}/Cu$_B^{I}$ catalyst (in the absence of phenol, which houses three electrons, two from haemFe^{II} and one from Cu$_B^{I}$) generates ~13% PROS, implying that since electron transfer is slow, some PROS can leak out before the fourth electron required for the complete reduction of O_2 can be supplied. For the haemFe^{II}/Cu$_B^{I}$/phenol catalyst, a negligible amount of PROS is formed (~4%),

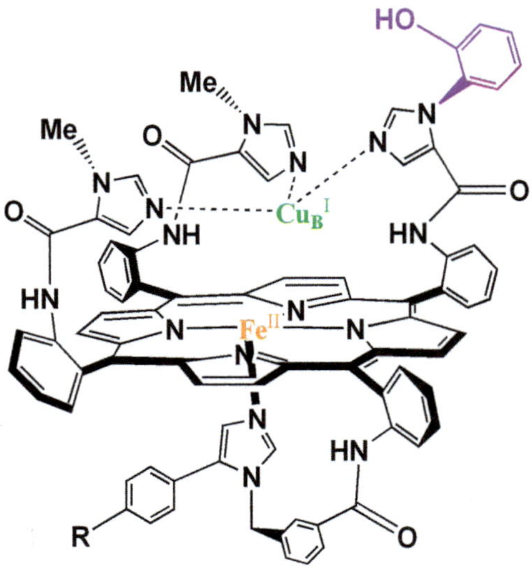

Figure 9.21 Synthetic model of cytochrome *c* oxidase.

indicating that the phenol acts as a source of the fourth electron for O_2 reduction under conditions of slow electron supply. The four electron reduction of O_2 to H_2O by the same CcO model using cytochrome *c*, the biological reductant of CcO, has also been studied. Biochemical CcO models have been reported by Lu *et al.* using site-directed mutagenesis of Mb. Three histidine residues have been introduced in the distal pocket, which can bind a distal Cu ion. These have provided valuable insights into the role of distal Cu and protons in the CcO active site.

To address the important question of why the active site of CcO is not severely inhibited by the diffusing hormone NO produced by neighbouring NO synthase, the reaction of NO with the functional CcO model was investigated.[102] When 1 equivalent of NO is added to the reduced haemFe^{II} CcO model complex (in the absence of Cu_B and phenol), there are distinct changes in the Soret and Q band regions of the absorption spectrum (Figure 9.22) relative to the parent complex, indicating the likely formation of a Fe–nitrosyl complex. The EPR spectrum shows the formation of a low-spin, six-coordinate $S = 1/2$ Fe^{II}–nitrosyl species, with multiple ^{14}N superhyperfine features in the A_y region, which is perturbed by isotopically enriched ^{15}NO (Figure 9.23), similarly to the data observed for native CcO and six-coordinate iron nitrosyl with imidazole axial ligands.[103] Thus, NO exposure to haemFe^{II} results in the formation of a haemFe^{II}–NO complex, which is stable under O_2 atmosphere, in contrast to enzymatic and non-enzymatic haem–nitrosyl species, which are extremely air sensitive. This has later been stated to be a general property of six-coordinate $S = 1/2$ haemFe^{II}–NO species.[104] When haemFe^{II}–NO is exposed to an equivalent of superoxide, the absorption spectrum resembles that of haemFe^{II}, and the characteristic low-spin, six-coordinate $S = 1/2$ Fe^{II}–NO EPR signal is lost.

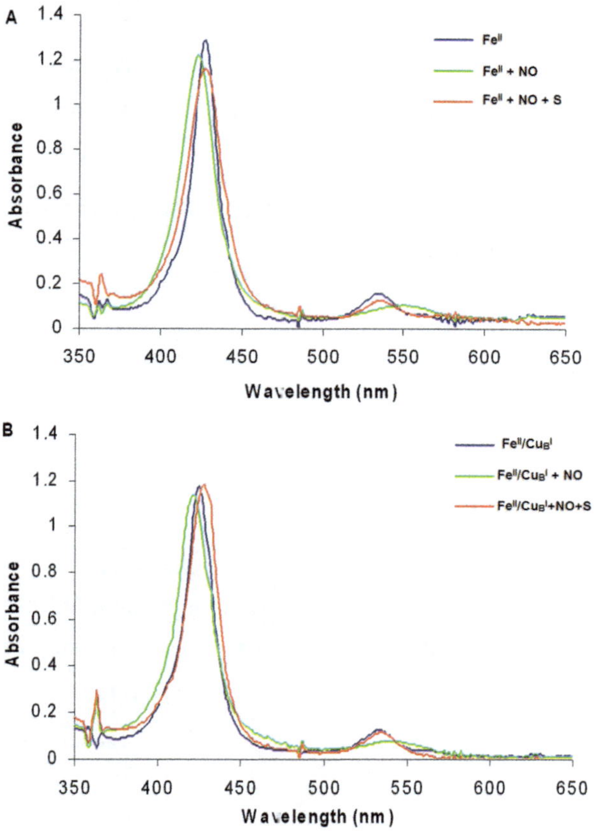

Figure 9.22 Absorption spectra of the nitric oxide (NO) derivatives and their reaction with O_2 and superoxide (S) for (a) haemFe^{II} and (b) haem$Fe^{II}/Cu_B{}^{I}$ complexes. Reproduced from ref. 37 with permission from the Royal Society of Chemistry.

Absorption and EPR data confirm that when NO is added to the haem$Fe^{II}/Cu_B{}^{I}$ complex (in the absence of phenol), it forms haemFe^{II}–$NO/Cu_B{}^{I}$. Note that unlike O_2, NO did not oxidize the FeCu centre, consistent with results obtained in CcO under strictly anaerobic conditions, indicating that NO may not be a substrate of aa_3-type CcO as previously thought. Later, a biochemical CcO model demonstrated the reduction of NO to N_2O mediated by a five-coordinate Fe^{II}–NO species similar to the ba_3-and baa_3-type CcOs.[29,77]

The absorption spectrum reflects that when the haemFe^{II}–$NO/Cu_B{}^{I}$ complex is exposed to O_2, it results in the formation of haemFe^{II} species (Figure 9.22b). There is loss of the characteristic Fe^{II}–NO EPR signal, along with the concomitant appearance of a new $S = 1/2$ type II Cu^{II} signal (Figure 9.23). In the presence of the distal $Cu_B{}^{I}$, superoxide is generated *in situ* in the presence of O_2, which reacts with the haemFe^{II}–NO to generate a ferrous species (the potential of the Cu site is ~0 mV,[18] which can reduce O_2 to O_2 which is around -100 mV,

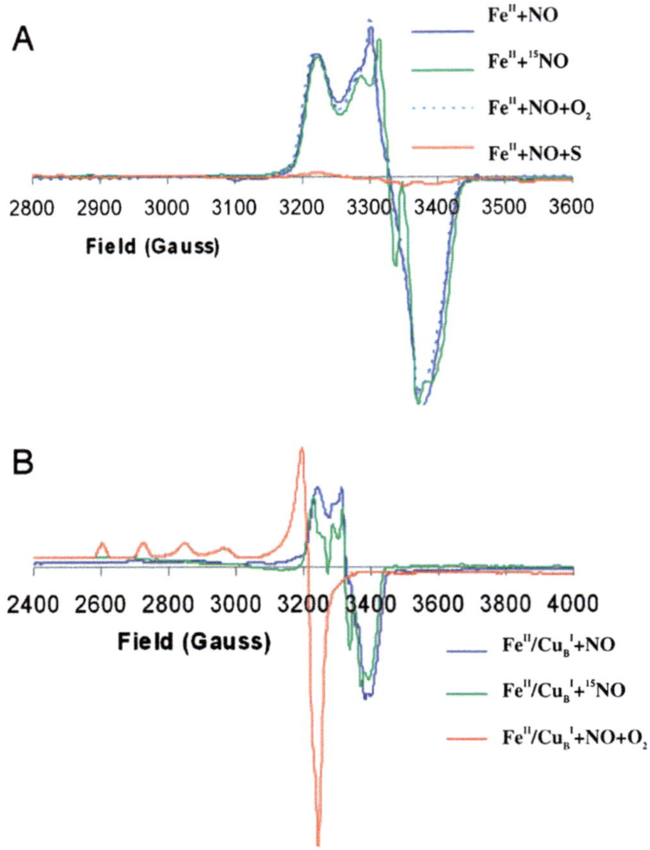

Figure 9.23 Electron paramagnetic resonance spectra of the nitric oxide (NO) derivatives and their reaction with O_2 and superoxide (S) for (a) haem-Fe^{II} and (b) haemFe^{II}/Cu_B^{I} complexes. Reproduced from ref. 37 with permission from the Royal Society of Chemistry.

in presence of excess O_2), similarly to the haemFe^{II} complex. This reaction probably proceeds *via* a peroxynitrite-type intermediate that forms by the reaction between superoxide and the NO complex, which will readily isomerise into nitrate in organic solvents. The destabilisation of a haemFe^{II}–NO species in the presence of the distal Cu and O_2 provides a possible recovery mechanism of CcO from inhibition by NO. These results have been schematically represented in Scheme 9.9.

Carbon monoxide (CO) and cyanide (CN^-) are molecules that are either present in the environment or in our diets and are potent inhibitors of CcO.[105–107] These small molecules have also been shown to inhibit the electrocatalytic O_2 reduction by these functional CcO models. The haemFe^{II} complex, on exposure to CO and CN^-, forms into haemFe^{II}–CO and haemFe^{II}–CN complexes, respectively, as reflected by the absorption data (Figure 9.24a). On exposing these haemFe^{II}–CO and haemFe^{II}–CN complexes to NO, the

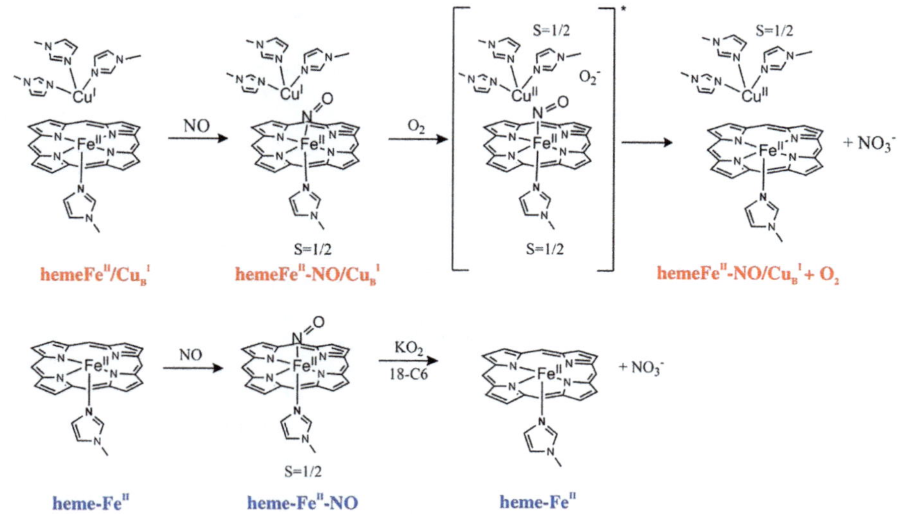

Scheme 9.9 Proposed reaction mechanism of removal of nitric oxide (NO) from cytochrome *c* oxidase using the haemFe^{II} and haemFe^{II}/Cu$_B^I$ catalysts. Reproduced from ref. 37 with permission from the Royal Society of Chemistry.

CO and CN^- ligands get replaced by NO, as demonstrated by their resultant absorption and EPR spectra (Figure 9.24a and b).

It is a well-known fact that amyl nitrite (AmN) is used as a treatment for accidental CN^- exposure.[108,109] To study its effect on the CcO model, AmN is added to the CN^--bound form of haemFe^{II}–CN/Cu$_B^I$. The resultant EPR spectrum is characteristic of the presence of an $S = 1/2$ Cu^{II} species and an $S = 1/2$ haemFe^{II}–NO species (total spin integration accounts for the presence of two paramagnetic centres) (Figure 9.24b). In contrast, no reaction is observed between haemFe^{II}–CN and AmN. It has been proposed that AmN is reduced by Cu$_B^I$ into amyl alkoxide (AmO^-) and NO, generating Cu^{II} (an oxidative addition reaction). The NO generated from AmN subsequently displaces CN^-, forming the haemFe^{II}–NO species. This illustrates a possible recovery mechanism of CcO from CN^- poisoning (Scheme 9.10). This proposed role of NO in reducing the antagonistic behaviour of CN^- was later established in CcO.[110,111]

Thus, NO may possibly act as a natural defence of CcO against inhibitors such as CN^- and CO. Because of its extremely high binding affinity for reduced haem a_3, NO can easily replace either of these ligands, forming a stable Fe^{II}–NO complex, which could then be oxidised by superoxide generated *in situ*, thereby regenerating the active enzyme. This *in situ* generation of superoxide should also provide defence against inhibitors that target the oxidised haem a_3 site (anionic ligands such as CN^-, N_3, *etc.*) by reducing the oxidised site to the corresponding ferrous forms, which have much higher NO affinities. Indeed, an increased concentration of superoxide is observed

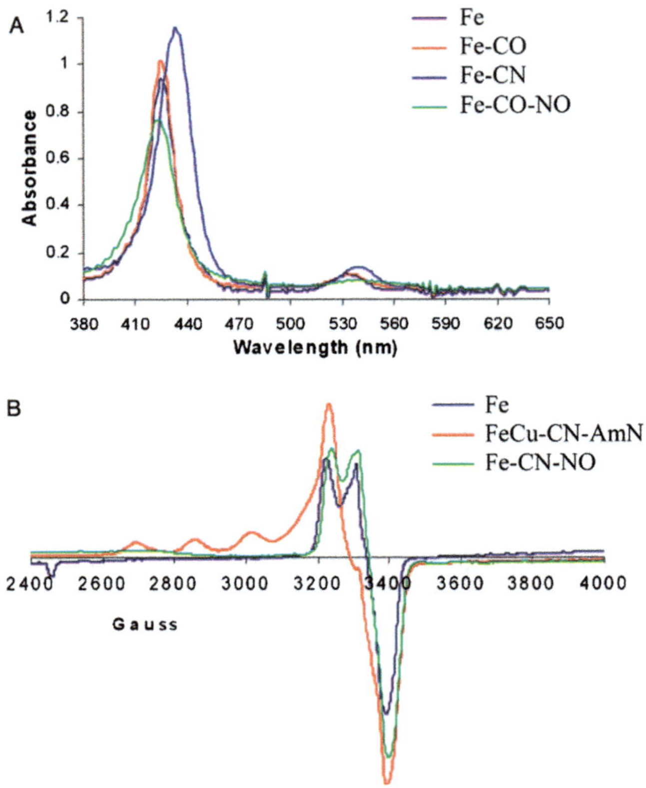

Figure 9.24 (a) Absorption and (b) electron paramagnetic resonance spectra of CO and CN derivatives of haemFe^{II} and haem-Fe^{II}/Cu$_B$^{I} and their reactions with nitric oxide (NO) and amyl nitrite (AmN), respectively. Reproduced from ref. 37 with permission from the Royal Society of Chemistry.

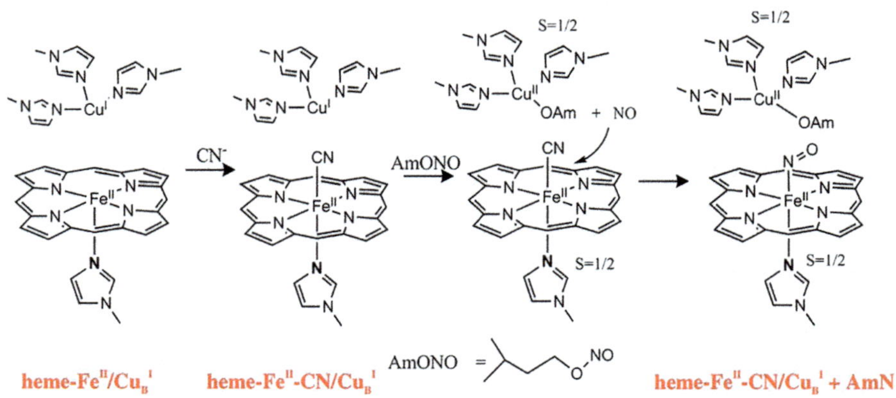

Scheme 9.10 Proposed mechanism for the protection of cytochrome *c* oxidase by amyl nitrite from cyanide poisoning. Reproduced from ref. 37 with permission from the Royal Society of Chemistry.

in CcO during turnover when the haem a_3 site is inhibited by ligand binding.[112] Thus, a symbiotic interaction between a potent inhibitor (NO) and a reactive oxygen species by-product (O superoxide) should protect CcO from external inhibitors. While NO defends the reduced site, superoxide protects the oxidised site.

Thus, functional models of CcO that reduce O_2 efficiently were not found to be capable of reducing NO to N_2O. It is likely that the distance between the Fe and the Cu centre is too large in these complexes, where bridging peroxide intermediates have not been observed, but azide bridging has been observed (*i.e.*, Fe–Cu ~6 Å). However, the synthetic model of NOR only varies in the distal metal, and the Fe–Fe distances in the NOR model are likely to be similar to the Fe–Cu distances in the CcO model. Similarly, the functional model of NOR, without the glutamate, is as good at reducing O_2 as the model for CcO. This presents an apparent conundrum to a rational mind, and exotic proposals based on enzyme quaternary structures and not fundamental inorganic chemistry have to be invoked in order to explain the selectivity of NOR for NO and CcO for O_2. However, if one considers the electronic structure of a hyponitrite intermediate, it is quite clear that such an intermediate will need stabilisation *via* hydrogen bonding or solvation. In the active site of CcO, hydrogen bonding is offered by the tyrosine residue, albeit weakly. However, in the active site of NOR, the two glutamates can provide substantial stabilisation to this species. In a synthetic system that is exposed to the solvent, the stabilisation likely does not require the hydrogen bonding interaction, which is why the model without the carboxylate works quite well. Similarly, the biosynthetic model, in which the active site is not well solvated, requires these glutamate residues to stabilise the hyponitrite intermediate by the hydrogen bonding interaction. *Ergo*, it is not surprising that the biosynthetic model with the two glutamates shows NOR activity and the one without any does not.

9.5 Concluding Remarks

The haemCu oxidase superfamily of enzymes represents one of the most important group of enzymes. Obtaining a solid understanding of the structure–function correlations of these enzymes is not only important to elucidating their biochemistry, which is significant in the field of medicine, but these enzymes also, in principle, perform reactions that are important for the environment. The reduction of O_2 to water by CcO also happens to be the cathodic reaction of a H_2/O_2 fuel cell. Similarly, the generation of N_2O has significance to energy (oxidant) and medicine (laughing gas). The synthetic functional models have been beneficial for this purpose, particularly considering the difficulties in investigating these membrane-bound enzymes. These investigations suggested that a *trans* haemFe^{II}–NO/$Fe_B{}^{II}$–NO bis-nitrosyl species is the probable intermediate in the reductive coupling step, that NORs could also act as good oxygen reduction catalysts, that CN^- and CO poisoning could be treated with NO and they have provided a probable mechanism of reversible inhibition of CcO by NO. All of these predictions, some made against predictions from computational investigations, have later been evaluated and proved to be correct in the natural enzyme. Similarly, the biosynthetic models have played a crucial role in evaluating the roles played by the glutamate residues in NOR activity. The results obtained using the synthetic models clearly suggest that the difference in reactivity between NOR and CcO likely stems from the hydrogen bonding stabilisation of the bridging hyponitrite intermediate by the two glutamate residues present in the active site. It remains to be seen if this proposition survives the test of time. Let the games begin!

References

1. P. M. H. Kroneck and W. G. Zumft, *FEMS Symp.*, 1990, **56**, 1.
2. W. G. Zumft, *Arch. Microbiol.*, 1993, **160**, 253.
3. W. G. Zumft, A. Dreusch, A. Glockner and P. M. H. Kroneck, *Bioinorganic Chemistry*, Wiley-VCH Verlag GmbH, 1997, p. 397.
4. W. G. Zumft, *Microbiol. Mol. Biol. Rev.*, 1997, **61**, 533.
5. S. Al-Attar and S. de Vries, *FEBS Lett.*, 2015, **589**, 2050.
6. T. M. Stevanin, J. W. B. Moir and R. C. Read, *Infect. Immun.*, 2005, **73**, 3322.
7. J. L. Edwards, *Infect. Immun.*, 2010, **78**, 1202.
8. M. J. Prather and J. Hsu, *Science*, 2010, **330**, 952.
9. A. R. Ravishankara, J. S. Daniel and R. W. Portmann, *Science*, 2009, **326**, 123.
10. D. E. Canfield, A. N. Glazer and P. G. Falkowski, *Science*, 2010, **330**, 192.
11. G. C. Brown and A. Vilalta, *Brain Res.*, 2015, **1628**, 288.
12. R. Banerjee, *J. Biol. Chem.*, 2015, **290**, 30726.
13. P. C. Ford, J. C. M. Pereira and K. M. Miranda, *Struct. Bonding*, 2014, **154**, 99.
14. P. C. Ford, *Inorg. Chem.*, 2010, **49**, 6226.
15. P. C. Ford and I. M. Lorkovic, *Chem. Rev.*, 2002, **102**, 993.
16. P. C. Ford, D. A. Wink and D. M. Stanbury, *FEBS Lett.*, 1993, **326**, 1.
17. J. A. McCleverty and M. D. Ward, *Comprehensive Coordination Chemistry II*, Elsevier Ltd., 2004, vol. 2, p. 743.
18. A. N. Osipov, G. G. Borisenko and Y. A. Vladimirov, *Biochemistry*, 2007, **72**, 1491.
19. P. K. Allan and R. E. Morris, *Struct. Bonding*, 2014, **154**, 225.
20. N. Lehnert, W. R. Scheidt and M. W. Wolf, *Struct. Bonding*, 2014, **154**, 155.
21. A. P. Hunt and N. Lehnert, *Acc. Chem. Res.*, 2015, **48**, 2117.
22. C. A. Brown, M. A. Pavlosky, T. E. Westre, Y. Zhang, B. Hedman, K. O. Hodgson and E. I. Solomon, *J. Am. Chem. Soc.*, 1995, **117**, 715.
23. Y. Zhang, M. A. Pavlosky, C. A. Brown, T. E. Westre, B. Hedman, K. O. Hodgson and E. I. Solomon, *J. Am. Chem. Soc.*, 1992, **114**, 9189.
24. T. E. Nappier, R. D. Feltham, J. H. Enemark, A. Kruse and M. Cooke, *Inorg. Chem.*, 1975, **14**, 806.

25. Y. Matsumoto, T. Tosha, A. V Pisliakov, T. Hino, H. Sugimoto, S. Nagano, Y. Sugita and Y. Shiro, *Nat. Struct. Mol. Biol.*, 2012, **19**, 238.
26. L. A. Schurig-Briccio, P. Venkatakrishnan, J. Hemp, C. Bricio, J. Berenguer and R. B. Gennis, *Proc. Natl. Acad. Sci. U. S. A.*, 2013, **110**, 12613.
27. C. Bricio, L. Alvarez, M. San Martin, L. A. Schurig-Briccio, R. B. Gennis and J. Berenguer, *Appl. Environ. Microbiol.*, 2014, **80**, 4871.
28. S. Khatua and A. Majumdar, *J. Inorg. Biochem.*, 2015, **142**, 145.
29. V. Daskalakis, T. Ohta, T. Kitagawa and C. Varotsis, *Biochim. Biophys. Acta, Bioenerg.*, 2015, **1847**, 1240.
30. J. D. Caranto, A. Weitz, M. P. Hendrich and D. M. Kurtz, *J. Am. Chem. Soc.*, 2014, **136**, 7981.
31. A. G. Duarte, C. M. Cordas, J. J. G. Moura and I. Moura, *Biochim. Biophys. Acta, Bioenerg.*, 2014, **1837**, 375.
32. B. Kramos, D. K. Menyhard and J. Olah, *J. Phys. Chem. B*, 2012, **116**, 872.
33. A. B. McQuarters, N. E. Wirgau and N. Lehnert, *Curr. Opin. Chem. Biol.*, 2014, **19**, 82.
34. S. de Vries and I. Schröder, *Biochem. Soc. Trans.*, 2002, **30**, 662.
35. F. L. Sousa, R. J. Alves, M. A. Ribeiro, J. B. Pereira-Leal, M. Teixeira and M. M. Pereira, *Biochim. Biophys. Acta, Bioenerg.*, 2012, **1817**, 629.
36. T. Hino, Y. Matsumoto, S. Nagano, H. Sugimoto, Y. Fukumori, T. Murata, S. Iwata and Y. Shiro, *Science*, 2010, **330**, 1666.
37. S. G. Dey and A. Dey, *Dalton Trans.*, 2011, **40**, 12633.
38. P. Sarti, E. Forte, D. Mastronicola, A. Giuffre and M. Arese, *Biochim. Biophys. Acta, Bioenerg.*, 2012, **1817**, 610.
39. P. Sarti, M. Arese, E. Forte, A. Giuffre and D. Mastronicola, *Adv. Exp. Med. Biol.*, 2012, **942**, 75.
40. M. Brunori, E. Forte, M. Arese, D. Mastronicola, A. Giuffre and P. Sarti, *Biochim. Biophys. Acta*, 2006, **1757**, 1144.
41. M. Brunori, A. Giuffre, E. Forte, D. Mastronicola, M. C. Barone and P. Sarti, *Biochim. Biophys. Acta*, 2004, **1655**, 365.
42. A. Loullis and E. Pinakoulaki, *Chem. Commun.*, 2015, **51**, 17398.
43. G. Stubauer, A. Giuffre, M. Brunori and P. Sarti, *Biochem. Biophys. Res. Commun.*, 1998, **245**, 459.
44. T. Fujiwara and Y. Fukumori, *J. Bacteriol.*, 1996, **178**, 1866.
45. A. Giuffre, M. C. Barone, M. Brunori, E. D'Itri, B. Ludwig, F. Malatesta, H.-W. Muller and P. Sarti, *J. Biol. Chem.*, 2002, **277**, 22402.
46. L. L. Pearce, A. J. Kanai, L. A. Birder, B. R. Pitt and J. Peterson, *J. Biol. Chem.*, 2002, **277**, 13556.
47. X.-J. Zhao, V. Sampath and W. S. Caughey, *Biochem. Biophys. Res. Commun.*, 1995, **212**, 1054.
48. F. Tani, M. Matsu-Ura, S. Nakayama and Y. Naruta, *Coord. Chem. Rev.*, 2002, **226**, 219.
49. J. P. Collman, R. Boulatov, C. J. Sunderland and L. Fu, *Chem. Rev.*, 2003, **104**, 561.
50. E. Kim, E. E. ChufÃ¡n, K. Kamaraj and K. D. Karlin, *Chem. Rev.*, 2004, **104**, 1077.
51. P. Venkateswara Rao and R. H. Holm, *Chem. Rev.*, 2004, **104**, 527.

52. G. Henkel and B. Krebs, *Chem. Rev.*, 2004, **104**, 801.
53. L. M. Mirica, X. Ottenwaelder and T. D. P. Stack, *Chem. Rev.*, 2004, **104**, 1013.
54. E. A. Lewis and W. B. Tolman, *Chem. Rev.*, 2004, **104**, 1047.
55. E. Y. Tshuva and S. J. Lippard, *Chem. Rev.*, 2004, **104**, 987.
56. J. H. Enemark, J. J. A. Cooney, J.-J. Wang and R. H. Holm, *Chem. Rev.*, 2004, **104**, 1175.
57. F. A. Walker, *Chem. Rev.*, 2004, **104**, 589.
58. C. J. Reedy and B. R. Gibney, *Chem. Rev.*, 2004, **104**, 617.
59. O. Farver, N. M. Marshall, S. Wherland, Y. Lu and I. Pecht, *Proc. Natl. Acad. Sci. U. S. A.*, 2013, **110**, 10536.
60. M. G. Savelieff and Y. Lu, *JBIC, J. Biol. Inorg. Chem.*, 2010, **15**, 461.
61. J. Liu, S. Chakraborty, P. Hosseinzadeh, Y. Yu, S. Tian, I. Petrik, A. Bhagi and Y. Lu, *Chem. Rev.*, 2014, **114**, 4366.
62. J. P. Collman, Y.-L. Yan, J. Lei and P. H. Dinolfo, *Inorg. Chem.*, 2006, **45**, 7581.
63. J. P. Collman, Y. Yang and R. A. Decreau, *Org. Lett.*, 2007, **9**, 2855.
64. J. P. Collman and S. Ghosh, *Inorg. Chem.*, 2010, **49**, 5798.
65. J. P. Collman, R. R. Gagne, C. Reed, T. R. Halbert, G. Lang and W. T. Robinson, *J. Am. Chem. Soc.*, 1975, **97**, 1427.
66. J. P. Collman, M. Broering, L. Fu, M. Rapta, R. Schwenninger and A. Straumanis, *J. Org. Chem.*, 1998, **63**, 8082.
67. J. P. Collman, M. Broering, L. Fu, M. Rapta and R. Schwenninger, *J. Org. Chem.*, 1998, **63**, 8084.
68. J. P. Collman, R. A. Decreau and C. Zhang, *J. Org. Chem.*, 2004, **69**, 3546.
69. J. P. Collman, Y.-L. Yan, J. Lei and P. H. Dinolfo, *Org. Lett.*, 2006, **8**, 923.
70. J. P. Collman, Y. Yang, A. Dey, R. A. Decreau, S. Ghosh, T. Ohta and E. I. Solomon, *Proc. Natl. Acad. Sci. U. S. A.*, 2008, **105**, 15660.
71. J. Wang, M. P. Schopfer, A. A. N. Sarjeant and K. D. Karlin, *J. Am. Chem. Soc.*, 2009, **131**, 450.
72. J. Wang, M. P. Schopfer, S. C. Puiu, A. A. N. Sarjeant and K. D. Karlin, *Inorg. Chem.*, 2010, **49**, 1404.
73. J. van der Oost, A. P. N. de Boer, J.-W. L. de Gier, W. G. Zumft, A. H. Stouthamer and R. J. M. van Spanning, *FEMS Microbiol. Lett.*, 1994, **121**, 1.
74. J. P. Collman, A. Dey, Y. Yang, R. A. Decréau, T. Ohta and E. I. Solomon, *J. Am. Chem. Soc.*, 2008, **130**, 16498.
75. J. P. Collman, Y. Yang, A. Dey, R. A. Decréau, S. Ghosh, T. Ohta and E. I. Solomon, *Proc. Natl. Acad. Sci. U. S. A.*, 2008, **105**, 15660.
76. M. R. A. Blomberg and P. E. M. Siegbahn, *Biochemistry*, 2012, **51**, 5173.
77. H. Matsumura, T. Hayashi, S. Chakraborty, Y. Lu and P. Moenne-Loccoz, *J. Am. Chem. Soc.*, 2014, **136**, 2420.
78. M. Shoji, K. Hanaoka, D. Kondo, A. Sato, H. Umeda, K. Kamiya and K. Shiraishi, *Mol. Phys.*, 2014, **112**, 393.
79. T. Hayashi, K. D. Miner, N. Yeung, Y.-W. Lin, Y. Lu and P. Moenne-Loccoz, *Biochemistry*, 2011, **50**, 5939.

80. N. Yeung, Y.-W. Lin, Y.-G. Gao, X. Zhao, B. S. Russell, L. Lei, K. D. Miner, H. Robinson and Y. Lu, *Nature*, 2009, **462**, 1079.
81. N. Yeung and Y. Lu, *Chem. Biodiversity*, 2008, **5**, 1437.
82. Y.-W. Lin, N. Yeung, Y.-G. Gao, K. D. Miner, S. Tian, H. Robinson and Y. Lu, *Proc. Natl. Acad. Sci. U. S. A.*, 2010, **107**, 8581.
83. Y.-W. Lin, N. Yeung, Y.-G. Gao, K. D. Miner, L. Lei, H. Robinson and Y. Lu, *J. Am. Chem. Soc.*, 2010, **132**, 9970.
84. E. Pinakoulaki, T. Ohta, T. Soulimane, T. Kitagawa and C. Varotsis, *J. Am. Chem. Soc.*, 2005, **127**, 15161.
85. E. Pinakoulaki and C. Varotsis, *Biochemistry*, 2003, **42**, 14856.
86. S. Chakraborty, J. Reed, M. Ross, M. J. Nilges, I. D. Petrik, S. Ghosh, S. Hammes-Schiffer, J. T. Sage, Y. Zhang, C. E. Schulz and Y. Lu, *Angew. Chem., Int. Ed.*, 2014, **53**, 2417.
87. A. A. A. Attia and R. Silaghi-Dumitrescu, *J. Mol. Model.*, 2015, **21**, 1.
88. L. M. Blomberg, M. R. A. Blomberg and P. E. M. Siegbahn, *Biochim. Biophys. Acta*, 2006, **1757**, 240.
89. M. R. A. Blomberg and P. E. M. Siegbahn, *Biochim. Biophys. Acta*, 2013, **1827**, 826.
90. S. Lu, Suharti, S. De Vries and P. Moeenne-Loccoz, *J. Am. Chem. Soc.*, 2004, **126**, 15332.
91. F. H. Thorndycroft, G. Butland, D. J. Richardson and N. J. Watmough, *Biochem. J.*, 2007, **401**, 111.
92. U. Flock, N. J. Watmough and P. Aedelroth, *Biochemistry*, 2005, **44**, 10711.
93. G. Butland, S. Spiro, N. J. Watmough and D. J. Richardson, *J. Bacteriol.*, 2001, **183**, 189.
94. L. Salomonsson, J. Reimann, T. Tosha, N. Krause, N. Gonska, Y. Shiro and P. Aedelroth, *Biochim. Biophys. Acta, Bioenerg.*, 2012, **1817**, 1914.
95. J. ter Beek, N. Krause, J. Reimann, P. Lachmann and P. Aedelroth, *J. Biol. Chem.*, 2013, **288**, 30626.
96. C. M. Cordas, A. G. Duarte, J. J. G. Moura and I. Moura, *Biochim. Biophys. Acta*, 2013, **1827**, 233.
97. J. P. Collman, N. K. Devaraj, T. P. A. Eberspacher and C. E. D. Chidsey, *Langmuir*, 2006, **22**, 2457.
98. J. P. Collman, N. K. Devaraj, R. A. Decréau, Y. Yang, Y.-L. Yan, W. Ebina, T. A. Eberspacher and C. E. D. Chidsey, *Science*, 2007, **315**, 1565.
99. J. P. Collman, R. Boulatov, C. J. Sunderland and L. Fu, *Chem. Rev.*, 2004, **104**, 561.
100. J. P. Collman, A. Dey, Y. Yang, S. Ghosh and R. A. Decreau, *Proc. Natl. Acad. Sci. U. S. A.*, 2009, **106**, 10528.
101. J. M. Burke, J. R. Kincaid, S. Peters, R. R. Gagne, J. P. Collman and T. G. Spiro, *J. Am. Chem. Soc.*, 1978, **100**, 6083.
102. J. P. Collman, A. Dey, R. A. Decreau, Y. Yang, A. Hosseini, E. I. Solomon and T. A. Eberspacher, *Proc. Natl. Acad. Sci. U. S. A.*, 2008, **1–5**, 5.
103. M. Hoshino, L. Laverman and P. C. Ford, *Coord. Chem. Rev.*, 1999, **187**, 75.

104. L. E. Goodrich, S. Roy, E. E. Alp, J. Zhao, M. Y. Hu and N. Lehnert, *Inorg. Chem.*, 2013, **52**, 7766.
105. M. T. Wilson, G. Antonini, F. Malatesta, P. Sarti and M. Brunori, *J. Biol. Chem.*, 1994, **269**, 24114.
106. M. W. Neil, *J. Forensic Med.*, 1957, **4**, 173.
107. W. W. Wainio and J. Greenlees, *Arch. Biochem. Biophys.*, 1960, **90**, 18.
108. K. K. Chen, C. L. Rose and G. H. A. Clowes, *J. Indiana State Med. Assoc.*, 1944, **37**, 344.
109. A. Buzzo and R. E. Carratala, *Sem. Med.*, 1933, 1772.
110. L. K. Cambal, M. R. Swanson, Q. Yuan, A. C. Weitz, H.-H. Li, B. R. Pitt, L. L. Pearce and J. Peterson, *Chem. Res. Toxicol.*, 2011, **24**, 1104.
111. L. L. Pearce, E. Lopez Manzano, S. Martinez-Bosch and J. Peterson, *Chem. Res. Toxicol.*, 2008, **21**, 2073.
112. B. S. Zuckerbraun, B. Y. Chin, M. Bilban, J. de Costa d'Avila, J. Rao, T. R. Billiar and L. E. Otterbein, *FASEB J.*, 2007, **21**, 1099.

CHAPTER 10

Model Compounds of Copper-Containing Enzymes Involved in Bacterial Denitrification

BRITTANY J. JOHNSON[a] AND NEAL P. MANKAD*[a]

[a]Department of Chemistry, University of Illinois at Chicago, 845 W. Taylor St. MC 111, Chicago, IL 60607, USA
*E-mail: npm@uic.edu

10.1 Introduction

Two copper enzymes are involved in bacterial denitrification.[1] First, one of the two nitrite reductases is the copper nitrite reductase (CuNiR), and its role is to catalyse the one-electron reduction of nitrite (NO_2^-) to nitric oxide (NO).[2] Second, the final step in denitrification involves nitrous oxide reductase (N_2OR), a copper-dependent enzyme whose role is to catalyse the two-electron reduction of nitrous oxide (N_2O) to dinitrogen (N_2).[3] CuNiR and N_2OR both feature two types of copper sites: an electron transfer (ET) site and a catalytic site for each enzyme. In the case of CuNiR, ET is mediated by a mononuclear type-1 (T1) copper site, while NO_2^- binding and reduction catalysis occurs at a mononuclear type-2 (T2) copper site. In the case of N_2OR, ET is mediated by a dinuclear Cu_A site, while N_2OR binding and reduction catalysis occurs at a tetranuclear site with either 4Cu : 2S (Cu_Z) or 4Cu : 1S (Cu_Z*) stoichiometry. The presence of these multiple copper sites—one for ET and one

RSC Metallobiology Series No. 9
Metalloenzymes in Denitrification: Applications and Environmental Impacts
Edited by Isabel Moura, José J. G. Moura, Sofia R. Pauleta and Luisa B. Maia

Published by the Royal Society of Chemistry, www.rsc.org

for catalysis—in each enzyme has complicated spectroscopic studies due to spectral overlap under some conditions. The ability of CuNiR to also catalyse reductive coupling of excess NO to N_2O[4] and the difficulty in isolating pure samples of N_2OR with respect to Cu_Z *vs.* Cu_Z* content[1,5,6] are examples of other complications to *in vitro* studies of the native systems. These and other issues have motivated inorganic model studies in which copper chemistry that is relevant to the catalytic processes can be examined in a controlled fashion. The aim of this chapter is to highlight structural and/or functional models of the catalytic copper sites involved in bacterial denitrification: the T2 site of CuNiR and the Cu_Z/Cu_Z* sites of N_2OR. Model studies that are relevant to the T1[7–9] and Cu_A[10–14] ET sites have also been conducted, but will not be covered here.

10.2 Models Relevant to Nitrite Reductase

10.2.1 Key Aspects of the Enzymatic T2 Site to Model

While the mechanistic aspects of nitrite reduction at the T2 site of CuNiR have been debated extensively,[15,16] the preponderance of experimental and computational data regarding the enzymatic system has allowed for a (largely) consensus mechanistic view to emerge (Figure 10.1).[1] The resting Cu^{II} state features a solvent-derived ligand in the reactive coordination site. This ligand is displaced (and lost through protonation as H_2O) by the

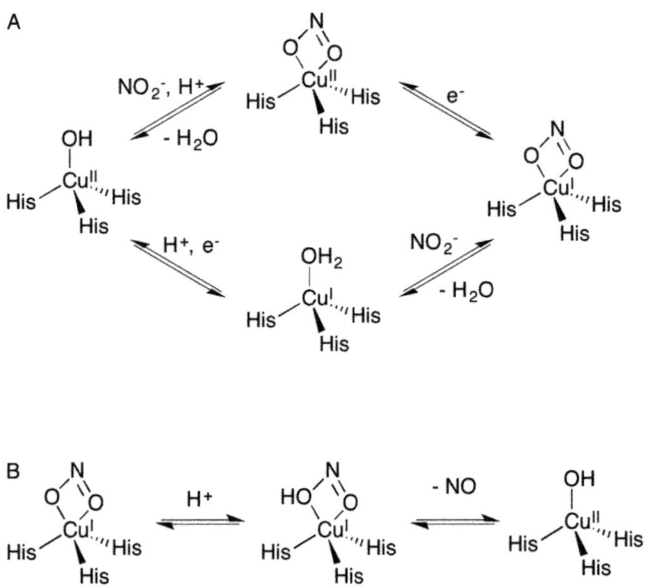

Figure 10.1 Plausible catalytic mechanism for copper nitrite reductase. (a) Dual pathways for the resting type-2 site undergoing reduction and substrate binding. (b) Proposed proton-triggered N–O bond cleavage.

NO_2^- substrate, in concert with the T2 site being reduced (through ET from the T1 site) to the Cu^I oxidation state. The sequence of these events (*i.e.*, whether substrate binding occurs before or after reduction by T1) has been contested, but is now thought to be random, with both pathways being catalytically viable depending on pH, nitrite concentration and other factors (Figure 10.1a).[17,18] The key Cu^I–NO_2^- intermediate then undergoes N–O bond cleavage (Figure 10.1b). Again, while the mechanism of this process has been debated, the consensus view is that protonation of the bound nitrite triggers direct NO loss without formation of an intermediate Cu–NO species of any type.[19] The role of experimentally observed NO-bound states of CuNiR remains to be clarified: such species could be off-cycle resting states, product-inhibited states or catalytic intermediates in the reductive NO coupling process *in vitro*, but not in the NO_2^- reduction process *in vivo*.

Structural and spectroscopic snapshots of several of the proposed intermediates in Figure 10.1 have been obtained specifically for the resting state (both oxidised and reduced), the nitrite-bound state (both oxidised and reduced) and the NO-bound state. The resting T2 site features a single copper centre bound by *three histidine residues*. In oxidised form, the copper is also ligated by a solvent-derived hydroxyl ligand,[20] completing a tetrahedral coordination geometry. Upon reduction, the resting state dissociates the solvent-derived ligand without any significant perturbation to the Cu–His linkages.[21] Several crystal structures feature substrate bound to the oxidised T2 site, with the nitrite anion having displaced the solvent-derived ligand.[1,22] In all cases, nitrite binds as a *bidentate* ligand, with *asymmetric binding* observed between the copper centre and each of the two oxygen atoms (intramolecular Cu–O distances differ by 0.2–0.6 Å). No significant changes to the Cu–NO_2^- binding were observed in the single report of a crystal structure featuring nitrite bound to the reduced T2 site.[21] Lastly, an unusual *side-on NO* binding mode was characterised crystallographically for the NO-bound state of the T2 site.[23] This unusual structure is now thought to represent a kinetically trapped local energy minimum that is metastable in the solid state,[24] with a *bent (~160°) end-on* Cu^I–$NO^•$ linkage having been characterised spectroscopically in solution under an NO atmosphere.[25]

10.2.2 Structural Models Relevant to the Nitrite-Bound T2 Site

Synthetic mononuclear copper–nitrite complexes are commonly known to take on one of three different binding modes (Figure 10.2): η^1-N, η^1-O and η^2-O,O. It is clear that nitrite binds to the T2 site in CuNiR in the η^2-O,O binding mode preferentially, regardless of oxidation state. All three binding modes are accessible for typical copper complexes, as exemplified by the simultaneous presence of all three in the structure $[Cu(NO_2)_6]^{4-}$ and related simple ions.[26–28] Structurally characterised model complexes for the nitrite-bound T2 site are known and adopt η^1-N,[29,30] η^1-O[29–53] and η^2-O,O[29,44,46,54–80] binding modes. These different binding modes are likely to be very close in energy in

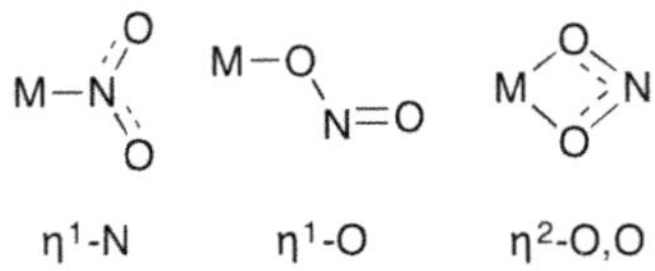

Figure 10.2 Three possible binding modes in mononuclear metal–nitrite complexes.

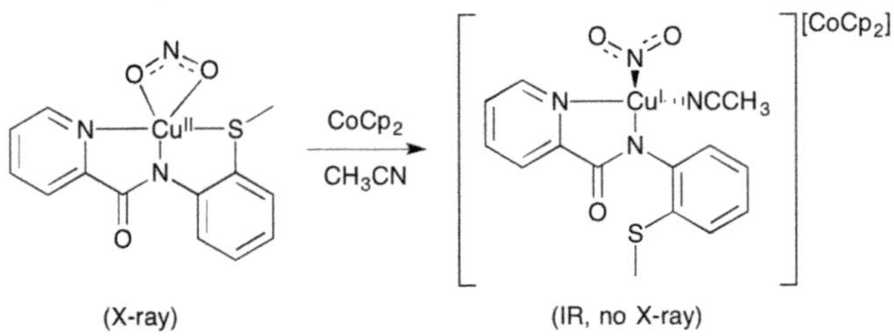

Figure 10.3 Redox-induced change in nitrite binding mode observed by Patra and co-workers.[29]

many cases, as exemplified by a tris(pyridylamine)-supported complex that switches between the η^1-N and η^1-O modes depending on the crystallisation conditions.[30,52] In another example, one-electron reduction of a $Cu^{II}(NO_2)$ complex with asymmetric η^2-O,O binding results in a reduced $Cu^{I}(NO_2)$ complex with η^1-N binding (Figure 10.3).[29] This observation indicates that, in the absence of secondary and tertiary protein structure effects, the copper oxidation state is a key factor determining the NO_2^- binding mode when all else is equal.

Most of the examples cited above do not duplicate the $N_3Cu(NO_2)$ (N = nitrogen donor) motif found in CuNiR. It is particularly instructive to examine the subset of model complexes that do fit this $N_3Cu(NO_2)$ motif. Indeed, within this smaller dataset, all Cu^{II} complexes feature the same η^2-O,O binding that is preferred by the T2 site of CuNiR. On the other hand, all known Cu^{I} nitrito complexes within this smaller dataset feature the η^1-N binding mode. The latter still serve as functional CuNiR models, as discussed further below.

In the Cu^{II} oxidation state, perhaps the most well-studied $N_3Cu(NO_2)$ models are the various tris(pyrazolyl)borate complexes, $(Tp^{R,R'})Cu(NO_2)$, synthesised by the groups of Tolman (R,R' = Me,Me; Ph,Ph; CF_3,CH_3; ^{t}Bu,H)[81–83] and Fujisawa (R,R' = ^{i}Pr,^{i}Pr; ^{t}Bu,^{i}Pr)[84] (Figure 10.4a). Related triazolyl derivatives have also been synthesised by the group of Papish.[85] Karlin and co-workers synthesised a tris(amine)-ligated $Cu^{II}(NO_2)$ complex (Figure 10.4b),[86] and Casella and co-workers has reported related N-heterocycle derivatives.[73,87] In all of these cupric species, the nitrite ligand adopts the η^2-O,O binding mode with varying degrees of asymmetry. The family of tris(pyrazolyl) borate complexes is particularly instructive in this regard, as the degree of

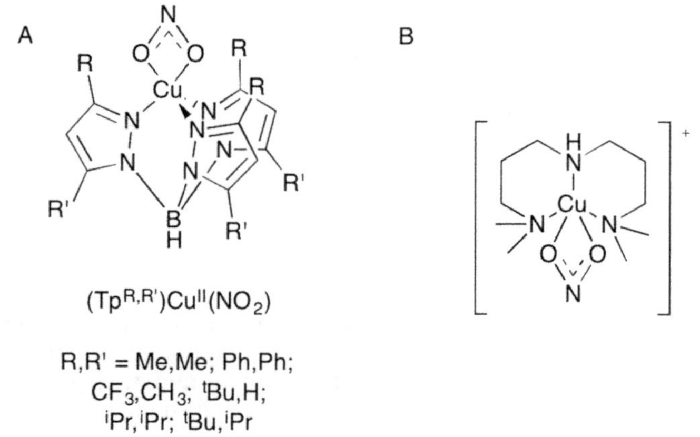

Figure 10.4 Examples of structurally characterised copper(II) nitrite model complexes within the $N_3Cu(NO_2)$ motif. (a) The family of $(Tp^{R,R'})Cu(NO_2)$ complexes. (b) A triaminocopper(II) nitrite complex.

Table 10.1 Asymmetric Cu–O distances for various $(Tp^{R,R'})Cu^{II}(NO_2)$ complexes shown in Figure 10.4a as a function of tris(pyrazolyl)borate substituents.

R,R'	Me,Me	CF_3,CH_3	^{i}Pr,^{i}Pr	^{t}Bu,H	^{t}Bu,^{t}Bu
d(Cu–$O_{proximal}$)	2.025 Å[a]	2.003 Å[a]	2.022Å	1.976Å	1.987Å
d(Cu–O_{distal})	2.025 Å[a]	2.003 Å[a]	2.031Å	2.169Å	2.195Å
Difference	0 Å	0 Å	0.009Å	0.193Å	0.208Å

[a]Crystallographically imposed symmetry.

asymmetry in the η^2-O,O binding can be correlated to the steric constraints of the $(Tp^{R,R'})$ ligand. As seen in Table 10.1, there is a rough relationship by which the difference between Cu–$O_{proximal}$ and Cu–O_{distal} distances gets larger as the $(Tp^{R,R'})$ ligand gets bulkier. This observation indicates that the asymmetric η^2-O,O found in CuNiR represents a strained geometry induced by secondary coordination sphere interactions with the protein environment. The varying degree of asymmetry in the η^2-O,O binding in these $(Tp^{R,R'})Cu(NO_2)$ complexes is also reflected in the shape of the electron paramagnetic resonance (EPR) signals coming from a change in electronic ground states,[84] although other spectroscopic methods such as infrared spectroscopy are less sensitive to this effect.[2] The singly occupied molecular orbital (SOMO) for a model $(Tp^{R,R'})Cu(NO_2)$ system has been computed by Lehnert *et al.*[84] and closely matches the redox-active molecular orbital calculated for the CuNiR active site.[19] This SOMO features antibonding overlap between the Cu $3d_{x2-y2}$ orbital and a nitrite π-orbital, and it can be thought of as modelling the destination of one-electron ET from the T1 site that triggers proton-coupled NO loss in CuNiR.

Several Cu^I models in the $N_3Cu(NO_2)$ motif have also been characterised. Among the most well studied are complexes supported by triazacyclononane

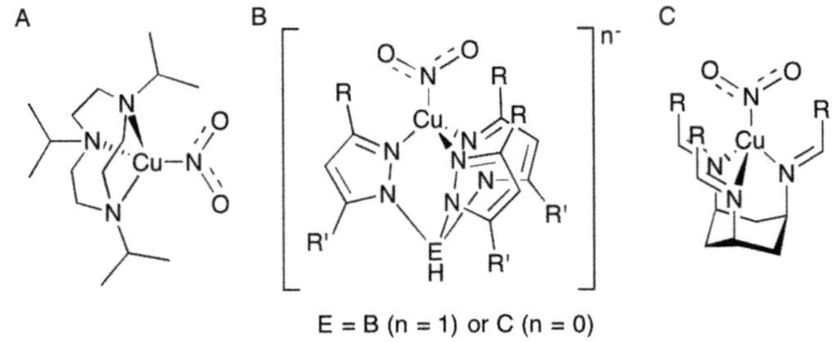

Figure 10.5 Examples of structurally characterised copper(I) nitrite model complexes within the $N_3Cu(NO_2)$ motif, supported by (a) a triazacyclononane ligand, (b) tris(pyrazolyl)borate and tris(pyrazolyl)methane ligands and (c) a tris(imino)cyclohexane ligand.

ligands, such as ($^iPr_3$TACN)Cu(NO_2) (Figure 10.5a), examined by the group of Tolman[88–90] and later by Fujii and co-workers.[91,92] A related species was synthesised by Itoh and colleagues.[93] Anionic tris(pyrazolyl)borate-supported $Cu^I(NO_2)$ complexes and their neutral tris(pyrazolyl)methane analogues have been studied by Hsu *et al.*[94] and Fujii and co-workers (Figure 10.5b),[92] as have related tris(imidazolyl)carbinol complexes.[92] Lastly, Walton and colleagues synthesised a tris(imino)cyclohexane-ligated $Cu^I(NO_2)$ species (Figure 10.5c).[95] All of these $Cu^I(NO_2)$ complexes feature the η^1-N binding mode for their nitrite ligands. In fact, the only structurally characterised example of a $Cu^I(NO_2)$ complex that does not exhibit η^1-N binding is the bis(phosphine) complex $(ArPh_2P)_2Cu(NO_2)$,[55] which instead features η^2-O,O binding as in CuNiR (Ar = 2-MeOC$_6$H$_4$).

10.2.3 Structural Models Relevant to the NO-Bound T2 Site

Whatever role (or lack thereof) that the NO-bound T2 site plays in nitrite reduction, its structure is fascinating. The side-on binding of NO to copper observed for the T2 site by crystallography[23] is unprecedented in the synthetic copper literature. In fact, for transition metal nitrosyl complexes generally, the side-on binding mode is typically observed only as a transient excited-state linkage isomer of an end-on bound ground state (Figure 10.6).[96] Another unusual feature of the side-on NO-bound T2 site is its EPR signature, which allowed for its initial assignment as a Cu^{II}–NO^- species rather than a Cu^I–$NO^\bullet$ species. As is evident by the model chemistry discussed below, Cu^I-nitrosyls are quite stable, while Cu^{II}-nitrosyl complexes are rare and unlikely to be stable under biological conditions. Subsequent studies of CuNiR revealed that the NO-bound state in solution is an end-on Cu^I–$NO^\bullet$ species with ∠Cu–N–O = ~160°,[25] more in line with observations from the model systems discussed here.

The most well-studied model copper–nitrosyls are the tris(pyrazolyl)borate complexes shown in Figure 10.7a. These complexes can be considered as Cu^I–$NO^\bullet$ complexes, or $\{CuNO\}^{11}$ complexes in Enemark–Feltham notation.[97]

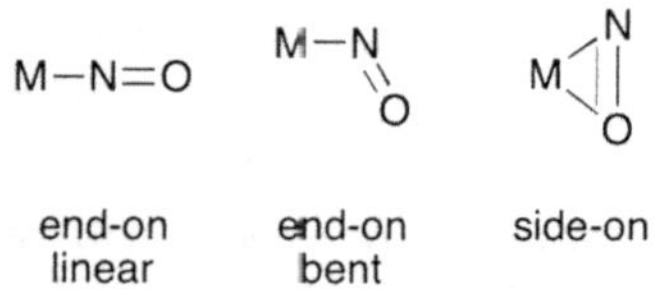

Figure 10.6 Three possible binding modes in mononuclear metal–nitrosyl complexes.

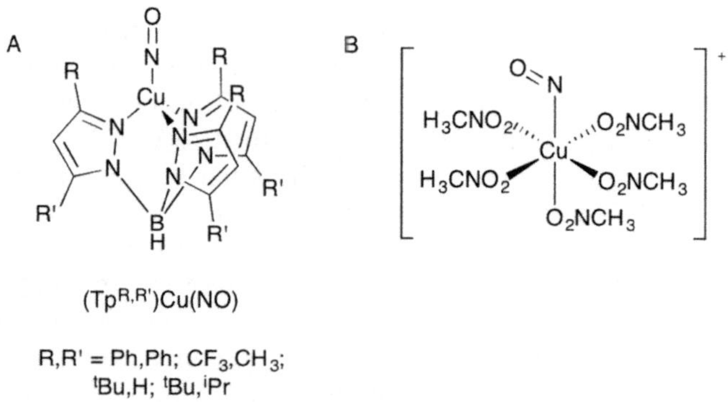

Figure 10.7 Structurally characterised examples of copper–nitrosyls. (a) $\{CuNO\}^{11}$ complexes supported by tris(pyrazolyl)borate ligands. (b) A rare $\{CuNO\}^{10}$ complex supported by nitromethane ligands.

Pioneering synthesis and structural characterisation studies on the first $(Tp^{R,R'})$CuNO complexes were conducted by Tolman and colleagues.[81,98,99] Further bulkier derivatives, as well as a cationic tris(pyrazolyl)methane analogue, were reported by Fujisawa and co-workers[100] and analysed spectroscopically by Lehnert.[100] Structural data for these model systems reveal end-on linear nitrosyl coordination, with $163° < \angle Cu–N–O < 176°$ across the known crystal structures, largely consistent with solution characterisation of the CuNiR system.[25] These $\{CuNO\}^{11}$ models systems are quite thermally robust.

On the other hand, Cu^{II}–$NO^{\bullet}$ complexes, or $\{CuNO\}^{10}$ complexes in Enemark–Feltham notation, are quite unstable in most cases. While simple $CuX_2(NO)$ salts have been known since the early 1960s,[101,102] the only structurally characterised $\{CuNO\}^{10}$ complex to date is $[Cu(CH_3NO_2)_5(NO)]^+$ reported by Hayton and co-workers in 2010 (Figure 10.7b).[103] Here, the end-on nitrosyl ligand is clearly bent with $\angle Cu–N–O = 121°$, and it readily dissociates upon exposure to vacuum. The use of unreactive nitromethane solvent molecules as the only available ligands in Hayton's system is crucial, as typically $\{CuNO\}^{10}$ complexes are observed only as transient intermediates[104] capable of nitrosylating most potential supporting ligands, including amines and alcohols.[105–107] In fact, Lippard and colleagues have exploited this behaviour to develop turn-on fluorescent probes for detection of NO *in vitro* and *in vivo*.

For Lippard's probes, the mechanism for chemical sensing depends upon unstable $\{CuNO\}^{10}$ intermediates demetalating to eliminate fluorescent ligands in nitrosylated form.[108–110] Clearly, a potential $\{CuNO\}^{10}$ form of the T2 site in CuNiR would rapidly nitrosylate nearby amino acid resides and therefore is unlikely to be involved in productive biochemistry.

10.2.4 Functional Models with Nitrite Reductase Activity

Among the $Cu^{I}(NO_2)$ complexes cited above, many examples are capable of nitrite reductase activity (*i.e.*, participation in the reaction described in eqn (10.1)).[29,33,52,55–57,73,83,87,89–92,94,111,112] In many cases, the stoichiometry of the reaction has been established convincingly, and in all cases it has been established by structural data that nitrite binds to the Cu^{I} centre prior to protonation and dehydration. Electrocatalysis has even been demonstrated successfully for selected examples.[29,52] However, it has proven difficult to probe the mechanism in most cases, for various reasons. First, initial protonation of Cu^{I}-bound nitrite is often the rate-determining step, making all subsequent electron transfer and N–O fragmentation processes difficult or impossible to probe experimentally. Second, in some cases, neither the NO nor the HO^-/H_2O produced in the reaction remain bound to copper, thereby leaving no trace of the processes that occurred at the metal centre. Third, in many cases, the Cu^{II} product is capable of reacting further with NO to produce N_2O and/or other NO_x products, further complicating the analysis and reaction setup. Only in a few cases, discussed below, have significant mechanistic details emerged from kinetics analysis and/or low-temperature detection of intermediates.

$$L_nCu^{I} + NO_2^{-} + 2H^{+} \rightarrow L_nCu^{II} + NO + H_2O \quad (10.1)$$

Casella and co-workers studied the nitrite reductase kinetics of Cu^{I} complexes ligated by bis(pyrazole)amine and bis(benzimidazole)amine ligands, with the ligand shown in Figure 10.8a giving the highest activity in the presence of HBF_4.[87] The reaction was found to be first order in the copper complex, with rate increasing with increasing $[H^+]/[Cu^+]$ up to a value of $[H^+]/[Cu^+] = 2.0$. Saturation kinetics were observed with respect to $[NO_2^-]$, indicating a nitrite binding pre-equilibrium occurring prior to rate-determining protonation. The observed rate constant was modelled according to eqn (10.2), where k is a fundamental rate constant proportional to the ET rate and K_a is the acid constant of the protonated $Cu^{I}(NO_2)$ species (*i.e.*, a $Cu^{I}[HNO_2]$ adduct). The observed rate constant being a composite parameter dependent on both ET rate and $Cu^{I}(NO_2)$ basicity indicates that nitrite reductase rates may not always correlate with Cu^{II}/Cu^{I} reduction potentials, but rather will have complex relationships with both reduction potential and basicity and need to be evaluated on a case-by-case basis. Casella and co-workers spectroscopically detected the presence of nitrous acid, HNO_2, as an intermediate in the nitrite reduction at low temperature, providing the first evidence for the

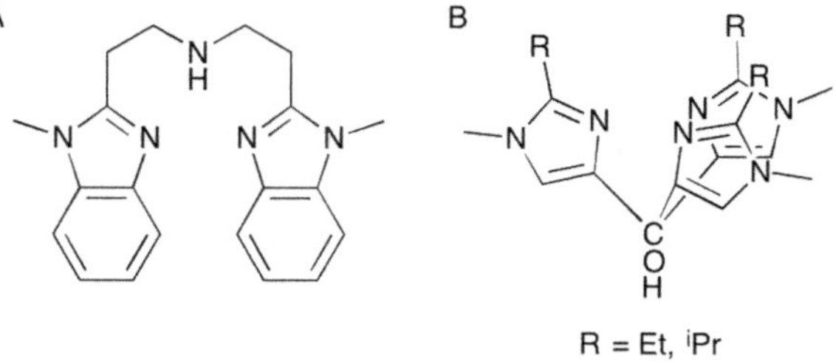

Figure 10.8 Selected N_3 ligands used in mechanistic studies of nitrite reductase reactivity by (A) Casella and (B) Fujii.

intermediacy of a HNO_2 complex along a synthetic nitrite reductase pathway. Yamaguchi and co-workers studied a closely related bis(pyridine)amine system and determined activation parameters of $\Delta H^{\ddagger} = 25$ kJ mol^{-1} and $\Delta S^{\ddagger} = -51$ J mol^{-1} K^{-1} for the synthetic nitrite reductase reaction.[57] Kujime and Fujii[91] revisited the nitrite reductase reaction of (^{i}Pr$_3$TACN)Cu(NO_2) (Figure 10.5a) in the presence of trifluoroacetic acid or acetic acid, a reaction originally reported by Tolman and colleagues.[89,90] The initial protonation of the bound nitrite was found to be rate determining, but two distinct intermediates were detected by monitoring of the trifluoroacetic acid reaction at low temperatures. The first intermediate was assigned as a Cu^I(HNO_2) complex (*i.e.*, the first known metal complex of nitrous acid). The identity of the second intermediate remains unknown, but it was demonstrated to arise from a second proton transfer step and could be either a transient copper-nitrosyl adduct or a Cu^{II} complex of H_2O that decays to the ultimate cupric product of the reaction, (^{i}Pr$_3$TACN)Cu(O_2CCF_3)$_2$. Collecting these observations, the overall mechanistic scheme shown in Figure 10.9 can be constructed.[15] Caution should be exercised in relating this mechanism to chemistry operative in the CuNiR system, as it was proposed based on Cu^I(NO_2) complexes with η^1-N rather than η^2-O,O binding modes.

$$k_{obs} = \frac{k\left[H^+\right]}{K_a} \tag{10.2}$$

Fujii and co-workers later compared the nitrite reductase activity of the ^{i}Pr$_3$TACN system to that of Cu^I(NO_2) complexes supported by tris(pyrazolyl)methane ligands (Figure 10.5b, E = C) and tris(4-imidazolyl)carbinol ligands (Figure 10.8b) of varying steric profiles.[92] Several important trends were noted. First, the tris(4-imidazolyl)carbinol system exhibited the highest nitrite reductase activity, consistent with the presence of three histidine-derived imidazole ligands supporting the T2 site in CuNiR. This increase in activity was rationalised as resulting from the energy-raising of the Cu^I(NO_2) highest occupied molecular orbital, the frontier orbital responsible for reacting with H^+ in the rate-determining step, by electron-rich imidazole ligands compared to less-donating pyrazole ligands. Second, the reaction rate was

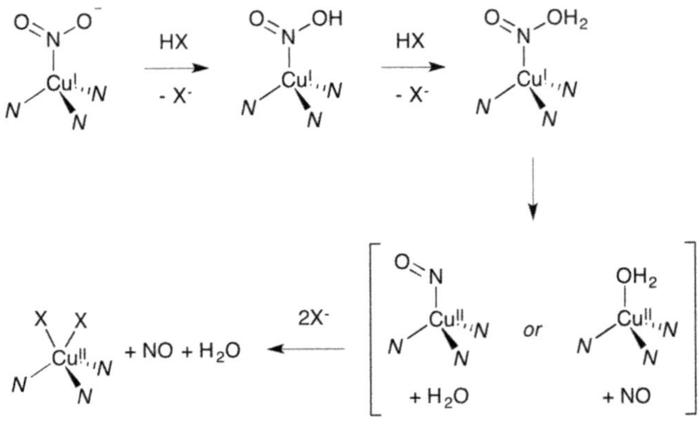

Figure 10.9 Hypothetical mechanism for the dehydration of η^1-N-bound nitrite.

observed to increase with decreasing steric bulk. This increase in activity was rationalised as resulting from the less crowded systems having shorter Cu–$N_{nitrite}$ bond distances, resulting in more efficient Cu → NO_2^- π-backbonding to weaken N–O bonding. Again, caution should be exercised in relating these observations, based on η^1-N nitrite complexes, to the behaviour of CuNiR, which is thought to utilise η^2-O,O nitrite binding due to the participation of secondary coordination sphere effects.[16]

Only two functional models of CuNiR are thought to proceed through O-bound nitrite ligands. First, protonation of Hsu's $(ArPh_2P)_2Cu(NO_2)$ complex with excess acetic acid resulted in quantitative NO formation, and the $Cu^I(NO_2)$ species was characterised crystallographically as having η^2-O,O binding (Figure 10.10a).[55] Again, initial protonation of bound nitrite was found to be rate determining. The measured activation parameters, $\Delta H^\ddagger$ = 41 kJ mol^{-1} and $\Delta S^\ddagger = -158$ J mol^{-1} K^{-1}, are dramatically different from those typical of η^1-N copper–nitrite systems (see above). The larger enthalpy of activation and more negative entropy of activation may be characteristic of the protonation of η^2-O,O Cu^I–nitrite complexes. Second, Moore and Szymczak studied the acid-free nitrite reductase activity of Cu^I ligated by the proton-responsive ligand shown in Figure 10.10b.[112] Although the $Cu^I(NO_2)$ intermediate was not characterised experimentally, it was modelled computationally as having η^1-O binding stabilised by hydrogen bonding in the secondary coordination sphere. Upon proton/electron transfer from the metal–ligand complex, a Cu^{II} aquo complex was isolated as the ultimate product, where the H_2O ligand was stabilised in the binding pocket by hydrogen bonding interactions in the secondary coordination sphere. This system highlights the importance of second-sphere hydrogen bonding both to position the nitrite substrate relative to the copper centre and to facilitate proton and electron transfers. Similar hydrogen bonding interactions are likely to be key to the activity of CuNiR itself and to form the basis of leading mechanistic hypotheses, including that of Suzuki and co-workers.[16]

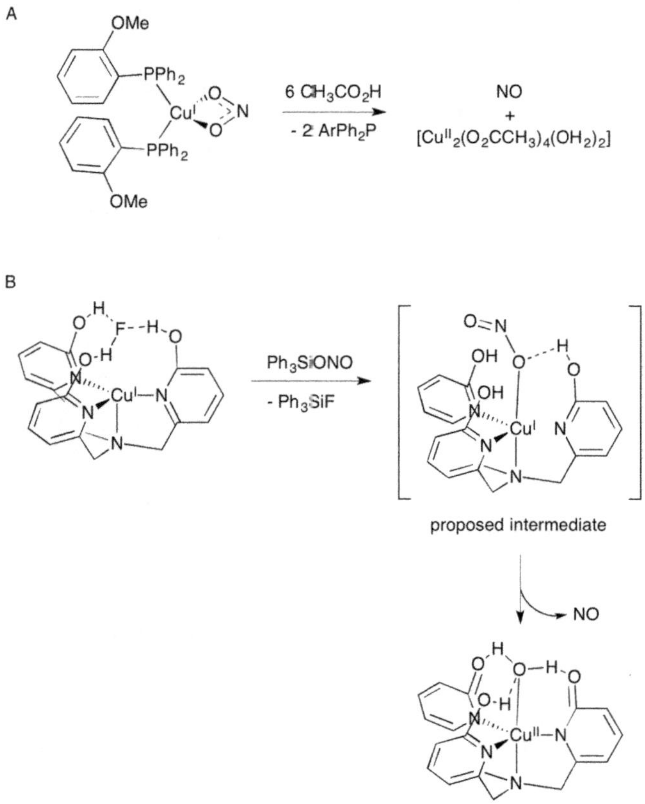

Figure 10.10 Model nitrite reductase reactions proceeding through O-bound nitrite ligands, studied by (A) Hsu and (B) Szymczak.

10.3 Models Relevant to N_2OR

10.3.1 Key Aspects of the Cu_Z Sites to Model

Several structural features are common to all forms and all redox states of the catalytic $[(His)_7Cu_4S_n]$ (n = 1 or 2) sites in N_2OR that have been characterised. The catalytic site is invariably a *tetracopper* site supported by *only nitrogen ligands* (*i.e.*, seven histidine residues), and the tetracopper core is consistently bridged by a μ_4-S^{2-} *ligand*. Most structural studies, including the seminal crystallographic study by Brown *et al.*,[113] were conducted on aerobically purified enzymes and feature the so-called Cu_Z* catalytic site (previously referred to as Cu_Z in some literature). In this form, the catalytic site has a 4Cu : 1S stoichiometry (Figure 10.11a). In the conventional labelling scheme, the Cu_{IV} centre has a low coordination number by virtue of being supported by one histidine rather than two. Bridging the Cu_I–Cu_{IV} edge is a solvent-derived ligand that is thought to be labile and, therefore, the site of N_2O binding. This hypothesis is given weight by a crystal structure of N_2OR

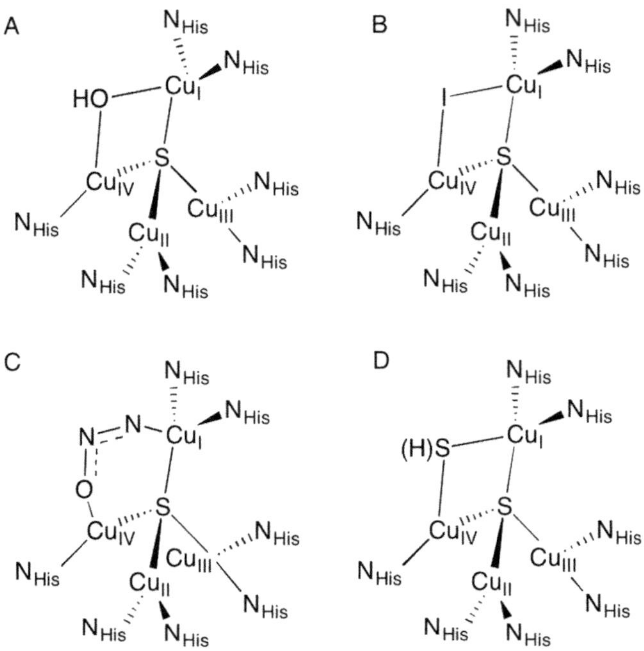

Figure 10.11 The catalytic site of nitrous oxide reductase. (a) The Cu_Z* form with a solvent-derived ligand bound. (b) The Cu_Z* form with an iodide inhibitor bound. (c) Proposed binding of N_2O to Cu_Z*. (d) The Cu_Z form.

incubated with iodide,[114] a known enzyme inhibitor, which shows I^- having replaced the solvent-derived ligand along the Cu_I–Cu_{IV} edge (Figure 10.11b) and therefore blocking N_2O from accessing the putative binding site. Computational studies have indicated that N_2O binds to the Cu_I–Cu_{IV} edge in a *μ-1,3 binding mode* (Figure 10.11c) along the Cu_I–Cu_{IV} edge, with ∠N–N–O = 139°,[115] although no experimental evidence is available for this proposal. The resting state of Cu_Z* under aerobic conditions is the formal $3Cu^I:1Cu^{II}$ ("1-hole", $S = 1/2$) redox state, while activity towards N_2O requires reduction to the $4Cu^I$ ("fully reduced", $S = 0$) state.[116] The proposed mechanism for N_2O reduction also involves intermediates in the formal $2Cu^I:2Cu^{II}$ redox state,[115] and so *three redox states* are relevant for model complexes. It is also important to note that *secondary-sphere hydrogen bonding* from a nearby lysine residue has been observed with the solvent-derived ligand in the resting state.[117] This hydrogen bonding is also thought to be key in stabilisation of the N_2O-bound intermediate and in lowering the kinetic barrier towards N–O cleavage (by roughly 9–13 kcal mol^{-1}).[118]

Anaerobic preparations of N_2OR have long been known to give distinct redox properties and spectroscopic features from aerobic preparations,[1,3] and a crystallographic study by Einsle and co-workers in 2011 indicated that these changes are due to a different form of the catalytic site being present.[119]

This catalytic site, now known as Cu_Z, features a 4Cu : 2S stoichiometry with a μ_2-sulphur ligand found along the Cu_I–Cu_{IV} edge (Figure 10.11d). The resting state of Cu_Z under aerobic conditions is the $2Cu^I$: $2Cu^{II}$ ("2-hole", $S = 0$) state, and the 1-hole state can be accessed upon reduction. Recent studies by Solomon and co-workers have shown that the 2-hole state of Cu_Z features a μ_2-S^{2-} ligand along the Cu_I–Cu_{IV} edge, but that the sulphur is protonated to give a μ_2-SH^- ligand in that position upon reduction to the 1-hole state.[6] A potential mechanism for aerobic conversion of Cu_Z to Cu_Z*, involving reductive loss of the second sulphur ligand by protonation, was suggested on this basis. This is in line with recent studies by Moura, Solomon and co-workers indicating that only the Cu_Z* form of the catalytic site is kinetically competent to perform N_2O reduction under catalytic conditions.[5] Therefore, *the most relevant cluster types to model are 4Cu : 1S complexes supported by nitrogen ligands in the* $4Cu^I$, $3Cu^I$: $1Cu^{II}$ *and* $2Cu^{II}$: $2Cu^I$ *redox states.*

For comparison to model complexes described below, some key structural and spectroscopic features of the Cu_Z* site are summarised here. Structurally, the $[Cu_4S]$ core of Cu_Z* is approximately C_2-symmetric, with the μ_4-S ligand being roughly coplanar with Cu_I, Cu_{II} and Cu_{IV}. The geometry of the four-coordinate sulphur atom can be described by $\tau_4 = 0.66$, where $\tau_4 = 1$ for a perfect tetrahedron and $\tau_4 = 0$ for a perfect square planar geometry.[120] The Cu–Cu distances fall into two categories, with two of the distances (Cu_{II}–Cu_{IV} and Cu_{II}–Cu_{III}) being short (~2.6 Å) and two of the distances (Cu_I–Cu_{IV} and Cu_I–Cu_{III}) being long (~3.4 Å). The one-electron redox potential of Cu_Z* has been measured as +60 mV *versus* NHE.[121] Finally, the open-shell forms of Cu_Z* and Cu_Z clusters feature characteristic optical features as denoted in Table 10.2, which have been assigned as μ_4-$S^{2-} \rightarrow$ Cu charge transfer bands.[1]

10.3.2 Structural Models Relevant to the Cu_Z* Site

A rich variety of copper–sulphur structures have been characterised using many types of supporting ligands.[122–127] Clusters as small as $[Cu_2S]$[128,129] and as large as $[Cu_{136}S_{56}]$[130] have been characterised. The vast majority of these clusters were synthesized by self-assembly, making rational control of Cu : S stoichiometry a challenge. In this respect, recent studies by the groups of Hillhouse[131] and Murray[132] represent landmarks. In the Hillhouse synthesis, a carbene-supported $[Cu_3S]$ cluster in the $3Cu^I$ redox state (Figure 10.12a) was

Table 10.2 Characteristic S → Cu CT transitions for the catalytic sites of nitrous oxide reductase.

Site	Redox state	λ (nm)	ε (M^{-1} cm^{-1})
Cu_Z*	$3Cu^I$: $1Cu^{II}$	640	4000
Cu_Z	$3Cu^I$: $1Cu^{II}$	670	4000
Cu_Z	$2Cu^{II}$: $2Cu^I$	629(sh),[a] 560	4000[b]

[a]Shoulder.
[b]For main feature.

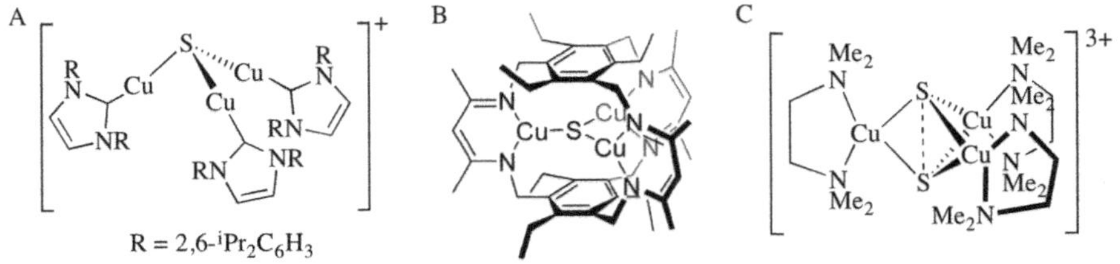

Figure 10.12 Selected tricopper model complexes synthesised by the groups of (a) Hillhouse, (b) Murray and (c) Tolman. The image in (b) was reprinted with permission from ref. 132. Copyright 2014 American Chemical Society.

constructed through rational design by introducing each of the core atoms sequentially, aided by a protection/deprotection strategy. In the Murray synthesis, ketiminate-supported $[Cu_3S]$ clusters in the $2Cu^{I}:1Cu^{II}$ and $1Cu^{I}:2Cu^{II}$ redox states were constructed using a cyclophane cage structure to control nuclearity and orientate the Cu reactive sites appropriately (Figure 10.12b). Prior to these recent reports, the most well-studied models for Cu_Z/Cu_Z* were synthesised by Tolman's group and featured $[Cu_3S_2]$ cores.[133] Although originally assigned as copper–sulphide clusters, subsequent spectroscopic and computational studies have revealed that these complexes feature significant S–S bonding and are best thought of as tricopper(II) complexes of S_2^{3-} (Figure 10.12c).[134–136]

The first synthetic $[Cu_4S]$ cluster, $[(dppm)_4Cu_4(\mu_4\text{-}S)]^{2+}$ (dppm = bis[diphenylphosphino]methane), was initially reported by Yam and co-workers in 1993 (Figure 10.13a).[137] The first synthesis of this complex involved self-assembly of the $[Cu_4S]$ cluster initiated by addition of Na_2S to the dicopper(I) precursor $[Cu_2(\mu\text{-dppm})_2(CH_3CN)_2]^{2+}$, although other methods have been reported subsequently.[138] This $[Cu_4S]$ complex and its derivatives featuring Ag, Au and Se have been studied extensively for their photophysical properties,[139–141] but the relationship with Cu_Z* was not appreciated initially because the first crystal structure of N_2OR had not been reported yet. The structure of this complex bears some resemblance to the Cu_Z* site. The complex has C_{2v} symmetry, and the sulphur geometry is characterised by τ_4 = 0.59. The Cu–Cu positions do follow the pattern of two short (~2.9 Å) and two long (~3.1 Å) distances, although the asymmetry is not as dramatic as that seen in Cu_Z*. Cyclic voltammetry measurements of Yam's complex showed only irreversible redox events, indicating that this cluster is stable only in the $4Cu^{I}$ state.

In 2014, Johnson *et al.* revisited the Yam-type compounds and explored variations in the bridging diphosphine ligands.[142] Increasing the steric bulk by using dcpm in place of dppm (dcpm = bis[dicyclohexylphosphino]methane) resulted in $[(dcpm)_3Cu_3(\mu_3\text{-}S)]^{+}$ instead of a tetracopper cluster, indicating that sterics play an important role in copper–sulphide nuclearity during self-assembly. Use of dppa in place of dppm (dppa = bis[diphenylphosphino]amine) did result in the tetracopper cluster $[(dppa)_4Cu_4(\mu_4\text{-}S)]^{2+}$. One interesting feature of this molecule is that the N–H groups in the backbone of

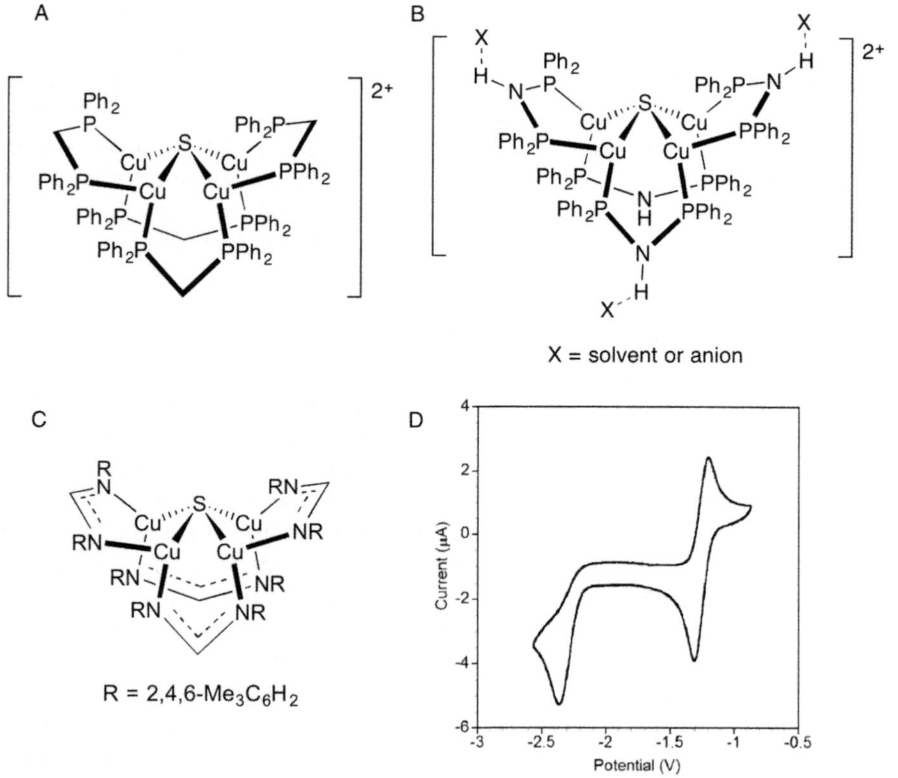

Figure 10.13 Diphosphine-supported $[Cu_4(\mu_4\text{-}S)]$ model complexes synthesised by the groups of (a) Yam and (b) Mankad. (c) An amidinate-supported $[Cu_4(\mu_4\text{-}S)]$ model complex and (d) its cyclic voltammogram. The image in (d) was reproduced from ref. 143 with permission from the Royal Society of Chemistry.

the dppa ligand act as hydrogen bond donors akin to the lysine residue near the Cu_Z* cluster. Evidence for hydrogen bonding to solvent molecules and anions was reported both in solution and in the solid state (Figure 10.13b). Another interesting feature of this molecule is that its cyclic voltammogram features a pseudo-reversible one-electron oxidation at −0.12 V *vs.* $[FeCp_2]^{+/0}$, indicating that both the $4Cu^I$ and $3Cu^I:1Cu^{II}$ states are stable on the electrochemical timescale, although synthesis of the oxidised species has not been reported. Collectively, these observations highlight the roles of both primary and secondary coordination sphere effects in tuning the nuclearities, substrate binding pathways and redox properties of $[Cu_4S]$ clusters. Although none of these $4Cu^I$ complexes are known to react with N_2O, the dppa-supported cluster was shown to react with both azide and iodide anions, resulting in the breakdown of the tetracopper clusters to yield tricopper products featuring the anionic substrates. Its particular affinity for iodide mimics the inhibition behaviour of N_2OR.

The first [Cu_4S] cluster supported only by nitrogen ligands was reported by Johnson *et al.* in 2015.[143] This complex, supported by bridging amidinate ligands (Figure 10.13c), was constructed by adding neutral sulphur atom donors—either S_8 or Ph_3SbS—to the dicopper(I) precursor $Cu_2(NCN)_2$ to yield the $2Cu^I:2Cu^{II}$ complex $(NCN)_4Cu_4(\mu_4\text{-}S)$ (NCN = $[2,4,6\text{-}Me_3C_6H_2N]_2CH$). The complex was found to be isostructural to Yam's complex, with a sulphur geometry characterised by $\tau_4 = 0.76$. The Cu–Cu contacts were much more faithful to the Cu_Z* site, with an alternating pattern of distinct short (~2.4 Å) and long (~3.0 Å) distances. The deep purple complex featured a characteristic optical feature at 561 nm with a shoulder at ~470 nm, closely matching the optical bands for the 2-hole state of Cu_Z. Also like the 2-hole Cu_Z cluster, the ground state of the synthetic $2Cu^I:2Cu^{II}$ species was found to be a singlet. The cyclic voltammogram (Figure 10.13d) featured a fully reversible reduction event at −1.3 V *vs.* $[FeCp_2]^{+/0}$, indicating that an anionic $3Cu^I:1Cu^{II}$ complex, $[(NCN)_4Cu_4(\mu_4\text{-}S)]^-$, should be highly stable and isolable. DFT analysis indicated that the redox-active molecular orbital for this process features significant S 3p character in addition to 3d contributions from all four Cu centres, highlighting the highly covalent nature of the [Cu_4S] core. Further reduction to the dianionic $4Cu^I$ state at −2.4 V *vs.* $[FeCp_2]^{+/0}$ was irreversible by cyclic voltammetry, indicating that it cannot be stabilised by this particular amidinate ligand. Between the diphosphine systems and the amidinate systems, all three catalytically relevant [Cu_4S] redox states have been modelled. However, no single ligand system has been identified in synthetic systems that can stabilise all three redox states as seen in N_2OR.

The only complexes featuring a structural connection between one or more Cu centres and N_2O were reported in 2013 by Severin and co-workers.[144] As shown in Figure 10.14, these molecules feature N_2O units sandwiched between Lewis acidic Cu^I or Cu^{II} centres and Lewis basic N-heterocyclic carbenes (NHCs). It should be noted that these complexes did not result from the direct reaction between a Cu species and gaseous N_2O. Rather, the N_2O molecule required pre-activation by adduct formation with the NHC[145] in order to enable reaction with the Cu species.

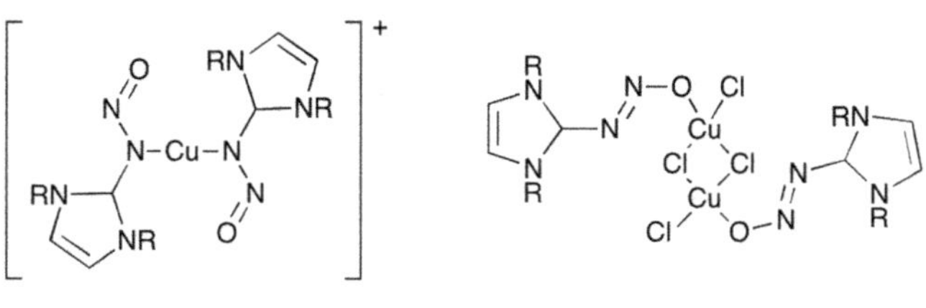

Figure 10.14 Severin's Cu complexes of N_2O that had been pre-activation by adduct formation with *N*-heterocyclic carbenes.

10.3.3 Functional N_2O Reactivity of Molecular Copper Complexes

Although no true copper–sulphide complex is known to react with gaseous N_2O, a handful of other molecular Cu species do exhibit such activity, and some of them feature other types of copper–sulphur interactions (*i.e.*, thiolates and disulphides) as key structural elements. As an aside, it should be noted that simple Cu salts[146] and heterogeneous Cu-containing materials[147–150] also are known to activate gaseous N_2O, but will not be explored in detail here.

The first molecular Cu complex to show any activity towards N_2O was reported in 2009 by Tolman and co-workers.[151] The Me_3TACN-supported tricopper species shown in Figure 10.15a was characterised at low temperature. Assigned as a mixed-valent $2Cu^I{:}1Cu^{II}$ complex bridged by a S_2^{2-} ligand, this species exhibited a Cu/S_2^{2-} charge transfer band at 631 nm, similar to that for the 1-hole Cu_Z* site. Exposure of this [Cu_3S_2] cluster to excess N_2O at −80 °C resulted in stoichiometric generation of N_2, measured to be quantitative by GC/MS. No oxygen-containing by-product was identified in the study. Based on the fact that excess $[(Me_3TACN)Cu(CH_3CN)]^+$ inhibited the N_2O reaction, it was proposed that the tricopper species was in equilibrium with the mixed-valent dicopper disulphide cluster shown in Figure 10.15b, which was proposed to be the active species for N_2O reduction. Computational modelling identified μ–O binding of N_2O to the dicopper cluster in the transition state (Figure 10.15b), distinct from the μ-N,O binding proposed for N_2OR. An analogous binding mode has been proposed for N_2O activation by a Cu-containing zeolite material,[149] and it is likely that the two bimetallic N_2O binding modes are relatively close in energy for many systems, including Cu_Z*.[152]

One common way to trace the fate of the oxygen atom during an N_2O reduction process is to employ an oxygen atom acceptor such as PPh_3. Two molecular Cu systems are known to reduce N_2O to N_2 with concomitant O atom transfer to PPh_3 to generate $OPPh_3$ as a by-product. The mixed-valent dicopper complex shown in Figure 10.16a was reported by Beloglazkina and co-workers in 2013 to be a catalyst for oxygen atom transfer from N_2O to PPh_3.[153] Quantitative conversion of PPh_3 to $OPPh_3$ was observed with 5 mol% catalyst loading at room temperature, whereas only 8% conversion was observed in the absence of the catalyst. The same report disclosed electrocatalytic activation of N_2O upon two-electron reduction of the catalyst and resulting surface deposition of Cu-containing material onto the electrode surface, but no attempts were made to detect the nitrogen- or oxygen-containing products. In 2014, Jayarathne *et al.* reported that the heterobimetallic Cu–Fe species shown in Figure 10.16b also reduces N_2O to N_2 at room temperature.[154] Stoichiometric (86%) formation of $OPPh_3$ was observed when the reaction was conducted in the presence of excess PPh_3. No oxygen atom transfer was noted when one of the two metal sites was omitted and Cu-only or Fe-only analogues were tested. Catalysis was not observed in this

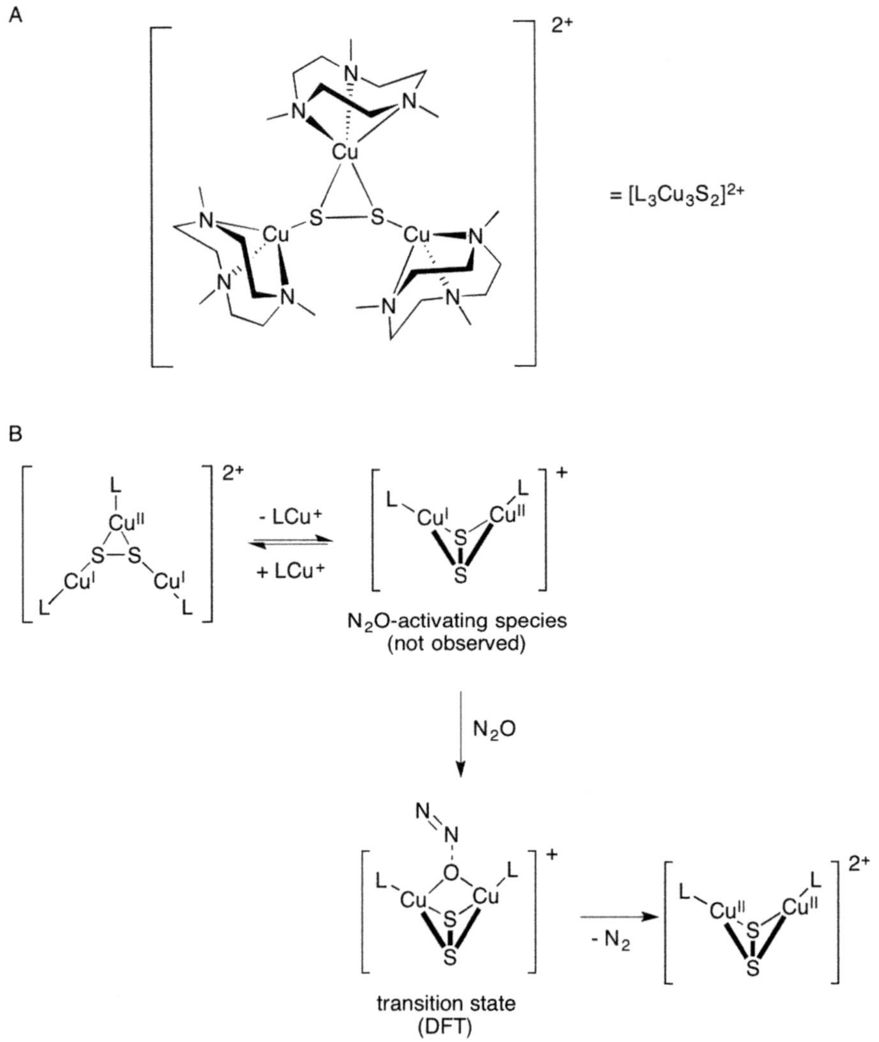

Figure 10.15 (a) Tolman's N_2O-activating tricopper–disulphide cluster. (b) Proposed mechanism for N_2 loss through N_2O activation by a dicopper–disulphide intermediate.

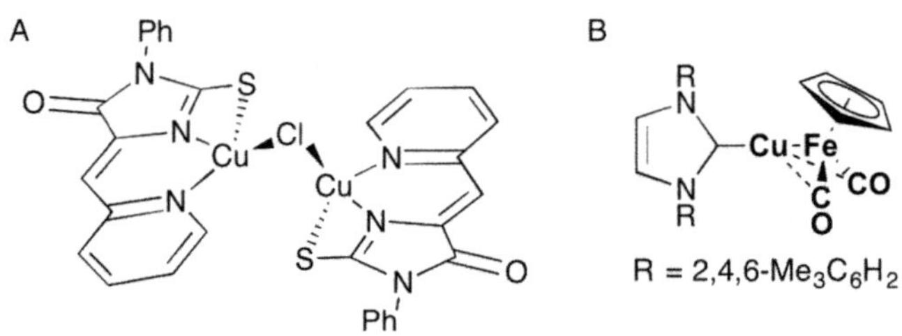

Figure 10.16 Two molecular Cu complexes that are active for oxygen atom transfer from N_2O to PPh_3, studied by (A) Beloglazkina and (B) Mankad.

case due to the rapid precipitation of an insoluble Fe-containing material upon exposure of the heterobimetallic complex to N_2O.

In 2010, Torelli and co-workers demonstrated that dinucleating poly(pyridine) ligand scaffolds could be used to construct unique mixed-valent dicopper mono(thiolate) complexes,[155] whose $[Cu_2S]$ cores with formal Cu^ICu^{II} redox states are relevant to the $[Cu_I–S–Cu_{IV}]$ edge of Cu_Z*, where N_2O is thought to bind. Although their first-generation system was not reactive towards N_2O due to its coordinatively saturated Cu centres, the second-generation system shown in Figure 10.17a did exhibit N_2O reduction activity, as reported in 2014.[156] Assigned as a fully delocalised mixed-valent species, the adjacent reactive sites of the two Cu centres were found to bind non-equivalent ligands, triflate and water. Exposure of this dissymmetric species to N_2O produced 0.4 equivalents of N_2 and resulted in a dicopper(II) hydroxide species with a unique $[Cu_2(\mu\text{-}SR)(\mu\text{-}OH)]$ core. A reaction stoichiometry consisting of 0.5 equivalents of N_2O consumed per dicopper molecule was proposed, consistent with the two-electron reduction of N_2O requiring two Cu^ICu^{II} electron donors. Evidence for an N_2O adduct was obtained by ^{19}F nuclear magnetic resonance spectroscopy, although no experimental data regarding the N_2O binding mode were obtained. Several different binding modes featuring both bridging and monodentate N_2O ligands were modelled

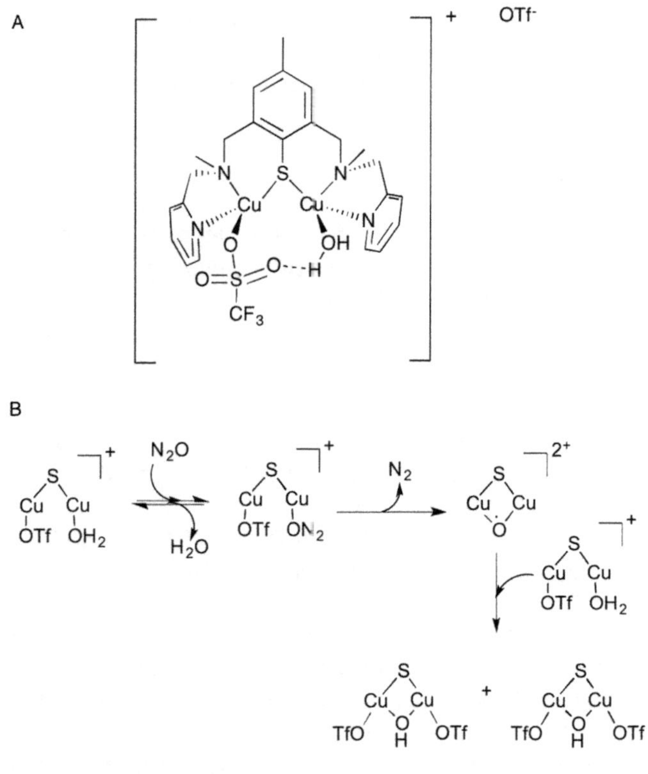

Figure 10.17 (a) Torelli's N_2O-reducing dicopper thiolate complex. (b) Proposed mechanism for N_2O reduction.

computationally, and the authors proposed that the intermediate features a monodentate N_2O ligand (either N-bound or O-bound, separated by ~2 kcal mol^{-1}) on the basis of its spectral similarities to the aquo species. The proposed mechanism for N_2O reduction is shown in Figure 10.17b. First, N_2O displaces the labile H_2O ligand. Then, loss of N_2 provides a formally $Cu^{II}Cu^{III}(\mu\text{-}O)$ intermediate that likely has $Cu^{II}Cu^{II}(\mu\text{-}O^{\bullet})$ character. Finally, reduction of this oxidised intermediate by an equivalent of the starting dicopper complex (coupled to proton transfers) provides two equivalents of the dicopper(II) hydroxide species. This series of results highlights the requirement for a labile coordination site with a weakly coordinated ligand in order for productive N_2O activation to occur. Furthermore, it is noteworthy that four Cu centres are required to reduce a single N_2O molecule in this case, mimicking the stoichiometry of a single N_2OR turnover.

Considering these four examples of molecular Cu complexes that are active for N_2O reduction, it is clear that cooperative effects are a requirement for N_2O activation by Cu. All four examples highlighted here employ bimetallic cooperativity and feature polarised Cu–M interactions, either from homobimetallic Cu/Cu mixed valency or from heterobimetallic Cu/Fe bonding. This view is congruent with the proposed binding of N_2O to the Cu_Z* site, which is thought to utilise two adjacent Cu centres that have different primary coordination spheres and therefore are likely to have different partial charges, even in the fully reduced $4Cu^I$ state. It is also in line with an emerging trend in N_2O activation outside of Cu chemistry, wherein Lewis acid–base cooperativity is a productive strategy for N_2O capture and utilisation.[157–164]

Acknowledgements

The authors are grateful for support from the Department of Chemistry and the College of Liberal Arts & Sciences at the University of Illinois at Chicago. NPM is an Alfred P. Sloan Research Fellow.

References

1. E. I. Solomon, D. E. Heppner, E. M. Johnston, J. W. Ginsbach, J. Cirera, M. Qayyum, M. T. Kieber-Emmons, C. H. Kjaergaard, R. G. Hadt and L. Tian, *Chem. Rev.*, 2014, **114**, 3659.
2. A. C. Merkle and N. Lehnert, *Dalton Trans.*, 2012, **41**, 3355.
3. S. R. Pauleta, S. Dell'Acqua and I. Moura, *Coord. Chem. Rev.*, 2013, **257**, 332.
4. S. Ghosh, A. Dey, O. M. Usov, Y. Sun, V. M. Grigoryants, C. P. Scholes and E. I. Solomon, *J. Am. Chem. Soc.*, 2007, **129**, 10310.
5. E. M. Johnston, S. Dell'Acqua, S. Ramos, S. R. Pauleta, I. Moura and E. I. Solomon, *J. Am. Chem. Soc.*, 2014, **136**, 614.
6. E. M. Johnston, S. Dell'Acqua, S. R. Pauleta, I. Moura and E. I. Solomon, *Chem. Sci.*, 2015, **6**, 5670.
7. P. L. Holland and W. B. Tolman, *J. Am. Chem. Soc.*, 1999, **121**, 7270.

8. B. A. Jazdzewski, P. L. Holland, M. Pink, V. G. Young, D. J. E. Spencer and W. B. Tolman, *Inorg. Chem.*, 2001, **40**, 6097.
9. N. P. Mankad, W. E. Antholine, R. K. Szilagyi and J. C. Peters, *J. Am. Chem. Soc.*, 2009, **131**, 3878.
10. R. Houser, V. Young and W. Tolman, *J. Am. Chem. Soc.*, 1996, **118**, 2101.
11. S. B. Harkins and J. C. Peters, *J. Am. Chem. Soc.*, 2004, **126**, 2885.
12. N. P. Mankad, E. Rivard, S. B. Harkins and J. C. Peters, *J. Am. Chem. Soc.*, 2005, **127**, 16032.
13. S. B. Harkins, N. P. Mankad, A. J. M. Miller, R. K. Szilagyi and J. C. Peters, *J. Am. Chem. Soc.*, 2008, **130**, 3478.
14. N. P. Mankad, S. B. Harkins, W. E. Antholine and J. C. Peters, *Inorg. Chem.*, 2009, **48**, 7026.
15. B. A. Averill, *Angew. Chem., Int. Ed.*, 1994, **33**, 2057.
16. K. Kataoka, H. Furusawa, K. Takagi, K. Yamaguchi and S. Suzuki, *J. Biochem.*, 2000, **127**, 345.
17. H. J. Wijma, L. Jeuken and M. P. Verbeet, *J. Biol. Chem.*, 2006, **281**, 16340.
18. R. H. Goldsmith, L. C. Tabares and D. Kostrz, *Proc. Nat. Acad. Sci. U. S. A.*, 2011, **108**, 17269.
19. S. Ghosh, A. Dey, Y. Sun, C. P. Scholes and E. I. Solomon, *J. Am. Chem. Soc.*, 2009, **131**, 277.
20. J. W. Godden, S. Turley, D. C. Teller, E. T. Adman, M. Y. Liu, W. J. Payne and J. LeGall, *Science*, 1991, **253**, 438.
21. M. E. Murphy, S. Turley and E. T. Adman, *J. Biol. Chem.*, 1997, **272**, 28455.
22. S. V. Antonyuk, R. W. Strange, G. Sawers, R. R. Eady and S. S. Hasnain, *Proc. Natl. Acad. Sci. U. S. A.*, 2005, **102**, 12041.
23. E. I. Tocheva, F. I. Rosell, A. G. Mauk and M. E. P. Murphy, *Science*, 2004, **304**, 867.
24. A. C. Merkle and N. Lehnert, *Inorg. Chem.*, 2009, **48**, 11504.
25. O. M. Usov, Y. Sun, V. M. Grigoryants, J. P. Shapleigh and C. P. Scholes, *J. Am. Chem. Soc.*, 2006, **128**, 13102.
26. A. Ozarowski and R. Allmann, *Z. Anorg. Allg. Chem.*, 1991, **592**, 187.
27. K. A. Klanderman, W. C. Hamilton and I. Bernal, *Inorg. Chim. Acta*, 1977, **23**, 117.
28. E. C. Lingafelter and D. L. Cullen, *Inorg. Chem.*, 1971, **10**, 1264.
29. R. C. Maji, S. K. Barman, S. Roy, S. K. Chatterjee, F. L. Bowles, M. M. Olmstead and A. K. Patra, *Inorg. Chem.*, 2013, **52**, 11084.
30. N. Komeda, H. Nagao, Y. Kushi and G. Adachi, *Bull. Chem. Soc. Jpn.*, 1995, **68**, 581.
31. G. Steel, A. Mustapha, J. Reglinski and A. R. Kennedy, *Polyhedron*, 2014, **67**, 360.
32. T. Tano, K. Mieda, H. Sugimoto, T. Ogura and S. Itoh, *Dalton Trans.*, 2014, **43**, 4871.
33. J. G. Woollard-Shore, J. P. Holland, M. W. Jones and J. R. Dilworth, *Dalton Trans.*, 2010, **39**, 1576.
34. D. Maiti, D.-H. Lee, A. A. Narducci Sarjeant, M. Y. M. Pau, E. I. Solomon, K. Gaoutchenova, J. Sundermeyer and K. D. Karlin, *J. Am. Chem. Soc.*, 2008, **130**, 6700.

35. K. Fujisawa, R. Kanda, Y. Miyashita and K.-I. Okamoto, *Polyhedron*, 2008, **27**, 1432.
36. F. A. Mautner, R. Vicente and S. S. Massoud, *Polyhedron*, 2006, **25**, 1673.
37. U. Mukhopadhyay, I. Bernal, S. S. Massoud and F. A. Mautner, *Inorg. Chim. Acta*, 2004, **357**, 3673.
38. M. Scarpellini, A. Neves, E. E. Castellano, E. F. de Almeida Neves and D. W. Franco, *Polyhedron*, 2004, **23**, 511.
39. K.-Y. Choi, M.-A. Whang, H.-H. Lee, K.-C. Lee and M.-J. Kim, *Transition Met. Chem.*, 2004, **29**, 792.
40. P. J. Arnold, S. C. Davies, M. C. Durrant, D. V. Griffiths, D. L. Hughes and P. C. Sharpe, *Inorg. Chim. Acta*, 2003, **348**, 143.
41. S. Youngme, N. Chaichit and N. Koonsaeng, *Inorg. Chim. Acta*, 2002, **335**, 36.
42. G. L. Breneman, C. W. Howald and O. J. Parker, *Acta Cryst.*, 2001, **E57**, 286.
43. R. Gupta and R. Mukherjee, *Polyhedron*, 2001, **20**, 2545.
44. M. Beretta, E. Bouwman, L. Casella and B. Douziech, *Inorg. Chim. Acta*, 2000, **310**, 41.
45. L. M. Berreau, J. A. Halfen, V. G. Young and W. B. Tolman, *Inorg. Chim. Acta*, 2000, **297**, 115.
46. Y. M. Lee, S. N. Choi, I. H. Suh and R. D. Bereman, *Acta Cryst.*, 1998, **C54**, 1582.
47. F. Hueso-Ureña and A. L. Peñas-Chamorro, *Polyhedron*, 1998, **18**, 351.
48. M. Harata, K. Jitsukawa and H. Masuda, *Bull. Chem. Soc. Jpn.*, 1998, **71**, 637.
49. T. K. Lal, J. F. Richardson, M. S. Mashuta and R. M. Buchanan, *Polyhedron*, 1997, **16**, 4331.
50. S. J. Hill, P. Hubberstey and W. S. Li, *Polyhedron*, 1997, **16**, 2447.
51. F. Jiang, R. R. Conry, L. Bubacco, Z. Tyeklar, R. R. Jacobson, K. D. Karlin and J. Peisach, *J. Am. Chem. Soc.*, 1993, **115**, 2093.
52. N. Komeda, H. Nagao, G. Adachi, M. Suzuki and A. Uehara, *Chem. Lett.*, 1993, 1521.
53. R. Allman, S. Kremer and D. Kucharzcyk, *Inorg. Chim. Acta*, 1984, **85**, L19.
54. A. Kalita, P. Kumar, R. C. Deka and B. Mondal, *Chem. Commun.*, 2012, **48**, 1251.
55. W.-J. Chuang, I.-J. Lin, H.-Y. Chen, Y.-L. Chang and S. C. N. Hsu, *Inorg. Chem.*, 2010, **49**, 5377.
56. C.-S. Chen and W.-Y. Yeh, *Chem. Commun.*, 2010, **46**, 3098.
57. H. Yokoyama, K. Yamaguchi, M. Sugimoto and S. Suzuki, *Eur. J. Inorg. Chem.*, 2005, **2005**, 1435.
58. C.-S. Chen, H.-F. Dai, C.-H. Chen and W.-Y. Yeh, *Inorg. Chim. Acta*, 2011, **376**, 396.
59. A. Indra, S. M. Mobin, S. Bhaduri and G. K. Lahiri, *Inorg. Chim. Acta*, 2011, **374**, 415.
60. R.-Q. Zhu, *Acta Cryst.*, 2011, **E67**, 937.
61. B. Sarkar, S. Konar, C. J. Gómez-García and A. Ghosh, *Inorg. Chem.*, 2008, **47**, 11611.

62. S. Chattopadhyay, M. G. B. Drew and A. Ghosh, *Inorg. Chim. Acta*, 2006, **359**, 4519.
63. Y. B. Shim, S. N. Choi, S. J. Lee, S. K. Kang and Y. M. Lee, *Acta Cryst.*, 2004, **E60**, 1573.
64. B. Murphy, G. Roberts, S. Tyagi and B. J. Hathaway, *J. Mol. Struct.*, 2004, **698**, 25.
65. S. Youngme, P. Phuengphai, S. Booncoon and N. Chaichit, *Polyhedron*, 2004, **23**, 1685.
66. A. A. Holder, R. T. Stibrany, N. Bolotina, M. Hall, V. C. R. Payne, K. Kirschbaum, A. A. Pinkerton and A. M. Newton, *J. Chem. Crystallogr.*, 2004, **34**, 383.
67. É. B. Hergovich, Z. Gréczi and J. Kaizer, *Eur. J. Inorg. Chem.*, 2002, 1687.
68. S. Youngme, N. Chaichit, C. Pakawatchai and S. Booncoon, *Polyhedron*, 2002, **21**, 1279.
69. A. Camus, N. Marsich, A. Lanfredi and F. Ugozzoli, *Inorg. Chim. Acta*, 2000, **309**, 1.
70. G. L. Breneman, C. W. Howald and O. J. Parker, *Acta Cryst.*, 1999, **C55**, 1.
71. R. Kuhlman, G. L. Schimek and J. W. Kolis, *Polyhedron*, 1999, **18**, 1379.
72. A. J. Blake, S. J. Hill and P. Hubberstey, *Chem. Commun.*, 1998, 1587.
73. L. Casella, O. Carugo, M. Gullotti, S. Doldi and M. Frassoni, *Inorg. Chem.*, 1996, **35**, 1101.
74. R. T. Stibrany, J. A. Potenza and H. J. Schugar, *Inorg. Chim. Acta*, 1996, **243**, 33.
75. J. A. Halfen and W. B. Tolman, *Acta Cryst.*, 1995, **C51**, 215.
76. C. J. Simmons, B. J. Hathaway, K. Amornjarusiri, B. D. Santarsiero and A. Clearfield, *J. Am. Chem. Soc.*, 1987, **109**, 1947.
77. C. J. Simmons, K. Seff, F. Clifford and B. J. Hathaway, *Acta Cryst.*, 1983, **C39**, 1360.
78. C. J. Simmons, A. Clearfield, W. Fitzgerald, S. Tyagi and B. J. Hathaway, *Inorg. Chem.*, 1983, **22**, 2463.
79. F. S. Stephens, *J. Chem. Soc. A*, 1969, 2081.
80. I. M. Procter and F. S. Stephens, *J. Chem. Soc. A*, 1969, 1248.
81. J. L. Schneider, S. M. Carrier, C. E. Ruggiero, V. G. Young and W. B. Tolman, *J. Am. Chem. Soc.*, 1998, **120**, 11408.
82. C. E. Ruggiero and S. M. Carrier, *Angew. Chem., Int. Ed. Engl.*, 1994, **33**, 895.
83. W. B. Tolman, *Inorg. Chem.*, 1991, **30**, 4877.
84. N. Lehnert, U. Cornelissen, F. Neese, T. Ono, Y. Noguchi, K.-I. Okamoto and K. Fujisawa, *Inorg. Chem.*, 2007, **46**, 3916.
85. M. Kumar, N. A. Dixon, A. C. Merkle, M. Zeller, N. Lehnert and E. T. Papish, *Inorg. Chem.*, 2012, **51**, 7004.
86. G. Y. Park, S. Deepalatha, S. C. Puiu, D.-H. Lee, B. Mondal, A. A. Narducci Sarjeant, D. del Rio, M. Y. M. Pau, E. I. Solomon and K. D. Karlin, *J. Biol. Inorg. Chem.*, 2009, **14**, 1301.
87. E. Monzani, G. J. Anthony, A. Koolhaas, A. Spandre, E. Leggieri, L. Casella, M. Gullotti, G. Nardin, L. Randaccio, M. Fontani, P. Zanello and J. Reedijk, *J. Biol. Inorg. Chem.*, 2000, **5**, 251.

88. J. A. Halfen, S. Mahapatra, M. M. Olmstead and W. B. Tolman, *J. Am. Chem. Soc.*, 1994, **116**, 2173.
89. J. A. Halfen and W. B. Tolman, *J. Am. Chem. Soc.*, 1994, **116**, 5475.
90. J. A. Halfen, S. Mahapatra, E. C. Wilkinson, A. J. Gengenbach, V. G. Young Jr, L. Que Jr and W. B. Tolman, *J. Am. Chem. Soc.*, 1996, **118**, 763.
91. M. Kujime and H. Fujii, *Angew. Chem., Int. Ed. Engl.*, 2006, **45**, 1089.
92. M. Kujime, C. Izumi, M. Tomura, M. Hada and H. Fujii, *J. Am. Chem. Soc.*, 2008, **130**, 6088.
93. A. Kunishita, M. Z. Ertem, Y. Okubo, T. Tano, H. Sugimoto, K. Ohkubo, N. Fujieda, S. Fukuzumi, C. J. Cramer and S. Itoh, *Inorg. Chem.*, 2012, **51**, 9465.
94. S. C. N. Hsu, Y.-L. Chang, W.-J. Chuang, H.-Y. Chen, I.-J. Lin, M. Y. Chiang, C.-L. Kao and H.-Y. Chen, *Inorg. Chem.*, 2012, **51**, 9297.
95. A. K. Nairn, S. J. Archibald, R. Bhalla, C. J. Boxwell, A. C. Whitwood and P. H. Walton, *Dalton Trans.*, 2006, 1790.
96. B. Cormary, I. Malfant, M. Buron-Le Cointe, L. Toupet, B. Delley, D. Schaniel, N. Mockus, T. Woike, K. Fejfarova, V. Petricek and M. Dusek, *Acta Cryst.*, 2009, **B65**, 612.
97. J. H. Enemark and R. D. Feltham, *Coord. Chem. Rev.*, 1974, **13**, 339.
98. S. M. Carrier, C. E. Ruggiero and W. B. Tolman, *J. Am. Chem. Soc.*, 1992, **114**, 4407.
99. C. E. Ruggiero, S. M. Carrier, W. E. Antholine, J. W. Whittaker, C. J. Cramer and W. B. Tolman, *J. Am. Chem. Soc.*, 1993, **115**, 11285.
100. K. Fujisawa, A. Tateda, Y. Miyashita, K.-I. Okamoto, F. Paulat, V. K. K. Praneeth, A. Merkle and N. Lehnert, *J. Am. Chem. Soc.*, 2008, **130**, 1205.
101. R. Fraser, *J. Inorg. Nucl. Chem.*, 1961, **17**, 265.
102. R. T. M. Fraser and W. E. Dasent, *J. Am. Chem. Soc.*, 1960, **82**, 348.
103. A. M. Wright, G. Wu and T. W. Hayton, *J. Am. Chem. Soc.*, 2010, **132**, 14336.
104. M. Sarma and B. Mondal, *Inorg. Chem.*, 2011, **50**, 3206.
105. M. Sarma, A. Kalita, P. Kumar, A. Singh and B. Mondal, *J. Am. Chem. Soc.*, 2010, **132**, 7846.
106. M. Sarma, A. Singh, S. G. Gupta, G. Das and B. Mondal, *Inorg. Chim. Acta*, 2010, **363**, 63.
107. D. Tran, B. W. Skelton, A. H. White, L. E. Laverman and P. C. Ford, *Inorg. Chem.*, 1998, **37**, 2505.
108. M. H. Lim, D. Xu and S. J. Lippard, *Nat. Chem. Biol.*, 2006, **2**, 375.
109. M. H. Lim, B. A. Wong, W. H. Pitcock, D. Mokshagundam, M.-H. Baik and S. J. Lippard, *J. Am. Chem. Soc.*, 2006, **128**, 14364.
110. Z. J. Tonzetich, L. E. McQuade and S. J. Lippard, *Inorg. Chem.*, 2010, **49**, 6338.
111. P. P. Paul and K. D. Karlin, *J. Am. Chem. Soc.*, 1991, **113**, 6331.
112. C. M. Moore and N. K. Szymczak, *Chem. Sci.*, 2015, **6**, 3373.
113. K. Brown, M. Tegoni, M. Prudêncio, A. S. Pereira, S. Besson, J. J. Moura, I. Moura and C. Cambillau, *Nat. Struct. Biol.*, 2000, **7**, 191.

114. K. Paraskevopoulos, S. V. Antonyuk, R. G. Sawers, R. R. Eady and S. S. Hasnain, *J. Mol. Biol.*, 2006, **362**, 55.
115. S. I. Gorelsky, S. Ghosh and E. I. Solomon, *J. Am. Chem. Soc.*, 2006, **128**, 278.
116. S. Ghosh, S. I. Gorelsky, P. Chen, I. Cabrito, J. G. G. Moura, I. Moura and E. I. Solomon, *J. Am. Chem. Soc.*, 2003, **125**, 15708.
117. T. Haltia, K. Brown, M. Tegoni, C. Cambillau, M. Saraste, K. Mattila and K. Djinovic-Carugo, *Biochem. J.*, 2003, **369**, 77.
118. S. Ghosh, S. I. Gorelsky, S. D. George, J. M. Chan, I. Cabrito, D. M. Dooley, J. J. G. Moura, I. Moura and E. I. Solomon, *J. Am. Chem. Soc.*, 2007, **129**, 3955.
119. A. Pomowski, W. G. Zumft, P. M. H. Kroneck and O. Einsle, *Nature*, 2011, **477**, 234.
120. L. Yang, D. R. Powell and R. P. Houser, *Dalton Trans.*, 2007, 955.
121. T. Rasmussen, B. Berks, J. Butt and A. Thomson, *Biochem. J.*, 2002, **364**, 807.
122. J. T. York, I. Bar-Nahum and W. B. Tolman, *Inorg. Chim. Acta*, 2008, **361**, 885.
123. Y. Lee, A. A. N. Sarjeant and K. D. Karlin, *Chem. Commun.*, 2006, 621.
124. J. T. York, I. Bar-Nahum and W. B. Tolman, *Inorg. Chem.*, 2007, **46**, 8105.
125. C. B. Khadka, B. K. Najafabadi, M. Hesari, M. S. Workentin and J. F. Corrigan, *Inorg. Chem.*, 2013, **52**, 6798.
126. S. Dehnen, A. Schäfer, D. Fenske and R. Ahlrichs, *Angew. Chem., Int. Ed.*, 1994, **33**, 746.
127. A. Eichhöfer, G. Buth, S. Lebedkin, M. Kühn and F. Weigend, *Inorg. Chem.*, 2015, **54**, 9413.
128. S. Delgado, P. J. Sanz Miguel, J. L. Priego, R. Jiménez-Aparicio, C. J. Gómez-García and F. Zamora, *Inorg. Chem.*, 2008, **47**, 9128.
129. J. Zhai, A. S. Filatov, G. L. Hillhouse and M. D. Hopkins, *Chem. Sci.*, 2016, **7**, 589.
130. M.-L. Fu, I. Issac, D. Fenske and O. Fuhr, *Angew. Chem., Int. Ed.*, 2010, **49**, 6899.
131. J. Zhai, M. D. Hopkins and G. L. Hillhouse, *Organometallics*, 2015, **34**, 4637.
132. G. N. Di Francesco, A. Gaillard, I. Ghiviriga, K. A. Abboud and L. J. Murray, *Inorg. Chem.*, 2014, **53**, 4647.
133. E. C. Brown, J. T. York, W. E. Antholine, E. Ruiz, S. Alvarez and W. B. Tolman, *J. Am. Chem. Soc.*, 2005, **127**, 13752.
134. R. Sarangi, L. Yang, S. G. Winikoff, L. Gagliardi, C. J. Cramer, W. B. Tolman and E. I. Solomon, *J. Am. Chem. Soc.*, 2011, **133**, 17180.
135. J. F. Berry, *Chem.–Eur. J.*, 2010, **16**, 2719.
136. S. Alvarez, R. Hoffmann and C. Mealli, *Chem.–Eur. J.*, 2009, **15**, 8358.
137. V. W.-W. Yam, W.-K. Lee and T.-F. Lai, *J. Chem. Soc., Chem. Commun.*, 1993, 1571.

138. R. N. Yang, Y. A. Sun, Y. M. Hou, X. Y. Hu and D. M. Jin, *Inorg. Chim. Acta*, 2000, **304**, 1.
139. V. Wing-Wah Yam, K. Kam-Wing Lo, W. Kit-Mai Fung and C. R. Wang, *Coord. Chem. Rev.*, 1998, **171**, 17.
140. C. R. Wang, K. K. W. Lo, W. K. M. Fung and V. W. W. Yam, *Chem. Phys. Lett.*, 1998, **296**, 505.
141. W. H. Lam, E. C.-C. Cheng and V. W.-W. Yam, *Inorg. Chem.*, 2006, **45**, 9434.
142. B. J. Johnson, S. V. Lindeman and N. P. Mankad, *Inorg. Chem.*, 2014, **53**, 10611.
143. B. J. Johnson, W. E. Antholine, S. V. Lindeman and N. P. Mankad, *Chem. Commun.*, 2015, **51**, 11860.
144. A. G. Tskhovrebov, B. Vuichoud, E. Solari, R. Scopelliti and K. Severin, *J. Am. Chem. Soc.*, 2013, **135**, 9486.
145. A. G. Tskhovrebov, E. Solari, M. D. Wodrich, R. Scopelliti and K. Severin, *Angew. Chem., Int. Ed.*, 2011, **51**, 232.
146. G. Kiefer, L. Jeanbourquin and K. Severin, *Angew. Chem., Int. Ed.*, 2013, **52**, 6302.
147. J. S. Woertink, P. J. Smeets, M. H. Groothaert, M. A. Vance, B. F. Sels, R. A. Schoonheydt and E. I. Solomon, *Proc. Natl. Acad. Sci. U. S. A.*, 2009, **106**, 18908.
148. P. J. Smeets, R. G. Hadt, J. S. Woertink, P. Vanelderen, R. A. Schoonheydt, B. F. Sels and E. I. Solomon, *J. Am. Chem. Soc.*, 2010, **132**, 14736.
149. M.-L. Tsai, R. G. Hadt, P. Vanelderen, B. F. Sels, R. A. Schoonheydt and E. I. Solomon, *J. Am. Chem. Soc.*, 2014, **136**, 3522.
150. P. Vanelderen, B. E. R. Snyder, M.-L. Tsai, R. G. Hadt, J. Vancauwenbergh, O. Coussens, R. A. Schoonheydt, B. F. Sels and E. I. Solomon, *J. Am. Chem. Soc.*, 2015, **137**, 6383.
151. I. Bar-Nahum, A. K. Gupta, S. M. Huber, M. Z. Ertem, C. J. Cramer and W. B. Tolman, *J. Am. Chem. Soc.*, 2009, **131**, 2812.
152. W. B. Tolman, *Angew. Chem., Int. Ed.*, 2010, **49**, 1018.
153. A. G. Majouga, E. K. Beloglazkina, A. A. Moiseeva, O. V. Shilova, E. A. Manzheliy, M. A. Lebedeva, E. S. Davies, A. N. Khlobystov and N. V. Zyk, *Dalton Trans.*, 2013, **42**, 6290.
154. U. Jayarathne, S. R. Parmelee and N. P. Mankad, *Inorg. Chem.*, 2014, **53**, 7730.
155. S. Torelli, M. Orio, J. Pécaut, H. Jamet, L. Le Pape and S. Ménage, *Angew. Chem., Int. Ed.*, 2010, **49**, 8249.
156. C. Esmieu, M. Orio, S. Torelli, L. Le Pape, J. Pécaut, C. Lebrun and S. Ménage, *Chem. Sci.*, 2014, **5**, 4774.
157. E. Otten, R. C. Neu and D. W. Stephan, *J. Am. Chem. Soc.*, 2009, **131**, 9918.
158. R. C. Neu, E. Otten, A. Lough and D. W. Stephan, *Chem. Sci.*, 2010, **2**, 170.
159. A. G. Tskhovrebov, E. Solari, M. D. Wodrich, R. Scopelliti and K. Severin, *J. Am. Chem. Soc.*, 2012, **134**, 1471.

160. A. G. Tskhovrebov, E. Solari, R. Scopelliti and K. Severin, *Organometallics*, 2012, **31**, 7235.
161. G. Ménard, J. A. Hatnean, H. J. Cowley, A. J. Lough, J. M. Rawson and D. W. Stephan, *J. Am. Chem. Soc.*, 2013, **135**, 6446.
162. A. G. Tskhovrebov, E. Solari, R. Scopelliti and K. Severin, *Inorg. Chem.*, 2013, **52**, 11688.
163. A. G. Tskhovrebov, L. C. E. Naested, E. Solari, R. Scopelliti and K. Severin, *Angew. Chem., Int. Ed.*, 2014, **54**, 1289.
164. Z. Mo, E. L. Kolychev, A. Rit, J. Campos, H. Niu and S. Aldridge, *J. Am. Chem. Soc.*, 2015, **137**, 12227.

CHAPTER 11

Electron Transfer and Molecular Recognition in Denitrification and Nitrate Dissimilatory Pathways

RUI M. ALMEIDA[a], SIMONE DELL' ACQUA[b], ISABEL MOURA[a], SOFIA R. PAULETA[a] AND JOSÉ J. G. MOURA*[a]

[a]UCIBIO, REQUIMTE, Dept. Química, FCT-UNL, Campus da Caparica, 2829-516 Caparica, Portugal; [b]Department of Chemistry, University of Pavia, Via Taramelli 12, 27100, Pavia, Italy
*E-mail: jose.moura@fct.unl.pt

11.1 Introduction

Denitrification is no longer a "black box" process starting with the reduction of nitrate and ending with the release of dinitrogen gas. In fact, the enzymes that catalyse each step, their substrates and consequently their products were identified and their 3D structures were solved, and the catalytic mechanisms and electron transfer pathways are now being unravelled. These enzymes are complex with multiple redox centres and intricate intramolecular electron transfer routes. The electron transfer reactions and the formation of the transient electron transfer complexes have been characterised using different biophysical and spectroscopic techniques by several groups in recent years.

RSC Metallobiology Series No. 9
Metalloenzymes in Denitrification: Applications and Environmental Impacts
Edited by Isabel Moura, José J. G. Moura, Sofia R. Pauleta and Luisa B. Maia

Published by the Royal Society of Chemistry, www.rsc.org

This chapter reviews the state of the art of the electron transfer mechanisms in the enzymes involved in the four sequential steps of denitrification according to eqn (11.1):

$$2NO_3^- \rightarrow 2NO_2^- \rightarrow 2^{\bullet}NO \rightarrow N_2O \rightarrow N_2 \quad (11.1)$$

catalysed by nitrate reductase, nitrite reductase (copper-containing nitrite reductase [Cu-NiR] or cytochrome cd_1 nitrite reductase [cd_1-NiR]), nitric oxide reductase (NOR) and by the terminal enzyme of the process, nitrous oxide reductase (N_2OR). The dissimilatory reduction of nitrate to ammonia starting with the periplasmic nitrate reductase (Nap) and followed by nitrite reductase (NrfA), which reduces nitrite to ammonium, will also be included in this review.

A brief overview of each system is given, since more detailed descriptions can be found in the other chapters of this book, with the exception of the dissimilatory nitrate reduction pathway.

The two main aspects discussed here are the intra- and inter-molecular electron transfer pathways and the molecular recognition processes involved in the formation of the transient complexes between the redox partners, in which either soluble or membrane-bound proteins take part.

11.2 Electron Transfer Involved in Nitrate Reduction

Respiratory nitrate reductase (NaR) is the first enzyme of the denitrification pathway, which catalyses the reduction of nitrate to nitrite, with concomitant release of one water molecule and the consumption of two protons and two electrons (according to eqn (11.2)):

$$NO_3^- + 2H^+ + 2e^- \rightarrow NO_2^- + H_2O \quad (11.2)$$

The same reaction is catalysed by the Nap in the first step of dissimilatory nitrate reduction to ammonia,[1] and by the prokaryotic cytoplasmic assimilatory nitrate reductase (Nas).[2] These three enzymes belong to the DMSO reductase family of molybdenum-containing enzymes,[3] wherein one Mo ion is coordinated by the *cis*-dithiolene group of two pyranopterin guanidine dinucleotide (Mo-bis PGD). In this chapter, only the first two will be discussed, namely dissimilatory Nap (Section 11.2.1) and respiratory membrane-bound NaR (Section 11.2.2). Assimilatory nitrate reductases are sometimes associated with inflammatory processes in the human gut, as the use of nitrate as a respiratory substrate increases the fitness of intracellular pathogens.[4,5]

Some prokaryotes encode all three classes of nitrate reductases, such as the *Paracoccus* (*Pa.*) species, with each taking part in a specific cell function: NarR (NarGHI) is thought to generate ATP under anaerobic growth, while Nap (NapAB) participates in non-electrogenic pathways and is proposed to dissipate excess reductant under aerobic growth,[6] and NasBCG, located in the cytoplasm, is involved in the two-step assimilatory pathway of nitrate to ammonium (nitrate and nitrite reductase [NiR]), with the electrons flowing from the NADH pool *via* FAD-NasB and a ferredoxin to nitrate reductase, NasC.[2]

Some eukaryotes (*e.g.*, plants, algae or fungi) also encode nitrate reductase, which in this case belongs to the sulphite oxidase family of molybdenum-containing enzymes and participates in the assimilative pathway.[7,8]

11.2.1 Periplasmic Dissimilatory Nitrate Reductase

Nap catalyses the first step of the catabolic dissimilation of nitrate and is restricted to the periplasmic space.[9] Nap was isolated for the first time from *Paracoccus* species in the early 1990s (*Pa. pantotrophus*,[10–12] previously known as *Thiosphaera pantotropha*, and *Pa. denitrificans*[13]), and since then, these enzymes have been found to be commonly encoded in the genomes of α-, β-, γ-, δ- and ε-proteobacteria.

Naps can be purified either as monomers or as heterodimers, the common subunit being the one containing the catalytic centre, NapA, while the other is the electron-transferring subunit, NapB (which is present in the majority of the Nap-encoding organisms, but not all). The monomeric NapA has been isolated from *Desulfovibrio* (*D.*) *desulfuricans* and *Escherichia* (*E.*) *coli*,[14,15] while heterodimeric NapAB was characterised from *Pa. denitrificans*,[13] *Shewanella* (*S.*) *oneidensis*,[16] *Wolinella* (*Wo.*) *succinogenes*,[17] *Rhodobacter* (*Rh.*) *sphaeroides*[18] and *Cupriavidus* (*C.*) *necator*.[19,20] These bacteria, with the exception of the latter three, do not perform complete denitrification, though in some cases, have the *nosZ* gene (encoding N_2OR) in their genome.

There are several genes involved in periplasmic nitrate reduction that are usually clustered together, which in some cases can form an operon. Some of these namely, *napABC*, are considered to be structural genes, while others, such as *napE*, *napK* and *napF*, are of unknown function (first two) or are proposed to be involved in the assembly of the Fe–S centre in NapA (*napF*). On the other hand, *napH*- and *napG*-encoded proteins have been associated with electron transfer from menaquinol to NapA and are required for its membrane attachment.[21]

The electron transfer pathway towards the active centre of NapA, which is buried within the protein core, ~15 Å away from the protein surface, begins in the quinone pool, and thus electron transfer proteins, namely NapC and/or NapGH, will oxidise quinol (menaquinol or ubiquinol), and subsequently reduce NapB (see Figure 11.1).[19,20] In organisms lacking NapB, such as *D. desulfuricans*, it is proposed that NapC, NapG or a putative redox partner, NapM, will transfer electrons directly to NapA.[22,23]

NapG and NapH are two putative ferredoxins encoded usually in the *nap* operon of organisms that do not encode NapC (namely ε-proteobacteria); the latter is an integral membrane protein facing the cytoplasm with two Fe–S binding motifs, and the former is a periplasmic-facing, membrane-associated protein with four Fe–S binding motifs. Both of these proteins

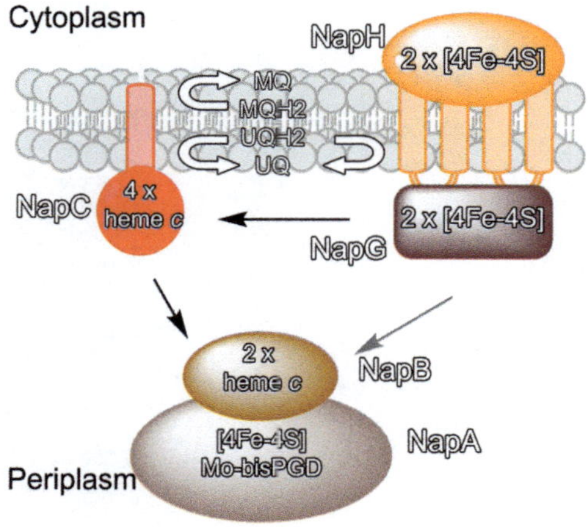

Figure 11.1 Flow of electrons from the quinone pool towards NapA. Ubiquinol (UQH2) is capable of reducing membrane-spanning protein NapH, which in turns transfers electrons to NapG. NapC can oxidise both ubiquinol and menaquinol (MQH2). NapC may also be reduced by periplasmic NapG, and both can transfer electrons to NapB. The electrons are then transferred to NapA. In organisms lacking NapB, electrons are proposed to be transferred directly from NapC to NapA (not shown), or *via* an alternative redox partner NapM, which is not depicted here.

are proposed to receive electrons from the quinone pool (acting as quinol dehydrogenases), transferring them directly to the NapAB active enzyme in a process that also translocates protons across the membrane into the periplasm (see Figure 11.1).[21,24] Although no direct experimental evidence exists of the electron transfer between NapC, NapG or NapH and NapAB, a series of *E. coli nap* gene knockout mutants established that their absence precludes nitrate reduction.[25,26] The structure and detailed function of NapC, NapG and NapH remain unclear, as the proteins are yet to be isolated and biochemically characterised.

In most cases, the redox partner NapB forms with NapA as a heterodimer. NapB contains two low-spin *c*-type haems, stacked in parallel, with potentials ranging from +50 to +160 mV,[19,27] which are able to transfer electrons to the [4Fe–4S] cluster of NapA, whose midpoint reduction potential is slightly negative (−15 mV). The electrons are then transferred to the catalytic centre (the distance between the Mo atom and the nearest Fe atom is ~12 Å), and two alternative routes have been proposed. One involves the conserved lysine and tyrosine residues (namely K56 and Y58 in *C. necator* numbering),[19,20] and the other involves two water molecules that bridge the Fe–S centre and the pterin (see Figure 11.2).[10,11]

In microorganisms that lack NapB, other redox proteins have been characterised as putative physiological electron donors. In *D. desulfuricans*, NapA has alternative redox partners that can fulfil this role, such as NapM[22] and NapC. NapC is a *c*-type tetrahaemic transmembrane protein belonging to the NapC/NirT/NrfH superfamily (Section 11.3.1) that receives electrons from the

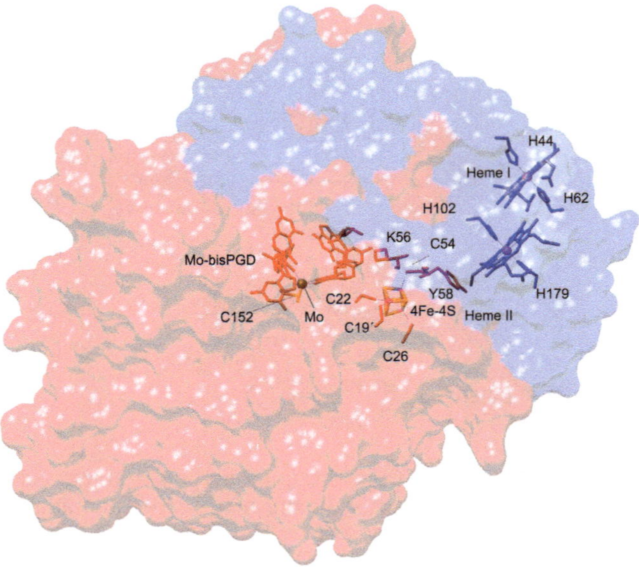

Figure 11.2 *Cupriavidus necator* NapA (red) and NapB (blue) with the metal centres in evidence (protein backbone not shown). The Mo atom is bound to six sulphur atoms and is represented as a brown sphere. The NapA residues K56 and Y58, proposed to be essential in the electron transfer pathway between NapB and NapA, are also depicted. Residues involved in the coordination of the metal centres are also identified. Image created using UCSF Chimera and Protein Data Bank identifier 3ML1.

oxidation of menaquinone, transferring them directly to NapA (see Figure 11.1), and NapM, a periplasmic *c*-type tetrahaemic cytochrome, is proposed to receive electrons from the oxidation of formate or hydrogen, which are subsequently transferred to the catalytic subunit of NapA.[22]

In *S. oneidensis*, the main electron transfer partner of NapAB seems to be CymA, a multipurpose electron transfer protein that also belongs to the NapC/NirT/NrfH superfamily (see Section 11.3.1) and feeds electrons to a number of other respiratory systems, including NiR of the ammonification pathway (see Section 11.3.1).[28]

Interestingly, organisms that do not form a functional NapAB heterodimer (*e.g.*, *D. desulfuricans*) also seem to lack the tyrosine residue that bridges haem 2 from NapB and the [4Fe–4S] cluster in NapA. Additionally, important charge distribution patterns on NapA's surface seem to be dependent on the interaction with NapB or other electron transfer partners. Jepson *et al.* found that Nap enzymes forming a NapAB heterodimer tend to have more polar and anionic residues at their surface, while the ones that do not tend to have more apolar or even cationic residues.[23] Furthermore, NapA–NapB interactions are dominated by hydrogen bonds involving conserved residues, which are less well conserved in NapA, forming transient interactions

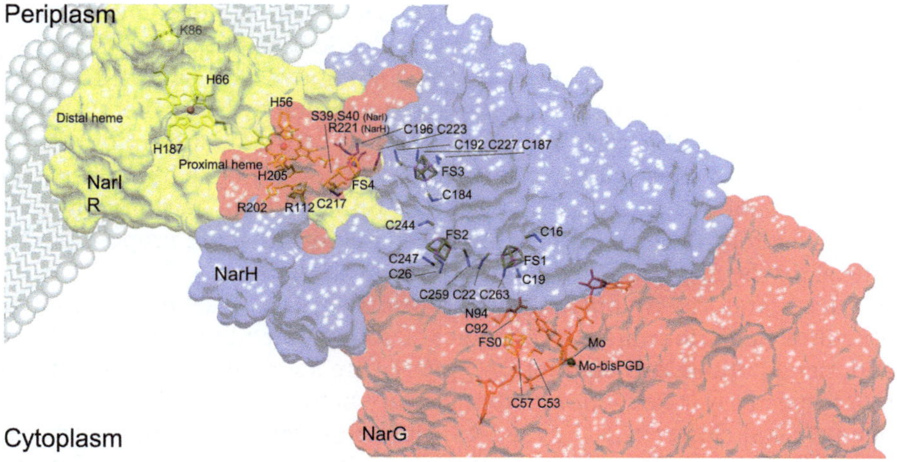

Figure 11.3 Representation of NarGHI in the cellular environment. Electrons flow from the ubiquinol or menaquinol bound close to the NarI distal haem *b*, towards the Mo-bis PGD active site *via* all other cofactors (proximal haem *b*, FS4, FS3, FS2, FS1 and FS0). NarI residue K86 is proposed to aid in the stabilisation of semiquinone binding, and NarG residues N94 and C92 are essential for efficient electron transfer between the NarH and NarG subunits. NarI, NarH and NarG are coloured yellow, blue and red, respectively. The metal centres are shown as sticks coloured by subunit, and the residues involved in their coordination are identified, as well as some other residues involved in the electron transfer pathway. Image created using UCSF Chimera and the coordinates are from Protein Data Bank identifier 1Q16.

with its partners, such as in *D. desulfuricans*.[23] The dissociation constant of *Rh. sphaeroides* heterodimeric NapAB has been calculated to be 0.5 nM, indicating a strong, non-transient complex.[18]

11.2.2 Respiratory Nitrate Reductase

A different type of nitrate reductase that is membrane bound and facing the cytoplasm has been isolated from organisms such as *E. coli*,[25] *Marinobacter* (*Ma.*) *hydrocarbonoclasticus* (previously known as *Pseudomonas nautica*)[29] and *Pa. pantotrophus*.[30,31] They form very large heterotrimers (see Figure 11.3) consisting of three different subunits: NarG, a 1246-residue-long polypeptide containing the catalytic, nitrate-reducing, Mo-bis PGD cofactor and one-electron transfer [4Fe–4S] centre (named FS0); NarH, a 512-residue-long polypeptide containing several ferredoxin-type Fe–S clusters (three [4Fe–4S], named FS1–3, and one [3Fe–4S], named FS4); and NarI, a membrane-spanning polypeptide with 225 residues containing a quinol binding site and two *b*-type haem groups.[32]

In NaR, the relative geometry of the metal centres ensures an efficient electron relay from the quinol binding site to the catalytic Mo-bis PGD centre (see Figure 11.3). NarG and NarH interact with each other *via* an extensive buried surface (over 10 000 Å^2), wherein two α-helices and one long β-strand of the N-terminus of NarG, which are amphipatic in nature, form a series of favourable electrostatic and hydrophobic/aromatic residue stacking interactions with NarH. The β-hairpin of NarG forms several hydrogen bonds with a C-terminal β-strand of NarI, and these, together with a β-hairpin from NarH, form a twisted β-sheet with an extensive network of hydrogen bonds between the three subunits. Residues Y213, R216 and R222 from NarI are involved in the electrostatic and hydrogen-bond interactions that are proposed to be crucial for the formation of the NarGHI heterotrimer.[32]

The respiratory NaR operates on the oxidising side of the menaquinol or ubiquinol pools, for which the $E^{0\prime}$ values of ubiquinone/ubiquinol (UQ/UQH_2) and menaquinone/menaquinol (MQ/MQH_2) are approximately +60 and −80 mV, respectively.[33] Reduction potentials vary between the high potential of one [4Fe–4S] cluster and the [3Fe–4S] cluster of NarH (+130 and +180 mV, respectively) to the low potential of the remaining [4Fe–4S] centres (−55 to −420 mV).[34,35] The two NarI haem groups have potentials in the +20 to +120 mV range,[36] while the midpoint reduction potential of the Mo(IV)/Mo(V) and Mo(V)/Mo(VI) transitions of the NarG Mo-bis PGD cofactor are +100 and +200 mV, respectively.[37–39]

In NaR, the semiquinone radical (and also ubiquinol or menaquinol) is proposed to bind near the distal haem *b* of NarI, stabilised by hydrogen bonds to H66 (axial ligand of NarI distal haem *b*) and to K86 from NarI (see Figure 11.3).[40] From this point of entry, the electrons travel a total of 75 Å towards the Mo-bis PGD catalytic centre, with cluster edge-to-edge distances between 5.4 Å (distal haem *b* to proximal haem *b* in NarI) and 11.2 Å (FS1 of NarH to FS0 of NarG),[32] which are well within the upper distance limit proposed by Dutton and co-workers for efficient electron transfer (14 Å).[41] Moreover, the propionate groups of the proximal haem *b* of NarI form an extensive hydrogen bond network towards NarH, involving residues S39, S40, R112 and R202 (from NarI) and R221 (from NarH), and additionally, N94 and C92 (from NarG) are proposed to be involved in the electron transfer between the latter two subunits (from FS1 to FS0 and towards the catalytic site) (see Figure 11.3).[32]

In some organisms, such as *Thermus* (*T.*) *thermophilus*, an additional NaR subunit—NarC—has been described.[42] NarC belongs to the NapC/NirT/NrfH superfamily (see Section 11.3.1) and is thought to capture electrons from NarI when there is not enough nitrate available to feed the anaerobic respiration process but there is a sufficient quantity of other nitrogen oxides. Then, NarC would transfer these electrons directly to the more downstream enzymes of the denitrification pathway (including cd_1-NiR; see Section 11.3.3).[43]

11.3 Electron Transfer Involved in Nitrite Reduction

NiR, involved in denitrification, is an enzyme that can present different types of metal centres, either copper (Cu-NiR; see Section 11.3.2) or haem iron (cd_1-NiR; see Section 11.3.3), but still catalyse the same basic reaction: reduction of nitrite to NO, the second step of this metabolic pathway.

There is also another family of NiRs that participates in the dissimilatory nitrate ammonification *via* the reduction of nitrite to ammonium, which is the pentahaemic NiR (NrfA) (Section 11.3.1).[44]

11.3.1 Direct Conversion of Nitrite to Ammonium by Dissimilatory NrfA

NiR that belongs to the nitrate dissimilatory pathway catalyses the reaction (according to eqn (11.3)):

$$NO_2^- + 8H^+ + 6e^- \rightarrow NH_4^+ + H_2O \tag{11.3}$$

This *c*-type cytochrome NiR, commonly designated as NrfA, exhibits highly specific activity of >1000 μmol NO_2^- min^{-1} mg^{-1}.[45,46] Additionally, NrfA is also able to transform nitric oxide and hydroxylamine to ammonium, albeit with a relatively low specific activity.[47,48]

This enzyme is usually arranged as a $NrfA_2$ dimer (2 × 53 kDa) anchored to an additional subunit (NrfH) arranged as a heterotrimer ($NrfA_2H$), with the biological active form of this complex being a dimeric form of the heterotrimer,[49] $NrfA_4H_2$, as is the case in several δ- and ε-proteobacteria, including several *Desulfovibrio* species, *Wo. succinogenes* and *Campylobacter* (*Ca.*) *jejuni*.[17,50,51] In *S. oneidensis*, the functional form of the enzyme is a standalone $NrfA_2$ dimer.[52]

NrfH is a member of the widespread NapC/NirT/NrfH superfamily (*i.e.*, membrane-bound tetra- or penta-haem *c*-type cytochromes that are able to receive electrons from menaquinol).[53] NrfH contains four *c*-type haems (see Figure 11.4), stacked in pairs and numbered 1–4 according to their order in the sequence. Haem 1 has the unique feature of being high-spin penta-coordinated by a methionine side chain, from the haem binding motif –CXXCHXM–,[49] with the distal ligand position (but not coordinating) being occupied by an aspartate or a histidine side chain, and this is responsible for interacting with menaquinol. Haem 2 and haem 3 are bis-histidinyl hexa-coordinated, while haem 4, despite also being hexa-coordinated, has a lysine residue from the NrfA polypeptide chain as the distal ligand and transfers electrons to NrfA.[49] NrfA contains five *c*-type haems, of which four (numbered 2–5) are hexa-coordinated by two histidine side chains, while the fifth (haem 1), the catalytic centre, has an unusual lysine ligand as the proximal axial ligand, with the sixth position occupied by a weak ligand (*e.g.*, water).[54] The distal, weak ligand of haem 1 is exchangeable and may leave the coordination sphere upon reduction of the haem; otherwise, it is stabilised in place by a network of hydrogen bonds involving a pocket of conserved histidine, tyrosine and arginine residues (see Figure 11.4).[49,50,55,56] N–O bond cleavage is facilitated by hydrogen bonding to the distal pocket histidine residue, since it was observed that an *E. coli* NrfA mutant lacking H264 is unable to reduce nitrite, but maintains hydroxylamine reductase activity.[57] In NrfA, the midpoint reduction

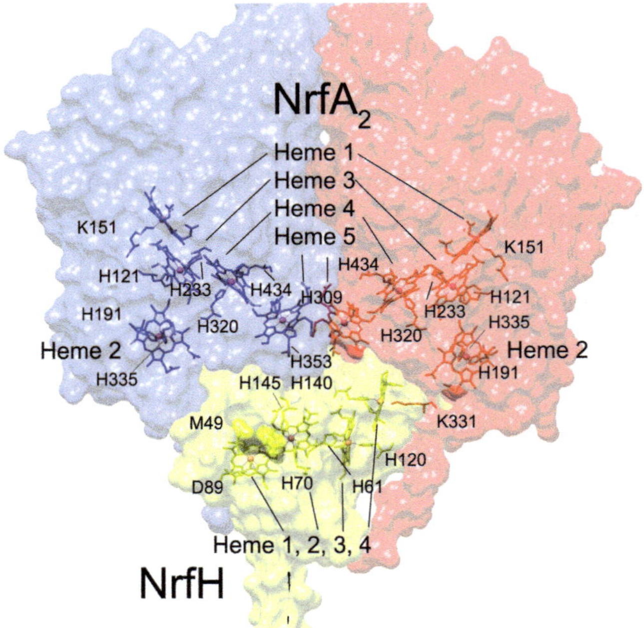

Figure 11.4 The arrangement of the haems in *Desulfovibrio vulgaris* $NrfA_2H$. NrfH haems, embedded in the membrane, are coloured light yellow, while the NrfA haems are coloured either red or blue, depending on the monomer. Electrons flow *via* the four haem cofactors of NrfH towards haem 2 of NrfA until they reach the catalytic centre (haem 1, coordinated by K151 and one water molecule in the crystal structure). Residues involved in the coordination of the metal centres are also identified. Image created using UCSF Chimera and Protein Data Bank identifier 2J7A.

potentials of the haem cofactors vary between approximately −300 and −35 mV at pH 6.0,[46,58] and in some cases going as low as −480 mV.[59] In NrfH, the penta-coordinated haem has a midpoint reduction potential above 0 mV, while for the other electron transfer haems, this value is close to −300 mV.[59] The functional $NrfA_4H_2$ is very stable, with each NrfH subunit having over 30% of its solvent-accessible surface covered by both NrfA monomers in an interaction with a strong electrostatic character (hydrogen bonding and salt bridges).[49]

In order to initiate nitrite reduction to ammonium, the iron atom of the catalytic haem has to be reduced, since in the oxidised state, the weakening of the N–O bond of nitrite is precluded.[60] Electrons are then transferred to NrfA by NrfH, either directly or indirectly,[44] as NrfH is able to collect electrons from menaquinol, which is present in the periplasmic side of the cytoplasmic membrane. Rapid intramolecular electron transfer within NrfA is guaranteed by the close packing of the five haem moieties, with iron–iron distances not exceeding 13 Å (see Figure 11.4), which aids the enzyme in attaining turnover numbers higher than 500 s^{-1}.[57] In fact, it is known that the proximity and potential of the adjacent redox states significantly influences the intramolecular electron transfer processes.[57,61,62]

Since the reaction mechanism entails the use of six electrons in order to reduce nitrite to ammonium, several studies addressed the topic on how the electrons flow towards the catalytic centre. In *Wo. succinogenes*, Einsle and co-workers found that electrons can be transferred either isolated or in pairs towards the active centre (the latter case in the reduction of HNO to hydroxylamine by electronically coupled haems).[54] In fact, electrons can be stored and released within the four non-catalytic haems of each NrfA protomer in pairs, so as to avoid the formation of radical intermediates during the six-electron reduction of nitrite to ammonium, and their intramolecular electron transfer kinetics are somewhat faster than what is found in other NiR families, with values in the range of 10^3 s^{-1} or higher.[63] However, this does not seem to be a universal mechanism, as in *S. oneidensis*, only single-electron transfer events have been described.[58]

Regarding the electron transfer partners of NrfA, in *S. oneidensis*, which lacks the NrfH-encoding gene, CymA, a quinol dehydrogenase from the NapC/NirT/NrfH superfamily and also containing four *c*-type haem moieties, is the main electron provider (see Section 11.2.1).[28] Enteric bacteria that also do not encode a NrfH homologue in their genome rely instead on an assembly consisting of NrfB, NrfC and NrfD, which are speculated to contain five *c*-type haems, one [4Fe–4S] cluster and one quinol binding site, respectively.[64] A putative pathway for the electron transfer predicts the oxidation of menaquinol by the NrfCD complex, with concomitant electron transfer to NrfB, which will then reduce NrfA.[53] The complex between NrfA and NrfB has a relatively strong dissociation constant (50 nM), which is an indication that these proteins remain tightly associated under physiological conditions, with NrfB being proposed to bind near NrfA haem 2.[65]

11.3.2 Cu-NiR

Cu-NiR, as well as cd_1-NiR (described in Section 11.3.3), catalyse the reduction of nitrite to nitric oxide (according to eqn (11.4)). This section deals with the former.

$$NO_2^- + e^- + 2H^+ \rightarrow {}^{\bullet}NO + H_2O \qquad (11.4)$$

Cu-NiR forms a heterogeneous family with the common characteristic being the presence of a type 1 copper (T1Cu) centre as the electron transferring centre and a type 2 copper (T2Cu) centre as the catalytic centre. Some members of this family contain additional domains harbouring other electron transfer centres. One example is the *Hyphomicrobium denitrificans* Cu-NiR, a hexameric enzyme with each monomer containing an N-terminal domain with a T1Cu binding centre (Cu_N) and a C-terminal domain with the canonical T1Cu (Cu_C) and T2Cu centres.[66] One other example is the Cu-NiR from *Ralstonia (Ra.) pickettii* with an additional *c*-type haem binding domain, which is proposed to preclude the need for a soluble electron donor.[67,68]

The canonical Cu-NiRs have been isolated from several α-, β- and γ-proteobacteria[69] as homotrimers containing only the T1Cu and T2Cu centres, and can be further divided into "blue" and "green" NiRs due to differences in their spectroscopic properties.[70] The T1Cu centre has its copper atom coordinated by two histidines, one cysteine and one methionine side chain, and is usually buried around 7 Å within the protein core, while the T2Cu centre, located between two monomers, is coordinated by three histidines from two polypeptide chains and a solvent ligand, and is buried 14 Å into the protein core, but is accessible *via* a substrate channel.[66,69] The distance between the two copper centres is approximately 12.5 Å (see Figure 11.5).[71]

It has been established that Cu-NiRs are regenerated *via* electron transfer from small electron donors, with these being either a *c*-type cytochrome (*e.g.*, cytochrome c_{550} or a cytochrome *c* domain located at the C-terminus of the enzyme itself) or a cupredoxin, directly to the T1Cu centre of NiR,[72,73] and from there *via* intramolecular electron transfer to the T2Cu centre.[74–76] Nevertheless, theoretical studies using DFT and model compounds on the reaction mechanism of Cu-NiRs failed to identify whether the T2Cu centre should be previously reduced in order to bind the substrate.[77]

In *Alcaligenes* (*Al.*) *faecalis* and *Achromobacter* (*Ac.*) *xylosoxidans* (previously known as *Al. xylosoxidans*) Cu-NiRs, single-molecule intramolecular electron

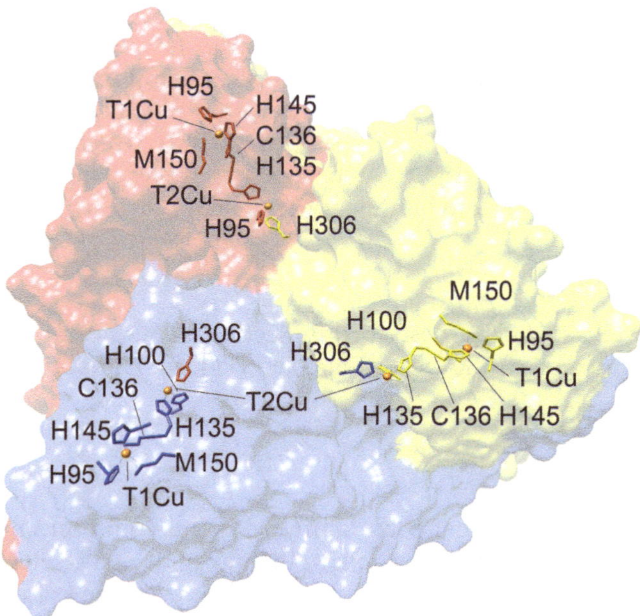

Figure 11.5 Electron transfer pathway in *Alcaligenes faecalis* homotrimeric copper-containing nitrite reductase. The electrons flow from the type-1 copper (T1Cu) centre to the type-2 copper (T2Cu) centre *via* the T1Cu ligand C136 to the T2Cu ligand H135. Note that one of the T2Cu centre ligands (H306) comes from a different subunit (indicated by the different colours red, blue and yellow). Residues involved in the coordination of the metal centres are also identified. Image created using UCSF Chimera and Protein Data Bank identifier 4YSE.

transfer studies estimated the reaction rate constant to be in the order of 4 ± 2 × 10^5 M^{-1} s^{-1} and the intramolecular electron transfer rate constants in the range of 14–21 s^{-1},[78] which are at least one to two orders of magnitude smaller than what had been described for other bulk systems using pulse radiolysis studies (450–2100 s^{-1}) in the absence of nitrite.[79] Nevertheless, it is of the same order of magnitude as the electron transfer reaction between pseudoazurin and Cu-NiR.[80] A different study using the *Ac. xylosoxidans* Cu-NiR found that intramolecular electron transfer rates are in the range of 200–450 s^{-1}, depending on factors such as pH (5.8–7.0),[71,81] and in the *Ac. cycloclastes* Cu-NiR, this rate can reach values exceeding 1000 s^{-1}.[82]

The residues involved in the intramolecular electron transfer include T1Cu-coordinating C136 and T2Cu-coordinating H135 (in *Al. faecalis* numbering), with the electrons proceeding from one centre to the other *via* 11 covalent bonds.[71] A different study also postulated that N90 (in *Ac. xylosoxidans*) is essential to forming a proton channel to the T2Cu site.[75]

The majority of the studies involving Cu-NiRs also found that the intramolecular rate of electron transfer is modulated by the rate of nitrite binding at the active site, which leads to conformational rearrangements that facilitate electron transfer (*e.g.*, by changing the reduction potential of the catalytic centre) in a concerted manner to avoid inactivation of the enzyme.[68] In fact, the mere difference in midpoint reduction potentials is insufficient to drive electron transfer *per se* in this enzyme.[81] It is also worth noting that intramolecular electron transfer from T1Cu and T2Cu centres seems to be coupled to proton transfer, which in some cases may result in a rate-limiting step.[76]

Pseudoazurin is a small cupredoxin that is proposed to be an electron donor to Cu-NiRs in the *Achromobacter* and *Alcaligenes* genera. In *Al. faecalis*, it was shown that the interaction between these two proteins is stabilised by complementary electrostatic forces between the negatively charged residues of Cu-NiR (E113, E197, D201, E204 and D205) and the lysine patch in pseudoazurin (K10, K38, K57 and K77).[83,84] Later studies revealed that the uncharged residues near the Cu-coordinating H81 (A15, A83, M84 and I110) of pseudoazurin are the most affected in a cross-saturation transfer nuclear magnetic resonance analysis of the pseudoazurin–Cu-NiR complex, with additional positively charged and polar residues aiding in the stabilisation of the complex (N9, K38, N61, K107 and K109).[85] Therefore, in this electron transfer complex, electrons are proposed to jump from the pseudoazurin residue H81 towards Cu-NiR's M84, and from there to the T1Cu ligand H145, or *via* a single through-space jump between pseudoazurin's H81 and Cu-NiR's P139 (see Figure 11.6). The Cu–Cu distance between the redox partners was established to be ~15 Å.[86] Notably, the redox state of pseudoazurin influences the binding mode, with reduced pseudoazurin having a higher affinity and longer complex lifetimes with Cu-NiR, with exchange times being far slower than the typical Cu-NiR turnover number, which might be due to redox state-dependent conformational changes.[87]

In the electron transfer complex between *Ac. xylosoxidans* Cu-NiR and its redox partner, cytochrome c_{551}, the electron transfer reaction is strongly

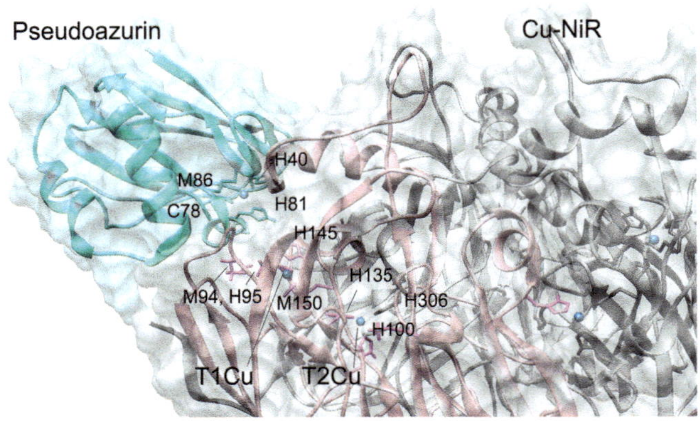

Figure 11.6 Productive electron transfer complex between pseudoazurin (blue ribbon) and copper-containing nitrite reductase (Cu-NiR; pink ribbon) from *Alcaligenes faecalis*. Cu-coordinating residues are identified in both proteins. Electron transfer is thought to proceed from pseudoazurin's H81 towards Cu-NiR's M94, and from there to the type-1 copper (T1Cu)-coordinating H145. Copper-to-copper distance ≈15 Å. The protein surface is shown in light grey and copper atoms are shown as spheres. Residues involved in the coordination of the metal centres are also identified. Image created using UCSF Chimera and Protein Data Bank identifier 2P80.

dependent on a network of hydrogen-bonded, bridging water molecules in the interface of the complex, which facilitate the electronic coupling between the *c*-type haem and the T1Cu centre.[68,88] In fact, *Ac. xylosoxidans* cytochrome c_{551} presents a hydrophobic patch, comprising residues S18, A19, V22, V28, P63, G67, T68 and A69, which interacts with a hydrophobic patch of Cu-NiR, composed of residues A86, M87, N109, G133, W138, T192, A194, E195, Y197 and G198. In this complex, the electron transfer centres are at a distance of 10.5 Å, with a maximum optimised intermolecular electron transfer rate in the order of magnitude of 10^6 s^{-1}, with the electron performing a through-space jump (4.0 Å) from the haem edge of cytochrome c_{551} to Cu-NiR's P88.[88]

The genome of *Ac. xylosoxidans*, in addition to cytochrome c_{551}, also encodes two different azurins. Azurin I was established to be a putative redox partner of Cu-NiR, together with cytochrome c_{551}, since the expression of its encoding gene is upregulated and transcribed along with *nirK* (the gene that encodes Cu-NiR; see Chapter 5 of this book for more details) under denitrifying conditions. The rate of Cu-NiR reduction *in vitro* by azurin I was estimated to be ~20 µmol NiR min^{-1} mg $azurin^{-1}$.[72]

In the *Ra. pickettii* Cu-NiR, which contains a *c*-haem in its additional C-terminal domain, the electron transfer centres are at a maximum distance of 10.4 Å, and the major intervenients in the electron path include the T1Cu ligands H143 and A138.[68]

In the *Hyphomicrobium denitrificans* Cu-NiR, which contains an additional T1Cu domain, electrons flow from the physiological electron donor, cytochrome c_{550}, to the copper atom of either T1Cu centre (Cu_N or Cu_C) with rate constants in the order of 10^5 M^{-1} s^{-1} for the former, and 10^3 M^{-1} s^{-1} for the latter, as determined by pulse radiolysis experiments. Intramolecular electron transfer will then drive the electron towards the catalytic centre (T2Cu). While electron transfer between $T1Cu_C$ and T2Cu centres is not rate limiting, the reduction of $T1Cu_C$ by $T1Cu_N$ is, owing to the fact that the distance between them is double that which exists between the $T1Cu_C$ and T2Cu centres.

Geobacillus (*Ge.*) *kaustophilus* HTA426 presents a Cu-NiR with an extended N-terminal region (an additional loop of 28 residues between T41 and G68) that is involved in the recognition of its redox partner, cytochrome c_{551}, and the intermolecular electron transfer rate was estimated by stopped-flow kinetics to be around 4 s^{-1}.[89]

NirK is a Cu-NiR isolated from *Neisseria* (*N.*) *gonorrhoeae* and *N. meningitidis*, which participates in a truncated denitrification pathway in these organisms.[90] Under certain conditions, such as low oxygen tension, NirK is activated and can receive electrons from either cytochrome c_N, cytochrome c_5 (both containing one *c*-type haem moiety) or CcoP (containing three *c*-type haem groups), which also donate electrons to cytochrome cbb_3 oxidase, in the aerobic respiratory pathway,[90] or from cytochrome c_2.[91] The two domains of cytochrome c_5 exhibited different electron transfer rates towards NirK, with second-order rate constants varying between values in the order of magnitude of 10^5 M^{-1} s^{-1}. In some *Neisseria* species, such as *N. weaverii*, NirK contains an additional haem-binding domain and thus might be able to bypass the need for an external electron transfer partner,[90] as in the case of *Ra. pickettii*.

11.3.3 *cd*$_1$-NiR

cd$_1$-NiR catalyses the reduction of nitrite to nitric oxide according to eqn (11.4), as mentioned before. This enzyme has been isolated from the periplasmic space of several denitrifying bacteria, such as from different *Paracoccus* and *Pseudomonas* (*Ps.*) species, as well as from the *Magnetospirillum* genera, among others.[92–96] This enzyme is a homodimer of approximately 120 kDa, with each monomer harbouring one *c*-type haem (the electron transferring centre) and one d_1-type haem (the catalytic centre) (see Figure 11.7) in two different domains. Although the Fe–Fe distances are around 20 Å and the distances between the haem edges are around 11 Å, which renders them suitable for intramolecular electron transfer,[97] both haems are hexacoordinated in the as-isolated state. Thus, the need for an activation mechanism can be envisaged.

The peculiar characteristics of the d_1-type haem compared to other iron porphyrins (saturation of two pyrrole bonds, addition of two keto and one acrylate groups) leads to a midpoint reduction potential that is more negative (by 15–40 mV), and lies in the region of +200 mV, when compared with iron porphyrins.[94]

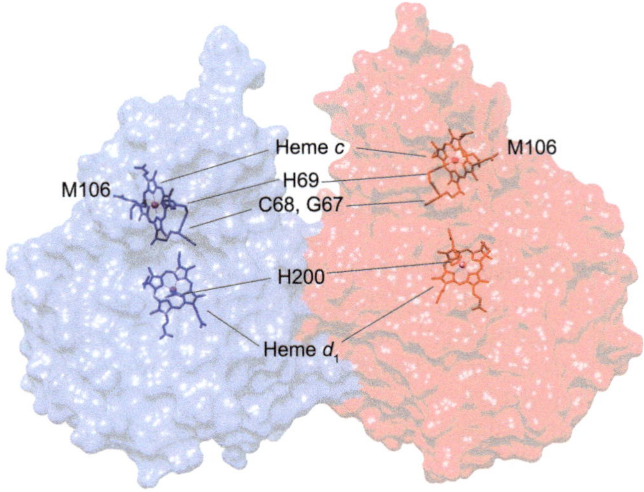

Figure 11.7 Electron transfer pathway between haem *c* and haem d_1 in the homodimeric *Paracoccus pantotrophus* cytochrome cd_1 nitrite reductase. The intramolecular electron transfer pathway involves C68 and G67 and a bridging water molecule (not shown). Residues involved in the coordination of the metal centres are also identified. Image created using UCSF Chimera and Protein Data Bank identifier 1H9X.

In the oxidised, inactive state, haem d_1 is coordinated by a histidine and a tyrosine side chain, but after reduction, allosteric conformational changes occur in the coordination sphere of both haems, with the tyrosine side chain being displaced from haem d_1, creating a vacant coordination site for the substrate, while haem *c*, which had a bis-histidinyl coordination in the oxidised form unready state, becomes coordinated by a methionine and a histidine side chain.[98,99]

Those conformational changes are brought about by the reduction of haem *c*, which leads to a decrease in the electronic coupling between the electron transfer centre and the catalytic centre, as hydrogen bonds bridging the two domains are disrupted, leading the haem *c* domain away from the catalytic domain, while the tyrosine residue coordinating haem d_1 is displaced.[100–102] Furthermore, conformational changes may have an effect on the hydrogen bonding network that enhances intramolecular electron transfer. The presence of water molecules between the two domains may increase intramolecular electron transfer rates by ~10-fold, as proposed in *Ps. aeruginosa* cd_1-NiR.[103] It was postulated that a similar justification would account for the increased electron transfer rate in *Ps. stutzeri* cd_1-NiR.[104]

Intramolecular electron transfer rates within cd_1-NiR may vary between 0.3 and 3 s^{-1} for the *Ps. aeruginosa* cd_1-NiR and 1400 s^{-1} for the *Pa. pantotrophus* enzyme.[103,105–108] An earlier study postulated a very fast electron transfer rate in *Ps. aeruginosa* cd_1-NiR from a laser flash photolysis experiment (in the range of 10^4 s^{-1}).[109] In the *Ps. stutzeri* cd_1-NiR, the rate is one to two orders of magnitude larger than in the *Ps. aeruginosa* cd_1-NiR (23 s^{-1}), due to conformational changes and to the end-product dissociation rate from the catalytic d_1-type haem.[105,110] The intramolecular electron transfer between the *c*- and d_1-type haems is thought to proceed *via* C50 and G49 (according to the *Ps. aeruginosa* cd_1-NiR primary sequence).

Small *c*-type cytochromes (*e.g.*, *Pa. pantotrophus* cytochrome c_{550}) and small T1Cu proteins (azurins or pseudoazurins) are the redox partners of cd_1-NiR,[111–114] as was found to be the case in Cu-NiR. As mentioned, these small electron transfer proteins present one patch that is rich in acidic residues and another patch that is rich in lysines, surrounding a predominantly hydrophobic surface (containing glycines, valines, alanines and prolines), from where the haem *c*'s edge protrudes,[114] or including the solvent-exposed imidazole ring of the histidine that coordinates the T1Cu centre. Cupredoxins were able to transfer electrons towards the electron transferring domain of cd_1-NiR with an observable rate constant in the order of the single digits to low tens per second. Thus, relatively slow intramolecular electron transfer avoids the formation of a dead-end product (NO bound to the reduced iron of the catalytic haem d_1).[104,105]

In *Paracoccus* species, both cytochrome c_{550} and pseudoazurin are proposed to be the electron transfer partners of cd_1-NiR, as well as of N_2OR (see Section 11.5.2), since the double-knockout strain for the genes that encode these small electron shuttle proteins proved to be unable to reduce nitrite.[115] Moreover, these two small proteins have been proposed to play an interchangeable role in electron transfer in the periplasm of *Paracoccus* species, for which the pseudospecificity of their charge surface plays an important role. In fact, they have different tertiary structures, but share a ring of positive charges around a hydrophobic patch that matches the negatively charged cd_1-NiR surface.[116] The hydrophobic patch includes the point of exit/entry for the electron, as mentioned before.

Zajicek *et al.* described that *Pa. pantotrophus* NapC (see Section 11.2.1) is also able to transfer electrons cooperatively to cd_1-NiR for its activation step *in vitro*. The same was true in *Ps. stutzeri*, where NirT, which belongs to the NapC/NirT/NrfH superfamily, was found to be a redox partner of this enzyme.[117] However, those authors also note that this interaction only occurs in the activation step in order to render the enzyme kinetically active. Catalytic turnover will then require electron transfer *via* the aforementioned cupredoxin or *c*-type cytochrome.

In the case of the *Ma. hydrocarbonoclasticus* cd_1-NiR, a study using electrochemical techniques established cytochrome c_{552} to be its physiological electron donor. Actually, this small electron shuttle protein also functions as an electron donor to the other enzymes of the denitrification pathway (NOR [see Section 11.4.2] and N_2OR [see Section 11.5.2]) in this organism. This first study in the denitrification pathway of *Ma. hydrocarbonoclasticus* showed that the formation of electron transfer complexes is not just driven by thermodynamics, but surface recognition must also play a role, since the other periplasmic *c*-type cytochromes—cytochrome c_{553}, cytochrome c_{549}, cytochrome

c' and cytochrome $c_{553(548)}$—were not able to transfer electrons in a catalytically efficient manner, despite having similar haem centres and reduction potentials.[107]

In fact, the electron transfer rates between cd_1-NiR and cytochrome c_{552} in *Ma. hydrocarbonoclasticus* follow a second-order rate constant in the region of $10^5\ M^{-1}\ s^{-1}$, which is similar to that observed in the *Ps. aeruginosa* enzyme interacting with either cytochrome c_{551} ($2.0 \times 10^6\ M^{-1}\ s^{-1}$) or azurin ($1.4 \times 10^5\ M^{-1}\ s^{-1}$).[112,118]

11.4 Electron Transfer Involved in Nitric Oxide Reduction

NOR catalyses the third step of the denitrification process, the reduction of nitric oxide to nitrous oxide (according to eqn (11.5)):

$$2NO + 2H^+ + 2e^- \rightarrow N_2O + H_2O \tag{11.5}$$

As described in Chapter 6, three different classes of NORs can be found in denitrifying bacteria, namely cytochrome *c*-dependent NOR (cNOR), quinol-dependent NOR (qNOR) and qCuNOR, which differ in the electron donors and in the number and type of electron transfer centres (see Chapter 6).[119–121]

The catalytic centre constitutes a binuclear iron centre (haem b_3–non-haem Fe_B) and is conserved in all three NOR classes. Moreover, in all cases, the catalytic centre receives electrons from a *b*-type haem, which acts as an electron transfer centre. Besides this centre, NOR can present another redox centre, acting as electron transferring centre, which is a *c*-type haem in cNOR and a dinuclear copper centre in qCuNOR, whereas qNOR receives electrons directly from membrane-bound ubiquinol or menaquinol. In the following paragraphs, the intramolecular electron transfer between the redox centres within the enzyme and the intermolecular electron transfer between the external electron donor and NOR will be described.

11.4.1 Intramolecular Electron Transfer

The intramolecular electron transfer in NOR includes both the electron transfer between different electron transfer centres, depending on the NOR class, and the electron transfer from the *b*-type haem to the catalytic centre, which is coupled with NO reduction. The discussion and current hypothesis regarding the mechanism of the catalytic cycle of NOR are addressed in Chapter 6 of this book, and these aspects are outside of the scope of the present chapter. However, it has been demonstrated that the electron transfer to the catalytic centre and the substrate binding are interconnected. In particular, a spectroscopic and electrochemical study using cNOR from *Pa. denitrificans* proves that electrons are delivered to the catalytic centre only after NO binding.[122] Similar conclusions were obtained by stopped-flow kinetic experiments of the single turnover rate of cNOR from *Pa. denitrificans*.[123] Further studies with mutated cNOR allowed the proposal of a specific channel for the proton transfer from the periplasm into the active site.[124]

In cNOR, the electron is first transferred from the periplasmic electron donor to the acceptor site, the *c*-type haem, and is subsequently transferred *via* a low-spin *b*-type haem to the binuclear active site (haem b_3–Fe_B) (see Figure 11.8). The electron transfer rate between *c*-type and *b*-type haems is about $3 \times 10^4\ s^{-1}$, while the rate of electron transfer between *b*-type and b_3-type haems is too fast to be resolved ($>10^6\ s^{-1}$).[125]

These reaction rates are in agreement with the interatomic distances found in the X-ray crystallographic structure.[126] In particular, the iron-to-iron distance between *c*-type haem and *b*-type haem is 20.5 Å, whereas the distance between *b*-type haem and b_3-type haem is 14.1 Å. Resemblance with cytochrome *c* oxidase (CcO) is obvious (haem *a*–haem a_3 distance), even if there is the replacement of the conserved phenylalanine by glycine (G348 according to the *Ps. aeruginosa* primary sequence) in cNOR (see Figure 11.8).[127]

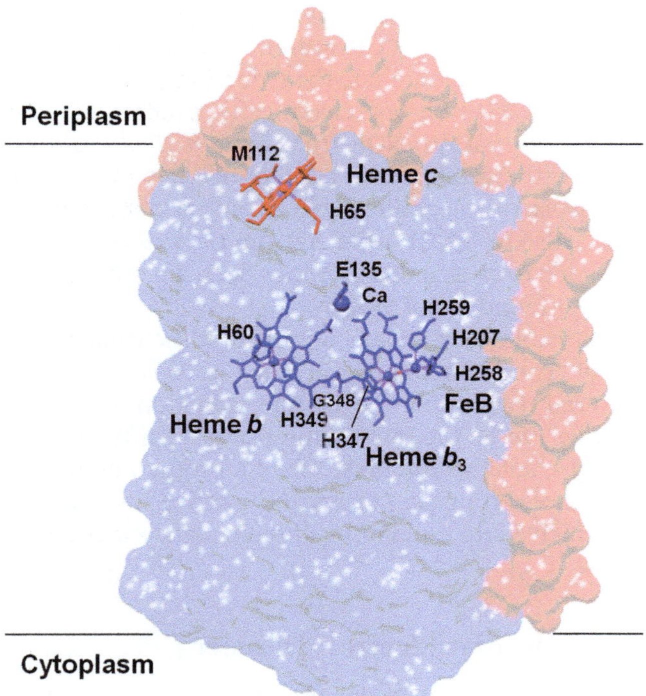

Figure 11.8 Proposed intramolecular electron transfer in cytochrome *c*-dependent nitric oxide reductase. Electrons are transferred from haem *c* to haem *b*, and finally to haem b_3. G348 is placed between haem *b* and haem b_3, replacing the well-conserved phenylalanine in cytochrome *c* oxidases. Residues involved in the coordination of the metal centres and calcium ions are also identified. Image created using UCSF Chimera and Protein Data Bank identifier 3O0R.

In CcO, the rate of intramolecular electron transfer has been determined by flash-photolysis techniques to be faster by about one order of magnitude (3×10^5 s^{-1})[128] than that which has been determined in NOR.

An electrochemical study demonstrated that all of the redox processes are one-electron transfers each, and that the reduction potential is higher for the *c*-type haem (about +208 mV) and becomes progressively lower for the other iron centres (haem *b*: +43 mV; haem b_3: −162 mV; Fe_B: −368 mV).[129]

In qNOR, the *c*-type haem electron transfer centre is missing and electrons are donated from menaquinol directly to the *b*-type haem and finally to haem b_3 of the catalytic centre. The X-ray crystallographic structure shows that the iron-to-iron distance is 13.7 Å.[130] However, direct measurements of intramolecular electron transfer rate for this enzyme class are lacking.

qCuNOR presents CuA- and *b*-type haem as the electron transfer centre, making this NOR type the closest homologue to CcO in terms of internal electron transfer pathways. However, direct measurements of electron transfer rate or structural crystallographic insights for this NOR class are also lacking.

Another important aspect related to intermolecular electron transfer is the possible generation of an electrogenic proton membrane gradient, as observed in the homologous CcO. However, while CcO stores energy by the generation of a membrane gradient in the reduction of oxygen,[131] cNOR catalyses the reduction of NO in a non-electrogenic process. However, recently it has been proposed that qCuNOR from *Bacillus* (*Ba.*) *azotoformans* has a higher homology with CcO than with other NOR types and it is able to generate a proton electrochemical gradient across the membrane, which is similar to that of CcO.[132] Also for qNOR isolated from *Ge. stearothermophilus*, a proton transfer coupled to electron transfer has been demonstrated by studying the single turnover reduction of oxygen.[133]

Finally, a remarkable difference is observed in the catalytic properties of cNOR and qNOR. The reduction of the redox centres is slow in qNOR and controls the overall reaction rate, while the reduction for cNOR is fast and therefore is unlikely to be the rate-limiting step.[134] Different structural and functional features, such as the presence of a proton and water channel in qNOR, could control certain properties of the active site, such as the reduction potential, which could explain the different catalytic properties of these two NOR classes.

11.4.2 Intermolecular Electron Transfer

NOR can receive electrons from redox proteins (*c*-type cytochromes or T1Cu proteins) in the case of cNOR or from quinols for qNOR, whereas for qCuNOR, both mechanisms have been proposed.[119,135]

cNOR, as reported in Chapter 6 of this book, has been isolated from *Pa. denitrificans*,[136] *Ps. stutzeri*,[137] *Ma. hydrocarbonoclasticus*,[138] *Halomonas halodenitrificans*[139] and *Ps. aeruginosa*.[140]

In *Pa. denitrificans*, both cytochrome c_{550} and pseudoazurin act as interchangeable electron donors for all of the enzymes involved in the denitrification pathway, including NOR.[141] In *Ps. stutzeri*, cytochrome c_{551} is the soluble periplasmic cytochrome that transfers electrons to the enzymes involved in the denitrification pathway.[142] However, direct measurements of the electron transfer between cytochrome c_{551} and NOR are lacking.

In *Ps. aeruginosa*, cytochrome c_{551} has been demonstrated to be the electron donor of cd_1-NiR.[114] Alternatively, azurin, a T1Cu protein, can also act as an electron donor of the enzymes involved in the denitrification process, even if its role is thought to be limited to stress conditions.[143] For electron transfer to cNOR from *Ps. aeruginosa*, experimental data are lacking, but some information regarding the electron transfer can be obtained from the X-ray crystallography studies.[126] cNOR was indeed co-crystallised with its monoclonal antibody Fab, which is bound to the surface area that presumably binds the electron donor *in vivo* (see Figure 11.9). This interaction is characterised by several water-mediated hydrogen bonds and hydrophobic interactions.

In the periplasm of *Ma. hydrocarbonoclasticus*, a series of *c*-type cytochromes can act as potential electron donors of cNOR. However, enzymatic

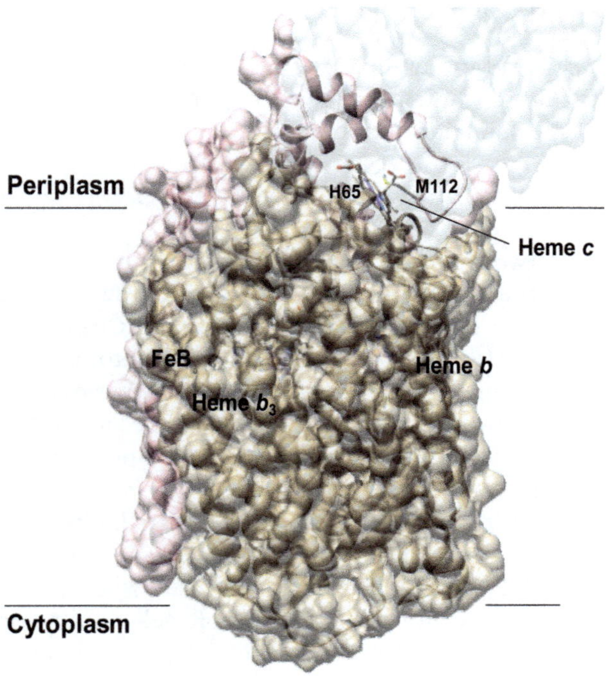

Figure 11.9 Proposed interaction surface of chrome *c*-dependent nitric oxide reductase (cNOR) with the electron donor. The two subunits of cNOR are coloured light pink (NorC) and light orange (NorB), with the surface of the monoclonal antibody Fab coloured light grey. It is evident that the antibody used in the crystallographic studies interacts with the surface containing the solvent-exposed haem *c* propionates. Image created using UCSF Chimera and Protein Data Bank identifier 3O0R.

activity assays show that only cytochrome c_{552} is able to donate electrons efficiently to the enzyme, indicating that this protein is the physiologic electron donor of cNOR from this organism.[138] The same was true for cd_1-NiR (see Section 11.3.2) and for N_2OR (see Section 11.5.2).

In qNOR, the electrons used in the NO reduction by qNOR are supplied from menaquinol to the binuclear centre through haem *b*.[144] Crystallographic analysis with the co-crystallised menaquinol analogue 2-heptylhydroxyquinoline N-oxide allowed the identification of a highly conserved area (D746, H328, E332, F336 and F337) in qNORs, but not in cNORs, which probably acts as binding site for menaquinol.[130]

The qCuNOR from *Ba. azotoformans* uses cytochrome c_{551}, a negatively charged periplasmic *c*-type cytochrome, as an electron donor.[145] Menaquinol had been proposed to also function as an electron donor in qCuNOR, but this electron transfer mechanism has recently been excluded for the enzyme from this bacterial source.[132]

11.5 Electron Transfer Involved in Nitrous Oxide Reduction

N_2OR is a copper-containing enzyme that catalyses the two-electron reduction of nitrous oxide to dinitrogen (according to eqn (11.6)),[135,146,147] which is the last step of the denitrification pathway.

$$N_2O + 2H^+ + 2e^- \rightarrow N_2 + H_2O \qquad (11.6)$$

N_2OR contains two multi-copper centres, the binuclear electron transfer centre, CuA, and the tetranuclear catalytic centre, the "CuZ centre", which can exist in two different forms: CuZ*(4Cu1S) and CuZ(4Cu2S). The "CuZ centre" as CuZ*(4Cu1S) presents the four copper atoms in a tetrahedron shape with an inorganic μ_4-sulphide bridge and a water-derived ligand coordinated at the Cu_I–Cu_{IV} edge, while the "CuZ centre" as CuZ(4Cu2S) has an extra sulphur atom instead of the water-derived ligand at the Cu_I–Cu_{IV} edge, as described in Chapter 7 of this book.[147]

The CuZ(4Cu2S) form of the "CuZ centre" is usually isolated when N_2OR is purified under anoxic conditions[148] or rapidly in the presence of oxygen,[149] while the CuZ*(4Cu1S) form is maximised when the purification is performed under oxic conditions from soluble cell extracts obtained from broken cells that have been stored frozen;[149,150] however, enzyme purifications typically yield N_2OR with a mixture of the two centres in different proportions.[149,151]

In the following paragraphs, the intramolecular electron transfer between the CuA centre and the catalytic centre in both forms—CuZ(4Cu2S) and CuZ*(4Cu1S)—and the intermolecular electron transfer between the periplasmic electron donor and CuA centre will be described.

11.5.1 Intramolecular Electron Transfer

The first three-dimensional structure of N_2OR[152,153] revealed the functional homodimeric conformation of this enzyme. Indeed, the distance within the same monomer between the electron transferring centre (CuA centre) and the catalytic centre ("CuZ centre") is around 40 Å, whereas the CuA–"CuZ" distance within the dimer is approximately 10 Å, which is considered suitable for efficient electron transfer (see Chapter 7 of this book).[41,152]

Experimental data regarding the electron transfer reaction between CuA and CuZ(4Cu2S) and CuZ*(4Cu1S) in N_2OR are still scarce. However, an analogous CuA centre is present in CcO,[127,154] for which several parameters and properties of the electron transfer between the metal cofactors within the enzyme have been determined.[146,155]

In CcO, the CuA centre receives electrons from a cytochrome *c*, which are then transferred to the electron transfer centre, haem *a*, and subsequently to the dinuclear catalytic centre, the haem a_3–CuB centre.[156]

The electron transfer between the CuA centre and haem *a* depends on the source of the enzyme and the experimental methodology used, but in general is characterised by a relatively high rate constant of between $2 \times 10^4\ s^{-1}$ and $10 \times 10^4\ s^{-1}$,[157–159] despite the distance of about 16 Å between the two metal centres. This range of reaction rate constant is in excellent agreement with the rate constant of $8.7 \times 10^4\ s^{-1}$ predicted from the structure by the Moser–Dutton equation.[160] The crucial role of a histidine residue that coordinates the CuA centre in the electron transfer pathway to haem *a* has been proven by fast kinetic experiments on *Rh. sphaeroides* CcO mutants.[161] The most probable electron transfer route, on the basis of a quantum mechanics/molecular mechanics study, involves the electron transfer from the histidine to two adjacent arginine residues that are connected to haem *a* propionates *via* hydrogen bonds.[162]

Regarding the electron transfer from the CuA centre to the catalytic centre in N_2OR, no kinetic parameters are available to date. However, a theoretical model based on the *Ma. hydrocarbonoclasticus* N_2OR crystal structure identified two possible electron transfer routes from the CuA to the CuZ*(4Cu1S) centre.[163] In the first putative pathway, the electron is transferred from W563, which is a CuA ligand, to the neighbour F564 and subsequently to the oxygen of the water-derived ligand bound in between Cu_I and Cu_{IV} of the CuZ*(4Cu1S) centre. The alternative route involves the electron transfer from the CuA centre to its H569 ligand, then to the adjacent M570 and finally to H128, which is a ligand of Cu_{II} of the CuZ*(4Cu1S) centre (see Figure 11.10).

An analogue analysis using the *Pa. denitrificans* N_2OR structure identified a pathway where the electron jumps directly from H639, the CuA ligand corresponding to H569 in *Ma. hydrocarbonoclasticus* N_2OR, to the water-derived ligand located near the CuZ*(4Cu1S) centre.[164]

The structure of N_2OR from *Ps. stutzeri* shows the presence of an additional sulphur atom in the "CuZ centre", thus suggesting a different electron

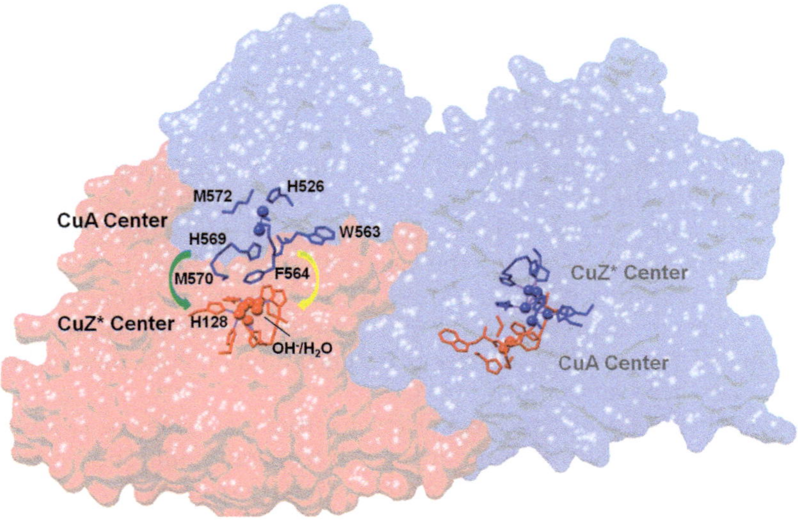

Figure 11.10 The two proposed intramolecular electron transfer pathways across the dimer interface between the CuA and CuZ*(4Cu1S) centres for the *Marinobacter hydrocarbonoclasticus* nitrous oxide reductase. The first route (yellow arrow) involves W563, which is a CuA ligand, and the neighbouring residue F564; the electron is then transferred to the water-derived ligand bound at Cu_I of the CuZ*(4Cu1S) centre. An alternative route (green arrow) involves electron transfer from the CuA centre to H569, then to M570 and subsequently to H128, which is a ligand of Cu_{II} of the CuZ*(4Cu1S) centre. Image created using UCSF Chimera and the coordinates are from Protein Data Bank identifier 1QNI.

transfer route from the CuA centre to CuZ(4Cu2S).[165] Einsle and co-workers also presented a further crystallographic structure in which a N_2O molecule is located in a cavity in between the two metal centres. The substrate is located in a binding pocket in which H626 (numbered according to the *Ps. stutzeri* N_2OR primary sequence), a ligand of the CuA centre, along with M627 and F621, assist in substrate orientation (see Figure 11.11). N_2O is therefore placed in the direct route for the electron transfer between the CuA centre and CuZ(4Cu2S) (the second route mentioned above for the *Ma. hydrocarbonoclasticus* N_2OR). Although these data were collected on a non-catalytic form of the enzyme,[166] Einsle and collaborators have raised the hypothesis that both centres are required for the reduction of nitrous oxide during catalysis and that they cooperate in nitrous oxide reduction.[165]

11.5.2 Intermolecular Electron Transfer

Small electron carrier proteins, either *c*-type cytochromes or T1Cu proteins, have been identified as electron donors of N_2OR. In the case of N_2OR purified from *Rh. capsulatus*,[167] *Rh. sphaeroides* f. sp. *denitrificans*,[168] *Ma.*

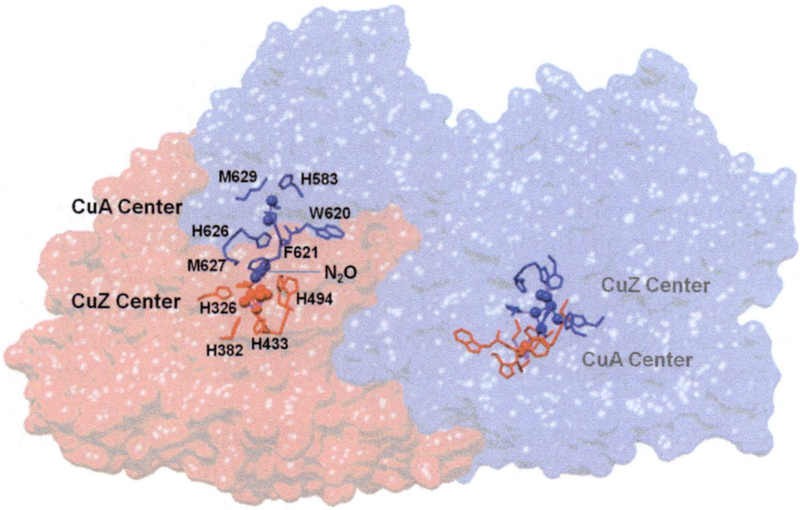

Figure 11.11 The proposed intramolecular electron transfer pathway across the dimer interface between the CuA and CuZ(4Cu2S) centres for the *Pseudomonas stutzeri* nitrous oxide reductase in the presence of nitrous oxide. The most favourable route involves electron transfer from the CuA centre to its ligand H626, subsequently to M627 and finally to nitrous oxide located in the proximity of the CuZ(4Cu2S) centre. Image created using UCSF Chimera and the coordinates are from Protein Data Bank identifier 3SBR.

hydrocarbonoclasticus[169] and *Pa. pantotrophus*,[170] the physiological donor is proposed to be a periplasmic *c*-type cytochrome. In the latter case, N_2OR can also accept electrons from a periplasmic T1Cu protein, pseudoazurin,[170] and from the mitochondrial horse heart cytochrome *c*.[171] In the case of N_2OR from *Ac. cycloclastes*, the physiological electron donor has been demonstrated to be a pseudoazurin,[172] but bovine heart cytochrome *c* is also able to reduce the enzyme, as an artificial electron donor, in *in vitro* assays.[173]

A peculiar case is represented by N_2OR from *Wo. succinogenes* that presents an additional domain in its C-terminal with the canonical *c*-type haem binding motif –CXXCH– acting as electron transfer centre from the periplasm to the CuA centre.[174] The three-dimensional structure of this enzyme is not available yet, and no experimental data regarding the electron transfer reaction rates have been evaluated. However, a model of the structure of N_2OR from *Wo. succinogenes* has been obtained through sequence homology and docking algorithms suggesting that this enzyme resembles the electron transfer complex between a N_2OR and a *c*-type cytochrome.[163] Moreover, N_2OR from *Wo. succinogenes* can accept electrons from a periplasmic *c*-type cytochrome isolated from the same organism, a putative electron donor.[175]

The electron transfer from these small electron donor proteins is specific to the CuA centre since the catalytic CuZ*(4Cu1S) centre is not reduced by these proteins.[169,171] Indeed, the as-isolated enzyme requires an activation process that includes the reduction of the CuZ*(4Cu1S) centre, and therefore most of the kinetic parameters obtained so far were determined using artificial electron donors. In particular, reduced dyes (methyl or benzyl viologen) were used, which are able to reduce both CuA and CuZ*(4Cu1S) centres.[146,169] Therefore, it has been proposed that these small dyes are able to interact directly with both CuZ*(4Cu1S) and CuA centres.

To date, more complete kinetics studies in the presence of the respective physiological electron donors have been reported for N_2ORs isolated from *Ma. Hydrocarbonoclasticus*,[169] *Pa. denitrificans*[171] and *Ac. cycloclastes*.[172]

The recognition and orientation of the encountered electron transfer complex between N_2OR and the small electron donor proteins is driven by the nature and composition of the surface of both proteins and the behaviour is different.

In the case of the N_2OR from *Ma. hydrocarbonoclasticus*, the ionic strength dependence of the catalytic activity indicated that the complex with the physiological donor is driven by hydrophobic interactions,[169] whereas for the N_2ORs from *Pa. denitrificans* and *Ac. cycloclastes*, the interaction is mainly governed by electrostatic effects,[163] as suggested by the experimental data.[171,176,177] This information has been used to screen the putative model structures of the electron transfer complexes between N_2ORs from different organisms and their respective physiological electron donors by docking simulation.[163]

In those model structures, the small electron donor (*Ma. hydrocarbonoclasticus* cytochrome c_{552}, both pseudoazurin and cytochrome c_{550} from *Pa. denitrificans* and pseudoazurin from *Ac. cycloclastes*) was proposed to bind at the N_2OR surface near the CuA centre,[163] in accordance with the direct electron transfer experiments.[164,169,170,172] In all models, the solvent-exposed edges of the electron donor protein (haem methyl of cytochrome *c* or the imidazolic ring of the histidine ligand of the copper centre of pseudoazurin) are positioned at about 5 Å from a conserved patch in the enzyme surface, which is proposed to be the electron delivery site (see Figure 11.12). The analysis of the model structures does not allow the identification of a single-electron transfer pathway in the enzyme, whereas a set of conserved residues has been suggested to be involved in the electron transfer route to the CuA centre.[163]

As in the case for the intermolecular electron transfer, it is useful to look at the case of CcO, given the similarity between the CuA centres of the two enzymes. In CcO, a surface-exposed tryptophan plays a key role in the electron transfer from a soluble cytochrome *c* to the CuA centre.[161,178] In N_2OR, the surface-exposed tryptophan residue is missing, but P496 and H566 (residues numbered according to the *Ma. hydrocarbonoclasticus* N_2OR primary sequence) occupy an analogous position to W121 in the *Pa. denitrificans* CcO, and thus have been suggested to be involved in the electron transfer pathway.[164,169] The conserved patch in the enzyme surface is composed of A495–P496, D519 and H566, in which the carboxyl group of D519 is hydrogen bonded to H526, the terminal ligand of the CuA centre.[163,164,169]

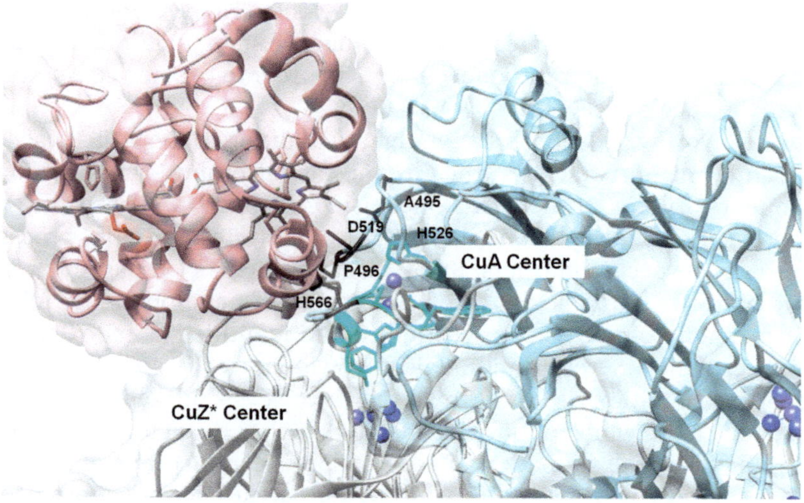

Figure 11.12 A top model structure for the electron transfer complex obtained by BiGGER is represented (probe 2, Table 1).[168] Nitrous oxide reductase (N_2OR) is coloured light grey and light blue and cytochrome c_{552} is coloured pink, with the haems and its axial ligands as sticks. Copper atoms are coloured blue, and N_2OR residues involved in the interaction and CuA ligands are shown as black and light blue sticks. Image created using UCSF Chimera.

Notably, in the recently determined structure of the fully oxidised *Ps. stutzeri* N_2OR, the side chain of this histidine residue is in fact not coordinating the CuA centre in the absence of substrate.[165] Therefore, Einsle and collaborators suggested that the H526 imidazole ring flips back, coordinating the CuA centre, upon substrate binding, and then its reduction is enabled by the electron donor protein.[165]

11.6 Concluding Remarks

Denitrification is a landmark in the study of electron transfer processes. A wide range of transition metals takes the spotlight in the direct catalysis of substrates that become products, but also substrates of the sequential reactions, with small electron shuttle proteins feeding electrons to this respiratory pathway.

The enzymes involved in denitrification are extensively described in this book, and here the electron transfer pathways were reviewed, with a main focus on the intra- and inter-molecular electron transfer pathways, as well as the molecular recognition mechanisms involving their redox partners, which are key players in the electron transfer process.

The enzymes involved in the four steps of denitrification and nitrate dissimilatory pathways contain multiple redox centres. In most cases, metal centres are assigned to specific roles: merely transferring electrons to the next redox centre or being the catalytic active site. The metal centre where catalysis occurs may be in a ready or in an inactive state, requiring activation that might involve conformational changes in order to become competent to catalyse the specific reaction.

Also noteworthy is the observation that, in addition to the structural complexity in terms of metal sites, the enzymes are also built as multi-subunit components, dimers, trimers, *etc.*, increasing the level of sophistication of the system. These catalytic units also have to face intermolecular processes and recognise specific partners. Small electron carriers (in particular soluble cytochromes, cupredoxins, membrane-bound metalloproteins, ubiquinol or menaquinol) have been scrutinised and identified as redox partners of these enzymes. Several methodologies evidenced the specificity of the molecular recognition process, which may involve the attraction of the two redox partners, in which oppositely charged patches may play an important role, followed by the formation of a more specific and competent electron transfer complex in which the hydrophobic patches have a more prominent role. These hydrophobic surfaces include, in most cases, the solvent-exposed haem edge in *c*-type cytochromes, or the imidazole ring of the histidine residue that coordinates the T1Cu centre in cupredoxins.[179] In addition, these interactions, as well as the electron transfer rates, are modulated by water molecules that stay at the interface of the transient complexes, since their formation is also driven by entropy (solvating water molecules being excluded from the interacting surface).[180,181]

Our level of knowledge of the intramolecular and intermolecular processes (rate constants, redox partners and key residues) in the denitrification steps varies between the sequential steps, and several gaps must (and will) be filled in the near future. It is clear that the emerging assembly of electron transfer data will soon enable a deeper discussion on electron transfer in the four steps mediating the reduction of nitrate to dinitrogen, enabling a better knowledge of the rate-limiting steps and the possibility of controlling the reactions involved, with relevant impacts on solving environmental problems.

In fact, the denitrification machinery that includes the enzymes and accessory proteins that might be important in maintaining the enzymes in an active state and are involved in their maturation or even as electron transfer proteins were found to be clustered together.[182] In this recent study in the *Ps. aeruginosa* denitrification pathway,[182] it was found that there is a supramolecular complex that involves NaR, cNOR, cd_1-NiR, N_2OR and the accessory factors of N_2OR. This supramolecular complex, which also interconnects with the formation of ATP (through ATP synthase) and enzymes of the TCA cycle, was named as nitrate respirasome. In such a complex, it is thus easy to conceive that a single small electron donor or, in some cases, two small electron donors (with an interchangeable role) might function as electron shuttle proteins between the enzymes, and this might be a very clever way for bacteria to use the gaseous product/substrates (nitric oxide, nitrous oxide and dinitrogen) that would not need to travel far to encounter the next active site.

Acknowledgements

The authors would like to thank Fundação para a Ciência e Tecnologia (FCT) for the financial support provided (PTDC/BIA-PRO/098882/2008 and PTDC/BIA-PRO/109796/2009 to SRP, PTDC/QUI-BIQ/116481/2010 and PTDC/BBB-BQB/0129/2014 to IM and SFRH/BPD/80293/2011 scholarship to RMA). This work was supported by the Unidade de Ciências Biomoleculares Aplicadas-UCIBIO, which is financed by national funds from FCT/MEC (UID/Multi/04378/2013) and co-financed by the ERDF under the PT2020 Partnership Agreement (POCI-01-0145-FEDER-007728). SRP is an IF fellow supported by FCT.

References

1. C. Coelho and M. J. Romao, *Protein Sci.*, 2015, **24**, 901.
2. A. J. Gates, V. M. Luque-Almagro, A. D. Goddard, S. J. Ferguson, M. D. Roldan and D. J. Richardson, *Biochem. J.*, 2011, **435**, 743.
3. P. J. Gonzalez, M. G. Rivas, C. S. Mota, C. D. Brondino, I. Moura and J. J. G. Moura, *Coord. Chem. Rev.*, 2013, **257**, 315.
4. A. Vazquez-Torres and A. J. Baumler, *Curr. Opin. Microbiol.*, 2015, **29**, 1.
5. J. O. Lundberg, E. Weitzberg, J. A. Cole and N. Benjamin, *Nat. Rev. Microbiol.*, 2004, **2**, 593.
6. D. J. Richardson, B. C. Berks, D. A. Russell, S. Spiro and C. J. Taylor, *Cell. Mol. Life Sci.*, 2001, **58**, 165.
7. W. H. Campbell, *Cell. Mol. Life Sci.*, 2001, **58**, 194.
8. K. Fischer, G. G. Barbier, H. J. Hecht, R. R. Mendel, W. H. Campbell and G. Schwarz, *Plant Cell*, 2005, **17**, 1167.
9. C. Sparacino-Watkins, J. F. Stolz and P. Basu, *Chem. Soc. Rev.*, 2014, **43**, 676.
10. L. C. Bell, D. J. Richardson and S. J. Ferguson, *FEBS Lett.*, 1990, **265**, 85.
11. L. C. Bell, D. J. Richardson and S. J. Ferguson, *J. Gen. Microbiol.*, 1992, **138**, 437.
12. B. C. Berks, D. J. Richardson, C. Robinson, A. Reilly, R. T. Aplin and S. J. Ferguson, *Eur. J. Biochem.*, 1994, **220**, 117.
13. H. J. Sears, S. J. Ferguson, D. J. Richardson and S. Spiro, *FEMS Microbiol. Lett.*, 1993, **113**, 107.
14. J. M. Dias, M. E. Than, A. Humm, R. Huber, G. P. Bourenkov, H. D. Bartunik, S. Bursakov, J. Calvete, J. Caldeira, C. Carneiro, J. J. Moura, I. Moura and M. J. Romao, *Structure*, 1999, **7**, 65.
15. B. J. Jepson, S. Mohan, T. A. Clarke, A. J. Gates, J. A. Cole, C. S. Butler, J. N. Butt, A. M. Hemmings and D. J. Richardson, *J. Biol. Chem.*, 2007, **282**, 6425.
16. P. J. Simpson, D. J. Richardson and R. Codd, *Microbiology*, 2010, **156**, 302.
17. J. Simon, M. Sanger, S. C. Schuster and R. Gross, *Mol. Microbiol.*, 2003, **49**, 69.

18. P. Arnoux, M. Sabaty, J. Alric, B. Frangioni, B. Guigliarelli, J. M. Adriano and D. Pignol, *Nat. Struct. Biol.*, 2003, **10**, 928.
19. C. Coelho, P. J. González, J. G. Moura, I. Moura, J. Trincão and M. João Romão, *J. Mol. Biol.*, 2011, **408**, 932.
20. C. Coelho, P. J. Gonzalez, J. Trincao, A. L. Carvalho, S. Najmudin, T. Hettman, S. Dieckman, J. J. Moura, I. Moura and M. J. Romao, *Acta Crystallogr., Sect. F: Struct. Biol. Cryst. Commun.*, 2007, **63**, 516.
21. M. Kern and J. Simon, *Microbiology*, 2009, **155**, 2784.
22. A. Marietou, D. Richardson, J. Cole and S. Mohan, *FEMS Microbiol. Lett.*, 2005, **248**, 217.
23. B. J. Jepson, A. Marietou, S. Mohan, J. A. Cole, C. S. Butler and D. J. Richardson, *Biochem. Soc. Trans.*, 2006, **34**, 122.
24. M. Kern and J. Simon, *Biochim. Biophys. Acta*, 2009, **1787**, 646.
25. T. H. Brondijk, A. Nilavongse, N. Filenko, D. J. Richardson and J. A. Cole, *Biochem. J.*, 2004, **379**, 47.
26. T. H. Brondijk, D. Fiegen, D. J. Richardson and J. A. Cole, *Mol. Microbiol.*, 2002, **44**, 245.
27. C. Coelho, J. Marangon, D. Rodrigues, J. J. G. Moura, M. J. Romão, P. M. Paes de Sousa and M. M. Correia dos Santos, *J. Electroanal. Chem.*, 2013, **693**, 105.
28. H. Gao, Z. K. Yang, S. Barua, S. B. Reed, M. F. Romine, K. H. Nealson, J. K. Fredrickson, J. M. Tiedje and J. Zhou, *ISME J.*, 2009, **3**, 966.
29. C. Correia, S. Besson, C. D. Brondino, P. J. Gonzalez, G. Fauque, J. Lampreia, I. Moura and J. J. Moura, *J. Biol. Inorg. Chem.*, 2008, **13**, 1321.
30. M. J. Ellington, W. L. Fosdike, R. G. Sawers, D. J. Richardson and S. J. Ferguson, *Arch. Microbiol.*, 2006, **184**, 298.
31. M. J. Ellington, G. Sawers, H. J. Sears, S. Spiro, D. J. Richardson and S. J. Ferguson, *Microbiology*, 2003, **149**, 1533.
32. M. G. Bertero, R. A. Rothery, M. Palak, C. Hou, D. Lim, F. Blasco, J. H. Weiner and N. C. Strynadka, *Nat. Struct. Biol.*, 2003, **10**, 681.
33. E. V. Morozkina and R. A. Zvyagilskaya, *Biochemistry (Moscow)*, 2007, **72**, 1151.
34. A. Magalon, M. Asso, B. Guigliarelli, R. A. Rothery, P. Bertrand, G. Giordano and F. Blasco, *Biochemistry*, 1998, **37**, 7363.
35. R. A. Rothery, A. Magalon, G. Giordano, B. Guigliarelli, F. Blasco and J. H. Weiner, *J. Biol. Chem.*, 1998, **273**, 7462.
36. A. Magalon, D. Lemesle-Meunier, R. A. Rothery, C. Frixon, J. H. Weiner and F. Blasco, *J. Biol. Chem.*, 1997, **272**, 25652.
37. R. A. Rothery, M. G. Bertero, R. Cammack, M. Palak, F. Blasco, N. C. Strynadka and J. H. Weiner, *Biochemistry*, 2004, **43**, 5324.
38. R. A. Rothery, F. Blasco, A. Magalon, M. Asso and J. H. Weiner, *Biochemistry*, 1999, **38**, 12747.
39. R. A. Rothery, F. Blasco and J. H. Weiner, *Biochemistry*, 2001, **40**, 5260.
40. P. Lanciano, A. Magalon, P. Bertrand, B. Guigliarelli and S. Grimaldi, *Biochemistry*, 2007, **46**, 5323.
41. C. C. Page, C. C. Moser, X. Chen and P. L. Dutton, *Nature*, 1999, **402**, 47.

42. F. Cava, O. Zafra and J. Berenguer, *Mol. Microbiol.*, 2008, **70**, 507.
43. L. Alvarez, C. Bricio, A. Blesa, A. Hidalgo and J. Berenguer, *Appl. Environ. Microbiol.*, 2014, **80**, 19.
44. J. Simon and P. M. Kroneck, *Met. Ions Life Sci.*, 2014, **14**, 211.
45. V. A. Bamford, H. C. Angove, H. E. Seward, A. J. Thomson, J. A. Cole, J. N. Butt, A. M. Hemmings and D. J. Richardson, *Biochemistry*, 2002, **41**, 2921.
46. J. D. Gwyer, H. C. Angove, D. J. Richardson and J. N. Butt, *Bioelectrochemistry*, 2004, **63**, 43.
47. P. Stach, O. Einsle, W. Schumacher, E. Kurun and P. M. Kroneck, *J. Inorg. Biochem.*, 2000, **79**, 381.
48. J. H. van Wonderen, B. Burlat, D. J. Richardson, M. R. Cheesman and J. N. Butt, *J. Biol. Chem.*, 2008, **283**, 9587.
49. M. L. Rodrigues, T. F. Oliveira, I. A. Pereira and M. Archer, *EMBO J.*, 2006, **25**, 5951.
50. C. A. Cunha, S. Macieira, J. M. Dias, G. Almeida, L. L. Goncalves, C. Costa, J. Lampreia, R. Huber, J. J. Moura, I. Moura and M. J. Romao, *J. Biol. Chem.*, 2003, **278**, 17455.
51. M. S. Pittman, K. T. Elvers, L. Lee, M. A. Jones, R. K. Poole, S. F. Park and D. J. Kelly, *Mol. Microbiol.*, 2007, **63**, 575.
52. M. Youngblut, E. T. Judd, V. Srajer, B. Sayyed, T. Goelzer, S. J. Elliott, M. Schmidt and A. A. Pacheco, *J. Biol. Inorg. Chem.*, 2012, **17**, 647.
53. J. Simon and M. G. Klotz, *Biochim. Biophys. Acta*, 2013, **1827**, 114.
54. O. Einsle, A. Messerschmidt, R. Huber, P. M. Kroneck and F. Neese, *J. Am. Chem. Soc.*, 2002, **124**, 11737.
55. O. Einsle, A. Messerschmidt, P. Stach, G. P. Bourenkov, H. D. Bartunik, R. Huber and P. M. Kroneck, *Nature*, 1999, **400**, 476.
56. O. Einsle, P. Stach, A. Messerschmidt, J. Simon, A. Kroger, R. Huber and P. M. Kroneck, *J. Biol. Chem.*, 2000, **275**, 39608.
57. C. W. Lockwood, B. Burlat, M. R. Cheesman, M. Kern, J. Simon, T. A. Clarke, D. J. Richardson and J. N. Butt, *J. Am. Chem. Soc.*, 2015, **137**, 3059.
58. E. T. Judd, M. Youngblut, A. A. Pacheco and S. J. Elliott, *Biochemistry*, 2012, **51**, 10175.
59. M. G. Almeida, S. Macieira, L. L. Goncalves, R. Huber, C. A. Cunha, M. J. Romao, C. Costa, J. Lampreia, J. J. Moura and I. Moura, *Eur. J. Biochem.*, 2003, **270**, 3904.
60. D. Bykov and F. Neese, *Inorg. Chem.*, 2015, **54**, 9303.
61. I. V. Kurnikov, M. A. Ratner and A. A. Pacheco, *Biochemistry*, 2005, **44**, 1856.
62. C. Lockwood, J. N. Butt, T. A. Clarke and D. J. Richardson, *Biochem. Soc. Trans.*, 2011, **39**, 263.
63. A. Bauss and T. Koslowski, *Phys. Chem. Chem. Phys.*, 2015, **17**, 4483.
64. J. Simon and M. Kern, *Biochem. Soc. Trans.*, 2008, **36**, 1011.
65. C. W. Lockwood, T. A. Clarke, J. N. Butt, A. M. Hemmings and D. J. Richardson, *Biochem. Soc. Trans.*, 2011, **39**, 1871.

66. M. Nojiri, Y. Xie, T. Inoue, T. Yamamoto, H. Matsumura, K. Kataoka, Deligeer, K. Yamaguchi, Y. Kai and S. Suzuki, *Proc. Natl. Acad. Sci. U. S. A.*, 2007, **104**, 4315.
67. C. Han, G. S. Wright, K. Fisher, S. E. Rigby, R. R. Eady and S. S. Hasnain, *Biochem. J.*, 2012, **444**, 219.
68. S. V. Antonyuk, C. Han, R. R. Eady and S. S. Hasnain, *Nature*, 2013, **496**, 123.
69. M. J. Ellis, J. G. Grossmann, R. R. Eady and S. S. Hasnain, *J. Biol. Inorg. Chem.*, 2007, **12**, 1119.
70. S. Suzuki, K. Kataoka and K. Yamaguchi, *Acc. Chem. Res.*, 2000, **33**, 728.
71. O. Farver, R. R. Eady, G. Sawers, M. Prudencio and I. Pecht, *FEBS Lett.*, 2004, **561**, 173.
72. R. L. Harris, R. R. Eady, S. S. Hasnain and R. G. Sawers, *Arch. Microbiol.*, 2006, **186**, 241.
73. S. Brenner, D. J. Heyes, S. Hay, M. A. Hough, R. R. Eady, S. S. Hasnain and N. S. Scrutton, *J. Biol. Chem.*, 2009, **284**, 25973.
74. M. J. Ellis, F. E. Dodd, G. Sawers, R. R. Eady and S. S. Hasnain, *J. Mol. Biol.*, 2003, **328**, 429.
75. N. G. Leferink, C. Han, S. V. Antonyuk, D. J. Heyes, S. E. Rigby, M. A. Hough, R. R. Eady, N. S. Scrutton and S. S. Hasnain, *Biochemistry*, 2011, **50**, 4121.
76. N. G. Leferink, C. R. Pudney, S. Brenner, D. J. Heyes, R. R. Eady, S. Samar Hasnain, S. Hay, S. E. Rigby and N. S. Scrutton, *FEBS Lett.*, 2012, **586**, 578.
77. S. Maekawa, T. Matsui, K. Hirao and Y. Shigeta, *J. Phys. Chem. B*, 2015, **119**, 5392.
78. S. Kuznetsova, G. Zauner, T. J. Aartsma, H. Engelkamp, N. Hatzakis, A. E. Rowan, R. J. Nolte, P. C. Christianen and G. W. Canters, *Proc. Natl. Acad. Sci. U. S. A.*, 2008, **105**, 3250.
79. S. Suzuki, K. Yamaguchi, K. Kataoka, K. Kobayashi, S. Tagawa, T. Kohzuma, S. Shidara, H. Iwasaki and Deligeer, *J. Biol. Inorg. Chem.*, 1997, **2**, 265.
80. Y. Astier, G. W. Canters, J. J. Davis, H. A. Hill, M. P. Verbeet and H. J. Wijma, *ChemPhysChem*, 2005, **6**, 1114.
81. N. G. Leferink, S. V. Antonyuk, J. A. Houwman, N. S. Scrutton, R. R. Eady and S. S. Hasnain, *Nat. Commun.*, 2014, **5**, 4395.
82. S. Wherland, O. Farver and I. Pecht, *ChemPhysChem*, 2005, **6**, 805.
83. M. Kukimoto, M. Nishiyama, M. Tanokura, E. T. Adman and S. Horinouchi, *J. Biol. Chem.*, 1996, **271**, 13680.
84. A. Impagliazzo, L. Krippahl and M. Ubbink, *ChemBioChem*, 2005, **6**, 1648.
85. A. Impagliazzo and M. Ubbink, *J. Am. Chem. Soc.*, 2004, **126**, 5658.
86. M. D. Vlasie, R. Fernandez-Busnadiego, M. Prudencio and M. Ubbink, *J. Mol. Biol.*, 2008, **375**, 1405.
87. A. Impagliazzo, A. J. Blok, M. J. Cliff, J. E. Ladbury and M. Ubbink, *J. Am. Chem. Soc.*, 2007, **129**, 226.

88. M. Nojiri, H. Koteishi, T. Nakagami, K. Kobayashi, T. Inoue, K. Yamaguchi and S. Suzuki, *Nature*, 2009, **462**, 117.
89. Y. Fukuda, H. Koteishi, R. Yoneda, T. Tamada, H. Takami, T. Inoue and M. Nojiri, *Biochim. Biophys. Acta*, 2014, **1837**, 396.
90. F. E. Aas, X. Li, J. Edwards, M. Hongro Solbakken, M. Deeudom, A. Vik, J. Moir, M. Koomey and M. Aspholm, *Environ. Microbiol.*, 2015, **17**, 2114.
91. A. C. Hopper, Y. Li and J. A. Cole, *J. Bacteriol.*, 2013, **195**, 2518.
92. T. Yamazaki, H. Oyanagi, T. Fujiwara and Y. Fukumori, *Eur. J. Biochem.*, 1995, **233**, 665.
93. E. Weegaerssens, W. S. Wu, R. W. Ye, J. M. Tiedje and C. K. Chang, *J. Biol. Chem.*, 1991, **266**, 7496.
94. S. Besson, C. Carneiro, J. J. Moura, I. Moura and G. Fauque, *Anaerobe*, 1995, **1**, 219.
95. V. Fulop, J. W. Moir, S. J. Ferguson and J. Hajdu, *Cell*, 1995, **81**, 369.
96. R. Timkovich, R. Dhesi, K. J. Martinkus, M. K. Robinson and T. M. Rea, *Arch. Biochem. Biophys.*, 1982, **215**, 47.
97. F. Cutruzzola, S. Rinaldo, F. Centola and M. Brunori, *IUBMB Life*, 2003, **55**, 617.
98. D. Nurizzo, F. Cutruzzola, M. Arese, D. Bourgeois, M. Brunori, C. Cambillau and M. Tegoni, *Biochemistry*, 1998, **37**, 13987.
99. D. Nurizzo, M. C. Silvestrini, M. Mathieu, F. Cutruzzola, D. Bourgeois, V. Fulop, J. Hajdu, M. Brunori, M. Tegoni and C. Cambillau, *Structure*, 1997, **5**, 1157.
100. S. Rinaldo, G. Giardina, N. Castiglione, V. Stelitano and F. Cutruzzola, *Biochem. Soc. Trans.*, 2011, **39**, 195.
101. J. W. Allen, N. J. Watmough and S. J. Ferguson, *Nat. Struct. Biol.*, 2000, **7**, 885.
102. C. D. Richter, J. W. Allen, C. W. Higham, A. Koppenhofer, R. S. Zajicek, N. J. Watmough and S. J. Ferguson, *J. Biol. Chem.*, 2002, **277**, 3093.
103. O. Farver, M. Brunori, F. Cutruzzola, S. Rinaldo, S. Wherland and I. Pecht, *Biophys. J.*, 2009, **96**, 2849.
104. M. Radoul, Y. Barak, S. Rinaldo, F. Cutruzzola, I. Pecht and D. Goldfarb, *Biochemistry*, 2012, **51**, 9192.
105. O. Farver, P. M. Kroneck, W. G. Zumft and I. Pecht, *Biophys. Chem.*, 2002, **98**, 27.
106. K. Kobayashi, A. Koppenhofer, S. J. Ferguson, N. J. Watmough and S. Tagawa, *Biochemistry*, 2001, **40**, 8542.
107. H. Lopes, S. Besson, I. Moura and J. J. Moura, *J. Biol. Inorg. Chem.*, 2001, **6**, 55.
108. K. Kobayashi, A. Koppenhofer, S. J. Ferguson and S. Tagawa, *Biochemistry*, 1997, **36**, 13611.
109. E. K. Wilson, A. Bellelli, S. Liberti, M. Arese, S. Grasso, F. Cutruzzola, M. Brunori and P. Brzezinski, *Biochemistry*, 1999, **38**, 7556.
110. O. Farver, P. M. Kroneck, W. G. Zumft and I. Pecht, *Proc. Natl. Acad. Sci. U. S. A.*, 2003, **100**, 7622.
111. I. M. Wasser, S. de Vries, P. Moenne-Loccoz, I. Schroder and K. D. Karlin, *Chem. Rev.*, 2002, **102**, 1201.

112. H. A. Pedroso, C. M. Silveira, R. M. Almeida, A. C. Almeida, S. Besson, I. Moura, J. J. Moura and M. G. Almeida, *Biochim. Biophys. Acta*, 2016, **1857**, 1412.
113. F. M. Ferroni, J. Marangon, N. I. Neuman, J. C. Cristaldi, S. M. Brambilla, S. A. Guerrero, M. G. Rivas, A. C. Rizzi and C. D. Brondino, *J. Biol. Inorg. Chem.*, 2014, **19**, 913.
114. F. Cutruzzolà, M. Arese, G. Ranghino, G. van Pouderoyen, G. Canters and M. Brunori, *J. Inorg. Biochem.*, 2002, **88**, 353.
115. I. V. Pearson, M. D. Page, R. J. van Spanning and S. J. Ferguson, *J. Bacteriol.*, 2003, **185**, 6308.
116. P. A. Williams, V. Fulop, Y. C. Leung, C. Chan, J. W. Moir, G. Howlett, S. J. Ferguson, S. E. Radford and J. Hajdu, *Nat. Struct. Biol.*, 1995, **2**, 975.
117. R. S. Zajicek, J. W. Allen, M. L. Cartron, D. J. Richardson and S. J. Ferguson, *FEBS Lett.*, 2004, **565**, 48.
118. E. Lojou, F. Cutruzzola, M. Tegoni and P. Bianco, *Electrochim. Acta*, 2003, **48**, 1055.
119. W. G. Zumft, *J. Inorg. Biochem.*, 2005, **99**, 194.
120. N. J. Watmough, S. J. Field, R. J. Hughes and D. J. Richardson, *Biochem. Soc. Trans.*, 2009, **37**, 392.
121. T. Hino, S. Nagano, H. Sugimoto, T. Tosha and Y. Shiro, *Biochim. Biophys. Acta*, 2012, **1817**, 680.
122. S. J. Field, M. D. Roldan, S. J. Marritt, J. N. Butt, D. J. Richardson and N. J. Watmough, *Biochim. Biophys. Acta*, 2011, **1807**, 451.
123. P. Lachmann, Y. Huang, J. Reimann, U. Flock and P. Ädelroth, *J. Biol. Chem.*, 2010, **285**, 25531.
124. J. ter Beek, N. Krause, J. Reimann, P. Lachmann and P. Ädelroth, *J. Biol. Chem.*, 2013, **288**, 30626.
125. J. H. M. Hendriks, A. Jasaitis, M. Saraste and M. I. Verkhovsky, *Biochemistry*, 2002, **41**, 2331.
126. T. Hino, Y. Matsumoto, S. Nagano, H. Sugimoto, Y. Fukumori, T. Murata, S. Iwata and Y. Shiro, *Science*, 2010, **330**, 1666.
127. S. Iwata, C. Ostermeier, B. Ludwig and H. Michel, *Nature*, 1995, **376**, 660.
128. A. Namslauer, M. Branden and P. Brzezinski, *Biochemistry*, 2002, **41**, 10369.
129. C. M. Cordas, A. G. Duarte, J. J. Moura and I. Moura, *Biochim. Biophys. Acta*, 2013, **1827**, 233.
130. Y. Matsumoto, T. Tosha, A. V. Pisliakov, T. Hino, H. Sugimoto, S. Nagano, Y. Sugita and Y. Shiro, *Nat. Struct. Mol. Biol.*, 2012, **19**, 238.
131. M. R. A. Blomberg and P. E. M. Siegbahn, *Biochim. Biophys. Acta*, 2013, **1827**, 826.
132. S. Al-Attar and S. de Vries, *FEBS Lett.*, 2015, **589**, 2050.
133. L. Salomonsson, J. Reimann, T. Tosha, N. Krause, N. Gonska, Y. Shiro and P. Ädelroth, *Biochim. Biophys. Acta*, 2012, **1817**, 1914.
134. E. Terasaka, N. Okada, N. Sato, Y. Sako, Y. Shiro and T. Tosha, *Biochim. Biophys. Acta*, 2014, **1837**, 1019.
135. P. Tavares, A. S. Pereira, J. J. Moura and I. Moura, *J. Inorg. Biochem.*, 2006, **100**, 2087.

136. J. Hoglen and T. C. Hollocher, *J. Biol. Chem.*, 1989, **264**, 7556.
137. D. H. Kastrau, B. Heiss, P. M. Kroneck and W. G. Zumft, *Eur. J. Biochem.*, 1994, **222**, 293.
138. C. G. Timóteo, A. S. Pereira, C. E. Martins, S. G. Naik, A. G. Duarte, J. J. Moura, P. Tavares, B. H. Huynh and I. Moura, *Biochemistry*, 2011, **50**, 4251.
139. N. Sakurai and T. Sakurai, *Biochemistry*, 1997, **36**, 13809.
140. H. Kumita, K. Matsuura, T. Hino, S. Takahashi, H. Hori, Y. Fukumori, I. Morishima and Y. Shiro, *J. Biol. Chem.*, 2004, **279**, 55247.
141. J. W. B. Moir and S. J. Ferguson, *Microbiology*, 1994, **140**, 389.
142. J. Lalucat, A. Bennasar, R. Bosch, E. García-Valdés and N. J. Palleroni, *Microbiol. Mol. Biol. Rev.*, 2006, **70**, 510.
143. E. Vijgenboom, J. E. Busch and G. W. Canters, *Microbiology*, 1997, **143**(9), 2853.
144. S. de Vries, M. J. Strampraad, S. Lu, P. Moenne-Loccoz and I. Schroder, *J. Biol. Chem.*, 2003, **278**, 35861.
145. Suharti, H. A. Heering and S. de Vries, *Biochemistry*, 2004, **43**, 13487.
146. W. G. Zumft and P. M. Kroneck, *Adv. Microb. Physiol.*, 2007, **52**, 107.
147. S. R. Pauleta, S. Dell'Acqua and I. Moura, *Coord. Chem. Rev.*, 2013, **257**, 332.
148. C. L. Coyle, W. G. Zumft, P. M. Kroneck, H. Korner and W. Jakob, *Eur. J. Biochem.*, 1985, **153**, 459.
149. S. Dell'Acqua, S. R. Pauleta, J. J. Moura and I. Moura, *Philos. Trans. R. Soc. London, Ser. B*, 2012, **367**, 1204.
150. M. Prudencio, A. S. Pereira, P. Tavares, S. Besson, I. Cabrito, K. Brown, B. Samyn, B. Devreese, J. Van Beeumen, F. Rusnak, G. Fauque, J. J. Moura, M. Tegoni, C. Cambillau and I. Moura, *Biochemistry*, 2000, **39**, 3899.
151. T. Rasmussen, B. C. Berks, J. N. Butt and A. J. Thomson, *Biochem. J.*, 2002, **364**, 807.
152. K. Brown, M. Tegoni, M. Prudencio, A. S. Pereira, S. Besson, J. J. Moura, I. Moura and C. Cambillau, *Nat. Struct. Biol.*, 2000, **7**, 191.
153. K. Brown, K. Djinovic-Carugo, T. Haltia, I. Cabrito, M. Saraste, J. J. Moura, I. Moura, M. Tegoni and C. Cambillau, *J. Biol. Chem.*, 2000, **275**, 41133.
154. T. Tsukihara, H. Aoyama, E. Yamashita, T. Tomizaki, H. Yamaguchi, K. Shinzawa-Itoh, R. Nakashima, R. Yaono and S. Yoshikawa, *Science*, 1996, **272**, 1136.
155. H. Beinert, *Eur. J. Biochem.*, 1997, **245**, 521.
156. V. R. I. Kaila, M. I. Verkhovsky and M. Wikström, *Chem. Rev.*, 2010, **110**, 7062.
157. J. R. Winkler, B. G. Malmström and H. B. Gray, *Biophys. Chem.*, 1995, **54**, 199.
158. L. M. Geren, J. R. Beasley, B. R. Fine, A. J. Saunders, S. Hibdon, G. J. Pielak, B. Durham and F. Millett, *J. Biol. Chem.*, 1995, **270**, 2466.
159. M. Ruitenberg, A. Kannt, E. Bamberg, B. Ludwig, H. Michel and K. Fendler, *Proc. Natl. Acad. Sci. U. S. A.*, 2000, **97**, 4632.
160. C. C. Moser, C. C. Page and P. L. Dutton, *Philos. Trans. R. Soc. London, Ser. B*, 2006, **361**, 1295.

161. K. Wang, L. Geren, Y. Zhen, L. Ma, S. Ferguson-Miller, B. Durham and F. Millett, *Biochemistry*, 2002, **41**, 2298.
162. M. F. Lucas, D. L. Rousseau and V. Guallar, *Biochim. Biophys. Acta*, 2011, **1807**, 1305.
163. S. Dell'acqua, I. Moura, J. J. Moura and S. R. Pauleta, *J. Biol. Inorg. Chem.*, 2011, **16**, 1241.
164. K. Mattila and T. Haltia, *Proteins*, 2005, **59**, 708.
165. A. Pomowski, W. G. Zumft, P. M. Kroneck and O. Einsle, *Nature*, 2011, **477**, 234.
166. E. M. Johnston, S. Dell'Acqua, S. Ramos, S. R. Pauleta, I. Moura and E. I. Solomon, *J. Am. Chem. Soc.*, 2014, **136**, 614.
167. D. J. Richardson, L. C. Bell, A. G. McEwan, J. B. Jackson and S. J. Ferguson, *Eur. J. Biochem.*, 1991, **199**, 677.
168. M. Itoh, K. Matsuura and T. Satoh, *FEBS Lett.*, 1989, **251**, 104.
169. S. Dell'acqua, S. R. Pauleta, E. Monzani, A. S. Pereira, L. Casella, J. J. Moura and I. Moura, *Biochemistry*, 2008, **47**, 10852.
170. B. C. Berks, D. Baratta, J. Richardson and S. J. Ferguson, *Eur. J. Biochem.*, 1993, **212**, 467.
171. T. Rasmussen, T. Brittain, B. C. Berks, N. J. Watmough and A. J. Thomson, *Dalton Trans.*, 2005, **21**, 3501.
172. K. Fujita, M. Hirasawa-Fujita, D. E. Brown, Y. Obara, F. Ijima, T. Kohzuma and D. M. Dooley, *J. Inorg. Biochem.*, 2012, **115**, 163.
173. K. Fujita, J. M. Chan, J. A. Bollinger, M. L. Alvarez and D. M. Dooley, *J. Inorg. Biochem.*, 2007, **101**, 1836.
174. S. Teraguchi and T. C. Hollocher, *J. Biol. Chem.*, 1989, **264**, 1972.
175. C.-s. Zhang and T. C. Hollocher, *Biochim. Biophys. Acta*, 1993, **1142**, 253.
176. M. Kukimoto, M. Nishiyama, T. Ohnuki, S. Turley, E. T. Adman, S. Horinouchi and T. Beppu, *Protein Eng.*, 1995, **8**, 153.
177. K. Kataoka, K. Yamaguchi, M. Kobayashi, T. Mori, N. Bokui and S. Suzuki, *J. Biol. Chem.*, 2004, **279**, 53374.
178. H. Witt, F. Malatesta, F. Nicoletti, M. Brunori and B. Ludwig, *J. Biol. Chem.*, 1998, **273**, 5132.
179. S. R. Pauleta, F. Guerlesquin, C. F. Goodhew, B. Devreese, J. Van Beeumen, A. S. Pereira, I. Moura and G. W. Pettigrew, *Biochemistry*, 2004, **43**, 11214.
180. S. R. Pauleta, A. Cooper, M. Nutley, N. Errington, S. Harding, F. Guerlesquin, C. F. Goodhew, I. Moura, J. J. Moura and G. W. Pettigrew, *Biochemistry*, 2004, **43**, 14566.
181. D. L. Cameron, J. Jakus, S. R. Pauleta, G. W. Pettigrew and A. Cooper, *J. Phys. Chem. B*, 2010, **114**, 16228.
182. J. M. Borrero-de Acuna, M. Rohde, J. Wissing, L. Jansch, M. Schobert, G. Molinari, K. N. Timmis, M. Jahn and D. Jahn, *J. Bacteriol.*, 2016, **198**, 1401.

CHAPTER 12

Channels and Transporters for Nitrogen Cycle Intermediates

SUSANA L. A. ANDRADE[a] AND OLIVER EINSLE*[a]

[a]Institut für Biochemie, Albert-Ludwigs-Universität Freiburg, Albertstrasse 21, 79104 Freiburg im Breisgau, Germany
*E-mail: einsle@biochemie.uni-freiburg.de

12.1 Bioenergetics of Denitrification

Bacterial denitrification is a dissimilatory metabolism, and denitrifiers utilise the free enthalpy of the four individual reaction steps for energy conservation in the form of a proton motive force (pmf) across their cytoplasmic membrane. In organismic metabolism, the energy yield of denitrification is second only to aerobic respiration, making it the preferred pathway in micro-oxic and anoxic environments, given the availability of nitrate.[1] This is of consequence for instance in the context of nitrogen fertilisation in agriculture, where nitrate salts are provided together with extensive irrigation that together create high-nitrogen/low-oxygen conditions in which denitrifiers thrive.[2] The extent of this problem is such that, according to current estimates, approximately 40–55% of all nitrogen fertilisers are lost to denitrifying bacteria without ever reaching the crop plants they were intended for.[3]

RSC Metallobiology Series No. 9
Metalloenzymes in Denitrification: Applications and Environmental Impacts
Edited by Isabel Moura, José J. G. Moura, Sofia R. Pauleta and Luisa B. Maia

Published by the Royal Society of Chemistry, www.rsc.org

12.1.1 The Compartmentalisation of the Nitrogen Cycle

As the substrates and intermediates of the denitrificatory enzymes are toxic or gaseous or both, most organisms localise the denitrification machinery outside of the cytoplasm, and in Gram-negative bacteria, this commonly means the periplasmic space.[1] In bacterial respiration, the enzymes of denitrification represent the reductive half-reaction of an electron transfer system originating from donors such as formate or hydrogen, and the actual electron acceptor in each of the reductions of nitrite (NO_3^-), nitrate (NO_2^-), nitric oxide (NO) and nitrous oxide (N_2O) is an oxygen atom that then yields water as a product (Figure 12.1a).[1,4] This formation of water requires protons, and if the reaction takes place in the periplasm, their consumption constitutes an effective reduction of the pmf. Such an energetic disadvantage could be avoided if the reaction steps were to occur in the cytoplasm, but in denitrification, nature has chosen this strategy only for the very first reaction: the reduction of nitrate to nitrite by the membrane-bound nitrate reductase (NarGHI).[5,6] The substrate and product of the nitrate reductase reaction are the only ionic intermediates of the pathway, so that their translocation across a lipid bilayer is not possible without the help of specific membrane-integral transport proteins. In denitrification, the reduction of nitrate is often transferred to the cytoplasmic side of the membrane, resulting in the consumption of two protons on the *n*-side of the membrane rather than on the *p*-side, thus increasing the protein gradient by four ions as compared to the periplasmic nitrate reductase (NapAB; Figure 12.1a).[7] Nevertheless, the following reduction of nitrite once more releases an oxygen atom as water, and thus is subject to the very same considerations of proton stoichiometry as nitrate reduction. However, although one of two distinct nitrite reductases is employed for denitrification (copper-containing or cytochrome cd_1),[8] both are exclusively found in the periplasmic space.[1] It can be assumed that evolution has not favoured the translocation of denitrificatory nitrite reduction to the cytoplasm due to the cytotoxicity of both nitrite and the reaction product NO. It can react with superoxide anions generated, for example, by xanthine oxidase to the highly reactive peroxynitrite, and the dangers of this chemistry seem to outweigh the energetic benefits.[9] Intracellular accumulation of nitrite is strictly avoided by most organisms, and the import of nitrite into the cytoplasm is therefore as tightly regulated as its generation by NarGHI. In denitrification, low nitrite levels in the cytoplasm are achieved simply and elegantly by coupling the import of nitrate to an export of nitrite, so that more substrate only becomes available when product is removed (Figure 12.1a).[10–12]

12.1.2 Translocation of Ions Across Membranes

Biological membranes structure the cell, and even in the simplest organisational motif of a bacterium, this gives rise to three distinct compartments with fundamentally different properties. The cytoplasm is a reducing,

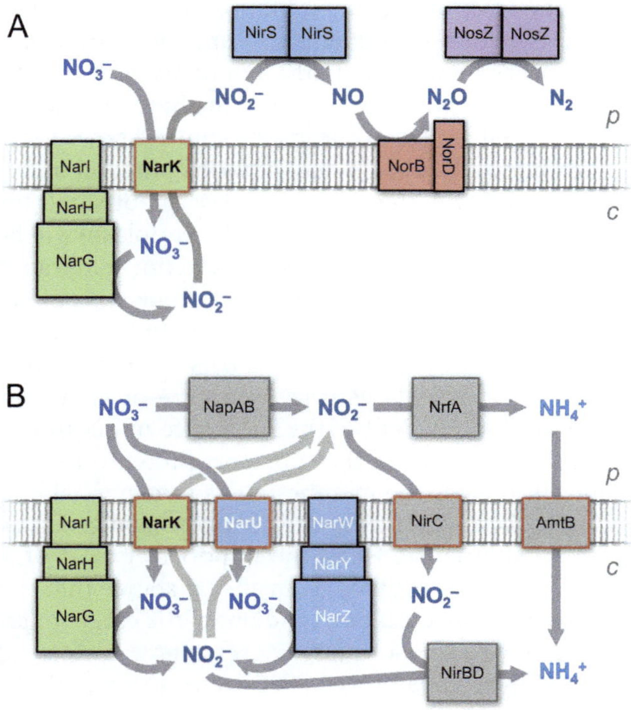

Figure 12.1 Arrangement of nitrogen cycle redox enzymes in the cytoplasmic membrane. (a) A typical denitrifier uses a NarGHI nitrate reductase that faces the cytoplasm so that the substrate NO_3^- must be imported and the product NO_2^- then re-exported. In order to avoid cytoplasmic accumulation of toxic nitrite, both transport functionalities are combined in the antiporter NarK. The subsequent steps of nitrite reductase (NirS), nitric oxide reductase (NorBD) and nitrous oxide reductase (NosZ) all occur in the periplasmic space. (b) The non-denitrifier *Escherichia coli* uses either a periplasmic (NapAB) or a cytoplasmic nitrate reductase, of which it contains two with different substrate affinities and regulations (NarGHI and NarZYW). Dissimilatory nitrite reduction is then completed by cytochrome *c* nitrite reductase NrfA, yielding ammonium as a product. For nitrogen assimilation, the organism can either uptake NO_2^- *via* the nitrite channel NirC, whose expression is tightly linked to that of cytoplasmic nitrite reductase NirBD, or directly import NH_4^+ with the ammonium transporter AmtB. © 2013 Wiley-VCH Verlag GmbH & Co. KGaA, Weinheim.[10]

hydrophilic environment, separated by the hydrophobic interior of the lipid bilayer from the once more hydrophilic, but oxidising periplasm. Ion gradients across energised membranes are generally directed inwards, meaning that respiratory processes lead to a depletion of H^+ or Na^+ at the inside of the membrane and an accumulation on the outside.[13] The permeability (*i.e.*, flux per area) of a membrane for ions is extremely low and is mediated exclusively through transport proteins *in vivo*.[14] It is their selectivity that largely determines the properties of a cell membrane, and two extreme cases of this are exemplified by the inner and outer membranes of Gram-negative bacteria, respectively. The outer membrane is the boundary of the cell. It contains lipopolysaccharides in its outer leaflet and is anchored to the peptidoglycan layer—the bacterial cell wall—on the inside, but it shows very little selectivity for small compounds. This is due to porins, integral outer membrane proteins of the β-barrel type that allow the passage of small molecules, so that the composition of most compounds in the periplasm will largely be the same as in the surrounding medium. Specific selection and transport instead occurs at the second boundary of the cell, the cytoplasmic membrane. This membrane is energised by an H^+ or Na^+ gradient, and its transport proteins mediate the uptake of nutrients and the extrusion of toxic or waste products either passively or actively. *Passive transport*—or channelling—equilibrates the concentration gradient of the substance in question and implies a specific selectivity filter within the transport protein to ensure that only the desired molecules cross the membrane. As in many habitats, nutrients are the objects of fierce competition, and so mechanisms of active transport have evolved as an alternative. Here, the import or export of a given substrate (or cargo) molecule is coupled to an energy source that is able to drive the process against the concentration gradient of the cargo. In *primary active transport*, this is achieved *via* a primary energy source such as light (bacteriorhodopsin) or the hydrolysis of ATP (ABC transporters). Alternatively, *secondary active transporters* utilise an existing ion gradient across an energised membrane to drive the energetically uphill translocation of the cargo. This mode of transport may function such that the energising ion and the cargo molecule are transported in the same direction, making the process a *symport*, while the mechanism resulting from opposite transport directions for both species is called an *antiport*. Channels characteristically achieve far higher transport rates of 10^6–10^7 ions s^{-1}, while transporters that require a mechanism of energy coupling are far slower, at 10–10^3 ions s^{-1}.[15] This is of relevance for the direct measurement of ion currents by electrophysiological techniques, where the slower transport processes remain below the detection limit unless the signals can be integrated over a large number of individual proteins. In the transport proteins of the biogeochemical nitrogen cycle, different mechanisms of transport are realised and will be discussed below.

12.2 Nitrate Transport

The two-electron reduction of nitrate to nitrite is catalysed by molybdenum-dependent nitrate reductases that obtain the required electrons from the quinol pool in the cytoplasmic membrane. Nitrate reductase activities can be localised either in the cytoplasm or in the periplasm, and in the context of denitrification, the latter appears to be the preferred solution at first glance, as all other steps of the pathway also take place in this compartment. However, the reduction of nitrate releases water as a by-product, and for its formation, two protons are consumed.

12.2.1 Types of Nitrate Transporters

As a major source of nitrogen and a substrate for energy metabolism, nitrate is a desired substrate, and consequently evolution has brought forth several different types of uptake permeases for NO_3^-.[16] These can either be members of the largest class of primary active transporters, the ABC transporters, or of the most important class of secondary active transport proteins, the major facilitator superfamily (MFS).[17] Nitrate-importing ABC transporters such as the prototypical NtrBCD are found in cyanobacteria and plants, where the free enthalpy of ATP hydrolysis is coupled to the active import of nitrate for assimilatory purposes.[18] In plants, these systems are primarily active in roots, where they are used to uptake nitrate as a nutrient from the soil.[19] As an alternative means of active nitrate uptake, major facilitators from two different families have evolved for the task,[20,21] both following the established *rocker-switch* model of transport that alternatively exposes a substrate-binding site to different sides of the membrane (Figure 12.2).[10] The two families in question differ in substrate affinity, with K_m values in the millimolar range for the low-affinity system NRT1 (or nitrate/peptide facilitator [NPF]) and in the micromolar range for the high-affinity system NRT2 (or nitrate/nitrite porter [NNP]). Primary structure similarity between both families is low, and—as recent structural

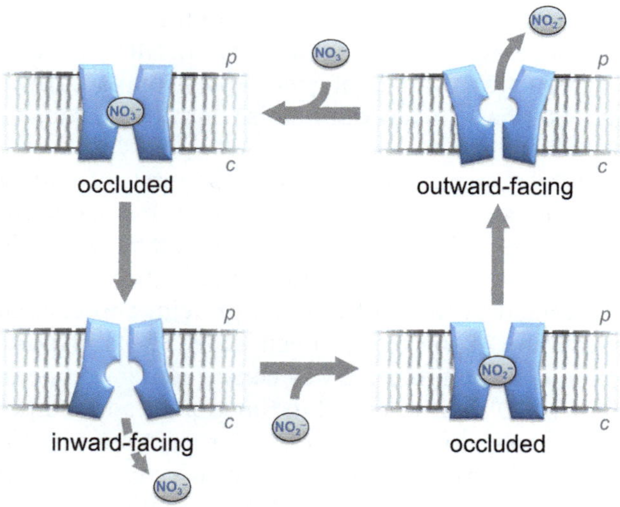

Figure 12.2 Transport mechanism of NarK. The typical "rocker-switch" model for secondary active transporters of the major facilitator superfamily includes two major states—inward-facing and outward-facing—that are connected through at least one further state in which the cargo molecule is sealed (occluded) within the transporter. The strict coupling of nitrate import to nitrite export in NarK implies that either substrate is only transported unidirectionally, meaning that it is only able to switch the protein from the occluded to either the inward- or the outward-facing state. © 2013 Wiley-VCH Verlag GmbH & Co. KGaA, Weinheim.[10]

analyses have revealed—their respective transport mechanisms are clearly distinct, justifying a separate discussion, albeit with a focus on the bacterial NRT2 systems that actually contribute to the denitrification pathway.

12.2.2 The NPF Family and the Plant Nitrate Transporter NRT1.1

As secondary active transporters, the members of the NPF family (NRT1/NPF)[22] are key mediators of nitrate uptake in plant roots, where they form part of the high-affinity uptake system for the metabolite.[16] The family as such exhibits a broad substrate spectrum that includes peptides, amino acids, dicarboxylates and other secondary metabolites, and it is generally assumed that the peptide transporters form the evolutionary root of this class of permeases.[23] The first plant nitrate transporter to be identified in higher plants was NRT1.1 from *Arabidopsis thaliana*,[22,23] also called NPF6.3 or CHL1. It is distinct from most other known NRT1 and NRT2 family members by the initially confusing property of having two different affinity values.[24–26] As shown through mutant studies, NRT1.1 is physiologically active in both the high- and low-affinity uptake pathways, with the switch of transport mode being mediated by the phosphorylation of a single threonine, T101.[27] Besides its function in nitrate assimilation, NRT1.1 also acts as a nitrate sensor, supporting the secondary role of the anion as a direct signal for the presence of bioavailable nitrogen.[28] Hereby, NRT1.1-mediated nitrate uptake directly regulates a variety of gene expression patterns in the plant, once more with a biphasic behaviour that is controlled through phosphorylation of T101.[29,30]

Two recent crystal structures of the inward-open state of *A. thaliana* NRT1.1 revealed a series of essential features of the ten-transmembrane helix protein, as well as some remarkable differences that may be rationalised through differences in the respective experimental setup.[31,32] In NRT1.1, the substrate binding site is located in the centre of the protein, approximately halfway across the membrane in the inward-open state. The transporter has high affinity to nitrate and the molecule was consequently well defined in both crystal structures. NRT1.1 is expected to function as a monomer, similar to other MFS transporters, although the crystal form obtained by Zheng's group shows a suggestive dimeric arrangement in which the crucial residue T101, located at the cytoplasmic end of helix TM3, is positioned close to the two-fold symmetry axis at the dimer interface.[32] In spite of the availability of only a single conformational state of NRT1.1, an alternating-access mechanism for nitrate transport was postulated by the Newstead group that involves the symport of two protons, one of which is used for change compensation of the anionic substrate during the passage of the hydrophobic barrier of the lipid bilayer (Figure 12.2).[31] The binding site residue H356 provides a suitable base for a transient protonation mechanism (see Section 12.3.4), and the absence of a compensating positive charge provided by the protein itself supports this suggestion, although direct electrophysiological evidence is yet to be provided.

12.2.3 NO_3^-/H^+ Symport *vs.* NO_3^-/NO_2^- Antiport

With a growing body of data on integral membrane transport proteins, it is becoming increasingly evident that the mechanism of transport can vary widely, even within one and the same family of permeases. Hereby, even subtle changes can effect a switch from antiport to symport, or to active transport through coupling to proton translocation. In the dissimilatory pathway of denitrification, energy is conserved through the generation of a pmf that drives ATP synthesis. The reduction of nitrate yields nitrite and water, effectively consuming two protons. Placing this reaction in the periplasm thus diminishes the H^+ gradient by these two electrons, while a cytoplasmic localisation of nitrate reductase has the opposite effect and is energetically favourable.[1] This, however, requires the possibility of transporting nitrate anions across the membrane. The task is well handled by the NRT1/NPF family described above, but while the proton symport found for this protein is well suited for a nitrate-assimilating plant, it would neutralise the benefits of relocating the enzyme to the cytoplasm for a denitrifying organism. It is thus bioenergetically reasonable that the NRT2/NNP proteins found in conjunction with NarGHI reductases refrain from symporting protons. This mechanistic alteration should require structural changes in the protein, in particular with respect to the lack of a compensating charge provided by the proton itself. In addition, denitrificatory nitrate dissimilation to dinitrogen should ideally split the downstream reaction into multiple steps that can be individually coupled to proton translocation, but this means that the cytotoxic intermediates nitrite and NO will be generated, which are best kept out of the cytoplasmic space. For denitrificatory NRT2 proteins, this has led to the elegant solution of exchanging the substrate nitrate of the cytoplasmic nitrate reductase with the reaction product nitrite in an obligatory antiport mechanism that reliably avoids the risk of intracellular nitrite accumulation (Figure 12.1a).

12.2.4 The NNP Family

Like the NRT1/NPF family, the NRT2/NNP family of transport proteins is a member of the general superfamily of major facilitators. A paradigmatic representative of the NNP family is the NarK protein, which is commonly encoded in an operon context with the genes for the dissimilatory, membrane-associated nitrate reductase NarGHI that catalyses the initial step of denitrification (Figure 12.1a).[33] The transporter is widely distributed throughout the organismic world, with NRT2 proteins occurring in bacteria, archaea, yeast, fungi and plants.[17] *Escherichia coli* contains two copies of the entire *nar* operon, each complete with its proprietary transporter, termed NarK and NarU, respectively (Figure 12.1b). Hereby, NarU is part of a constitutive nitrate respiration system that is employed during periods of nutrient deprivation, while NarK and its corresponding nitrate reductase NarGHI are highly expressed only during anaerobic growth on nitrate.[33] NarK and NarU were shown to also act in nitrite extrusion,[34,35] but their primary role is in the exchange of nitrate for nitrite during denitrification.

12.2.5 Structural Features of Bacterial NarK and NarU

The orthologous transporters NarK (463 aa) and NarU (462 aa) from *E. coli* share 75% sequence identity, and the two proteins were simultaneously characterised by X-ray crystallography, providing the first glimpse of members of the NRT2/NNP family of transport proteins. Most conveniently, although both structures obviously represented the inward-open state, NarK was solved with bound nitrite,[36] while NarU was characterised in complex with nitrate.[37] Consequently, a comparison of the two close orthologues provided essential hints towards understanding the selectivity mechanism of nitrate/nitrite exchangers (see Section 12.2.6).[10] The architectures of NarK and NarU show the typical features of MFS proteins and many other transporters, in that the peptide chain consists of two inverted repetitions of six transmembrane helices that form two structurally similar domains, interrelated through a pseudo-two-fold symmetry axis in the plane of the lipid bilayer (Figure 12.3).[38] Both halves are connected by a loop region between helices TM6 and TM7 that was not defined in the electron density maps of the initial analysis, indicating high structural flexibility. Working according to the canonical alternating-access model (Figure 12.2), the hinge region for the conformational change should be located in close proximity to the substrate binding site, but as both structures were inward facing, no information was available on the remaining conformational states of the two NNP transporters.

The gaps in the mechanistic picture were partially filled by a recent re-evaluation of *E. coli* NarK by Fukuda and co-workers,[39] who were able to obtain three further structures of the protein to complement the already available nitrite-bound inward-open state that was obtained by using an antibody F_{ab}

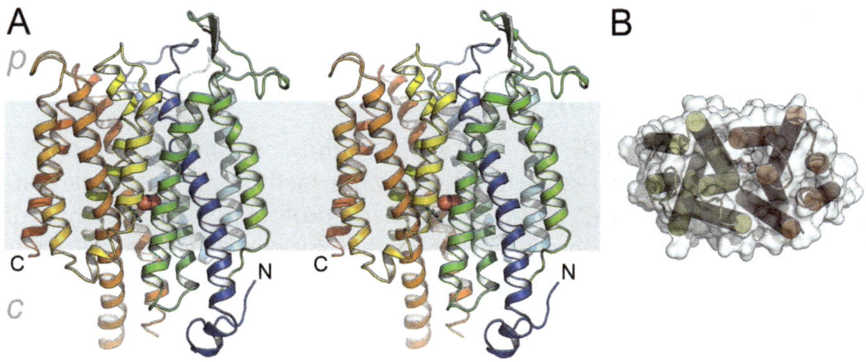

Figure 12.3 Structure of the nitrate/nitrite antiporter NarU from *Escherichia coli* (Protein Data Bank identifier 4IU8). (a) Stereo image of the protein in cartoon representation coloured from blue at the N-terminus to red at the C-terminus. Both termini are on the cytoplasmic side of the membrane. © 2013 Wiley-VCH Verlag GmbH & Co. KGaA, Weinheim.[10] (b) A top view of *E. coli* NarK from the periplasmic side. The transporter acts as a monomer, consisting of two pseudosymmetric half-channels (helices I–VI and VII–XII), as indicated by the colouring scheme.

fragment (Figure 12.4).[36] Among the available NarK structures, the inward-open apo form may serve as a reference for a resting state of the transporter, and a comparison with the nitrate-bound inward-facing form reveals only minor conformational changes (Figure 12.4a). This changes notably upon transition to an occluded state with nitrate being bound (Figure 12.4b), where the funnel-like opening to the cytoplasmic side is sealed, largely through a concerted movement of helices TM10 and TM11, while the rest of the protein structure remains virtually unchanged. As seen in other cases of MFS transporters, this emphasises that the "rocker-switch" model of alternative access does not simply imply a rigid-body movement of the two pseudosymmetric halves of the protein, but a rather subtler change that fulfils the structural requirements of closing off access to the transport channel from the cytoplasmic side. In a final structure, nitrite was found to bind to the very same binding site as nitrate (Figure 12.4c), and in this inward-open structure, the position of helices TM10 and TM11 is highly similar to the one in the nitrate-bound form (Figure 12.4a). With the slightly smaller ligand nitrite, however, the overall appearance of the transporter is more closed than with the larger nitrate, and this difference is reflected in the immediate vicinity of the central binding site.

12.2.6 Transport Mechanism of Nitrate/Nitrite Porters

The key mechanistic question regarding a nitrate/nitrite antiporter is how the two structurally very similar molecules NO_3^- and NO_2^- can be reliably distinguished at the molecular level to ensure that the former is exclusively

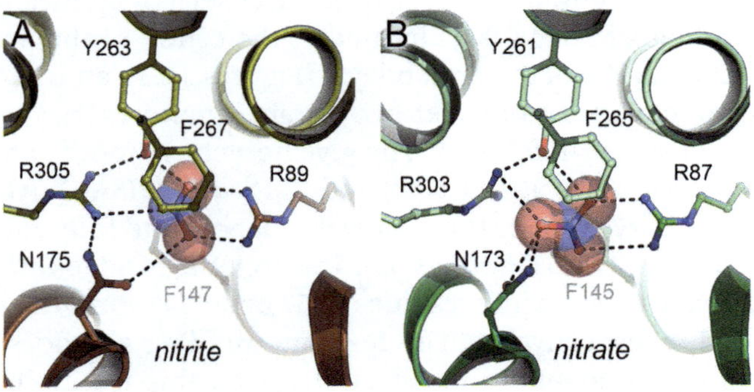

Figure 12.4 Binding of the cargo molecules in NarK and NarU. In the first crystal structures of the orthologous NarK and NarU that were published in parallel, the NarK structure that was obtained by co-crystallisation with an antibody F_{ab} fragment fortuitously contained a bound nitrite anion, while the NarU structure revealed the binding mode of nitrate. In conjunction, the two analyses showed the subtle differences that likely form the basis for substrate selectivity in NarK/U transporters. (a) In the nitrite-bound structure of NarK, a short (2.9 Å) hydrogen bond links the residues N175 and R305 that belong to two different halves of the protein. (b) Upon binding of nitrate, the presence of an additional oxygen atom in the cargo molecule disrupts this hydrogen bond, possibly allowing for a conformational transition of the transporter.

imported, while the translocation of the latter is always directed to the outside. Nitrate is larger than nitrite by a single oxygen atom, but it is anything but straightforward to rationalise how a binding site that is able to specifically bind nitrate should be able to exclude the nitrite molecule that—at the same total charge of −1—has its three atoms at almost precisely identical positions as three of the four atoms of the nitrate molecule. The situation is somewhat reminiscent of the case of the aminoacyl-tRNA synthetases, which face a similar challenge in exclusively coupling the correct amino acid residue to their corresponding tRNAs. These enzymes are the very point where the almost universal genetic code is actually realised, and thus must reliably discriminate even in cases such as serine and threonine that only differ by an exchange of a C_β proton for a methyl group. Threonyl-tRNA synthetase achieves this by employing two separate active sites: a coupling site recognises Thr, but is unable to exclude the highly similar Ser reliably, resulting in a non-negligible number of falsely loaded Ser-$tRNA^{Thr}$. Before leaving the enzyme, however, the aminoacylated 3′ end of the tRNA is scanned by a secondary active site in another domain of the protein that acts as a hydrolase. Its smaller access pathway now will only fit Ser, but not the larger Thr, resulting in the hydrolysis of Ser-$tRNA^{Thr}$, but not of correctly loaded Thr-$tRNA^{Thr}$.[40]

For a transporter such as NarK, this type of selectivity mechanism is not an option for a variety of reasons, including the detrimental effect on transport speed. With the structural data available (Figure 12.4), the exact binding modes of the monovalent anions nitrate and nitrite were revealed for the inward-facing conformations of NarK and NarU,[36,37,41] and their direct comparison provides essential clues regarding the selectivity mechanism. The substrate-binding site of NarK is located at the centre of the protein, quite precisely in the middle of the lipid bilayer (Figures 12.3b and 12.4). Early on, two conserved sequence motifs, termed nitrate signatures (NS), were defined as hallmarks of the NNP family.[42] The glycine-rich NS motifs are located in helices TM5 (^{164}GGALGLNGGLGN^{175}) and TM11 (^{408}GFISAIGAIGGFF^{420}) and form a central cavity lined by the positively charged and fully conserved R89 in TM2 and R305 in TM8 (Figure 12.5a). The arginine residues are juxtaposed and bind the substrates, while residues F267 and F147 form the bottom and lid of the cavity, respectively.[36] The resulting binding site does not have a rigid structure, but adapts to its substrates, and this feature enables it to convey reliable selectivity for nitrate and nitrite. For nitrite binding, R89 forms a tight salt bridge with the oxygen atoms of the NO_2^- molecule, while R305 is involved in a tight hydrogen-binding network with N175 and F267 that stabilise nitrite from the other side (Figure 12.5a). If nitrite is exchanged for nitrate, however, the salt bridge to R89 (R87 in NarU) remains, so that the additional oxygen atom protrudes on the other side, displacing not only the hydrogen-bonding interaction with R305 (R303 in NarU) itself, but also forcing apart R89 and the asparagine 175 (N173 in NarU) (Figure 12.5b). Within the binding pocket, this hydrogen bond is the only direct interaction between polar residues from the different halves of the transporters, and its

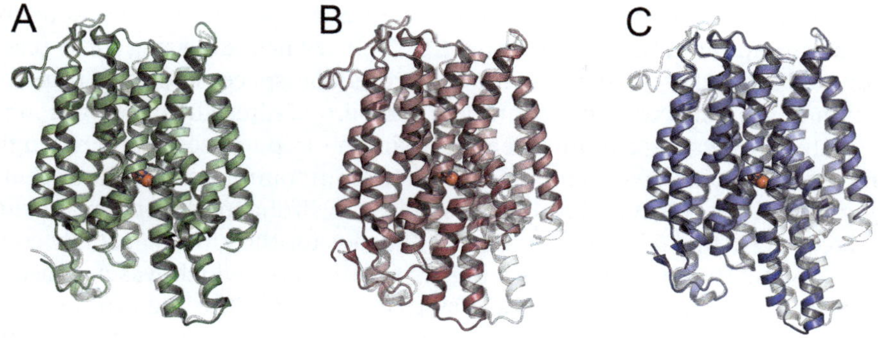

Figure 12.5 Known structures of the conformational states of *Escherichia coli* NarK. All panels show the inward-open apo form of the transporter in light grey (Protein Data Bank [PDB] identifier 4U4V) as a reference. (a) The nitrate-bound, inward-facing form (PDB identifier 4U4T) only shows minor differences from the apo state. (b) Upon transition to the nitrate-bound occluded state (PDB identifier 4U4W), helices TM10 and TM11 move inward to block access from the cytoplasmic side. (c) The inward-open nitrite-bound state (PDB identifier 4JRE) resembles the nitrate-bound form (a), but is slightly more closed, possibly due to the smaller ligand in the central binding pocket. © 2013 Wiley-VCH Verlag GmbH & Co. KGaA, Weinheim.[10]

disruption may be what makes a conformational change of the entire molecule possible. The polar residues of the binding pocket convey a degree of structural flexibility that not only allows for telling nitrite from nitrate, but also elegantly links the molecular recognition process to the transition of a different conformational state in an alternating-access mechanism.[10]

12.3 Nitrite Transport

As an intermediate of a series of metabolic pathways in the biogeochemical nitrogen cycle, the nitrite anion, NO_2^-, is constantly produced and consumed by different enzymatic reactions that can be assimilatory or dissimilatory by nature.

12.3.1 The Metabolic Role of Nitrite

The nitrite anion is characterised by a marked cytotoxicity in the cytoplasm, so that most denitrifiers employ mechanisms to prevent the accumulation of the molecule inside the cell. At the same time, it constitutes a metabolic centrepiece of the biogeochemical nitrogen cycle, where it is generated as the product of nitrate reduction in denitrification,[1] but also as the oxidation product of hydroxylamine, H_2NOH, in nitrification.[5] Two dissimilatory pathways emerge from nitrite, namely the one-electron reduction to NO in the further course of denitrification, or a single six-electron step catalysed by cytochrome *c* nitrite reductase in the ammonification pathway that directly yields ammonium, NH_4^+, as a product.[43] The enzymes catalysing these reactions are limited to the periplasmic or extracellular space, so that no translocation of the ion across the membrane is required. Alternatively, nitrite may be utilised as a source of nitrogen for assimilatory purposes, again through the conversion in a six-electron reduction to ammonium, but this time occurring directly in the cytoplasm catalysed by the sirohaem-dependent nitrite reductase, NirBD.[6] The *nir* operon that encodes for the enzyme usually contains a gene for an integral membrane protein, *nirC*, which was suggested early on to act as a specific uptake system for nitrite into the cytoplasm.[44] NirBD has evolved primarily for nitrogen assimilation from nitrite and offers a possibility to divert excess substrate from the energy metabolism of denitrification towards assimilation, including the nucleotide and protein biosynthesis that is required for cell proliferation. In the enteropathogenic γ-proteobacterium *Salmonella typhimurium*, an additional role for the Nir system was discovered recently. *S. typhimurium* has the ability to proliferate in the cytoplasm of macrophages that constitute a primary line of defence within the mammalian system of innate immunity. In order to survive inside their aggressive host cells, the bacteria must counteract the cytotoxins nitrite and peroxynitrite, $ONOO^-$, that macrophages generate from NO and peroxide or superoxide, the reaction products of inducible NO synthase and xanthine dehydrogenase, respectively.[45–47] Nitrogen assimilation seems less of an issue in the nutrient-rich environment of the mammalian gut, but the NiR system also serves very well as a defence mechanism by simply taking up nitrite or peroxynitrite for efficient cytoplasmic detoxification to ammonium and water by NirBD. In fact, a Δ*nirC* strain of *S. typhimurium* was strongly inhibited in its ability to proliferate in murine macrophages, while cellular NO levels were increased.[48]

12.3.2 The Formate/Nitrite Transporter Family of Ion Channels/Transporters

The pentameric NirC protein forms part of the family of formate/nitrite transporters (FNTs),[49] constituting transporter class 2.A.44 in the Saier nomenclature.[50] FNT proteins are widespread among enteric bacteria such as *Escherichia* and *Salmonella* species that may be either commensural or pathogenic. The first member of this family to be described in detail was the formate channel/transporter FocA that plays a central role in mixed-acid fermentation.[51] In the oxygen-limited but nutrient-rich environment of the mammalian gut, which is a common habitat for enterobacteria, this pathway is preferred in spite of its excessive and incomplete use of glucose, but the release of different acids, including lactate, acetate, succinate and formate, leads to a successive acidification of the medium that eventually becomes prohibitive for further bacterial proliferation.[51] Mixed-acid fermentation branches off the central metabolism in the reaction of pyruvate:-formate lyase, PflB, which generates formate and acetate in the cytoplasm from the glycolysis product pyruvate,[52] rather than following the oxidative carboxylation to acetyl-CoA and CO_2 catalysed by the mitochondrial pyruvate dehydrogenase complex.[53] In the normal metabolic state, the formate anion produced in the cytoplasm is exported by passive channelling *via* the anion channel FocA and is subsequently oxidised to CO_2 by the periplasmic formate dehydrogenase FdhABC.[54] However, once the continued release of acids causes a drop in the pH value of the surrounding medium below pH 6.8,[51] the FocA protein undergoes a striking functional switch from passive export to active import, reflected in a conformational change in the N-termini of the pentamer that faces the cytoplasmic side.[55–57] In this mode, formate is actively imported in a proton-driven symport mechanism to then undergo disproportionation to CO_2 and H_2 on the formate:hydrogen lyase complex at the cytoplasmic face of the membrane. Instead of acidifying their environment, the bacteria thus dispose of excess electrons and protons as H_2 gas.[58]

Additional orthologues of FocA were subsequently identified in the context of different bacterial operons, such as FdhC in the *fdh* operon that encodes formate dehydrogenase of *Methanobacterium formicium*[59] and the *nir* operon encoding the assimilatory nitrite reductase of *E. coli*.[33] To date, structural information is available for FocA from *E. coli*,[57] *Vibrio cholerae*[56] and *Salmonella typhimurium*,[55] as well as for the nitrite channel NirC from the same organism.[60] The most recent structural characterisation was of a hydrosulphide channel termed HSC (or AsrD) from the *asr* operon encoding assimilatory sulphite reductase.[61]

The FNT family thus constitutes a structurally conserved family of channels and transporters for monovalent anions. Functional studies of isolated proteins reconstituted in planar lipid bilayers revealed a rather low specificity for FocA, allowing it to serve as a universal exporter for the entire product range of mixed-acid fermentation.[62] Similarly, NirC was not only shown to transport nitrite anions, but also the chemically similar nitrate.[60,63]

12.3.3 The Nitrite Channel NirC

A three-dimensional structure of a NirC channel was obtained for the *S. typhimurium* protein at a resolution of 2.4 Å by X-ray crystallography.[60] NirC is a polytopic membrane protein with seven transmembrane helices per protomer (Figure 12.6a). In the crystal structure, almost the entire protein chain was well defined in the electron density maps, including in particular an N-terminal α-helix whose structural flexibility was an important mechanistic property of the functional switch from channel to transporter in FocA at low pH.[55] The same helix in NirC is tightly attached to the core domain of the protein *via* a series of hydrogen-bonding interactions, in line with the absence of a comparable functional switch or an alteration of the channel functionality at low pH.[60] The protein forms the pentamer that is observed for all members of the family, with an apparent transmembrane conduct in the central position that is not relevant for anion translocation, as it is blocked by detergent molecules *in vivo* (Figure 12.6b). This leaves a distinct

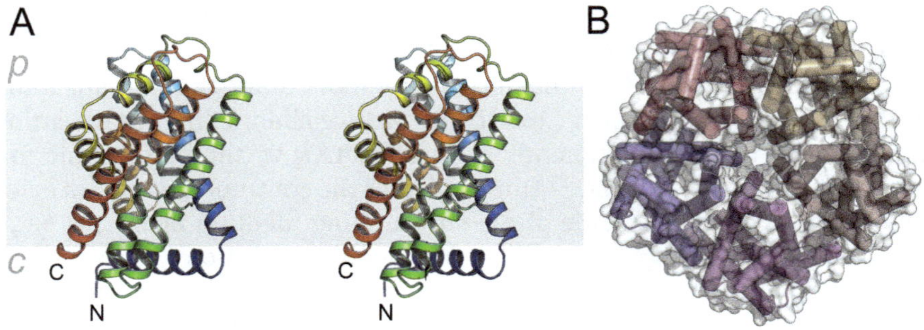

Figure 12.6 Structure of the nitrite channel NirC from *Salmonella typhimurium*. (a) A stereo representation of the monomer with the position of the membrane and its periplasmic (p) and cytoplasmic (c) side. NirC has both termini in the cytoplasm and its fold resembles those of aquaporins. (b) In contrast to the tetrameric structures of aquaporins, NirC forms a stable pentamer, but its electrophysiological transport characteristics do not provide evidence for cooperative transport.

channel in each protomer that shows the same hydrophobic constrictions around a central vestibule that was observed in the structure of FocA.[49]

12.3.4 The Transport Mechanism of NirC

The enterobacterial NirC channel is typically organised in an operon structure *nirBDCcysG*, together with the actual nitrite reductase NirBD and the CysG protein involved in the biogenesis of the cofactor sirohaem.[44] Cole and co-workers investigated the function of *E. coli* NirC *in vivo* and found the membrane protein to function as a passive anion channel, but possibly also as an active transporter by symporting nitrite with H^+.[34] The crystal structure of the *S. typhimurium* orthologue then revealed the precise architecture of the conducting channel in the NirC protomer that, in conjunction with the available *in vivo* data and electrophysiological studies on the protein reconstituted in planar lipid bilayers, allowed for the postulation of a transport mechanism in molecular detail.[49,60] Similar to other known members of the FNT family, the ion channel in NirC is tripartite, with funnel-like entrances from either side of the membrane terminating in tight constrictions consisting of rings of hydrophobic residues (Figure 12.7a). While these constrictions do not fully seal the transport pathway, the remaining passage is too narrow for even a water molecule, and it also prevents the conduction of protons as shown by NirC-containing proteoliposomes that retain a transmembrane proton gradient.[49,63] The structure of NirC—similar to the other available structures of FNT channels—thus likely represents a closed resting state that must undergo structural rearrangements in order to conduct any type of cargo. The two constrictions in the channel enclose a central hydrophobic vestibule of NirC in the very centre of the lipid bilayer, and the conserved arrangement includes three distinct, protonable sites that are very

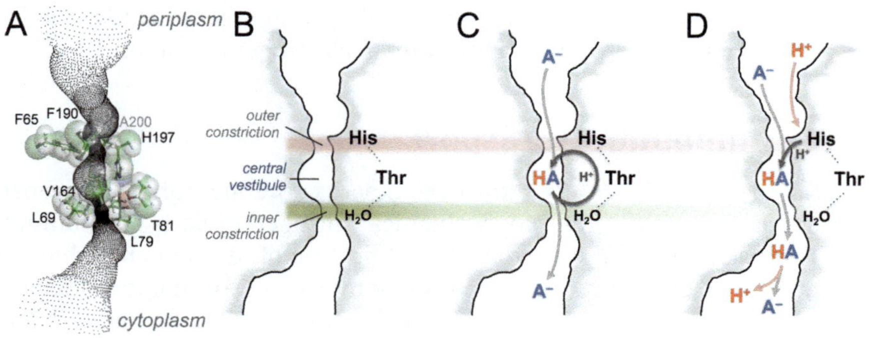

Figure 12.7 Ion channel and transport mechanism of NirC. (a) Each protomer of the nitrite channel NirC holds a narrow transport channel that is lined by hydrophobic residues, with the exception of a histidine and a threonine. (b) Within the channel, two hydrophobic constrictions separate a central vestibule from the environment. (c) This arrangement suggests a transient protonation that helps anions to overcome the hydrophobic barriers. The mechanism involves the cycling of a proton that is not transported, and this is in line with observed short (ion transport) and long (loss of the proton) bursts of gating. (d) Transient protonation is also in line with a facultative anion/H^+ symport mechanism that would switch the function of the protein from passive channelling to active transport, as observed in the pH-dependent gating of the formate channel/transporter FocA.

likely of mechanistic relevance for transport. These are a strictly conserved histidine (H197 in *S. typhimurium* NirC), a threonine (T81) and a tightly coordinated water molecule that constitutes a rather exceptional finding in the most hydrophobic region of the membrane (Figure 12.7b). In line with electrophysiological data that show the occurrence of a fast and a slow gating mode,[60] this arrangement suggests a transport mechanism that involves an internal proton relay across the three conserved sites (Figure 12.7c). In this model, anions can readily enter the charged funnels leading towards the hydrophobic constrictions. In NirC, the electrostatic surface charge distribution seems to favour anion import over export, in line with the physiological role of the protein.[60] In order to cross (and open) the hydrophobic constriction, however, the nitrite anion must be protonated to the uncharged form, which can be achieved *via* H197 (Figure 12.7c). In this state, the cargo can cross the central vestibule as well as the distal constriction, and here the charged environment of the exit funnel promotes deprotonation. The proton itself can then be transferred back to H197 *via* the H_2O molecule and T81 to be ready for the next anion, resulting in a net translocation of only the nitrite anion (Figure 12.7c). Interestingly, this mechanism can be seamlessly transferred into an active proton symport mechanism as found in FocA channels if the proton is co-transported instead of recycled and subsequently replenished from the *p*-side of the membrane (Figure 12.7d). A NO_2^-/H^+ symport mechanism would be electroneutral and therefore cannot be investigated by electrophysiological techniques, but Fendler and co-workers could show by solid-supported membrane electrophysiology that at least *in vitro* NirC is able to act as a nitrite/proton antiporter in an electrogenic fashion.[63]

12.4 Ammonium Transport

The final ionic species to get translocated across the membrane bilayer in the reaction network of the biogeochemical nitrogen cycle is ammonium, NH_4^+. It constitutes the most reduced modification of the element, which in this state can be directly assimilated into biomolecules for biosynthetic purposes, so that it serves as an important nutrient for bacteria and plants.[64] The availability of nitrogen frequently becomes the limiting factor for organismic growth, and a dedicated uptake system for the ion is thus almost ubiquitously found throughout all kingdoms of life.

12.4.1 Amt/Rh/Mep Transporters

Ammonium transport proteins are integral membrane proteins that form the Amt/Rh/Mep family of transporters, as they include the mammalian Rhesus (Rh) proteins and the fungal methylammonium permeases (Mep).[64] In spite of their wide distribution and evolutionary spread, the core structure of this class of proteins is highly conserved, consisting of two sets of five transmembrane helices related by a pseudo-two-fold symmetry axis in the membrane plane that are stabilised by a long 11th helix (Figure 12.8a). Amt proteins form stable trimers in the membrane that can be isolated after detergent solubilisation and do not dissociate fully, even after SDS treatment for gel electrophoresis.[65] Bacterial Amt homologues are frequently found in an operon context together with a small regulatory protein of

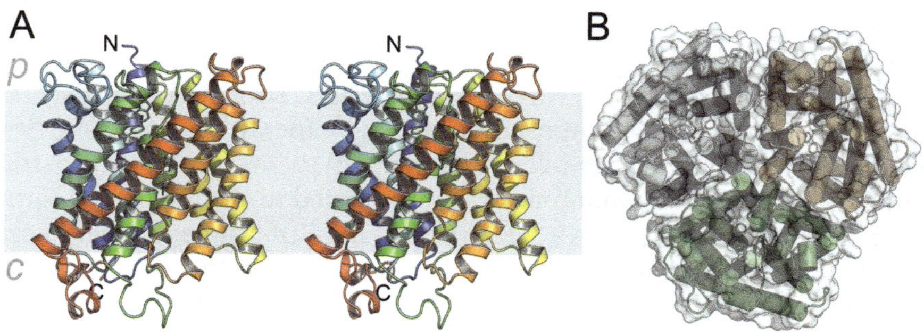

Figure 12.8 Structure of the ammonium transporter Amt-1 from *Archaeoglobus fulgidus*. (a) Stereo view of the Amt-1 monomer, highlighting the diagonal helix XI that stabilises the two symmetry-related halves of the integral membrane protein. (b) Ammonium transporters form stable trimers in the membrane, with distinct transport channels in each individual protomer. The arrangement of the transmembrane segment of this class of transporters is highly conserved throughout all domains of life.

the P_{II} family termed GlnK.[66,67] P_{II} proteins are able to respond to the cellular energy and nitrogen status *via* the effectors 2-oxoglutarate and ATP, and block Amt-mediated uptake of reduced nitrogen under high-nitrogen/low-energy conditions by the formation of a tight complex with the transporter.[68,69] The binding of ATP and ADP to GlnK proteins is cooperative,[70] and the binding of 2-oxoglutarate and MgATP leads to a conformational change that prevents the regulator from fitting its extended T-loops into the exit channels of the Amt trimer.[71] P_{II} proteins thus are finely tuned signal integrators for the metabolic state of the cell, and their importance for Amt functionality lies in their ability to block the active uptake of NH_4^+ when the level of 2-oxoglutarate, the immediate acceptor of NH_4^+, is low. Continuous ammonium transport in this case would result in intracellular accumulation of the cation, which on one side is cytotoxic by itself, and on the other side would eventually generate a backflow of uncharged ammonia, NH_3, across the membrane. With its high permeability, NH_3 would establish a futile cycle that would lead to a deterioration of the proton gradient across the cytoplasmic membrane without the concomitant benefit of taking up reduced nitrogen species.[71]

12.4.2 Passive *vs.* Active Transport of Reduced Nitrogen

With respect to cellular ammonium uptake, the seemingly small difference between NH_4^+ and NH_3 as a transported species has major functional consequences. Uncharged ammonia is a hydrophobic gas with a permeability coefficient $P_d = 10^{-3}$ cm s^{-1}, which is very similar to the value for pure water.[72] It has long been debated whether a dedicated transport system for ammonium across biological membranes would be required at all or whether passive diffusion might suffice, as it does in the cases of N_2, O_2 or H_2. However, several studies pointed towards an active mechanism for cellular ammonium uptake as early as the 1970s and 1980s.[72,73] It would be required at least at low external concentrations of ammonium, where passive diffusion of ammonia would not be sufficient for the cells.[74]

Historically, the functionality of Amt family proteins was monitored by following the uptake of ^{14}C-labelled methylammonium, $H_3CNH_3^+$, a presumably non-physiological alternative cargo molecule that was suggested to behave analogously to NH_4^+.[74] For yeast Mep proteins, an active and electrogenic uptake mechanism was found,[75,76] and the same was the case for the Amt proteins of *Arabidopsis thaliana*,[77] further supported by voltage-clamp electrophysiology on the protein produced in *Xenopus laevis* oocytes, in which the tomato protein LeAMT1; 1 and the human RhBG glycoprotein were studied.[78] This mechanistic hypothesis was challenged by Kustu and co-workers, who studied the Amt proteins of *E. coli*[79] and *S. typhimurium*[80] and did not observe methylammonium accumulation in the cytoplasm. They concluded that Amt proteins must be passive facilitators for NH_3 and termed the proteins ammonia channels rather than ammonium transporters.[80]

12.4.3 Structures of Ammonium Transporters

Based on the information that Amt proteins form trimers of extraordinary stability,[65] the transport protein family became an interesting target for structural analysis. Within a few years, a low-resolution structure obtained by cryo-electron microscopy and atomic force microscopy[81] was followed by two structures of *E. coli* AmtB solved independently by two laboratories[82,83] and the structure of Amt-1 from the hyperthermophilic archaeon *Archaeoglobus fulgidus*.[84] Subsequently, the structures of the Rh50 protein from *Nitrosomonas europaea*[85,86] and the human RhCG protein[87] shed light on the architecture of the Rh protein branch of the Amt/Mep/Rh family. All structures revealed the strict conservation of the transmembrane core of the family, including most of the amino acid residues identified as relevant for cation transport. In particular, these residues include a postulated recruitment site for ammonium cations on the extracellular face of the transporter that consists of a serine and a tryptophan residue. In conjunction, this arrangement can select for molecules that are both cationic (through cation-π interaction with Trp) and are able to form H-bonds (to Ser), thus efficiently excluding the similarly sized K^+ and H_2O, but allowing for the passage of ammonium and methylammonium.[64] Immediately below this recruitment site, the transport channel is sealed by two phenylalanine residues (the "*Phe-gate*") that must have some degree of conformational flexibility before it opens into a channel that leads to the cytoplasmic side and is lined by two histidines (the "*twin-His motif*").[64,88] The remarkable arrangement of these two residues that form a direct H-bonding interaction in all structures where they are present raised the possibility that here the transported NH_4^+ cation could be deprotonated, leading to an effective NH_3/H^+ symport after initial recruitment of NH_4^+ on the extracellular side.[64,84,89,90]

With the obtaining of this structural information, the question of the mechanism and cargo of ammonium transporters gained new urgency. After the first structural analysis of *E. coli* AmtB in 2004, Stroud and co-workers adopted the view of Kustu and spoke about "ammonia transport", presenting a functional assay in liposomes that showed internal basification upon AmtB-mediated uptake.[82] Although this experiment was later criticised,[91] the idea of an "ammonia channel" then spread rapidly and was reiterated by multiple groups, based largely on the finding that the conduction pore in the protein structures was perceived to be hydrophobic in character.[83,92–94] The idea of NH_3 channelling (or uniport), however, is bioenergetically unreasonable and was contested early on by us and others,[64,84,90] and the reason has largely caught on within the field over time.[91] Direct experimental evidence was nevertheless only obtained in the oocyte system, but not in a defined *in vitro* system with isolated protein reconstituted into an artificial membrane.

The important regulatory feature of complex formation between Amt proteins and the regulatory, trimeric GlnK proteins was also significantly advanced through structural studies. Already, the obvious complementarity between the Amt trimer and the GlnK proteins strongly suggested a mode of interaction along a shared three-fold axis of both components, as was readily postulated based on the structure of the *A. fulgidus* Amt-1.[84] Shortly thereafter, two groups presented crystal structure analyses of the AmtB–GlnK complex of *E. coli* that precisely verified this model, with the regulatory T-loops of the GlnK proteins inserting deeply into the substrate exit channels of the transporters.[69,95] An analysis of the structural changes induced by the binding of ATP and 2-oxoglutarate to *A. fulgidus* GlnK-2 then completed the picture by revealing that effector binding leads to an outward movement of the T-loops to a point that they are sterically hindered from inserting into the exit channels of the Amt trimer.[96]

12.4.4 Electrogenic NH_4^+ Transport by Amt Proteins

In spite of the unusually high resolution obtained with the crystal structures of different Amt proteins, the question concerning the nature of the transported substrate remained elusive, as the number of electrons, and thus the appearance in an electron density map, is exactly identical for NH_3, NH_4^+ and H_2O. Experimental evidence had been presented to support either electroneutral or electrogenic transport (see Section 12.4.2), and no direct data had been obtained from isolated proteins under defined *in vitro* conditions. The method of choice for the characterisation of electrogenic transport across biological membranes is direct electrophysiology, with its ability to immediately quantify the charges translocated through a given patch of protein-containing membrane. This can be carried out in planar lipid bilayers, as in the case of the nitrite channel NirC (see Section 12.3.3), yielding precise information about the actual conductivity of the system and its gating properties.[60] However, while isolation and reconstitution for Amt proteins is well established, no successful current measurements in planar lipid bilayers were obtained. This could have meant either that transport was not electrogenic in the first place or that indeed the transport rates for Amt proteins were slower than those to be expected for a channel. In order to specifically address the low conductance of ion transporters, further studies were carried out by solid-supported membrane electrophysiology.[97,98] Here, transport proteins are reconstituted into proteoliposomes that are then adsorbed in a reaction cuvette (Figure 12.9a) into a hybrid bilayer consisting of a thioalkane (Figure 12.9b, T) attached to a gold electrode (G) that is complemented by a lipid monolayer (L). The setup in the flow-through cuvette can be perfused with different solutions containing the transported ion (active) or not (non-active), and as it accommodates many proteoliposomes, the measured current is integrated over a large number of individual transporters, so that systems with low conductance become accessible (Figure 12.9a and b). For *Af*Amt-1, this approach was successful and allowed for the recording of specific currents of different substrates (Figure 12.9c). Here, the observed peak currents provided clear evidence that ammonium and methylammonium are transported species, while dimethylammonium and the monovalent cations Na^+ and K^+ are not (Figure 12.9d).[99] Amt-mediated transport thus is an

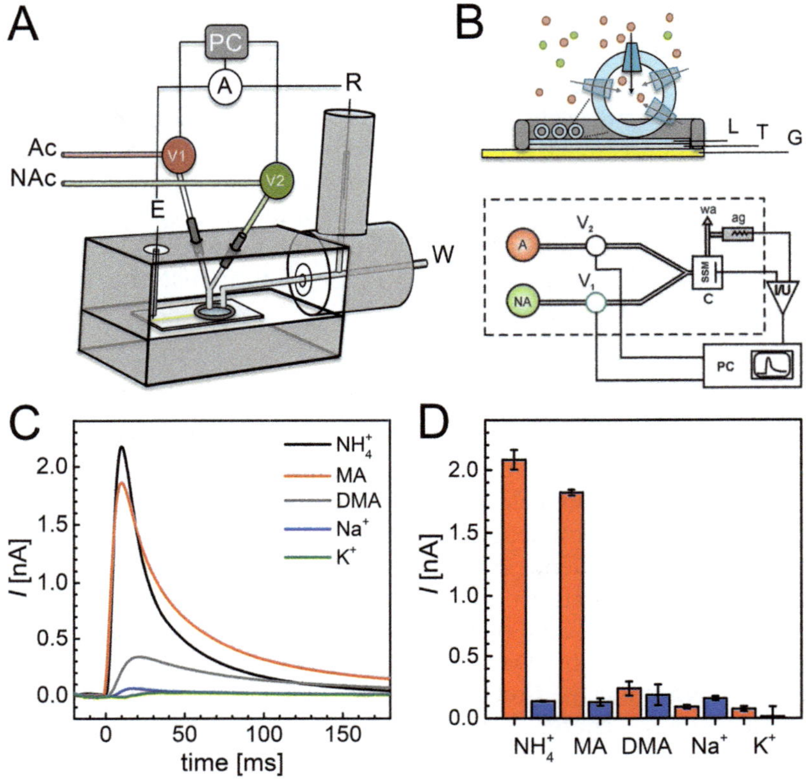

Figure 12.9 Electrogenic ammonium transport and mechanism of Amt proteins investigated by solid-supported membrane (SSM) electrophysiology. (a) Flow-through cuvette assembly for SSM measurements. A measurement electrode (E) and a reference (R) are connected to an amperemeter (A) and a computer (PC). The cuvette can be perfused with active (Ac) or non-active (NAc) solutions, devoid of or containing the ions in question, respectively. After passing the adsorbed liposomes, the solutions drain to a waste line (W). (b) Above: in the cuvette, proteoliposomes are adsorbed to a hybrid bilayer consisting of an outer leaflet of lipids (L) and an inner leaflet of octadecane thiol (T) bound to a gold electrode (G). Below: schematics for the cuvette (C) as shown in (a). (c) Transient currents recorded in a SSM experiment with proteoliposomes containing *Af*Amt-1. Active solutions with 300 mM of different cations evoke substantially different currents depending on the conductivity of the transporter for the respective species. (d) Evaluation of the experiment shown in (c) showing transient current maxima (red) and background currents (blue) for each ionic species. *Af*Amt-1 shows similar conductance for ammonium and methylammonium (MA), while dimethylammonium (DMA), sodium and potassium ions are efficiently excluded.

electrogenic uniport of NH_4^+, in line with the general bioenergetics considerations made above. This does not exclude an internal deprotonation of the cargo and a separate pathway for NH_3 and H^+ within the transport channel, but it rules out passive channelling of ammonia gas. Further studies will be required in order to show whether the electrogenic transport mechanism observed for the archaeal Amt-1 also holds true for other members of the Amt/Rh/Mep family of transport proteins.

12.5 Conclusions

The coupling of the enzymatic reactions of the biogeochemical nitrogen cycle to the cytoplasmic membrane is a necessary consequence of the requirement for storing the free enthalpy released in the catalytic sequence in the form of an electrochemical potential for energy conservation and ATP synthesis. Three ionic species—nitrate, nitrite and ammonium—require specific transport mechanisms, and the proteins known to perform these tasks span the entire range of known types of transporters. In the canonical denitrification pathway, only the reaction of the membrane-associated nitrate reductase NarGHI is located in the cytoplasm, while the three subsequent enzymatic steps occur in the periplasm. Only one family of nitrate transport proteins, the NRT2/NNP transporters, with its paradigmatic member NarK, have evolved to optimally support this sequence by stoichiometrically coupling the import of nitrate from the periplasm to the extrusion of the toxic product nitrite. This allows for the consumption of protons that accompanies the release of O^{2-} from nitrate to take place on the cytoplasmic side of the membrane, so that the reaction effectively adds to the transmembrane proton gradient. In contrast, the closely related NRT1/NPF family comprises nitrate uptake transporters and utilises a proton symport mechanism for a secondary active import of the anion. The structures of the two MFS transporters reveal characteristic differences in their anion binding sites that help to rationalise this remarkable mechanistic switch. If nitrite is allowed access to the cytoplasm, the metabolic context is assimilatory. The nitrite channel NirC is co-expressed with the cytoplasmic sirohaem nitrite reductase, and the underlying principle here is the prevention of an intracellular accumulation of nitrite as well. Finally, the ideal substrate for nitrogen assimilation is always ammonium that is directly fixed as glutamine or glutamate, and consequently the family of ammonium transporters is virtually ubiquitous.

References

1. O. Einsle and P. M. H. Kroneck, *Biol. Chem.*, 2004, **385**, 875.
2. G. J. Leigh, *The World's Greatest Fix: A History of Nitrogen in Agriculture*, Oxford University Press, Oxford, 2004.
3. V. Smil, *Nature*, 1999, **400**, 415.
4. D. E. Canfield, A. N. Glazer and P. G. Falkowski, *Science*, 2010, **330**, 192.
5. R. M. Martinez-Espinosa, J. A. Cole, D. J. Richardson and N. J. Watmough, *Biochem. Soc. Trans.*, 2011, **39**, 175.
6. B. C. Berks, S. J. Ferguson, J. W. B. Moir and D. J. Richardson, *Biochim. Biophys. Acta, Bioenerg.*, 1995, **1232**, 97.
7. P. Arnoux, M. Sabaty, J. Alric, B. Frangioni, B. Guigliarelli, J. M. Adriano and D. Pignol, *Nat. Struct. Biol.*, 2003, **10**, 928.

8. U. H. Hole, K. U. Vollack, W. G. Zumft, E. Eisenmann, R. A. Siddiqui, B. Friedrich and P. M. H. Kroneck, *Arch. Microbiol.*, 1996, **165**, 55.
9. P. Pacher, J. S. Beckman and L. Liaudet, *Physiol. Rev.*, 2007, **87**, 315.
10. S. L. A. Andrade and O. Einsle, *Angew. Chem., Int. Ed.*, 2013, **52**, 10422.
11. S. Noji, T. Nohno, T. Saito and S. Taniguchi, *FEBS Lett.*, 1989, **252**, 139.
12. J. A. DeMoss and P. Y. Hsu, *J. Bacteriol.*, 1991, **173**, 3303.
13. N. Lane and W. F. Martin, *Cell*, 2012, **151**, 1406.
14. I. Bernhardt, J. C. Ellory, J. S. Gibson and R. J. Wilkins, *Adv. Planar Lipid Bilayers Liposomes*, 2009, **10**, 1.
15. S. Noskov, *Biophys. J.*, 2011, **100**, 6.
16. B. G. Forde, *Biochim. Biophys. Acta, Biomembr.*, 2000, **1465**, 219.
17. S. S. Pao, I. T. Paulsen and M. H. Saier, *Microbiol. Mol. Biol. Rev.*, 1998, **62**, 1.
18. T. Omata, *Plant Cell Physiol.*, 1995, **36**, 207.
19. P. Nacry, E. Bouguyon and A. Gojon, *Plant Soil*, 2013, **370**, 1.
20. Y. F. Tsay, C. C. Chiu, C. B. Tsai, C. H. Ho and P. K. Hsu, *FEBS Lett.*, 2007, **581**, 2290.
21. Y. Y. Wang, P. K. Hsu and Y. F. Tsay, *Trends Plant Sci.*, 2012, **17**, 458.
22. S. Léran, K. Varala, J. C. Boyer, M. Chiurazzi, N. Crawford, F. Daniel-Vedele, L. David, R. Dickstein, E. Fernandez, B. Forde, W. Gassmann, D. Geiger, A. Gojon, J. M. Gong, B. A. Halkier, J. M. Harris, R. Hedrich, A. M. Limami, D. Rentsch, M. Seo, Y. F. Tsay, M. Y. Zhang, G. Coruzzi and B. Lacombe, *Trends Plant Sci.*, 2014, **19**, 5.
23. Y. F. Tsay, J. I. Schroeder, K. A. Feldmann and N. M. Crawford, *Cell*, 1993, **72**, 705.
24. N. C. Huang, C. S. Chiang, N. M. Crawford and Y. F. Tsay, *Plant Cell*, 1996, **8**, 2183.
25. R. C. Wang, D. Liu and N. M. Crawford, *Proc. Natl. Acad. Sci. U. S. A.*, 1998, **95**, 15134.
26. K. H. Liu, C. Y. Huang and Y. F. Tsay, *Plant Cell*, 1999, **11**, 865.
27. K. H. Liu and Y. F. Tsay, *EMBO J.*, 2003, **22**, 1005.
28. G. Krouk, B. Lacombe, A. Bielach, F. Perrine-Walker, K. Malinska, E. Mounier, K. Hoyerova, P. Tillard, S. Leon, K. Ljung, E. Zazimalova, E. Benkova, P. Nacry and A. Gojon, *Dev. Cell*, 2010, **18**, 927.
29. S. Munos, C. Cazettes, C. Fizames, F. Gaymard, P. Tillard, M. Lepetit, L. Lejay and A. Gojon, *Plant Cell*, 2004, **16**, 2433.
30. C. H. Ho, S. H. Lin, H. C. Hu and Y. F. Tsay, *Cell*, 2009, **138**, 1184.
31. J. L. Parker and S. Newstead, *Nature*, 2014, **507**, 68.
32. J. Sun, J. R. Bankston, J. Payandeh, T. R. Hinds, W. N. Zagotta and N. Zheng, *Nature*, 2014, **507**, 73.
33. S. Clegg, F. Yu, L. Griffiths and J. A. Cole, *Mol. Microbiol.*, 2002, **44**, 143.
34. W. J. Jia, N. Tovell, S. Clegg, M. Trimmer and J. Cole, *Biochem. J.*, 2009, **417**, 297.
35. J. J. Rowe, T. Ubbink-Kok, D. Molenaar, W. N. Konings and A. J. M. Driessen, *Mol. Microbiol.*, 1994, **12**, 579.
36. H. Zheng, G. Wisedchaisri and T. Gonen, *Nature*, 2013, **497**, 647.

37. H. Yan, W. Huang, C. Yan, X. Gong, S. Jiang, Y. Zhao, J. Wang and Y. Shi, *Cell Rep.*, 2013, **3**, 716.
38. S. Radestock and L. R. Forrest, *J. Mol. Biol.*, 2011, **407**, 698.
39. M. Fukuda, H. Takeda, H. E. Kato, S. Doki, K. Ito, A. D. Maturana, R. Ishitani and O. Nureki, *Nat. Commun.*, 2015, **6**, 7097.
40. A. C. Dock-Bregeon, R. Sankaranarayanan, P. Romby, J. Caillet, M. Springer, B. Rees, C. S. Francklyn, C. Ehresmann and D. Moras, *Cell*, 2000, **103**, 877.
41. M. Fukuda, H. Takeda, H. E. Kato, S. Doki, K. Ito, A. D. Maturana, R. Ishitani and O. Nureki, *Nat. Commun.*, 2015, **6**, 7097.
42. L. J. Trueman, A. Richardson and B. G. Forde, *Gene*, 1996, **175**, 223.
43. O. Einsle, *Methods Enzymol.*, 2011, **496**, 399.
44. T. Peakman, J. Crouzet, J. F. Mayaux, S. Busby, S. Mohan, N. Harborne, J. Wootton, R. Nicolson and J. A. Cole, *Eur. J. Biochem.*, 1990, **191**, 315.
45. Q. Xie and C. Nathan, *J. Leukocyte Biol.*, 1994, **56**, 576.
46. P. J. Brett, M. N. Burtnick, H. Su, V. Nair and F. C. Gherardini, *Cell. Microbiol.*, 2008, **10**, 487.
47. D. Chakravortty and M. Hensel, *Microbes Infect.*, 2003, **5**, 621.
48. P. Das, A. Lahiri, A. Lahiri and D. Chakravortty, *Microbiology*, 2009, **155**, 2476.
49. W. Lü, J. Du, N. J. Schwarzer, T. Wacker, S. L. A. Andrade and O. Einsle, *Biol. Chem.*, 2013, **394**, 715.
50. M. H. Saier Jr, B. H. Eng, S. Fard, J. Garg, D. A. Haggerty, W. J. Hutchinson, D. L. Jack, E. C. Lai, H. J. Liu, D. P. Nusinew, A. M. Omar, S. S. Pao, I. T. Paulsen, J. A. Quan, M. Sliwinski, T. T. Tseng, S. Wachi and G. B. Young, *Biochim. Biophys. Acta*, 1999, **1422**, 1.
51. B. Suppmann and G. Sawers, *Mol. Microbiol.*, 1994, **11**, 965.
52. G. Sawers, *FEMS Microbiol. Rev.*, 1998, **22**, 543.
53. J. Knappe and G. Sawers, *FEMS Microbiol. Rev.*, 1990, **75**, 383.
54. S. Leonhartsberger, I. Korsa and A. Böck, *J. Mol. Microbiol. Biotechnol.*, 2002, **4**, 269.
55. W. Lü, J. Du, T. Wacker, E. Gerbig-Smentek, S. L. A. Andrade and O. Einsle, *Science*, 2011, **332**, 352.
56. A. B. Waight, J. Love and D. N. Wang, *Nat. Struct. Mol. Biol.*, 2010, **17**, 31.
57. Y. Wang, Y. J. Huang, J. W. Wang, C. Cheng, W. J. Huang, P. L. Lu, Y. N. Xu, P. Y. Wang, N. Yan and Y. G. Shi, *Nature*, 2009, **462**, 467.
58. R. G. Sawers, *Biochem. Soc. Trans.*, 2005, **33**, 42.
59. W. B. White and J. G. Ferry, *J. Bacteriol.*, 1992, **174**, 4997.
60. W. Lü, N. J. Schwarzer, J. Du, E. Gerbig-Smentek, S. L. A. Andrade and O. Einsle, *Proc. Natl. Acad. Sci. U. S. A.*, 2012, **109**, 18395.
61. B. K. Czyzewski and D. N. Wang, *Nature*, 2012, **483**, 494.
62. W. Lü, J. Du, N. J. Schwarzer, E. Gerbig-Smentek, O. Einsle and S. L. Andrade, *Proc. Natl. Acad. Sci. U. S. A.*, 2012, **109**, 13254.
63. A. Rycovska, L. Hatahet, K. Fendler and H. Michel, *Biochim. Biophys. Acta, Biomembr.*, 2012, **1818**, 1342.
64. S. L. A. Andrade and O. Einsle, *Mol. Membr. Biol.*, 2007, **24**, 357.

65. D. Blakey, A. Leech, G. H. Thomas, G. Coutts, K. Findlay and M. Merrick, *Biochem. J.*, 2002, **364**, 527.
66. T. Arcondéguy, R. Jack and M. Merrick, *Microbiol. Mol. Biol. Rev.*, 2001, **65**, 80.
67. G. Thomas, G. Coutts and M. Merrick, *Trends Genet.*, 2000, **16**, 11.
68. S. L. Andrade, A. Dickmanns, R. Ficner and O. Einsle, *Proc. Natl. Acad. Sci. U. S. A.*, 2005, **102**, 14994.
69. M. J. Conroy, A. Durand, D. Lupo, X. D. Li, P. A. Bullough, F. K. Winkler and M. Merrick, *Proc. Natl. Acad. Sci. U. S. A.*, 2007, **104**, 1213.
70. S. Helfmann, W. Lü, C. Litz and S. L. A. Andrade, *J. Mol. Biol.*, 2010, **402**, 165.
71. S. Maier, P. Schleberger, W. Lü, T. Wacker, T. Pflüger, C. Litz and S. L. A. Andrade, *PLoS One*, 2011, **6**, e26327.
72. D. Kleiner, *Biochim. Biophys. Acta*, 1981, **639**, 41.
73. S. L. Hackette, G. E. Skye, C. Burton and I. H. Segel, *J. Biol. Chem.*, 1970, **245**, 4241.
74. R. Stevenson and S. Silver, *Biochem. Biophys. Res. Commun.*, 1977, **75**, 1133.
75. A. M. Marini, S. SoussiBoudekou, S. Vissers and B. Andre, *Mol. Cell. Biol.*, 1997, **17**, 4282.
76. A. M. Marini, S. Vissers, A. Urrestarazu and B. Andre, *EMBO J.*, 1994, **13**, 3456.
77. S. Gazzarrini, T. Lejay, A. Gojon, O. Ninnemann, W. B. Frommer and N. von Wirén, *Plant Cell*, 1999, **11**, 937.
78. U. Ludewig, N. von Wirén and W. B. Frommer, *J. Biol. Chem.*, 2002, **277**, 13548.
79. E. Soupene, L. H. He, D. L. Yan and S. Kustu, *Proc. Natl. Acad. Sci. U. S. A.*, 1998, **95**, 7030.
80. E. Soupene, H. Lee and S. Kustu, *Proc. Natl. Acad. Sci. U. S. A.*, 2002, **99**, 3926.
81. M. J. Conroy, S. J. Jamieson, D. Blakey, T. Kaufmann, A. Engel, D. Fotiadis, M. Merrick and P. A. Bullough, *EMBO Rep.*, 2004, **5**, 1153.
82. S. Khademi, J. O'Connell, J. Remis, Y. Robles-Colmenares, L. J. W. Miercke and R. M. Stroud, *Science*, 2004, **305**, 1587.
83. L. Zheng, D. Kostrewa, S. Bernèche, F. K. Winkler and X. D. Li, *Proc. Natl. Acad. Sci. U. S. A.*, 2004, **101**, 17090.
84. S. L. A. Andrade, A. Dickmanns, R. Ficner and O. Einsle, *Proc. Natl. Acad. Sci. U. S. A.*, 2005, **102**, 14994.
85. X. Li, S. Jayachandran, H. H. Nguyen and M. K. Chan, *Proc. Natl. Acad. Sci. U. S. A.*, 2007, **104**, 19279.
86. D. Lupo, X. D. Li, A. Durand, T. Tomizaki, B. Cherif-Zahar, G. Matassi, M. Merrick and F. K. Winkler, *Proc. Natl. Acad. Sci. U. S. A.*, 2007, **104**, 19303.
87. F. Gruswitz, S. Chaudhary, J. D. Ho, A. Schlessinger, B. Pezeshki, C. M. Ho, A. Sali, C. M. Westhoff and R. M. Stroud, *Proc. Natl. Acad. Sci. U. S. A.*, 2010, **107**, 9638.
88. A. Javelle, D. Lupo, L. Zheng, X. D. Li, F. K. Winkler and M. Merrick, *J. Biol. Chem.*, 2006, **281**, 39492.

89. R. T. Ullmann, S. L. A. Andrade and G. M. Ullmann, *J. Phys. Chem. B*, 2012, **116**, 9690.
90. F. C. Boogerd, H. W. Ma, F. J. Bruggeman, W. C. van Heeswijk, R. Garcia-Contreras, D. Molenaar, K. Krab and H. V. Westerhoff, *FEBS Lett.*, 2011, **585**, 23.
91. A. Javelle, D. Lupo, P. Ripoche, T. Fulford, M. Merrick and F. K. Winkler, *Proc. Natl. Acad. Sci. U. S. A.*, 2008, **105**, 5040.
92. S. Khademi and R. M. Stroud, *Physiology*, 2006, **21**, 419.
93. J. Thornton, D. Blakey, E. Scanlon and M. Merrick, *FEMS Microbiol. Lett.*, 2006, **258**, 114.
94. F. K. Winkler, *Pflügers Arch.: Eur. J. Physiol.*, 2006, **451**, 701.
95. F. Gruswitz, J. O'Connell 3rd and R. M. Stroud, *Proc. Natl. Acad. Sci. U. S. A.*, 2007, **104**, 42.
96. S. Maier, P. Schleberger, W. Lü, T. Wacker, T. Pflüger, C. Litz and S. L. Andrade, *PLoS One*, 2011, **6**, e26327.
97. P. Schulz, J. J. Garcia-Celma and K. Fendler, *Methods*, 2008, **46**, 97.
98. K. Seifert, K. Fendler and E. Bamberg, *Biophys. J.*, 1993, **64**, A402.
99. T. Wacker, J. J. Garcia-Celma, P. Lewe and S. L. A. Andrade, *Proc. Natl. Acad. Sci. U. S. A.*, 2014, **111**, 9995.

CHAPTER 13

Regulation of Denitrification

STEPHEN SPIRO[a]

[a]Department of Biological Sciences, University of Texas at Dallas, 800 W Campbell Road, Richardson, Texas 75080, USA
*E-mail: stephen.spiro@utdallas.edu

13.1 Introduction

The nitrogen oxyanions and oxides of the denitrification pathway (nitrate, nitrite, nitric oxide [NO] and nitrous oxide) function as oxidants in an electron transport pathway that is coupled to proton translocation to generate a proton motive force (PMF) across the cytoplasmic membrane. Thus, denitrification is classically viewed as an anaerobic respiration, the function of which is to provide a mechanism for PMF generation and ATP synthesis when oxygen is absent or limiting. Many denitrifying organisms are facultative; that is, they can also use molecular oxygen as a terminal electron acceptor. There is considerable biochemical diversity in the respiratory enzyme systems of denitrifying bacteria,[1] making it difficult to generalise about the bioenergetic efficiency of the pathway. Nevertheless, the ATP yield per two electrons transferred from a reductant to a terminal electron acceptor is higher when the acceptor is oxygen, providing an explanation for the observation that aerobic respiration is preferred over denitrification.[2] Thus, provided with a choice of oxygen and the substrates for denitrification, the denitrifiers preferentially reduce oxygen until it becomes limiting, at which point nitrate and nitrite reduction can initiate. It follows that there must be regulation of denitrification—in the presence of oxygen, either the enzymes and pathway are

RSC Metallobiology Series No. 9
Metalloenzymes in Denitrification: Applications and Environmental Impacts
Edited by Isabel Moura, José J. G. Moura, Sofia R. Pauleta and Luisa B. Maia

Published by the Royal Society of Chemistry, www.rsc.org

inactive, or the genes encoding the protein components of denitrification are not expressed, or both. There is evidence of post-transcriptional control of denitrification; one well-known example is the inhibition of nitrate transport by oxygen.[3] But the dominant (or at least most studied) mode of regulation is at the level of gene expression, which will therefore be the focus of this chapter.

The biochemical and genetic basis of denitrification has been well studied in a number of bacteria and Archaea that represent a broad swathe of phylogenetic diversity. For the most part, these are "environmental" organisms that are associated with soil, aquatic environments and sewage treatment plants (including *Paracoccus* [*Pa.*] *denitrificans*, *Pseudomonas* [*Ps.*] *stutzeri*, *Rhodobacter* [*Rh.*] *sphaeroides* and *Ralstonia* [*Ra.*] *eutropha*, which are discussed in detail in this chapter). But the pathway has also been studied in plant- and animal-associated microorganisms (*Ps. aeruginosa*, *Bradyrhizobium* [*Bra.*] *japonicum*, *Brucella* species and *Neisseria* species, which are discussed below). This is because metabolic processes of pathogenic bacteria are attracting increasing interest and because the intermediates of denitrification (especially nitrite and NO) have important roles to play in some host–pathogen and host–symbiont interactions. From the point of view of understanding how denitrification is regulated, the two most important model organisms have probably been *Pa. denitrificans* and *Ps. stutzeri*. Studies of other organisms have also made vital contributions, and the approach of studying multiple species has been very important, since there is considerable diversity in regulatory mechanisms.[4,5] For this reason, conclusions drawn from the studies of one organism cannot necessarily be extrapolated to other denitrifiers, and the study of many organisms has been very well justified.

In what follows, my goal is to review what is known about the regulation of denitrification (at the level of gene expression) in a selection of model organisms. The focus will be on the environmental signals that influence the expression of denitrification genes, the roles and mechanisms of specific transcriptional regulators and the topologies of regulatory networks. The diversity of protein participants and regulatory mechanisms will be illustrated, but some general overarching themes will also be described. The regulation of NO detoxification is also considered, since there are interesting similarities and differences in the regulators of NO detoxification and respiratory NO reduction.

13.2 Regulation of Denitrification in Model Organisms

13.2.1 *Pa. denitrificans*

Pa. denitrificans is a Gram-negative bacterium from the α-proteobacteria. Although often described as a soil organism, members of the genus *Paracoccus* seem to be almost ubiquitous, being detected in a wide variety of terrestrial and aquatic environments and sewage treatment plants and associated with eukaryotic hosts. Laboratory strains of *Pa. denitrificans* are vigorous denitrifiers; under anaerobic conditions in the presence of nitrate, growth is rapid and copious dinitrogen is generated. Since it is also biochemically and genetically tractable, *Pa. denitrificans* has proved to be an excellent model for studies of all aspects of denitrification.

Pa. denitrificans expresses periplasmic and membrane-associated respiratory nitrate reductases (NAP and NAR, respectively), a cytochrome cd_1-type nitrite reductase (NIR), a *c*-type NO reductase (NOR) that accepts electrons from small cytochromes and/or copper proteins and a canonical periplasmic N_2O reductase, NOS. The main physiological role of NAP appears to be to dispose of excess reductant;[6] therefore, NAR is assumed to be the enzyme that functions in denitrification in this organism.

The *nar*, *nir*, *nor* and *nos* genes encoding the enzymes of denitrification and their accessory proteins are expressed at low levels in cultures growing aerobically.[7] Subsequent studies have shown that the major signals regulating the expression of the denitrification genes are nitrate, oxygen and NO. Transcriptional control is exerted by three members of the FNR/CRP superfamily of transcriptional regulators, designated FnrP, NarR and NNR, which will be discussed in turn.

FnrP is believed to be an activator of the transcription of the *nar* and *nos* operons.[8–10] The FnrP protein contains four conserved cysteine residues that very likely coordinate a [4Fe–4S] cluster.[11] Thus, mechanistically, FnrP is almost certainly similar to the *Escherichia coli* FNR protein, which contains an oxygen-sensitive iron–sulphur cluster.[12] Reaction of oxygen with the cluster causes conversion to the [2Fe–2S] form, which is accompanied by a decreasing tendency of the protein to dimerise and bind to DNA. Thus, FNR targets are down-regulated in aerobic cultures. FNR-like proteins with this mechanism probably have different sensitivities to oxygen, since subtle changes in the protein environment of the cluster may alter its oxygen reactivity. For example, there is evidence that the CydR (FNR) protein of *Azotobacter vinelandii* is significantly more oxygen sensitive than the *E. coli* homologue.[13] The oxygen sensitivity of the *Pa. denitrificans* FnrP is not known.

The [Fe–S] cluster of FNR-like proteins is potentially reactive towards other small molecules, of which the most relevant is NO, an intermediate of denitrification. Indeed, there is evidence from *in vivo* and *in vitro* experiments to suggest that FNR can be inactivated by NO, leading to down-regulation of target genes.[13–16] Whether this is a physiologically relevant response to the low levels of NO that accumulate in steady-state denitrifying cultures is yet to be determined.

NarR is a little-studied protein that activates transcription of the *nar* operon in response to nitrate and/or nitrite.[17] There are no obvious sequence features in NarR that point to a signal-sensing mechanism, and the nature of the interaction of NarR with nitrate/nitrite is not known. Indeed, the initial report describing NarR suggested that it might respond to nitrate and nitrite indirectly.[17]

Besides oxygen, the other major signal regulating the expression of the denitrification genes is NO (Figure 13.1). In *Pa. denitrificans*, NO is sensed by another FNR/CRP family member, designated NNR, which activates transcription of the *nir* and *nor* genes encoding NIR and NOR, respectively.[8] More recent work suggests that NNR also regulates the transcription of *nos*,[10] although on the basis of proteomics data, *nos* does not belong to the NNR regulon.[9] Results from *in vivo* experiments suggest that NNR responds directly to NO.[18,19] The mechanism of NO sensing by NNR and its orthologues from other organisms has remained somewhat elusive. Early experiments suggested a haem-based sensing mechanism,[20,21] and structural and biochemical data suggest that DNR (from *Ps. aeruginosa*) is a haem protein.[22,23] More recent work has shown that the haem-nitrosyl form of DNR binds to DNA with high affinity.[24] NNR and DNR are in the same subfamily of regulators, but are more distantly related to other NNR-type (designated NnrR) proteins.[25] The details of the NO-sensing mechanism of NNR/DNR/NnrR proteins remain to be fully resolved, and it remains to be seen whether there is a universal mechanism of NO sensing in this family of proteins.

Regulation by multiple FNR/CRP-type proteins in the same organism raises some interesting mechanistic questions. Specifically, in *Pa. denitrificans*, it is evident that the three proteins have DNA binding specificities that are indistinguishable. Attempts to find sequence features that distinguish promoters activated by FnrP, NarR and NNR have proved unsuccessful. Indeed, it seems that, for example, FnrP and NarR can activate the *nar* promoter by, it

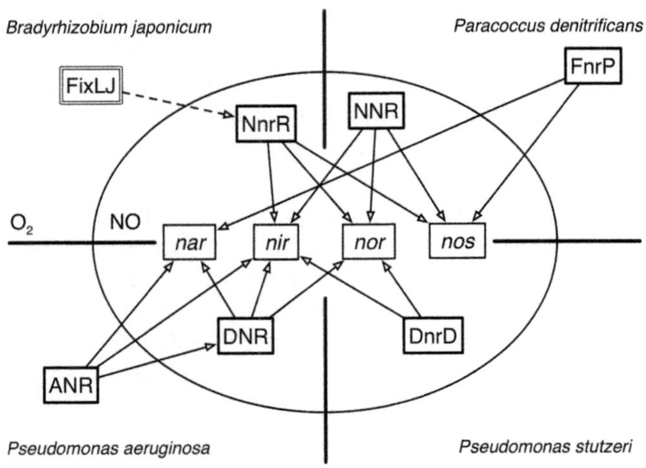

Figure 13.1 Summary of the regulatory roles of oxygen and nitric oxide (NO) sensors in four model denitrifying bacteria. Regulatory proteins that sense NO are shown inside the oval border; those that sense oxygen are outside the border. All of the regulators shown are FNR/CRP family members (Table 13.1), with the exception of FixLK, which is a two-component histidine kinase and response regulator (denoted by the double lined box). Solid arrows indicate the positive regulation of transcription; the dashed arrow indicates indirect regulation.

is thought, binding to the same *cis*-acting sequence.[17] The *nos* promoter is activated by FnrP and NNR[10] and contains two binding sites for FNR-related proteins.[9] The question of how three transcription factors with the same DNA binding specificities activate promoters containing one or more of the cognate binding site(s) is an interesting one that has not been satisfactorily answered.

Other *Pa. denitrificans* proteins have roles in the regulation of denitrification genes that are less well understood. NirI is a six-transmembrane helix membrane protein that likely has a periplasmic flavin-containing domain and cytoplasmic [Fe–S] clusters. NirI is required for the induction of the *nir* operon in response to oxygen limitation and denitrifying conditions.[26] NirI is co-expressed with NirX, a soluble protein of unknown function, and the *nirIX* transcription unit is in the NNR regulon. The signal transduction pathway that links the membrane-associated NirI protein to the activity of the *nir* operon promoter is not understood. NirI is similar in sequence and domain organisation to NosR, which is required for the transcription of the *nosZ* gene in *Ps. stutzeri* (see below). There is also a NosR protein in *Pa. denitrificans*, but its regulatory role has not been investigated.

NOS is a copper enzyme, and copper limitation causes down-regulation of the *nosRZDFYLX* genes that are required for the synthesis of an active NOS.[27] Repression of the *nos* operon by low copper requires a protein of unknown function (NosC) and NosR, but the mechanism by which high copper causes up-regulation of gene expression remains to be described. Interestingly, copper limitation causes accumulation of N_2O (because of reduced expression and lower activity of NOS), which in turn causes up-regulation of genes located downstream of vitamin B_{12} binding riboswitches.[27] The likely explanation is that N_2O interacts directly with vitamin B_{12} to control the activity of riboswitches. The genes that are up-regulated by this mechanism encode vitamin B_{12}-independent versions of enzymes that have vitamin B_{12}-dependent isoenzymes (methionine synthase and ribonucleotide reductase) and proteins involved in vitamin B_{12} scavenging.[27] It is tempting to speculate that vitamin B_{12} riboswitches could also regulate the expression of denitrification genes in response to N_2O, but no such mechanism has yet been described. Denitrification is a metal-demanding pathway (with a high requirement for iron, molybdenum and copper), so it is quite possible that other mechanisms exist for the metal-dependent regulation of denitrification genes.

The *Pa. denitrificans* denitrification system has been used as the platform for the development of a quantitative model that attempts to predict the behaviour of the system during the transition from aerobic to anaerobic growth.[28] While many biochemical details are currently missing (*e.g.*, the affinities of regulatory proteins for their cognate ligands), the model is nevertheless consistent with the gas and metabolite concentrations measured during the transition. The model also provides a satisfying rationale for the coordinate regulation of *nir* and *nor* expression by NNR (and therefore by NO). Previously, NO activation of the genes encoding both the NO-producing and the NO-consuming enzymes had been viewed as an atypical and slightly counterintuitive regulatory pattern. The suggested explanation (with no supporting evidence) was that this arrangement provided a mechanism to prevent the accumulation of the potentially toxic NO. The new view is that positive feedback regulation of *nir* expression mediated by NO drives the cell into making the transition from aerobic to anaerobic respiration. More surprisingly, best fits to the data are obtained if the model assumes that only a fraction of the population makes the transition. This is suggested to be a "bet-hedging" strategy: a fraction of the population does not synthesise the denitrification apparatus in order to be well prepared in the event that oxygen concentrations return to a level that supports aerobic respiration.[28] This behaviour is consistent with denitrifying bacteria occupying niches in the oxic/anoxic transition zone where oxygen tensions fluctuate, as opposed to, for example, the sulphate-reducing bacteria, which are active in strictly anoxic environments. In an extension of this model, fits to experimental data are obtained if it is assumed that both *nar* and *nir* are induced stochastically and are expressed in only a small proportion of cells, while *nosZ* is expressed in all cells.[29]

13.2.2 *Ps. stutzeri* and *Ps. aeruginosa*

Ps. stutzeri has long been used as a model system for biochemical and genetic studies of denitrification.[30] *Ps. aeruginosa* has also emerged as an important model, in part because of the interest in this organism as a human pathogen. As is the case for *Pa. denitrificans*, regulation of denitrification genes in *Pseudomonas* spp. is dominated by FNR/CRP family members (Figure 13.1), though there are important differences in the topologies of the regulatory networks and less well understood roles for additional regulators.

In *Ps. aeruginosa*, ANR is an oxygen sensor that is functionally and structurally equivalent to FNR of *E. coli* and FnrP of *Pa. denitrificans*.[31] A gene cluster encoding a nitrate transporter (NarK) and the respiratory nitrate reductase is positively regulated by ANR in response to low oxygen.[32] ANR also activates transcription of the *dnr* gene encoding a second FNR/CRP family member, DNR.[33] DNR functions as a sensor of NO using a haem-based mechanism.[23,24] In response to NO, DNR activates the transcription of the genes encoding NIR and NOR.[33] When over-expressed, DNR can also activate the genes encoding NarK and NAR,[33] and in a wild-type strain, it seems that ANR and DNR are co-activators of the *nar* promoter, with ANR playing the dominant role.[32] The *nir* promoter is activated by both ANR and DNR, with regulation by DNR dominating at this promoter.[34] *Ps. aeruginosa* is a metabolically versatile organism, and the ANR protein controls the expression of a large regulon of genes, some of which are unrelated to denitrification[35] and some of which have accessory roles in denitrification. An example of the latter is ANR regulation of the *azu* gene, which encodes a small copper protein that functions as an electron donor to NIR.[36] On the other hand, the DNR regulon is much smaller and is exclusively involved in denitrification.[35]

Ps. aeruginosa expresses a two-component system designated NarXL, which, by analogy to the *E. coli* system,[37] likely functions as a sensor of extracellular nitrate and nitrite. Expression of *narXL* is up-regulated by ANR and DNR, and the response regulator NarL co-activates the *nar* promoter together with ANR and DNR.[32] NarXL also positively regulate *dnr* transcription together with ANR, and NarL and DNR co-activate *nirQ*, which encodes a regulatory protein with a poorly defined role in denitrification.[32]

In *Ps. stutzeri*, the ANR homologue (called FnrA) does not control denitrification, since an *anr* mutant does not have a defect in denitrification, and NAR, NIR, NOR and NOS have wild-type activities in this background.[38] The lack of a role for FnrA is confirmed by a recent mutant fitness analysis, which did not identify *fnrA* as a gene contributing to fitness under denitrifying conditions.[39] The *nar* genes of *Ps. stutzeri* are up-regulated in response to nitrate by NarXL;[40] it is not known whether or not an oxygen-responsive regulator also contributes to the regulation of *nar* expression. The principal NO sensor of *Ps. stutzeri* is DnrD, which activates *nir* and *nor* transcription in response to NO.[41,42] Higher concentrations of NO cause the induction of the *nosZ* gene encoding NOS; this response is mediated by DnrD, perhaps indirectly through activation of *nosR*.[42]

NosR is an interesting and unusual protein that is required for the transcription of *nosZ*.[43] NosR is a membrane protein with six transmembrane helices and a periplasmic flavin-binding domain. Cytoplasmic domains of the protein contain two Cys–X_3–Cys–Pro motifs and two other clusters of cysteine residues with variable spacing.[44] Spectroscopic and biochemical characterisation of purified NosR confirmed the presence of flavin and of two [4Fe–4S] clusters that are likely coordinated by the aforementioned cysteine-rich regions.[44] The function of the Cys–X_3–Cys–Pro motifs is not known. Thus, NosR is a membrane protein with features suggesting a role in transmembrane electron transfer, and it does not have an easily recognisable DNA binding domain. That NosR is required for the transcription of the *nosZ* gene[43] is therefore surprising and has yet to be explained mechanistically. Perhaps most surprising is the observation that only the periplasmic flavin-containing domain of NosR is required for *nosZ* expression,[44] suggesting that NosR acts indirectly on the transcription of *nosZ*. NosR also has a role in NOS activity, perhaps by functioning as an electron donor to the enzyme.[44] In *Ps. aeruginosa*, NosR is not required for *nosZ* expression.[45] As mentioned above, the NosR of *Pa. denitrificans* is involved in the copper regulation of NOS expression.[27] As is also mentioned above, NosR is similar in sequence and domain organisation to the NirI protein of *Pa. denitrificans*. NirI is required for the transcription of the *nir* genes encoding NIR.[26] The mechanistic link between NirI and the transcription of the *nir* genes is also not understood, as for NosR it seems likely that NirI acts indirectly on transcription.

Ps. stutzeri has two additional regulators from the FNR/CRP family, designated DnrE and DnrS.[41] Expression of *dnrE* is activated by NarXL in response to nitrate, while *dnrS* is activated by FnrA in response to low oxygen.[41] Mutation of either *dnrE* or *dnrS* does not impact denitrification.[41] Currently, DnrE and DnrS have no known roles in the regulation of denitrification genes; rather, DnrS activates the expression of the gene encoding an anaerobic ribonucleotide reductase.[42]

Mutant fitness analysis has provided a genome-wide view of the role of accessory and regulatory genes in denitrification in *Ps. stutzeri*.[39] This study confirms the importance of DnrD and shows that the NO-sensing regulatory protein NorR is not involved in denitrification.[39] The explanation for the latter observation is likely that NorR regulates the expression of a NO-scavenging flavohaemoglobin (see below), as is the case in *Ps. aeruginosa*.[46] NnrS is a membrane-associated haem- and copper-containing protein that has been suggested to play a role in protection against the toxic effects of NO.[47] Evidence from the mutant fitness analysis suggests that NnrS may act as a NO sensor with a regulatory or repair function.[39] The mutant fitness analysis also implicates three regulatory proteins not previously associated with denitrification, including an anti-σ factor and the chemotaxis protein CheR.[39]

13.2.3 *Bra. japonicum*

Many of the plant-associated, symbiotic, nitrogen-fixing bacteria are capable of denitrification. Among these, the pathway and its regulation are best characterised in the soybean symbiont *Bra. japonicum*. In this organism, oxygen sensing is mediated by a cytoplasmic haem-containing histidine kinase designated FixL (note that in some related organisms FixL is a membrane protein). When oxygen tension is low, FixL auto-phosphorylates, then transfers the phosphate group to its partner protein, FixJ.[48] FixJ activates the expression of *fixK2*, which encodes an FNR/CRP family member, and FixK2 activates the expression of *fixK1*, which encodes a second FNR/CRP family member.[49] FixK2 also activates the expression of *nnrR*, the product of which is a likely NO sensor that activates the transcription of the *nir* and *nor* operons.[50] The FixK2 protein has a single reactive cysteine residue; exposure of the purified protein to oxidising agents *in vitro* causes the formation of inactive dimers that are cross-linked by a disulphide bridge, or inactive monomers with other types of cysteine modification.[51] Gene expression analysis suggests that FixK2 mediates a physiological response to exposure to hydrogen peroxide.[51] Successful crystallisation of FixK2 required substitution of the redox-active cysteine. The FixK2 structure shows that the redox-active cysteine is solvent exposed in the wild-type protein and suggests that modification of this cysteine by oxidation might disrupt DNA binding by stearic hindrance.[52] Thus, expression of *nir* and *nor* in *Bra. japonicum* is regulated by a complex cascade (Figure 13.1) with multiple sites for signal input: FixLJ (oxygen), FixK2 (reactive oxygen species) and NnrR (NO).

Recent work has added further complexity to the *Bra. japonicum* denitrification regulatory network by revealing roles for additional proteins. The oxygen-sensitive NifA protein that regulates genes involved in nitrogen fixation is also required for maximum expression of the *nap* (periplasmic nitrate reductase), *nir* and *nor* genes.[53] NifA belongs to the enhancer binding protein family of transcriptional activators, and as such, its activity is typically dependent on the alternative σ factor σ^N. Strangely, in the case of the denitrification genes, the requirement for NifA is independent of σ^N. Furthermore, there are no NifA binding sites detectable in the promoter regions of the denitrification genes, leading to the conclusion that the effect of NifA is indirect.[53] The RegBA proteins are a histidine kinase/response regulator pair that controls expression of a variety of genes in response to a redox signal.[54] RegB is a histidine kinase that is active under reducing conditions. There is evidence that RegB activity is controlled directly by the redox state of the quinone pool[55] and by the oxidation state of a conserved cysteine residue.[56] In *Bra. japonicum*, the RegBA proteins are designated RegSR. RegR is a regulator of the *nor* and *nos* promoters, and since purified RegR binds to these promoters, it can be inferred that this is direct regulation.[57] RegR activation of *nor* and *nos* requires anoxia and nitrate (though this may reflect co-dependence on another regulator, such as NnrR).[57] Interestingly, RegR activation of denitrification genes is independent of RegS, meaning that the identity of the cognate histidine kinase for RegR in this pathway has not been identified. One candidate is the sensor kinase of an orthologous sensor regulator pair in *Bra. japonicum*.[57] RegR also activates the expression of the gene encoding NifA; this activation is likewise independent of RegS.[58]

The regulatory cascade controlling denitrification in *Bra. japonicum* seems to be significantly more complex (in terms of the number of players and signal inputs) than those of other organisms. This may reflect the complex lifestyle of the organism as a root nodule symbiont and free-living soil bacterium. There is some evidence that denitrification occurs in root nodules and has a role to play in nodule formation.[59]

13.2.4 *Rh. sphaeroides*

Study of the regulation of denitrification in the photosynthetic organism *Rh. sphaeroides* is historically important, since this is the first species in which a NO-responsive transcriptional regulator was documented.[60,61] Strain 2.4.3 of *Rh. sphaeroides* expresses NAP, NIR and NOR and is therefore a *bona fide* denitrifier, although a frameshift mutation in *nosR* means that NOS is not active in this strain,[62] such that N_2O is the end product of denitrification. Expression of the *nir* and *nor* genes is activated by NnrR in response to NO. The *nir* promoter is co-dependent on NnrR and the PrrBA (RegBA) two-component system; the presence of PrrA binding sites in the *nirK* promoter is evidence of direct regulation by PrrA.[63]

13.2.5 *Ra. eutropha*

Ra. eutropha (formerly *Alcaligenes eutrophus*) is a metabolically versatile organism that has been studied for various aspects of its physiology, especially hydrogen oxidation. The respiratory systems of *Ra. eutropha* are quite well characterised; interestingly, some of the denitrification genes are duplicated.[64,65] In the context of the regulation of denitrification, the distinguishing feature of *Ra. eutropha* is the use of an enhancer binding protein, NorR, to regulate the expression of the NOR-encoding genes,[66] rather than the FNR/CRP-type regulators that have this role in other organisms. The binding site for NO in NorR (partly inferred from studies of the orthologous *E. coli* protein) is a mononuclear non-haem iron in the N-terminal GAF domain.[67] Results from site-directed mutagenesis experiments suggest that the iron is coordinated by the side chains of arginine, aspartate and cysteine residues.[68,69] Other regulatory mechanisms in this organism have not been studied, though there is evidence for roles for FNR- and DNR-type regulators.[64,65]

13.2.6 *Brucella* Species

Members of the genus *Brucella* are Gram-negative intracellular animal pathogens. *Brucella* is an α-proteobacterium and as such is quite closely related to some of the model denitrifiers discussed above (*Pa. denitrificans*, *Bra. japonicum* and *Rh. sphaeroides*). There is evidence that denitrification is required for the growth of *Brucella* in the intracellular host environment.[70–72] The advantage provided by denitrification can be rationalised in two ways: it provides a mechanism for ATP synthesis in microenvironments that might be depleted of O_2; and denitrification provides a pathway for the removal of NO, which is made by host cells as a mechanism to counteract invading pathogens. Thus, denitrification in *Brucella* and other pathogens has attracted considerable interest.

In *Brucella* (*Bru.*) *melitensis*, expression of the *nar* operon is regulated by an FNR/CRP family member (NarR), while a probable NO sensor designated NnrR regulates the expression of the *nir*, *nor* and *nos* genes.[73] NnrR is required for virulence in a mouse model and for intracellular resistance to NO. However, a *nor* mutant had a much less severe phenotype, suggesting that other unidentified members of the NnrR regulon are required for virulence.[73]

In *Bru. abortus*, a membrane-associated haem protein NtrY is a histidine kinase that is active at low O_2 tensions, transferring phosphate to its partner protein NtrX.[74] An *ntrYX* mutation prevents the up-regulation of the denitrification genes in response to low O_2 availability, suggesting that the NtrYX system is a regulator of denitrification genes.[74] It is not known whether the NtrYX regulation of denitrification genes is direct. The RegBA homologues of *Bru. abortus* (called PrrBA) directly regulate the transcription of denitrification genes and of *ntrYX* in response to microaerobic growth conditions.[75] The RegBA proteins of *Bru. suis* are also required for expression of the denitrification enzymes.[76]

13.2.7 *Neisseria* Species

Neisseria (*N.*) *meningitidis* and *N. gonorrhoeae* are human pathogens that have also attracted interest because of their ability to use nitrogen oxyanions for anaerobic respiration. Most pathogenic *Neisseria* species express a functional NIR and NOR and can reduce nitrite to N_2O under microaerobic conditions and are therefore *bona fide* denitrifiers; some strains also express NOS.[77,78] In *N. gonorrhoeae*, the NIR gene *aniA* is up-regulated by microaerobic growth in the presence of nitrite, and is co-induced by FNR and NarQP.[79,80] The nitrite effect does not require NarQP, which appear to be signal independent and constitutively active.[81] Rather, nitrite regulation of *aniA* requires the repressor NsrR, which is responsive to the product of nitrite reduction (NO), rather than to nitrite *per se*.[81] NsrR also represses the NOR gene *norB*, which is therefore up-regulated by sources of NO.[82] The *norB* gene is also activated by the iron-sensing repressor Fur by an indirect mechanism involving anti-repression.[82]

13.3 NO Detoxification

NO is somewhat toxic towards bacteria, and its toxicity is exploited by cells of the mammalian innate immune system, which synthesise NO in response to encounters with pathogenic microorganisms.[83] In the denitrifying bacteria, the activity of the respiratory NOR appears to be sufficient to avoid NO toxicity, and NO concentrations in growing cultures are in the nM range.[84] Strains that cannot express an active NOR (because of mutations in structural, accessory or regulatory genes) typically cannot grow or grow very poorly by denitrification, presumably because of the accumulation of NO.[85–87] The respiratory NOR is expressed by some non-denitrifying bacteria; for example, the cyanobacterium *Synechocystis* sp. strain PCC6803, where its role may be to protect against exogenous sources of NO.[88] The *norB* gene of this organism is divergently transcribed form *dnr*, encoding a transcriptional regulator related to the DNR/NNR-type NO sensors described above. However, the preliminary evidence suggests that expression of *norB* is not up-regulated by a source of NO in *Synechocystis* sp. strain PCC6803.[88]

Other enzymes with NO-scavenging activity have been described in both denitrifying and non-denitrifying bacteria. The flavohaemoglobin (Hmp) is a NO denitrosylase that oxidises NO in the presence of molecular oxygen.[89] This appears to be the major physiological function of Hmp, although a slower NOR activity (with N_2O as the reaction product) has been reported in the absence of oxygen.[90] Early studies of Hmp were in the denitrifying organism *Ra. eutropha*, where a minor role in denitrification was suggested.[91] In *Ps. aeruginosa*, Hmp appears not to have a role in denitrification, but rather serves to protect against exogenous NO in aerobic cultures.[92] In non-denitrifying organisms such as *E. coli*, Hmp similarly functions to detoxify NO from exogenous sources or synthesised endogenously as an adventitious by-product of nitrite reduction to ammonia. In *Salmonella*, Hmp is a virulence determinant because of its ability to protect against host-derived NO.[93]

Another important NO detoxifying activity is that of the flavorubredoxin FlRd, an NADH-dependent NOR. This enzyme has been well characterised in *E. coli*, where it protects against exogenous NO under anaerobic conditions.[94,95] FlRd is less widely distributed than Hmp and is not commonly expressed by organisms capable of denitrification.[4] The reason for this is presumably that anaerobic removal of NO is accomplished by the respiratory NOR, so organisms that express this enzyme have no need for the FlRd.

Expression of NO-scavenging activities is typically up-regulated by exposure to NO, being under the control of NO-sensing transcriptional regulators. In many organisms, including *E. coli*, transcription of *hmp* is (or is very likely to be) negatively regulated by NsrR.[4,96] NsrR coordinates an NO-sensitive [Fe–S] cluster[97] such that exposure to NO leads to de-repression of NsrR targets. In *Ps. aeruginosa*[92] and in *Vibrio cholerae*,[4,98] *hmp* is positively regulated by NorR, the same protein that activates the expression of the respiratory NOR gene in *Ra. eutropha*.[66] However, in other *Vibrio* species, *hmp* expression is regulated by NsrR.[4,99] In *E. coli*, NorR regulates the expression of the *norVW* genes encoding the FlRd and its redox partner.[100,101] This small selection of examples demonstrates that there is considerable flexibility in the wiring patterns that connect regulatory proteins and their target genes in different organisms; this is true both for genes encoding NO-scavenging enzymes and genes encoding the respiratory NORs involved in denitrification.[4] The existence of NO detoxification activities in denitrifying bacteria may seem paradoxical, given that these organisms express respiratory NORs that efficiently remove NO. However, NOR is typically expressed and active only under anaerobic conditions, so accessory scavenging enzymes presumably protect against exogenous sources of NO in cells that are growing in the presence of oxygen. The source of the exogenous NO depends upon the ecological niche occupied by the organism—it might, for example, be synthesised by cells of the host immune system, or by neighbouring organisms in complex microbial communities.

13.4 Emerging Themes: Regulators and Signals

As will now be clear, the expression of denitrification genes is controlled by a wide diversity of transcriptional regulators (Table 13.1). Particularly prominent roles are played by members of the FNR/CRP superfamily of regulators and two-component systems (histidine kinase and response regulator). In some organisms, regulators from other families play important roles (*e.g.*, NorR of *Ra. eutropha*), and there are important roles for regulators that are not well understood (NosR and NirI). These regulators respond to a variety of environmental and metabolic signals, the most important of which appear to be oxygen and NO (Figure 13.1). Although diverse, there is a limited set of regulators that controls expression of denitrification and NO detoxification genes. There is, however, considerable diversity in the specific regulator–target combinations and in the topology of regulatory networks in different organisms.[4,5]

There is considerable interest in the mechanisms by which transcriptional regulators interact with their signals. For the denitrification regulators, mechanisms based on [Fe–S] clusters (FnrP and ANR), haem (FixL and DNR) and non-haem iron (NorR) have been described, and are mentioned briefly

Table 13.1 Regulators of denitrification genes. This compilation is restricted to proteins that are known to regulate gene expression. The list is not intended to be exhaustive, but rather to show selected examples for which some functional and mechanistic information is available. In most cases, references are chosen to provide evidence of the signal-sensing mechanism; additional literature references can be found in the text.[a]

Protein	Organism	Mechanism	Signal	Reference
FNR/CRP family				
FnrP	*P. denitrificans*	[Fe–S]	O_2	Hutchings *et al.* (2002)[11]
ANR	*P. aeruginosa*	[Fe–S]	O_2 (and NO?)	Yoon *et al.* (2007)[16]
FixK1	*B. japonicum*	[Fe–S]?	O_2	Mesa *et al.* (2008)[49]
FixK2	*B. japonicum*	Reactive thiol	H_2O_2	Mesa *et al.* (2009)[51]
NNR	*P. denitrificans*	Haem?	NO (and O_2?)	Lee *et al.* (2006)[20]
DNR	*P. aeruginosa*	Haem	NO	Lobato *et al.* (2014)[24]
NarR	*P. denitrificans*	Not known	Nitrate/nitrite	Wood *et al.* (2001)[17]
Histidine kinase/response regulator				
NarXL	*P. aeruginosa*	Ligand binding	Nitrate/nitrite	Cheung and Hendrickson (2009)[114]
FixLJ	*B. japonicum*	Haem	O_2	Gilles-Gonzalez and Gonzalez (2005)[48]
RegBA	*R. sphaeroides* *B. abortus*	Reactive thiol; quinone binding	Redox	Wu and Bauer (2010)[55] Wu *et al.* (2013)[115]
Enhancer binding protein				
NorR	*R. eutropha*	Non-haem iron	NO	D'Autréaux *et al.* (2005)[67] Klink *et al.* (2007)[68]
Rrf2 family				
NsrR	*N. gonorrhoeae*	[Fe–S]	NO	Isabella *et al.* (2009)[107] Crack *et al.* (2015)[97]

[a]NO: nitric oxide.

in the relevant sections above. Mechanistically, the ultimate goal is to achieve a complete description of the way in which the output (usually, though not always, DNA binding) is stimulated or inhibited by interaction of the sensory protein or domain with its cognate signal. Extensive biochemical and structural data mean that the early events associated with ligand binding are well understood for the FixL proteins,[48,102] though a complete mechanistic description must await the availability of structures of full-length FixL proteins in active and inactive conformations. Also well understood are oxygen sensors of the FNR type, in which reaction of O_2 with a [4Fe–4S] cluster triggers a series of events that leads to dissociation of the protein dimer and loss of DNA binding.[12,103] The recent publication of the first FNR structure[104] promises to shed further light on the mechanism by which O_2 regulates DNA binding in this class of protein. Open questions are whether FNR homologues from different organisms (*e.g.*, ANR and FnrP) operate by the same mechanism and with the same sensitivity to oxygen. Indeed, there is growing evidence that FNR homologues have different oxygen sensitivities.[13,105]

Understanding of the haem-based NO sensors has lagged somewhat, but recent advances with DNR[23,24] suggest that rapid progress may soon be made. The DNR/NNR/NnrR-type NO sensors are quite diverse in sequence,[25] and so it is not clear whether they will turn out to share a common mechanism. In the case of the non-haem iron NO sensor NorR, the iron–nitrosyl complex has been defined spectroscopically,[67] and candidate ligands to the iron have been identified,[68,69] but the conformational changes that result from NO binding and trigger NorR activation remain to be elucidated. In the case of the NO sensor NsrR, early disagreement in the literature regarding the status of its iron–sulphur cluster as [2Fe–2S][106,107] or [4Fe–4S][108] appears to have been resolved in favour of the latter.[97] Earlier reports of [2Fe–2S] forms may be a consequence of oxygen- and thiol-mediated degradation of the [4Fe–4S] cluster.[97] As for other proteins, some caution should be exercised in extrapolating results for NsrR between different organisms. Thus, the (perhaps remote) possibility remains that NsrR proteins from different organisms contain different types of cluster. The protein ligands to the [4Fe–4S] cluster of the *Streptomyces coelicolor* NsrR are suggested to include three cysteines and a glutamate residue.[97]

Another interesting NO-sensing protein is the H-NOX protein or domain, which is related in sequence to the haem-containing regulatory domain of the NO-activated eukaryotic soluble guanylate cyclase.[109,110] Bacterial H-NOX proteins have been shown to interact with a histidine kinase[111] or with a bifunctional diguanylate cyclase/phosphodiesterase.[112] In both cases, the ultimate output of the signal transduction pathway is a change in the concentration of the second messenger cyclic di-GMP, which is responsible for the reciprocal regulation of sessile and planktonic growth in those organisms where this pathway has been studied.[111,112] H-NOX proteins are encoded in the genomes of some denitrifying bacteria, including *Pa. denitrificans*. On the basis of genome organisation, the *Pa. denitrificans* H-NOX is predicted to interact with a diguanylate cyclase that cyclises GTP to make cyclic di-GMP (in this case, there is no phosphodiesterase domain). Preliminary work in the author's laboratory has suggested that NO activation of the H-NOX/cyclic di-GMP signalling pathway regulates NO metabolism in this organism (S. Kumar and S. Spiro, unpublished work).

13.5 Conclusions and Future Prospects

Work over a number of years employing several model organisms has established the major environmental signals (O_2 and NO) and transcriptional regulators (FNR/CRP family members and sensor–regulator pairs) that control the expression of denitrification genes. Some regulators (*e.g.*, NorR) are important in only a subset of denitrifiers, and there are roles for proteins (such as NosR and NirI) that remain to be fully understood. A less well-explored, though potentially important, question is the extent to which metal ion availability regulates the expression of the denitrification apparatus. Expression of the gene encoding the copper-containing NOS is sensitive to copper availability;[27] less is known about regulatory mechanisms that respond to other metals involved in denitrification, particularly molybdenum and iron. Denitrification is a significant source of N_2O, which is an important contributor to anthropogenic climate forcing.[113] This fact alone is sufficient justification for the desire to understand fully the factors that regulate the abundance and activity of the denitrification enzymes. Much has been learned, but much remains still to be understood. Technical advances in many areas of microbiology offer new opportunities for furthering our understanding of this important group of organisms.

References

1. J. Simon and M. G. Klotz, *Biochim. Biophys. Acta*, 2013, **1827**, 114.
2. R. J. M. van Spanning, D. J. Richardson and S. J. Ferguson, in *Biology of the Nitrogen Cycle*, ed. H. Bothe, S. J. Ferguson and W. E. Newton, Elsevier, Amsterdam, 2007, ch. 1, pp. 3–20.
3. D. Hernandez and J. J. Rowe, *J. Biol. Chem.*, 1988, **263**, 7937.
4. D. A. Rodionov, I. L. Dubchak, A. P. Arkin, E. J. Alm and M. S. Gelfand, *PLOS Comput. Biol.*, 2005, **1**, e55.
5. S. Spiro, *Philos. Trans. R. Soc., B*, 2012, **367**, 1213.
6. D. J. Richardson, B. C. Berks, D. A. Russell, S. Spiro and C. J. Taylor, *Cell. Mol. Life Sci.*, 2001, **58**, 165.
7. B. Baumann, M. Snozzi, A. Zehnder and J. van der Meer, *J. Bacteriol.*, 1996, **178**, 4367.
8. R. J. van Spanning, A. P. de Boer, W. N. Reijnders, H. V. Westerhoff, A. H. Stouthamer and J. van der Oost, *Mol. Microbiol.*, 1997, **23**, 893.
9. P. Bouchal, I. Struhárová, E. Budinská, O. Sedo, T. Vyhlídalová, Z. Zdráhal, R. van Spanning and I. Kucera, *Biochim. Biophys. Acta*, 2010, **1804**, 1350.
10. L. Bergaust, R. J. M. van Spanning, A. Frostegård and L. R. Bakken, *Microbiology*, 2012, **158**, 826.
11. M. I. Hutchings, J. C. Crack, N. Shearer, B. J. Thompson, A. J. Thomson and S. Spiro, *J. Bacteriol.*, 2002, **184**, 503.
12. A. S. Fleischhacker and P. J. Kiley, *Curr. Opin. Chem. Biol.*, 2011, **15**, 335.
13. G. Wu, H. Cruz-Ramos, S. Hill, J. Green, G. Sawers and R. K. Poole, *J. Biol. Chem.*, 2000, **275**, 4679.
14. H. Cruz-Ramos, J. Crack, G. Wu, M. N. Hughes, C. Scott, A. J. Thomson, J. Green and R. K. Poole, *EMBO J.*, 2002, **21**, 3235.
15. S. T. Pullan, M. D. Gidley, R. A. Jones, J. Barrett, T. M. Stevanin, R. C. Read, J. Green and R. K. Poole, *J. Bacteriol.*, 2007, **189**, 1845.
16. S. S. Yoon, A. C. Karabulut, J. D. Lipscomb, R. F. Hennigan, S. V. Lymar, S. L. Groce, A. B. Herr, M. L. Howell, P. J. Kiley, M. J. Schurr, B. Gaston, K. H. Choi, H. P. Schweizer and D. J. Hassett, *EMBO J.*, 2007, **26**, 3662.
17. N. J. Wood, T. Alizadeh, S. Bennett, J. Pearce, S. J. Ferguson, D. J. Richardson and J. W. Moir, *J. Bacteriol.*, 2001, **183**, 3606.
18. R. J. van Spanning, E. Houben, W. N. Reijnders, S. Spiro, H. V. Westerhoff and N. Saunders, *J. Bacteriol.*, 1999, **181**, 4129.

19. M. I. Hutchings, N. Shearer, S. Wastell, R. J. van Spanning and S. Spiro, *J. Bacteriol.*, 2000, **182**, 6434.
20. Y.-Y. Lee, N. Shearer and S. Spiro, *Microbiology*, 2006, **152**, 1461.
21. N. Castiglione, S. Rinaldo, G. Giardina and F. Cutruzzolà, *Microbiology*, 2009, **155**, 2838.
22. S. Rinaldo, N. Castiglione, G. Giardina, M. Caruso, A. Arcovito, S. D. Longa, P. D'Angelo and F. Cutruzzolà, *Antioxid. Redox Signaling*, 2012, **17**, 1178.
23. F. Cutruzzolà, A. Arcovito, G. Giardina, S. della Longa, P. D'Angelo and S. Rinaldo, *Biometals*, 2014, **27**, 763.
24. L. Lobato, L. Bouzhir-Sima, T. Yamashita, M. T. Wilson, M. H. Vos and U. Liebl, *J. Biol. Chem.*, 2014, **289**, 26514.
25. H. Korner, H. J. Sofia and W. G. Zumft, *FEMS Microbiol. Rev.*, 2003, **27**, 559.
26. N. F. Saunders, E. N. Houben, S. Koefoed, W. N. Reijnders, H. V. Westerhoff, A. P. de Boer and R. J. van Spanning, *Mol. Microbiol.*, 1999, **34**, 24.
27. M. J. Sullivan, A. J. Gates, C. Appia-Ayme, G. Rowley and D. J. Richardson, *Proc. Natl. Acad. Sci. U. S. A.*, 2013, **110**, 19926.
28. J. Hassan, L. L. Bergaust, I. D. Wheat and L. R. Bakken, *PLoS Comput. Biol.*, 2014, **10**, e1003933.
29. J. Hassan, Z. Qu, L. L. Bergaust and L. R. Bakken, *PLoS Comput. Biol.*, 2016, **12**, e1004621.
30. W. G. Zumft, *Microbiol. Mol. Biol. Rev.*, 1997, **61**, 533.
31. M. Galimand, M. Gamper, A. Zimmermann and D. Haas, *J. Bacteriol.*, 1991, **173**, 1598.
32. K. Schreiber, R. Krieger, B. Benkert, M. Eschbach, H. Arai, M. Scobert and D. Jahn, *J. Bacteriol.*, 2007, **189**, 4310.
33. H. Arai, T. Kodama and Y. Igarashi, *Mol. Microbiol.*, 1997, **25**, 1141.
34. M. Kuroki, Y. Igarashi, M. Ishii and H. Arai, *Environ. Microbiol. Rep.*, 2014, **6**, 792.
35. K. Trunk, B. Benkert, N. Quäck, R. Münch, M. Scheer, J. Garbe, L. Jänsch, M. Trost, J. Wehland, J. Buer, M. Jahn, M. Schobert and D. Jahn, *Environ. Microbiol.*, 2010, **12**, 1719.
36. E. Vijgenboom, J. E. Busch and G. W. Canters, *Microbiology*, 1997, **143**, 2853.
37. E.-B. Goh, P. J. Bledsoe, L. L. Chen, P. Gyaneshwar, V. Stewart and M. M. Igo, *J. Bacteriol.*, 2005, **187**, 4890.
38. H. Cuypers and W. G. Zumft, *J. Bacteriol.*, 1993, **175**, 7236.
39. B. J. Vaccaro, M. P. Thorgersen, W. A. Lancaster, M. N. Price, K. M. Wetmore, F. L. Poole, A. Deutschbauer, A. P. Arkin and M. W. Adams, *Appl. Environ. Microbiol.*, 2015, **82**, 51.
40. E. Hartig, U. Schiek, K. U. Vollack and W. G. Zumft, *J. Bacteriol.*, 1999, **181**, 3658.
41. K. U. Vollack, E. Hartig, H. Korner and W. G. Zumft, *Mol. Microbiol.*, 1999, **31**, 1681.
42. K. U. Vollack and W. G. Zumft, *J. Bacteriol.*, 2001, **183**, 2516.

43. H. Cuypers, A. Viebrock-Sambale and W. G. Zumft, *J. Bacteriol.*, 1992, **174**, 5332.
44. P. Wunsch and W. G. Zumft, *J. Bacteriol.*, 2005, **187**, 1992.
45. H. Arai, M. Mizutani and Y. Igarashi, *Microbiology*, 2003, **149**, 29.
46. T. Koskenkorva, N. Aro-Kärkkäinen, D. Bachmann, H. Arai, A. D. Frey and P. T. Kallio, *Arch. Microbiol.*, 2008, **189**, 385.
47. A. M. Stern, B. Liu, L. R. Bakken, J. P. Shapleigh and J. A. Zhu, *J. Bacteriol.*, 2013, **195**, 4702.
48. M. A. Gilles-Gonzalez and G. Gonzalez, *J. Inorg. Biochem.*, 2005, **99**, 1.
49. S. Mesa, F. Hauser, M. Friberg, E. Malaguti, H.-M. Fischer and H. Hennecke, *J. Bacteriol.*, 2008, **190**, 6568.
50. S. Mesa, E. J. Bedmar, A. Chanfon, H. Hennecke and H.-M. Fischer, *J. Bacteriol.*, 2003, **185**, 3978.
51. S. Mesa, L. Reutimann, H.-M. Fischer and H. Hennecke, *Proc. Natl. Acad. Sci. U. S. A.*, 2009, **106**, 21860.
52. M. Bonnet, M. Kurz, S. Mesa, C. Briand, H. Hennecke and M. G. Grütter, *J. Biol. Chem.*, 2013, **288**, 14238.
53. E. Bueno, S. Mesa, C. Sanchez, E. J. Bedmar and M. J. Delgado, *Environ. Microbiol.*, 2010, **12**, 393.
54. S. Elsen, L. R. Swem, D. L. Swem and C. E. Bauer, *Microbiol. Mol. Biol. Rev.*, 2004, **68**, 263.
55. J. Wu and C. E. Bauer, *mBio*, 2010, **1**, e00272.
56. L. R. Swem, B. J. Kraft, D. L. Swem, A. T. Setterdahl, S. Masuda, D. B. Knaff, J. M. Zaleski and C. E. Bauer, *EMBO J.*, 2003, **22**, 4699.
57. M. J. Torres, M. Argandona, C. Vargas, E. J. Bedmar, H. M. Fischer, S. Mesa and M. J. Delgado, *PLoS One*, 2014, **9**, e99011.
58. E. Bauer, T. Kaspar, H.-M. Fischer and H. Hennecke, *J. Bacteriol.*, 1998, **180**, 3853.
59. S. Mesa, J. de Dios Alché, E. Bedmar and M. J. Delgado, *Physiol. Plant.*, 2004, **120**, 205.
60. A. V. Kwiatkowski and J. P. Shapleigh, *J. Biol. Chem.*, 1996, **271**, 24382.
61. A. V. Kwiatkowski, W. P. Laratta, A. Toffanin and J. P. Shapleigh, *J. Bacteriol.*, 1997, **179**, 5618.
62. A. Hartsock and J. P. Shapleigh, *J. Bacteriol.*, 2010, **192**, 903.
63. W. P. Laratta, P. S. Choi, I. E. Tosques and J. P. Shapleigh, *J. Bacteriol.*, 2002, **184**, 3521.
64. R. Cramm, *J. Mol. Microbiol. Biotechnol.*, 2009, **16**, 38.
65. Y. Kohlmann, A. Pohlmann, E. Schwartz, D. Zühlke, A. Otto, D. Albrecht, C. Grimmler, A. Ehrenreich, B. Voigt, D. Becher, M. Hecker, B. Friedrich and R. Cramm, *J. Proteome Res.*, 2014, **13**, 4325.
66. A. Pohlmann, R. Cramm, K. Schmelz and B. Friedrich, *Mol. Microbiol.*, 2000, **38**, 626.
67. B. D'Autréaux, N. P. Tucker, R. Dixon and S. Spiro, *Nature*, 2005, **437**, 769.
68. A. Klink, B. Elsner, K. Strube and R. Cramm, *J. Bacteriol.*, 2007, **189**, 2743.

69. N. P. Tucker, B. D'Autréaux, F. K. Yousafzai, S. A. Fairhurst, S. Spiro and R. Dixon, *J. Biol. Chem.*, 2008, **283**, 908.
70. S. Kohler, V. Foulongne, S. Ouahrani-Bettache, G. Bourg, J. Teyssier, M. Ramuz and J. P. Liautard, *Proc. Natl. Acad. Sci. U. S. A.*, 2002, **99**, 15711.
71. S.-H. Baek, G. Rajashekara, G. A. Splitter and J. P. Shapleigh, *J. Bacteriol.*, 2004, **186**, 6025.
72. S. Loisel-Meyer, M. P. Jiménez de Bagüés, E. Basseres, J. Dornand, S. Köhler, J. P. Liautard and V. Jubier-Maurin, *Infect. Immun.*, 2006, **74**, 1973.
73. V. Haine, M. Dozot, J. Dornand, J.-J. Letesson and X. de Bolle, *J. Bacteriol.*, 2006, **188**, 1615.
74. M. del Carmen Carrica, I. Fernandez, M. A. Martí, G. Paris and F. A. Goldbaum, *Mol. Microbiol.*, 2012, **85**, 39.
75. M. del Carmen Carrica, I. Fernandez, R. Sieira, G. Paris and F. A. Goldbaum, *Mol. Microbiol.*, 2013, **88**, 222.
76. E. Abdou, A. Deredjian, M. P. Jiménez de Bagüés, S. Köhler and V. Jubier-Maurin, *Infect. Immun.*, 2013, **81**, 2053.
77. J. D. Rock, M. R. Mahnane, M. F. Anjum, J. G. Shaw, R. C. Read and J. W. Moir, *Mol. Microbiol.*, 2005, **58**, 800.
78. K. R. Barth, V. M. Isabella and V. L. Clark, *Microbiology*, 2009, **155**, 4093.
79. T. C. Householder, W. A. Belli, S. Lissenden, J. A. Cole and V. L. Clark, *J. Bacteriol.*, 1999, **181**, 541.
80. S. Lissenden, S. Mohan, T. Overton, T. Regan, H. Crooke, J. A. Cardinale, T. C. Householder, P. Adams, C. D. O'Conner, V. L. Clark, H. Smith and J. A. Cole, *Mol. Microbiol.*, 2000, **37**, 839.
81. T. W. Overton, R. Whitehead, Y. Li, L. A. Snyder, N. J. Saunders, H. Smith and J. A. Cole, *J. Biol. Chem.*, 2006, **281**, 33115.
82. V. Isabella, L. F. Wright, K. Barth, J. M. Spence, S. Grogan, C. A. Genco and V. L. Clark, *Microbiology*, 2008, **154**, 226.
83. F. C. Fang, *Nat. Rev. Microbiol.*, 2004, **2**, 820.
84. J. Goretski, O. Zafiriou and T. Hollocher, *J. Biol. Chem.*, 1990, **265**, 11535.
85. C. Braun and W. G. Zumft, *J. Biol. Chem.*, 1991, **266**, 22785.
86. R. J. van Spanning, A. P. de Boer, W. N. Reijnders, S. Spiro, H. V. Westerhoff, A. H. Stouthamer and J. van der Oost, *FEBS Lett.*, 1995, **360**, 151.
87. A. P. de Boer, J. van der Oost, W. N. Reijnders, H. V. Westerhoff, A. H. Stouthamer and R. J. van Spanning, *Eur. J. Biochem.*, 1996, **242**, 592.
88. A. Busch, B. Friedrich and R. Cramm, *Appl. Environ. Microbiol.*, 2002, **68**, 668.
89. A. Hausladen, A. Gow and J. S. Stamler, *Proc. Natl. Acad. Sci. U. S. A.*, 2001, **98**, 10108.
90. S. O. Kim, Y. Orii, D. Lloyd, M. N. Hughes and R. K. Poole, *FEBS Lett.*, 1999, **445**, 389.
91. R. Cramm, R. A. Siddiqui and B. Friedrich, *J. Biol. Chem.*, 1994, **269**, 7349.
92. H. Arai, M. Hayashi, A. Kuroi, M. Ishii and Y. Igarashi, *J. Bacteriol.*, 2005, **187**, 3960.

93. I.-S. Bang, L. Liu, A. Vazquez-Torres, M. L. Crouch, J. S. Stamler and F. C. Fang, *J. Biol. Chem.*, 2006, **281**, 28039.
94. C. M. Gomes, A. Giuffre, E. Forte, J. B. Vicente, L. M. Saraiva, M. Brunori and M. Teixeira, *J. Biol. Chem.*, 2002, **277**, 25273.
95. A. M. Gardner and P. R. Gardner, *J. Biol. Chem.*, 2002, **277**, 8166.
96. D. M. Bodenmiller and S. Spiro, *J. Bacteriol.*, 2006, **188**, 874.
97. J. C. Crack, J. Munnoch, E. L. Dodd, F. Knowles, M. M. Al Bassam, S. Kamali, A. A. Holland, S. P. Cramer, C. J. Hamilton, M. K. Johnson, A. J. Thomson, M. I. Hutchings and N. E. Le Brun, *J. Biol. Chem.*, 2015, **290**, 12689.
98. A. M. Stern, A. J. Hay, Z. Liu, F. A. Desland, J. Zhang, Z. Zhong and J. Zhu, *mBio*, 2012, **3**, e00013.
99. Y. Wang, A. K. Dunn, J. Wilneff, M. J. McFall-Ngai, S. Spiro and E. G. Ruby, *Mol. Microbiol.*, 2010, **78**, 903.
100. M. I. Hutchings, N. Mandhana and S. Spiro, *J. Bacteriol.*, 2002, **184**, 4640.
101. A. M. Gardner, C. R. Gessner and P. R. Gardner, *J. Biol. Chem.*, 2003, **278**, 10081.
102. K. R. Rodgers and G. S. Lukat-Rodgers, *J. Inorg. Biochem.*, 2005, **99**, 963.
103. J. C. Crack, J. Green, A. J. Thomson and N. E. Le Brun, *Curr. Opin. Chem. Biol.*, 2012, **16**, 35.
104. A. Volbeda, C. Darnault, O. Renoux, Y. Nicolet and J. C. Fontecilla-Camps, *Sci. Adv.*, 2015, **1**, e1501086.
105. S. A. Ibrahim, J. C. Crack, M. D. Rolfe, J. M. Borrero de Acuna, A. J. Thomson, N. E. Le Brun, M. Schobert, M. R. Stapleton and J. Green, *J. Biol. Chem.*, 2015, **290**, 16812.
106. N. P. Tucker, M. G. Hicks, T. A. Clarke, J. C. Crack, G. Chandra, N. E. Le Brun, R. Dixon and M. I. Hutchings, *PLoS One*, 2008, **3**, e3623.
107. V. M. Isabella, J. D. Lapek, E. M. Kennedy and V. L. Clark, *Mol. Microbiol.*, 2009, **71**, 227.
108. E. T. Yukl, M. A. Elbaz, M. M. Nakano and P. Moënne-Loccoz, *Biochemistry*, 2008, **47**, 13084.
109. L. M. Iyer, V. Anantharaman and L. Aravind, *BMC Genomics*, 2003, **4**, 5.
110. L. Plate and M. A. Marletta, *Trends Biochem. Sci.*, 2013, **38**, 566.
111. L. Plate and M. A. Marletta, *Mol. Cell*, 2012, **46**, 449.
112. N. Liu, Y. Xu, S. Hossain, N. Huang, D. Coursolle, J. A. Gralnick and E. M. Boon, *Biochemistry*, 2012, **51**, 2087.
113. D. Richardson, H. Felgate, N. Watmough, A. Thomson and E. Baggs, *Trends Biotechnol.*, 2009, **27**, 388.
114. J. Cheung and W. A. Hendrickson, *Structure*, 2009, **17**, 190.
115. J. Wu, Z. Cheng, K. Reddie, K. Carroll, L. A. Hammad, J. A. Karty and C. E. Bauer, *J. Biol. Chem.*, 2013, **288**, 4755.

CHAPTER 14

Denitrification in Fungi

HIROFUMI SHOUN*[a] AND SHINYA FUSHINOBU[a]

[a]The University of Tokyo, 1-1-1 Yayoi, Bunkyo-ku, Tokyo 113-8657, Japan
*E-mail: hirofumishoun@yahoo.co.jp

14.1 Introduction

The nitrogen cycle performed by microorganisms is composed of three processes: nitrogen fixation, nitrification and denitrification. The cycle is very important for life and the global environment, providing nitrogen to life as nutrition and maintaining the homeostasis of Earth. Denitrification is the reverse reaction of nitrogen fixation in the sense that it carries fixed nitrogen back to the atmosphere. The major source of global nitrous oxide (N_2O; a potent greenhouse gas) emissions is the microbial activities of nitrification and denitrification. Therefore, the control and understanding of microbial denitrification is crucial for reducing N_2O emissions. The features of bacterial denitrifying systems are well characterised at the molecular level.[1–3] The bacterial denitrifying system is composed of four reducing steps:

$$NO_3^- \rightarrow NO_2^- \rightarrow NO \rightarrow N^2O \rightarrow N^2 \quad (14.1)$$

Each of these steps is catalysed by a dissimilatory nitrate reductase (dNar), dissimilatory nitrite reductase (dNir), nitric oxide reductase (Nor) and nitrous oxide reductase (Nos), respectively. The reducing equivalents for these reactions are provided from the respiratory chain coupled to ATP synthesis, and thus, bacterial denitrification functions as anaerobic respiration. Previously,

RSC Metallobiology Series No. 9
Metalloenzymes in Denitrification: Applications and Environmental Impacts
Edited by Isabel Moura, José J. G. Moura, Sofia R. Pauleta and Luisa B. Maia

Published by the Royal Society of Chemistry, www.rsc.org

organisms involved in the nitrogen cycle were thought to be restricted to bacteria (prokaryotes). More than two decades ago, Shoun *et al.* showed that many fungi and yeasts (eukaryotes) also exhibit distinct denitrifying activities.[4–6] Before their finding, there were many papers reporting that fungi can generate a small amount (recovering up to 15% of the nitrogen in the substrate nitrite) of N_2O from, in most cases, nitrite, and thus they may exhibit denitrifying activity.[7] However, these papers only reported a simple observation without providing any evidence that the small amount of N_2O was generated by a biological reaction in fungi. Nitrite is chemically unstable as compared with nitrate and thus it is possible that N_2O is formed by some chemical reaction during the incubation of nitrite with fungi. On the other hand, those studies have characterised the denitrifying system in fungi at the cellular and molecular levels (identifying both proteins and genes), mainly employing two fungal species, *Fusarium* (*F.*) *oxysporum* strain MT811 (JCM11502) and *Cylindrocarpon* (*C.*) *tonkinense* IFO (NBRC) 30 561. The most characteristic feature of the fungal denitrifying system is the involvement of cytochrome P450 (P450) as a Nor (P450nor).[8,9] Since then, many papers from other groups have also shown that fungal denitrification functions in nature as a major process in the nitrogen cycle.[10–13]

14.2 Serendipitous P450

More than three decades ago, Shoun *et al.* isolated a haem protein from the soluble fraction of *F. oxysporum* fungal cells, and the protein exhibited a quasi-lipoxygenase activity. Although the physiological roles of the lipoxygenase activity were not known, the haem protein was identified as a type of P450.[14] At that time, P450 took centre stage as a group of haem proteins involved in mammalian drug metabolism and steroidogenesis, whereas little was known about the microbial P450, except for bacterial P450cam in *Pseudomonas putida.* During the investigation of fungal P450 physiological functions, researchers attempted to isolate NADPH-P450 reductase (P450 reductase), the redox partner of P450, from *F. oxysporum.* P450 reductase can be traced by NADPH-cytochrome *c* reductase activity as its substitute.[15] Then, they isolated the activity and found that the active fraction exhibited an unusual absorption spectrum, characteristic of a sirohaem. These authors submitted a paper regarding the NADPH-cytochrome *c* reductase activity, which was unsuccessful. However, during the review process, a reviewer described an important comment about assimilatory nitrite reductase, which also contains a sirohaem and exhibits the same activity as NADPH-cytochrome *c* reductase. They were shocked at the comment because it seemed correct. Therefore, they checked the growth medium and found that it contained nitrate, which was intended to support the fungal growth as a nitrogen source. As such, they thought that the sirohaem-dependent NADPH-cytochrome *c* reductase activity would disappear when nitrate in the growth medium was replaced with another nitrogen source (ammonium). The replacement caused quite an unexpected change! The P450 that was produced in large amounts in the soluble fraction disappeared (Figure 14.1).[16]

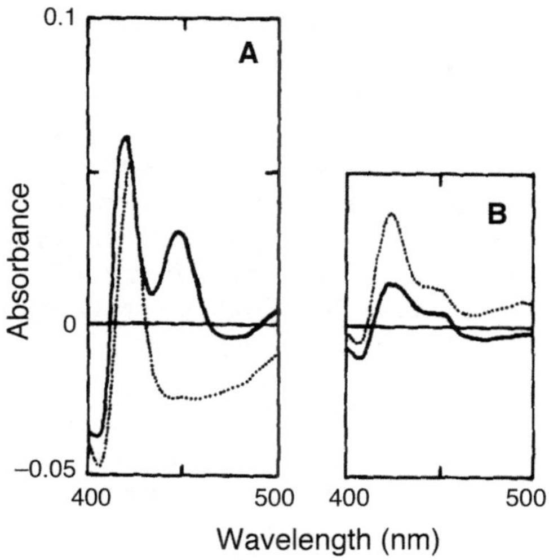

Figure 14.1 The CO difference spectra of (a) soluble and (b) microsome fractions of *Fusarium oxysporum*. Each fraction was diluted ten-fold with 50 mM potassium phosphate buffer containing 20% glycerol, 2 mM 2-mercaptoethanol and 0.1 mM EDTA (pH 7.2). To avoid interference by cytochrome oxidase, the reference was not reduced with dithionite. Solid line: nitrate-grown cells; dotted line: ammonium-grown cells. Reprinted from *FEBS Letters*, 244, Hirofumi Shoun, Wakako Suyama, Tsuneo Yasui, Soluble, nitrate/nitrite-inducible cytochrome P-450 of the fungus, *Fusarium oxysporum*, 11–14, Copyright 1989 with permission from Elsevier.

Furthermore, the soluble P450 was not produced even in the presence of nitrate when the aeration during cell growth was increased. The conditions for producing the P450—presence of nitrate and restriction of aeration—seemed quite strange because P450s were known to function as monooxygenases (aerobic enzymes) against lipophilic, organic substrates. These findings concerning inducing conditions led to the understanding that "this fungus is denitrifying". After 3–4 years, the research showed that the fungus *F. oxysporum* exhibits a denitrifying activity because its main denitrifying product was not N_2 but N_2O, unlike the textbook description. Finally, the denitrifying activity of *F. oxysporum* was shown, which stoichiometrically (with 100% yield) converts NO_3^- to N_2O.[4] Since then, many other ascomycetous and basidiomycetous soil fungi[5] and yeasts[6] have been shown to exhibit denitrifying activity when forming N_2O from NO_3^- or NO_2^-.

14.3 Fungal Denitrifying System

After the discovery of fungal denitrification, the nitrate-inducible P450 was shown to be a Nor and thus termed P450nor.[8,9] A NirK protein was also isolated from *F. oxysporum*, which showed a close resemblance to the bacterial counterpart, copper-containing dNir.[17] The fungal denitrifying system has been extensively studied in *F. oxysporum* MT811 (JCM 11502) and *C. tonkinense* IFO (currently, NBRC) 30 561 (Figure 14.2).

NirK (copper-containing dNir) and P450nor comprise the minimal couple for denitrification. *F. oxysporum* MT811 also contains dNar, which resembles the bacterial counterpart, NarGHI.[1–3] The mitochondrial dNir (*C. tonkinense*) and dNar (*F. oxysporum*) activities were shown to be associated with the respiratory chain coupled to ATP synthesis.[18] This was the first evidence of anaerobic respiration occurring in mitochondria.[19] The fungal system seems to lack Nos and thus the final product is N_2O. The fungal system of *F. oxysporum* MT811 is also unique in that dNar is supported by a ubiquinone-dependent formate dehydrogenase (UQFdh).[20,21] The coupling of dNar and UQFdh in *Escherichia coli* is well characterised, but has not been identified in denitrifying bacteria. Formate, the electron donor for the dNar–UQFdh couple, is provided by pyruvate *via* a pyruvate-formate lyase (PfL). One study suggests that the same electron transport system comprising PfL, UQFdh and dNar is also functioning in *F. oxysporum* MT811.[22] The importance of formate as the electron donor to the fungal denitrifying system in

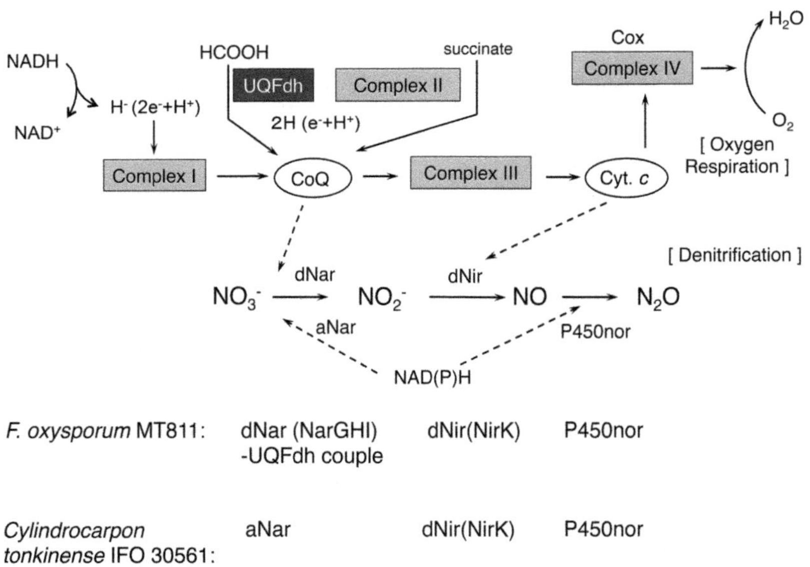

Figure 14.2 The mitochondrial denitrifying system of *Fusarium oxysporum* MT811 and *Cylindrocarpon tonkinense* sharing the respiratory chain with oxygen respiration. *F. oxysporum* contains the dNar (NarGHI)–UQFdh couple, NirK (dNir) and P450nor as the terminal reductases (oxidases). The dNar–UQFdh couple is known to occur among bacteria performing ammonification such as *Escherichia coli*, but is not known among denitrifying bacteria. *C. tonkinense* also contains NirK and P450nor, but not dNar. Instead, *C. tonkinense* can denitrify nitrate by utilising aNar. Reproduced from Figure 1 in Shoun *et al.*, *Phil. Trans. R. Soc. B*, 2012, **367**, 1186, with permission.

natural environments was recently demonstrated.[11] Fungal denitrification requires a minimal amount of oxygen supply.[23] Under such conditions, oxygen respiration and denitrification occur simultaneously (hybrid respiration).[21] In more anaerobic conditions, denitrification in *F. oxysporum* is replaced by ammonia fermentation (eqn (14.2)), in which aNar and aNir are involved.[23]

$$NO_3^- \rightarrow NO_2^- \rightarrow NH_4^+ \tag{14.2}$$

Figure 14.3 shows transmission electron microscopy images of *F. oxysporum* cells grown under aerobic, hypoxic and anaerobic conditions in the presence of nitrate. The mitochondria in the denitrifying cells seem intact, similar to aerobic cells, showing a sharp contrast to the presence of vacant mitochondria formed under anaerobic conditions for ammonia fermentation.[23] The results showed that denitrification requires intact mitochondria, which is consistent with the above observation that the fungal denitrifying system is localised in the mitochondria, whereas ammonia fermentation is performed only in the cytosol.

In contrast to the involvement of dNar in *F. oxysporum*, the denitrifying system of *C. tonkinense* does not contain dNar. However, it is extremely interesting that *C. tonkinense* can denitrify nitrate under certain conditions, using an assimilatory nitrate reductase (aNar) to reduce nitrate to nitrite.[24] The assimilatory nitrate-reducing system is ubiquitous in plants and microorganisms, removing nitrogen atoms from nitrate to provide nutrition for life. Assimilatory and dissimilatory nitrate-reducing systems were previously understood to function independently of each other. Therefore, the involvement of aNar or an assimilatory nitrate-reducing system (aNar and aNir) in denitrification[24] and ammonia fermentation[23] demonstrated the first use of an assimilatory system for a dissimilatory purpose (producing ATP).

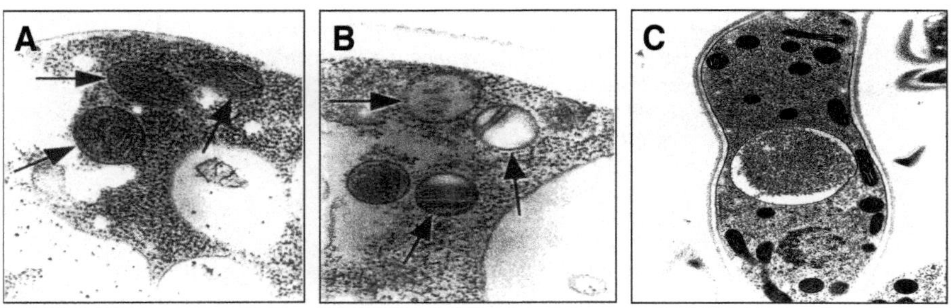

Figure 14.3 Photographs by transmission electron microscopy of *Fusarium oxysporum* cells. Cells cultured under aerobic (a), anoxic (b) and denitrification (c) conditions are shown. Magnification: × 36 000. Arrows indicate mitochondria. Panel (a) is reprinted from Figure 4a in ref. 23 with permission from ASBMB.

14.4 Eukaryotic *nirK* Genes: Originating from the Protomitochondrion?

Recent genome analyses revealed the presence of *nirK* (the gene for NirK) homologue genes in many eukaryotes (fungi, protozoa and green algae), a finding that is consistent with previous results on the isolation of the fungal NirK protein.[17,25] We also isolated a *nirK* homologue gene from *F. oxysporum* MT811 and showed that its recombinant protein exhibits the same properties as the isolated NirK protein.[22] The results showed that the *nirK* gene homologue from *F. oxysporum* encodes the NirK protein involved in fungal denitrification and thus is the first eukaryotic orthologue of bacterial *nirK* genes.

A phylogenetic relationship between NirK proteins and NirK homologues of prokaryotes and eukaryotes based on the deduced amino acid sequences was constructed (Figure 14.4),[25] which showed that eukaryotic NirK homologues including FoNir (dNir from *F. oxysporum*) cluster together. Bacterial

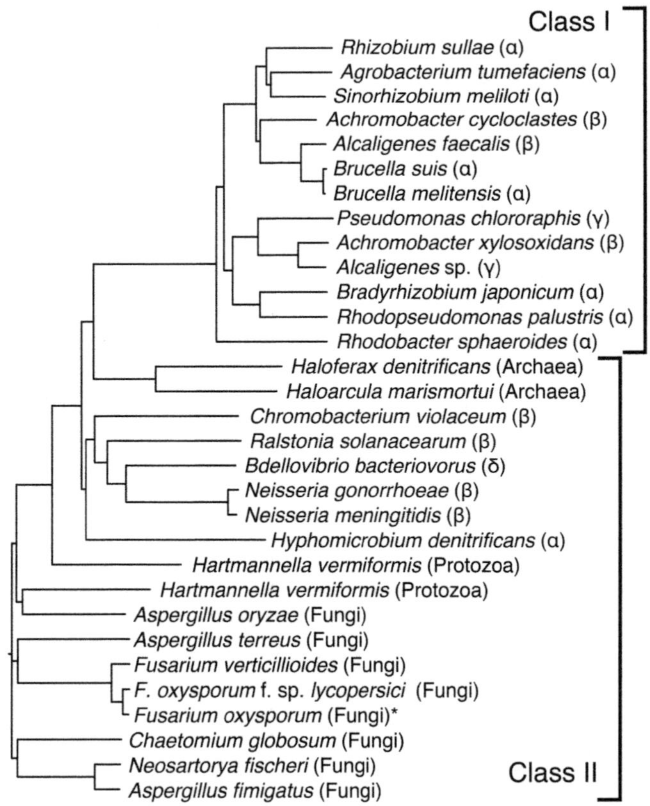

Figure 14.4 The phylogenetic tree of NirK proteins. The tree was generated using the Clustal Omega server. Each NirK protein is shown by its origin (organism), and the classification of each organism (subclass of proteobacteria, archaea or eukaryotes) is shown in parentheses. The NirK protein from *F. oxysporum* is shown with an asterisk. The protozoal homologue sequences are partial.

NirK proteins are divided into two major groups: classes I and II.[2] The phylogenetic tree shows that classes I and II diverged first, and then, the class II NirK proteins further split into archaeal, bacterial and eukaryotic groups. The eukaryotic branch was further divided into fungi and protozoa. The random distribution of NirK proteins among proteobacteria suggests that horizontal transfer of *nirK* genes frequently occurred between these bacteria. By contrast, the eukaryotic NirK homologues were systematically distributed in the same branch, suggesting that they emerged from the same ancestor.

There is another distinct dNir species—cytochrome cd_1-type dNir (NirS)—which is also widely distributed among denitrifying bacteria.[1–3] Denitrifying bacteria usually contain either NirK or NirS, but not both. No correlation seems to exist between the phylogeny of bacteria and the type of dNir (NirK or NirS). For example, both types are distributed equally among α-proteobacteria; *Hyphomicrobium denitrificans* and *Bradyrhizobium* have NirK, whereas *Paracoccus denitrificans* contain NirS. Therefore, the distribution of dNir (NirK and NirS) among proteobacteria seems random, showing a sharp contrast to the systematic distribution of *nirK* homologues among eukaryotes. Furthermore, only NirK homologues (and not NirS) have been found in eukaryotic genomes. These results indicate that eukaryotic *nirK* homologues evolved from a single ancestor and diverged with each host that harboured the gene.

It is generally accepted that the mitochondrion of eukaryotes evolved as the result of endosymbiosis of an α-proteobacterium with an anaerobic host, and that the mitochondria of all eukaryotes have a common origin.[19] The results presented above suggest that eukaryotic *nirK* homologues originated from the protomitochondrion (the endosymbiont that gave rise to the mitochondrion), which harboured the NirK-type (but not the NirS-type) dNir. Although eukaryotic *nirK* homologues are localised in the nuclear genome, a mitochondrial targeting signal can be found in most of them. Some eukaryotes, in particular those harbouring both NirK and P450nor, such as *F. oxysporum*, have conserved the denitrification system derived from the protomitochondrion, although it has been modified (an adoption of P450nor).

14.5 Co-Denitrification

Fungal denitrification is often accompanied by a unique phenomenon, co-denitrification, in which a hybrid N_2 or N_2O species is formed by combining nitrogen atoms from nitrite and other nitrogen compounds (nitrogen donor).[5,26] A similar phenomenon was later found in the anaerobic ammonium oxidation reaction.[27] The ratio of denitrification and co-denitrification varies depending on the conditions (fungal strains and nitrogen donors). The co-denitrification product (N_2 or N_2O) varies depending on the redox state of the nitrogen donor. Amines provide N_2,[5] whereas imines or azide form N_2O as the co-denitrification product,[26] as shown below:

$$^{15}NO_2^- + R^{14}NH_2 \rightarrow {}^{15}N^{14}N \quad (14.3)$$

$$^{15}NO_2^- + R^{14}NHOH \rightarrow {}^{15}N^{14}NO \quad (14.4)$$

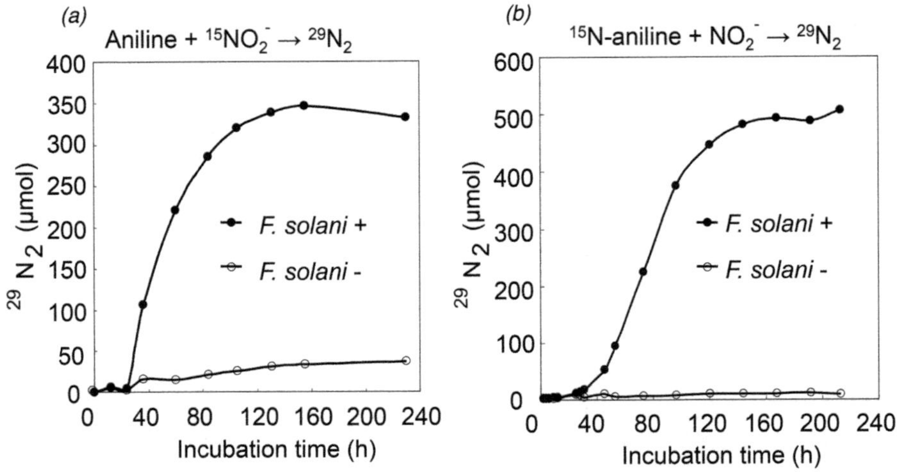

Figure 14.5 Co-denitrification by *Fusarium solani*. *F. solani* IFO 9425 was cultured in an Erlenmeyer flask containing the medium, which was composed of 2% glucose, 5 mM potassium nitrite (KNO_2; (a) ^{15}N-labelled or (b) non-labelled), 5 mM aniline ((a) ^{15}N-labelled or (b) non-labelled), 0.2% peptone and inorganic salts (pH 7.5), on a rotary shaker at 150 rpm, at 30 °C. The gas phase of the flask was replaced with helium to attain anaerobic conditions. Hybrid N_2 species ($^{14}N^{15}N$) were determined by gas chromatography–mass spectrometry. Reproduced from Figure 2 in Shoun *et al.*, *Phil. Trans. R. Soc. B*, 2012, **367**, 1186, with permission.

P450nor was shown to catalyse the co-denitrification reaction, forming N_2O and N_2 from NO and azide.[28] An external electron donor such as NADH, which is required for the Nor reaction by P450nor, is not necessary, indicating that the nitrogen donor also functions as an internal electron donor to reduce NO. Therefore, the direct reactant is NO rather than nitrite in co-denitrification reactions. In co-denitrification, nitrite would be reduced to NO by dNir (NirK), followed by the co-denitrification reaction by P450nor.[5]

F. solani IFO (NBRC) 9425 exhibits potent co-denitrification activity.[5] Among the three fungal strains tested (*F. oxysporum* MT811, *C. tonkinense* and *F. solani* IFO 9425), *F. solani* exhibited the highest co-denitrification activity against the nitrogen donor (aniline) (Figure 14.5). The recovery of nitrogen atoms from nitrite and aniline into N_2 is high (more than 50%). Inhibitors of NirK (diethyldithiocarbamate and cyanide) strongly inhibited co-denitrification, suggesting the involvement of NirK in the reaction. The same product (N_2) was formed when nitrite was replaced with NO, again suggesting that the direct reactant is NO rather than nitrite.[5,28] Therefore, the fungal denitrifying system can produce N_2 as the co-denitrification product, although it cannot form N_2 by denitrification.

14.6 P450nor (Fungal Nor)

Gene isolation showed that the nitrate-inducible haem protein belongs to the P450 superfamily, with the family number 55 (CYP55) being identified.[29] Interestingly, despite its eukaryotic origin, CYP55 exhibits a closer relationship

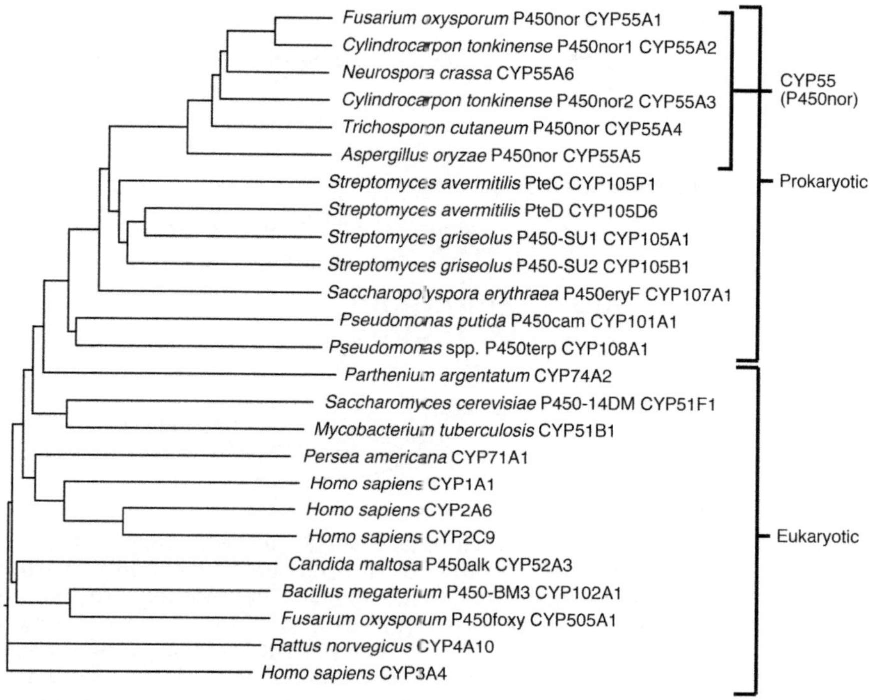

Figure 14.6 The phylogenetic tree of cytochrome P450s. The tree was generated using the Clustal Omega server.

to bacterial P450s than to eukaryotic P450s (Figure 14.6). The amino acid sequence of CYP55 shows sequence identities similar to the bacterial (actinomycetes) CYP105 members of up to 40%. Therefore, the fungus may have acquired the P450 gene from actinomycetes by horizontal gene transfer.[29]

P450 usually functions as a monooxygenase. Two electrons are required in the reaction to activate molecular oxygen (O_2), which are supplied by NAD(P)H (Figure 14.7). NAD(P)H releases its two electrons simultaneously as hydride ions (H^-), whereas the haem of P450 is a one-electron redox centre and thus can never receive the two electrons directly. Therefore, the two electrons transported from NAD(P)H to P450 are supported by electron transport systems containing flavoproteins, and there are two types in the electron transport system. Bacterial and mitochondrial P450s are supplied with electrons by the coupled ferredoxin reductase and ferredoxin, whereas eukaryotic (microsomal) P450s are supplied by a P450 reductase containing FAD and FMN. Surprisingly, however, P450nor was shown to receive two electrons directly from NAD(P)H.[8] This phenomenon opposed the central dogma of physiological electron transfer. P450nor is an exceptional P450[15] that does not require a redox partner (direct electron transfer from NAD(P)H).

The direct electron transfer from NADH to the haem of P450nor was conclusively demonstrated by kinetic analysis[30] (Figure 14.8) and by determining

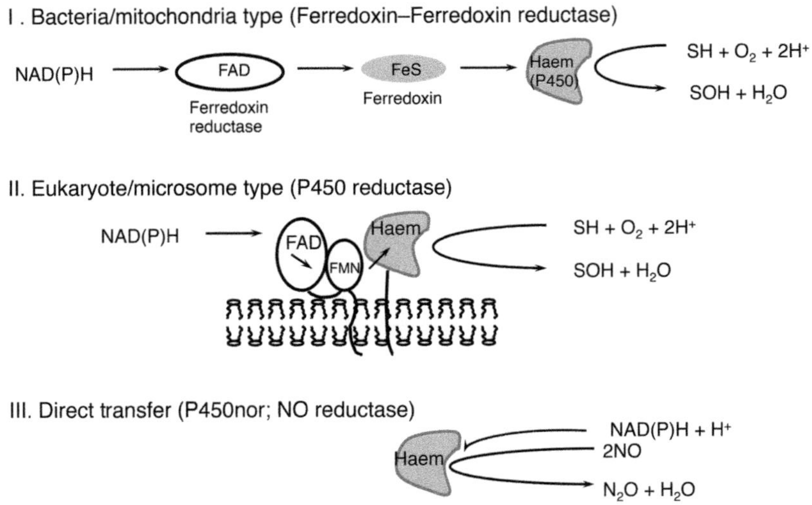

Figure 14.7 Classifications of cytochrome P450 (P450) depending on the type of redox partner. Electrons of NAD(P)H are transferred to P450 as indicated by arrows *via* each component. P450nor (type III) receives electrons directly from NAD(P)H and thus does not require redox partners. Components of the bacterial P450 system are water soluble. Mitochondrial P450s are membrane bound (inner membrane), whereas mitochondrial ferredoxin reductase and ferredoxin are soluble in the matrix. Components of type II (eukaryotic, microsomal type) are membrane bound. P450nor is water soluble, localised to both mitochondria and cytosol. SH represents the organic substrate to be hydroxylated by the P450 (monooxygenase) reaction. Reproduced from Figure 4 in Shoun *et al.*, *Phil. Trans. R. Soc. B*, 2012, **367**, 1186, with permission.

the crystal structure of a P450nor (GG mutant) complexed with an NADH analogue (nicotinic acid adenine dinucleotide [NAAD]).[31] The structure of the P450nor–NAAD complex was compared with the ferric–NO complex of P450nor[32] in Figure 14.9. Little difference was observed between the structures of P450nor in the ferric resting state[33] and in the ferric–NO complex.[32] Conversely, a remarkable conformational change in the protein was induced upon the binding of NAAD; the entrance gate of the haem distal pocket was closed. Two Arg residues, Arg_{64} and Arg_{174}, play a key role in the binding[34] by putting the pyrophosphate moiety of NAAD between them. Glu_{71}, Arg_{64} and Asp_{88} form a salt bridge network to stabilise the protein structure.[30,35] The interaction between Arg_{64} and Asp_{88} is broken once NAAD binds to destabilise the protein. This provides a driving force to rapidly remove a NAD^+ molecule from the active site after electron transfer is finished. One of the propionate side chains of the haem moves up to accompany the movement of Arg_{292}, which fixes the nicotinamide ring stereochemically. The conserved Thr residue (Thr_{243}) in the I-helix interacts with the carboxyl of the nicotinic acid ring to fix it stereochemically. This interaction of Thr_{243}, along with the

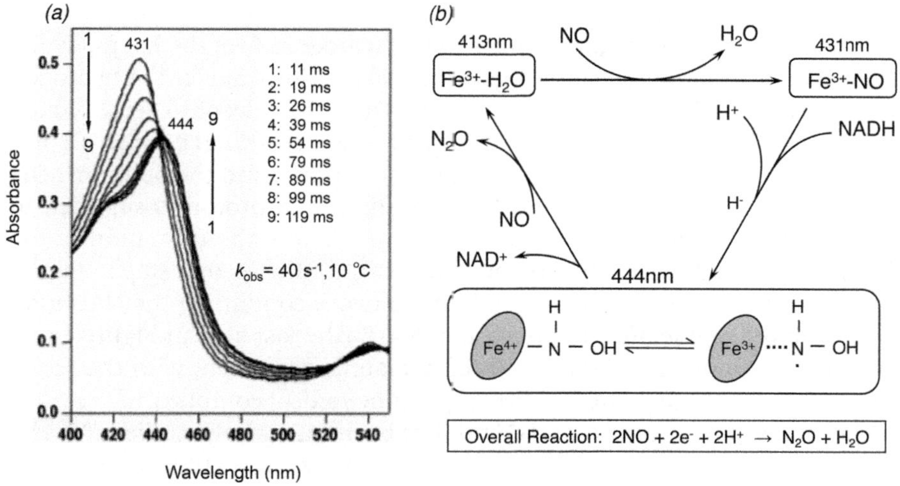

Figure 14.8 The reaction mechanism of P450nor. (a) Spectral changes in the bound haem during anaerobic reduction of the ferric (Fe_3^{+})–nitric oxide (NO) complex (431 nm species) with NADH to form a specific intermediate, I (444 nm species), observed by rapid scan analyser.[37] (b) Reaction scheme of P450nor. The structure of I (444 nm species) as a Fe_3^{+}–hydroxylamine radical complex was proposed by Daiber *et al.*[36] Reproduced from Figure 5 in Shoun *et al.*, *Phil. Trans. R. Soc. B*, 2012, **367**, 1186, with permission.

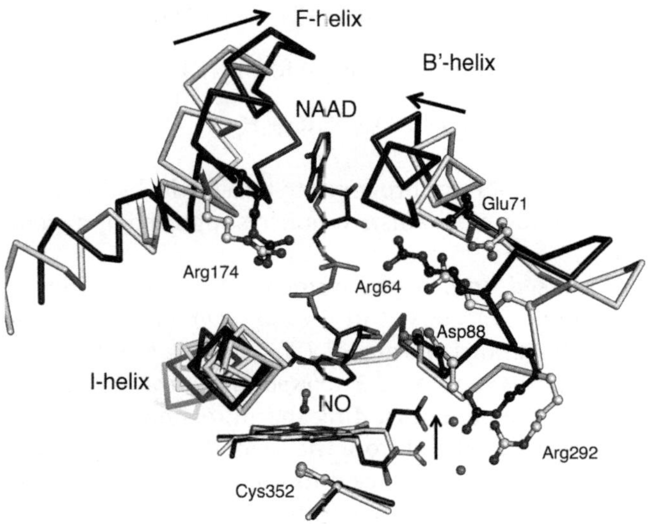

Figure 14.9 The movement of P450nor during cofactor binding. The superposition of the ferric–nitric oxide (NO) complex (white) and NAAD complex (black). Reproduced from Figure 6 in Shoun *et al.*, *Phil. Trans. R. Soc. B*, 2012, **367**, 1186, with permission.

propionate of haem moving upward, restricts the conformation of the nicotinic acid ring so that the pro-*R* sides of C4-hydrogens face the haem, which is consistent with the pro-*R* hydrogen-specific hydride transfer.[36] A hydrogen bond network is formed to deliver a proton from the solvent to Ser_{286}, which is located near the haem.[32] However, the network is rearranged to form a proton channel once NADH (NAAD) binds to the protein, and the bound NADH is itself involved in the network,[31] suggesting that a proton is supplied to the enzymatic reaction *via* the proton channel before the formation of the intermediate (with an absorption band at 444 nm) (the second step in Figure 14.8). This is because the hydrogen bond network containing the NAD molecule would degrade after the release of NAD^+ (the last step in Figure 14.8). The proton supplied to form the intermediate is consistent with the structure of the intermediate (ferric–hydroxylamine radical complex).[5]

The reaction mechanism of P450nor has been extensively studied.[36,37] The turnover of the overall reaction (eqn (14.5)) is very rapid (1000 s^{-1} at 10 °C) and thus should be of the order of 10^5 min^{-1} or more at 25 °C.[37]

$$2NO + NADH + H^+ \rightarrow N_2O + H_2O + NAD^+ \tag{14.5}$$

The overall reaction can be divided into three steps (Figure 14.8). The first substrate (NO) binds to ferric (Fe^{3+}) P450nor to form a ferric–NO complex:

$$Fe_3^+ + NO \rightarrow Fe_3^+ - NO \tag{14.6}$$

Fe^{3+}–NO is then reduced by NADH to form a specific intermediate (I) with a Soret absorption band at 444 nm:

$$Fe_3^+ - NO + NADH + H^+ \rightarrow I + NAD^+ \tag{14.7}$$

Finally, I interacts with the second NO to form N_2O:

$$I + NO \rightarrow N_2O + H_2O \tag{14.8}$$

The chemical entity of I was proposed to be a ferric–hydroxylamine radical complex.[36] The reaction mechanism was also supported by a quantum chemical calculation.[38]

P450nor is localised both in the mitochondria and in the cytoplasm of fungal cells. P450nor is localised in *F. oxysporum* and *C. tonkinense* in different ways. Two P450nor isoforms of *F. oxysporum* are derived from a single gene (*CYP55A1*). P450norA of *F. oxysporum* is translated from the first initiation codon of the gene including the mitochondrial targeting signal, whereas P450norB is translated from the second initiation codon below the targeting signal and thus is localised to the cytosol.[39,40] *C. tonkinense* contains two P450nor genes: P450nor1 (*CYP55A2*) and P450nor2 (*CYP55A3*).[41] *CYP55A2* contains a sequence for a mitochondria-targeting signal, whereas *CYP55A3* does not. P450nor1 specifically employs NADH as the electron donor, while P450nor2 prefers NADPH to NADH, although NADH can provide sufficient activity.[42] The electron donor specificity depends on the amino acid residues at two positions in the B′-helix (73rd and 75th positions in the case of CYP55A1).[34,43] Steric hindrance due to side chains of Ser_{73} and Ser_{75} in CYP55A1 excludes the 2′-phosphate moiety of NADPH from the site. In P450nor2 of *C. tonkinense*, Ser_{75} is replaced with a Gly residue, which is able to accommodate NADPH. Double mutations at these P450nor sites in *F. oxysporum* (S73G/S75G; GG mutant) greatly improved the specificity for NADPH.[31]

14.7 Occurrence of Fungal Denitrification and Co-Denitrification in Ecosystems

N_2O is a potent greenhouse gas with a 100-year global warming potential approximately 300-times that of CO_2, and it has also become the dominant stratosphere ozone-depleting substance.[44] N_2O is a product of two microbial activities: nitrification and denitrification.[45] With the rise of agricultural activities and the associated use of fertilisers, atmospheric N_2O levels (325 ppb in 2012, 19% higher than preindustrial levels) are currently increasing at a rate of 0.25% per year.[46] Therefore, it is of great interest how fungal denitrification contributes to the nitrogen cycle in the natural environment, because the main product of fungal denitrification is N_2O.

About a decade after our reports on fungal denitrification and co-denitrification, Laughlin and Stevens reported fungal dominance of denitrification and co-denitrification in grassland soil.[10] They showed that denitrification and co-denitrification by fungi in grassland soil may be responsible for up to 90% of the N_2O produced. More recently, Maeda *et al.* indicated that N_2O production is a common trait in the fungal taxa that are frequently abundant in soils.[45] They inoculated 15 fungal strains into previously sterilised arable, forest and grassland soils and showed that fungi contribute to soil N_2O emission, with fluxes potentially as high as those reported in natural soils. The importance of fungal denitrification and co-denitrification in ecosystems was further shown by numerous reports.[46–53] Long *et al.* showed that fungal co-denitrification is the primary process contributing to N_2 production in North Carolina soil.[49] It can be concluded from these results that fungal denitrification and co-denitrification are involved in the nitrogen cycle as one of the major pathways in ecosystems.

14.8 Concluding Remarks

Most of the fungal denitrifying systems are composed of two reducing steps (eqn (14.9)), containing NirK and P450nor as essential components.

$$NO_2^- \rightarrow NO \rightarrow N_2O \qquad (14.9)$$

These two genes are the minimum pair required for denitrification from nitrite to N_2O. Some fungi further utilise dNar (NarGHI type in the case of

F. oxysporum MT811) and/or aNar, which also enable the denitrification of nitrate. NirK and dNar are associated with the mitochondrial respiratory chain, coupled to ATP synthesis. This is the first example of anaerobic respiration in mitochondria.[19] Because the dNar (NarGHI) homologue gene is rarely observed in fungal genomes, the system in *C. tonkinense* is representative of fungal denitrification systems, which denitrifies nitrate by utilising aNar. P450nor and aNar receive electrons directly from NAD(P)H and thus are not associated with the respiratory chain and function as an electron sink under anoxic conditions. Only the NirK process contributes to ATP production in *C. tonkinense*. Thus, the ATP-producing metabolism, with its low efficiency of ATP production, reflects the fungal strategy for survival under anoxic conditions where the preference is for the speed of metabolism (degradation of carbohydrates such as glycolysis) over energy efficiency. The involvement of aNar or an assimilatory nitrate-reducing system (aNar and aNir) in denitrification[24] and ammonia fermentation[23] was the first to show the use of an assimilatory system for a dissimilatory purpose (ATP production).

Eukaryotic *nirK* and its homologue genes originated from the same ancestor, possibly the protomitochondrion, which harboured NirK-type (but not NirS-type) dNir.[25] Thus, the fungal and bacterial denitrifying systems share the same origin. On the other hand, P450nor is unique to the fungal system. Consequently, the mitochondrial denitrifying system replaced the original Nor protein with P450nor, whose gene was initially obtained from bacteria (actinomycetes) *via* horizontal gene transfer. The prototype P450 gene would have encoded the usual monooxygenase, whereas fungi would have modulated the gene to produce Nor activity because P450nor is not currently found in bacteria.

P450 proteins, belonging to the P450 superfamily, are among the most diversified enzyme proteins. However, even among such diversified P450 proteins, the function of P450nor is peculiar.[15] The function of P450nor is thus atypical of the most diversified P450s. The mechanism of the stereospecific transfer of H^- from NADH to the haem of P450nor has been elucidated. In addition to Nor activity, P450nor will catalyse the co-denitrification reaction. We have also found that P450nor exhibits NADH-peroxidase activity (Nakaya *et al.*, unpublished).

$$H_2O_2 + NADH + H^+ \rightarrow 2H_2O + NAD^+ \qquad (14.10)$$

P450nor is therefore a multifunctional detoxifying enzyme. P450nor is also related to the pathogenicity of a fungus.[54]

Co-denitrification is the first process to show the formation of a hybrid N_2 or N_2O species.[5,27] This depends on the reaction of NO with a nitrogen donor, which is possibly catalysed by P450nor and thus is characteristic of fungal denitrification. Co-denitrification products (N_2 or N_2O) vary depending on the redox state of the nitrogen donor: N_2 is formed from amines and N_2O is formed from imines or azide. This suggests that the nitrogen donors also act as internal electron donors. Therefore, the co-denitrification process is considered to be a kind of Nor reaction employing amines or imines as internal electron donors (and nitrogen donors). The molecular mechanism of co-denitrification needs further elucidation. The mechanism of co-denitrification by the bacterium *Streptomyces antibioticus*,[55] in which a very small amount of a hybrid N_2 species is formed, needs to be investigated further because P450nor is not found in bacteria.

In addition to the differences between the denitrifying systems of *F. oxysporum* and *C. tonkinense*, their carbon sources are also different. In *F. oxysporum*, denitrification is repressed and hetero-lactic acid fermentation dominates when glucose is available under anoxic conditions.[56] In *C. tonkinense*, denitrification is not repressed by glucose, yet works in parallel with glycolysis *via* the pentose phosphate shunt.[57] The NADPH-specific P450nor isozyme (P450nor2), which is localised in the cytosol, functions as an electron sink for the pentose phosphate shunt.

Genome analyses have also revealed that not only *nirK* homologue genes, but also the genes homologous to *CYP55* (P450nor) and *nap* (periplasmic nitrate reductase), are found in many fungal genomes. Our blast results indicated that out of 72 fungal genomes, 19 contained *nirK* homologies, 16 contained *CYP55* homologues and 15 contained *nap* homologues (26.4–20.8%), and many of these genomes contained both *nirK* and *CYP55*. The high rate of the appearance of *nirK* and *CYP55* homologues suggests that the denitrifying system comprising NirK and P450nor is widely distributed among fungi. By contrast, no genome contained a *narGHI* homologue, suggesting that the fungal denitrifying system that contains NarGHI, such as *F. oxysporum* MT811, is minor. The significance of the presence of *nap* homologues in many fungal genomes remains to be understood.

Fungal denitrification and co-denitrification are involved in the nitrogen cycle as major pathways in nature. This is supported by the distribution of *nirK* (dNir) and *CYP55* (P450nor) gene homologues in many fungal genomes (more than 20%) and recent papers identifying the predominance of fungal activities in various environments, such as arable fields, grasslands and forests. Because the final product of fungal denitrification is N_2O, fungal denitrification is one of major sources of N_2O emissions. Acidification of environments (*e.g.*, by acid rain and excess use of ammonia fertiliser) will promote fungal activity and result in increased N_2O emissions.

Acknowledgements

This work was a collaboration of the author (HS) with many researchers and students at the University of Tsukuba, The University of Tokyo, Riken, University of Konstanz, Osaka Prefecture University and Chiba University. This work was supported by grants-in-aid for scientific research from the Japan Society for the Promotion of Science and the Research and Development Program for New Bio-Industry Initiatives. We thank the staff at Photon Factory and SPring-8 for X-ray data collection.

References

1. B. C. Berks, S. J. Ferguson, J. W. Moir and D. J. Richardson, *Biochim. Biophys. Acta*, 1995, **1232**, 97.
2. W. G. Zumft, *Microbiol. Mol. Biol. Rev.*, 1997, **61**, 533.
3. S. J. Ferguson, *Curr. Opin. Chem. Biol.*, 1998, **2**, 182.
4. H. Shoun and T. Tanimoto, *J. Biol. Chem.*, 1991, **266**, 11078.
5. H. Shoun, D. H. Kim, H. Uchiyama and J. Sugiyama, *FEMS Microbiol. Lett.*, 1992, **73**, 277.
6. S. Tsuruta, N. Takaya, L. Zhang, H. Shoun, K. Kimura, M. Hamamoto and T. Nakase, *FEMS Microbiol. Lett.*, 1998, **168**, 105.
7. B. A. Averill, *Chem. Rev.*, 1996, **96**, 2951.
8. K. Nakahara, T. Tanimoto, K. Hatano, K. Usuda and H. Shoun, *J. Biol. Chem.*, 1993, **268**, 8350.
9. N. Takaya and H. Shoun, *Mol. Gen. Genet.*, 2000, **263**, 342.
10. R. J. Laughlin and R. J. Stevens, *Soil Sci. Soc. Am. J.*, 2002, **66**, 1540.
11. W. K. Ma, R. E. Farrell and S. D. Siciliano, *Appl. Environ. Microbiol.*, 2008, **74**, 6690.
12. R. L. Sutka, G. C. Adams, N. E. Ostrom and P. H. Ostrom, *Rapid Commun. Mass Spectrom.*, 2008, **22**, 3989.
13. S. J. Cathrine and C. Raghukumar, *Mycol. Res.*, 2009, **113**, 100.
14. H. Shoun, Y. Sudo, Y. Seto and T. Beppu, *J. Biochem.*, 1983, **94**, 1219.
15. T. Omura, *J. Biochem.*, 2010, **147**, 297.
16. H. Shoun, W. Suyama and T. Yasui, *FEBS Lett.*, 1989, **244**, 11.
17. M. Kobayashi and H. Shoun, *J. Biol. Chem.*, 1995, **270**, 4146.
18. M. Kobayashi, Y. Matsuo, A. Takimoto, S. Suzuki, F. Maruo and H. Shoun, *J. Biol. Chem.*, 1996, **271**, 16263.
19. A. G. Tielens, C. Rotte, J. J. van Hellemond and W. Martin, *Trends Biochem. Sci.*, 2002, **27**, 564.
20. H. Uchimura, H. Enjoji, T. Seki, A. Taguchi, N. Takaya and H. Shoun, *J. Biochem.*, 2002, **131**, 579.
21. N. Takaya, S. Kuwazaki, Y. Adachi, S. Suzuki, T. Kikuchi, H. Nakamura, Y. Shiro and H. Shoun, *J. Biochem.*, 2003, **133**, 461.
22. S. Kuwazaki, N. Takaya, A. Nakamura and H. Shoun, *Biosci., Biotechnol., Biochem.*, 2003, **67**, 937.
23. Z. Zhou, N. Takaya, A. Nakamura, M. Yamaguchi, K. Takeo and H. Shoun, *J. Biol. Chem.*, 2002, **277**, 1892.
24. T. O. Watsuji, N. Takaya, A. Nakamura and H. Shoun, *Biosci., Biotechnol., Biochem.*, 2003, **67**, 1115.
25. S. W. Kim, S. Fushinobu, S. Zhou, T. Wakagi and H. Shoun, *Appl. Environ. Microbiol.*, 2009, **75**, 2652.
26. T. Tanimoto, K. Hatano, D. H. Kim, H. Uchiyama and H. Shoun, *FEMS Microbiol. Lett.*, 1992, **93**, 177.
27. M. Strous, J. A. Fuerst, E. H. Kramer, S. Logemann, G. Muyzer, K. T. van de Pas-Schoonen, R. Webb, J. G. Kuenen and M. S. Jetten, *Nature*, 1999, **400**, 446.

28. F. Su, N. Takaya and H. Shoun, *Biosci., Biotechnol., Biochem.*, 2004, **68**, 473.
29. H. Kizawa, D. Tomura, M. Oda, A. Fukamizu, T. Hoshino, O. Gotoh, T. Yasui and H. Shoun, *J. Biol. Chem.*, 1991, **266**, 10632.
30. M. Umemura, F. Su, N. Takaya, Y. Shiro and H. Shoun, *Eur. J. Biochem.*, 2004, **271**, 2887.
31. R. Oshima, S. Fushinobu, F. Su, L. Zhang, N. Takaya and H. Shoun, *J. Mol. Biol.*, 2004, **342**, 207.
32. H. Shimizu, E. Obayashi, Y. Gomi, H. Arakawa, S. Y. Park, H. Nakamura, S. Adachi, H. Shoun and Y. Shiro, *J. Biol. Chem.*, 2000, **275**, 4816.
33. S. Y. Park, H. Shimizu, S. Adachi, A. Nakagawa, I. Tanaka, K. Nakahara, H. Shoun, E. Obayashi, H. Nakamura, T. Iizuka and Y. Shiro, *Nat. Struct. Biol.*, 1997, **4**, 827.
34. T. Kudo, N. Takaya, S. Y. Park, Y. Shiro and H. Shoun, *J. Biol. Chem.*, 2001, **276**, 5020.
35. F. Su, S. Fushinobu, N. Takaya and H. Shoun, *Biosci., Biotechnol., Biochem.*, 2004, **68**, 1156.
36. A. Daiber, T. Nauser, N. Takaya, T. Kudo, P. Weber, C. Hultschig, H. Shoun and V. Ullrich, *J. Inorg. Biochem.*, 2002, **88**, 343.
37. Y. Shiro, M. Fujii, T. Iizuka, S. Adachi, K. Tsukamoto, K. Nakahara and H. Shoun, *J. Biol. Chem.*, 1995, **270**, 1617.
38. N. Lehnert, V. K. Praneeth and F. Paulat, *J. Comput. Chem.*, 2006, **27**, 1338.
39. K. Nakahara and H. Shoun, *J. Biochem.*, 1996, **120**, 1082.
40. N. Takaya, S. Suzuki, S. Kuwazaki, H. Shoun, F. Maruo, M. Yamaguchi and K. Takeo, *Arch. Biochem. Biophys.*, 1999, **372**, 340.
41. T. Kudo, D. Tomura, D. L. Liu, X. Q. Dai and H. Shoun, *Biochimie*, 1996, **78**, 792.
42. K. Usuda, N. Toritsuka, Y. Matsuo, D. H. Kim and H. Shoun, *Appl. Environ. Microbiol.*, 1995, **61**, 883.
43. L. Zhang, T. Kudo, N. Takaya and H. Shoun, *J. Biol. Chem.*, 2002, **277**, 33842.
44. A. R. Ravishankara, J. S. Daniel and R. W. Portmann, *Science*, 2009, **326**, 123.
45. K. Maeda, A. Spor, V. Edel-Hermann, C. Heraud, M. C. Breuil, F. Bizouard, S. Toyoda, N. Yoshida, C. Steinberg and L. Philippot, *Sci. Rep.*, 2015, **5**, 9697.
46. H. Yang, H. Gandhi, N. E. Ostrom and E. L. Hegg, *Environ. Sci. Technol.*, 2014, **48**, 10707.
47. O. Spott and C. F. Stange, *Rapid Commun. Mass Spectrom.*, 2007, **21**, 2398.
48. O. Spott, R. Russow and C. F. Stange, *Soil Biol. Biochem.*, 2011, **43**, 1995.
49. A. Long, J. Heitman, C. Tobias, R. Philips and B. Song, *Appl. Environ. Microbiol.*, 2013, **79**, 168.
50. S. F. Bender, F. Plantenga, A. Neftel, M. Jocher, H. R. Oberholzer, L. Kohl, M. Giles, T. J. Daniell and M. G. van der Heijden, *ISME J.*, 2014, **8**, 1336.
51. L. Rohe, T. H. Anderson, G. Braker, H. Flessa, A. Giesemann, D. Lewicka-Szczebak, N. Wrage-Monnig and R. Well, *Rapid Commun. Mass Spectrom.*, 2014, **28**, 1893.

52. W.-L. Liu, M. Guan, S.-Y. Liu, J. Wang, J. Chang, Y. Ge and C.-B. Zhang, *Ecol. Eng.*, 2015, **77**, 163.
53. K. K. Treseder and J. T. Lennon, *Microbiol. Mol. Biol. Rev.*, 2015, **79**, 243.
54. M. P. Nittler, D. Hocking-Murray, C. K. Foo and A. Sil, *Mol. Biol. Cell*, 2005, **16**, 4792.
55. Y. Kumon, Y. Sasaki, I. Kato, N. Takaya, H. Shoun and T. Beppu, *J. Bacteriol.*, 2002, **184**, 2963.
56. Z. Zhou, N. Takaya and H. Shoun, *Biosci., Biotechnol., Biochem.*, 2010, **74**, 2431.
57. T. O. Watsuji, N. Takaya, A. Nakamura and H. Shoun, *Biosci., Biotechnol., Biochem.*, 2003, **67**, 1109.

CHAPTER 15

Denitrification and Non-Denitrifier Nitrous Oxide Emission in Gram-Positive Bacteria

YI-HUA SUN[a], PAUL DE VOS[a] AND KIM HEYLEN*[a]

[a]Laboratory of Microbiology, Department of Biochemistry and Microbiology, University of Ghent, K.L. Ledeganckstraat 35, 9000 Ghent, Belgium
*E-mail: kimheylen@hotmail.com

15.1 Denitrification is Widespread but Underexplored in Gram-Positive Bacteria

Denitrification is a facultative anaerobic process involving the stepwise dissimilatory reduction of nitrate over nitrite to the cytotoxic nitric oxide (NO), the potent greenhouse gas nitrous oxide (N_2O) and the inert and harmless dinitrogen gas (N_2), coupled to electron transport phosphorylation.[1] It is a highly modular process, meaning that the four reduction steps are not necessarily linked in a microorganism and that the intricate metalloproteins involved in each reduction can occur separately or in any combination possible.[2,3] So, the presence of one or more of their encoding genes is not *per se* indicative of the capacity to denitrify. In addition, over the last decades, it has become generally accepted that not only denitrification, but also dissimilatory

RSC Metallobiology Series No. 9
Metalloenzymes in Denitrification: Applications and Environmental Impacts
Edited by Isabel Moura, José J. G. Moura, Sofia R. Pauleta and Luisa B. Maia

Published by the Royal Society of Chemistry, www.rsc.org

nitrate reduction to ammonium (DNRA) and nitrite detoxification, can produce nitrogen gases from nitrate or nitrite, albeit as side products and not intermediate or major end products, calling into question the classic phenotypic screening methods for denitrification.[4] We adhere to the most recent definition of denitrification[1] and consider those microorganisms that can at least convert nitrite to N_2O in stoichiometric amounts (with at least 80% N converted) while conserving energy for growth as a "true denitrifier".

Gram-positive bacteria stain purple with the classical Gram staining because of their distinctive cell wall structure with a typically thick, multilayered peptidoglycan lacking an outer membrane and periplasm, although a small "periplasm-like" space exists between the cytoplasmic membrane and the peptidoglycan. The latter feature is of importance as most denitrification enzymes are either periplasmic (periplasmic nitrate reductase, both NirS- and NirK-type nitrite reductases and N_2O reductase) or membrane associated with the active site in the periplasm (NO reductase), and so this might have consequences for the organisation of the denitrification proteome in Gram-positive denitrifiers (see Section 15.2). Gram-positive bacteria are phylogenetically separated into two phyla, *Firmicutes* and *Actinobacteria*; this subdivision was historically based on their guanine plus cytosine ratio in DNA of low GC (well below 50%) and high GC (well above 50%), respectively. Since the introduction of the term "denitrification" by Gayon and Dupetit in 1882 and the subsequent search for its aetiological agents, several Gram-positive genera have been described to contain denitrifiers, although these were always severely outnumbered by the description of their Gram-negative counterparts in the lists of denitrifying taxa compiled based on literature surveys.[5–8] In 2011, we scrutinised the described Gram-positive denitrifiers by assessing the experimental and molecular data available,[4] and more recently, Shapleigh[2] did a similar exercise focusing on those *Firmicutes* and *Actinobacteria*, for which whole-genome sequences were available. In summary, most Firmicute denitrifiers belong to endospore-forming species of the genus *Bacillus* or close relatives, termed "bacilli" in the remainder of this chapter, such as *Bacillus azotoformans*,[1,9] *Geobacillus thermodenitrificans*,[10] *Virgibacillus halodenitrificans*[11,12] and multiple *Paenibacillus* species,[13,14] while Actinobacterial denitrifiers are mostly limited to the *Actinomycetales* with both spore formers such as *Streptomyces*[15,16] and non-spore formers such as *Corynebacterium*.[17,18] As a general feature, they seem to have a truncated denitrification pathway, most often lacking a N_2O reductase, which suggests that they might be vigorous N_2O emitters in their habitats.

Although the denitrifying capacities of most bacilli have been known for a very long time, denitrification-related biochemistry, physiology, regulation and environmental surveys have almost exclusively focused on Gram-negative denitrifiers. As for other non-proteobacterial lineages like *Bacteroidetes*, the main reasons as to why this trait is underexplored in Gram-positive bacteria are its strain-dependent nature, the mismatches with primers targeting denitrification genes due to their high sequence divergence (for more detail, see Section 15.4) and the relatively uncommon nature of the trait in its model organisms such as *Bacillus subtilis*, *Bacillus cereus* or even in type strains of newly described species.[4] Nevertheless, our screening of a collection of 180 *Bacillus* strains, at that time covering half of the validly described taxonomic diversity of the genus, revealed that around 25% were able to produce stoichiometric amounts of nitrogen gas from nitrate and/or nitrite while supporting growth.[19] That study suggested that denitrification is potentially widespread and clearly has been underestimated in strains of the genus *Bacillus* and, by extrapolation, in Gram-positive bacteria. For the remainder of this chapter, we will describe the current understanding of denitrification in *B. azotoformans*, one of the few *Bacillus* species that contains the complete denitrification pathway and for which the genome sequence is available, including details on other bacilli when relevant. We will also highlight the old and more recent insights into N_2O emission by non-denitrifying *Bacillus* strains, as this might be another, highly undervalued trait in bacilli, often confused with denitrification. We will end the chapter by exploring the methodological issues hampering the assessment of the environmental abundance and importance of denitrifying bacilli.

15.2 Novel Features in the Denitrifier *B. azotoformans*

15.2.1 Membrane-Bound Enzymes and a Novel NO Reductase

B. azotoformans LMG 9581^T was originally isolated from garden soil and has been recognised as a true denitrifier for decades.[1,9,20] It is capable of the complete dissimilatory nitrate reduction to N_2. In Gram-negative denitrifiers, the canonical denitrification pathways are carried out by at least two periplasmic reductases, the copper- or cytochrome cd_1-dependent nitrite reductases (NirK or NirS, respectively) and the copper-dependent N_2O reductase (NosZ). Activity of all four reduction steps was shown to be membrane associated in *B. azotoformans* LMG 9581^T,[21] which was already previously demonstrated for NirK in *V. halodenitrificans*.[12] *In silico* sequence analyses of the *nirK* and *nosZ* genes of *B. azotoformans* and other bacilli[22] indeed predicted that they encode lipoproteins that covalently bind to the membrane, while being preceded by a Sec or Tat signal, respectively, for protein export, indicative of a periplasmic localisation of the processed protein. So, despite Gram-positive denitrifiers only having a small "periplasm-like" space, the periplasmic reductases indeed appear to be located there, but as membrane-bound variants.

In addition to the more common cytoplasmic membrane protein complex nitrate reductase (Nar), a periplasmic variant (Nap) is also known; both bind a molybdenum bis-molybdopterin guanine dinucleotide (Mo-bis-MGD) with a 4Fe–4S cluster at the catalytic subunit for electron transfer.[23] Surprisingly, for the first time in a Gram-positive bacterium, a complete *nap* operon was found in the genome of *B. azotoformans* LMG 9581^T,[22] which, based on *in silico* analyses, seemed to constitute a new variant in addition to the four known Nap systems.[24] This putative new organisational structure of Nap was described in detail previously[22] and is shown in Figure 15.1. The *nap* operon lacks genes for the cytoplasmic maturation factors NapF and NapL, as well as the quinol-oxidising membrane-bound NapC, but contains two gene copies for NapG, as well as a gene coding for one of the enzymes involved in Mo-bis-MGD biosynthesis. In this Nap system, NapA represents the catalytic subunit and NapD is involved in the post-translational assembly of NapA, which receives its electrons for nitrate reduction from the companion dihaem *c* protein NapB. With NapA having a distinct Tat signal and NapB possessing a Sec signal, it can be inferred that both of them are exported to the periplasm. As in other organisms,[25–27] NapH, a membrane-bound enzyme specifically oxidising menaquinol, and NapG2, a periplasmic adaptor protein delivering electrons from menaquinol oxidation,[27] most likely form a membrane-bound complex for the transfer of electrons. As expected, NapG2 has a Tat signal while NapG1 does not, so the latter might substitute for NapF found in other organisms at the cytoplasmic side. As such, *B. azotoformans* LMG 9581^T has a NapAB and a NapGH module, possibly assembled as one membrane-bound complex, for menaquinol-dependent nitrate reduction to nitrite (Figure 15.1). As is the case for other Nap systems, its topology and architecture suggest it will not contribute to the generation of a proton motive force.

All NO reductases are integral membrane proteins and three kinds have been described thus far, all belonging to the haem–copper oxidase superfamily: cytochrome *c*-oxidising cNor, the quinol-dependent qNor and the Cu_ANor with cytochrome c_{551} as an electron donor.[28] The latter Nor, encoded by the *cbaBA* operon,[22] was discovered in *B. azotoformans* LMG 9581^T by the research group

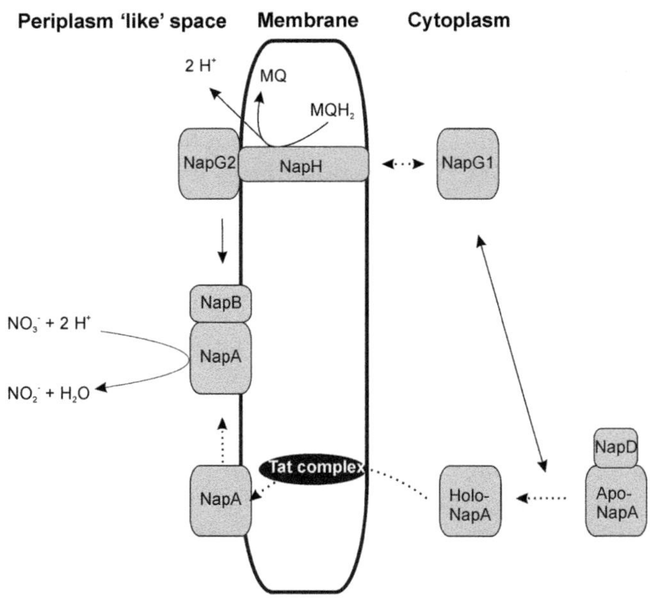

Figure 15.1 Proposed novel organisation of periplasmic nitrate reductase in *Bacillus azotoformans*. The scheme is based on the four Nap organisational structures described previously.[24]

of Simon de Vries and was previously thought to be bifunctional, being able to accept electrons from both menaquinol and cytochrome *c*.[21,29,30] The cNor is considered not to be electrogenic as it takes both electrons and protons from the periplasm.[31–33] In contrast, NO reduction by qNor could be electrogenic, as the crystal structure of the qNor from the Gram-positive *Geobacillus stearothermophilus*[34] revealed the presence of a putative proton transfer pathway between the cytoplasm and the binuclear metal centres. Recently, work with Cu_ANor reconstituted in closed liposomes provided the first experimental evidence of the formation of a proton electrochemical gradient across the membrane.[28] The proton electrochemical gradient is formed because protons are taken from the cytoplasm, causing a decrease in cytoplasmic charge and proton concentration; actual proton pumping has yet to be determined. This means that denitrifiers with Cu_ANor could exploit NO reduction for increased cellular ATP production (6.7% higher than those with cNor or even 13% if Cu_ANor would also pump two protons).[28] Thus far, all Gram-positive denitrifiers harbour either a gene for a Cu_ANor, a qNor or both, suggesting that they have the potential to conserve more energy from denitrification than their Gram-negative counterparts.

15.2.2 High Genetic Redundancy and Potential Metabolic Versatility

Genome analyses of *B. azotoformans* LMG 9581^T revealed a strikingly high gene redundancy for the canonical denitrification pathway.[22] This organism encodes two Nar and one Nap for nitrate reduction to nitrite, one NirK for nitrite reduction to NO, two qNor and two Cu_ANor for NO reduction to N_2O and three NosZ for N_2O reduction to N_2 (Figure 15.2). Hypothetically, 36 different gene combinations could result in a complete denitrification pathway. As Jeff Cole remarked for the alternative sets of nitrate reductase genes in *Escherichia coli*, "Either their co-existence is just an accident of evolution that arose after trivial gene duplication or acquisition event, or these sets of genes remained conserved long after they evolved because they are physiologically useful to the organism, otherwise they would have been eliminated by evolutionary selective pressures."[35] We concur with the latter hypothesis and assume that the various enzymes for the same N conversion will become expressed in different environmental contexts and/or may have slightly different structural roles. Indeed, even copies of the same enzyme differed quite substantially in amino acid sequence identity (NarG1–NarG2: 74.9%; qNor1–qNor2: 38%; NosZ1–NosZ2–NosZ3: 76.5–83.1%). The denitrification pathway might be completely modular or it can have preferential combinations of specific gene sets due to their regulation. Also, some denitrification genes might be constitutively expressed while others might depend on specific environmental triggers, or the proteome might contain multiple enzymes for the same reduction step at the same moment. In addition, it is plausible that under some conditions (*e.g.*, low pH), only truncated denitrification is carried out.[36,37]

To our surprise, the genome of *B. azotoformans* LMG 9581^T also contains the gene for the periplasmic pentahaem cytochrome *c* nitrite reductase

NrfA[38] involved in DNRA, which is also predicted to be a lipoprotein (Figure 15.2). DNRA is, like denitrification, a facultative anaerobic process involving nitrate reduction to nitrite followed by the six-electron reduction of nitrite to ammonium.[38] Their co-occurrence was very unexpected because, at that time, denitrification and DNRA were still considered to be mutually exclusive nitrate reduction pathways that were thought to occur in completely different microbial populations. Nevertheless, we observed the same for *Bacillus bataviensis*,[22] and at around the same time, the co-occurrence of both pathways was seen in several Gram-negative denitrifiers, like *Opitutus terrae* strain PB90-1 (DSM 11246), *Marivirga tractuosa* DSM 4126 and *Shewanella loihica* PV-4 (DSM 17748).[39] Since then, data mining of over 250 genomes unexpectedly revealed that over 20% of NirK-type denitrifiers also encode a NrfA gene (Decleyre and Heylen, unpublished) and have the genetic potential for

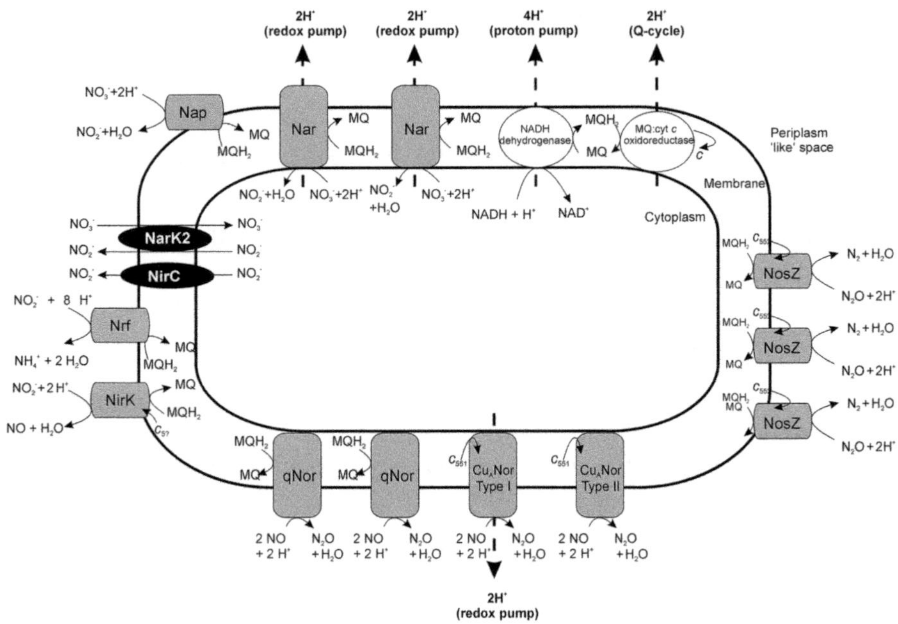

Figure 15.2 Schematic overview of the gene inventory related to dissimilatory nitrate reduction in *Bacillus azotoformans*. Menaquinol (MQH_2) donates electrons to a menaquinol:cytochrome *c* oxidoreductase (related to bc_1/b_6f), Nar, Nir, Nrf, qNor and NosZ. Membrane-bound *c*-type cytochromes subsequently donate electrons to their corresponding enzymes. The locations of the substrate-binding sites are hypothesised to be similar to other bacteria (see text for further explanation). Enzymes involved in respiratory nitrate reduction are depicted in dark grey. Nitrate/nitrite transport systems are in black. Menaquinol:cytochrome *c* oxidoreductase (related to bc_1/b_6f) and NADH dehydrogenase are in white. Cytochrome *c* lipoproteins are in light grey. Charge displacements contributing to the proton motive force are given. Deduced from genome data or taken from Suharti and de Vries.[21] N_2O: nitrous oxide; NO: nitric oxide.

both pathways. This is very exciting, as now the environmental drivers partitioning nitrate can be investigated without organism-dependent variation blurring the data, which could mean a giant leap forward for the design of predictive models of nitrate removal and N_2O emissions. Worth mentioning is that both *S. loihica* PV-4 and *B. azotoformans* LMG 9581^T lack the assimilatory nitrate and nitrite reductase genes. The concerted action of Nar (and/or Nap for *B. azotoformans*) and NrfA could compensate for this loss and still enable anaerobic nitrate assimilation, which might be an evolutionary reason for the co-occurrence of both pathways.

Actual functionality of both pathways has thus far only been demonstrated for *S. loihica* PV-4.[40,41] That work demonstrated that denitrification dominated at nitrate sufficiency and low carbon-to-nitrate (C/N) and nitrite-to-nitrate ratios, while high C/N and nitrite-to-nitrate ratios, pH 7.0, temperature of 30 °C and nitrite as the sole electron acceptor favoured DNRA. Until now, all nitrate or nitrite concentrations or C/N ratios experimentally tested for Gram-positive *B. azotoformans* LMG 9581^T only triggered denitrification (Sun and Heylen, unpublished), probably due to the use of batch cultures and their continuously changing growth conditions. Indeed, this was also seen for *S. loihica*, with both pathways being expressed under high C/N ratios in batch culture, while under the same conditions in chemostat, only DNRA was observed.[40] This might be the reason why DNRA bacteria are commonly studied and enriched in continuous culture.[40,42,43] So, currently scheduled chemostat experiments with *B. azotoformans* LMG 9581^T will either confirm the identified environmental drivers partitioning nitrate into both pathways for *Shewanella* or demonstrate taxon-dependent differences.

To our knowledge, *B. azotoformans* LMG 9581^T is the first denitrifier described with such a high genetic redundancy on functional gene and pathway levels, and therefore it is the ideal model organism to study redundancy and modularity of dissimilatory nitrate reduction by applying experimental, transcriptional, regulatory and mutational work in order to verify the above-mentioned theoretical considerations. Interesting to note is that another *B. azotoformans* strain, MEV2011, contains an identical dissimilatory nitrate reduction gene inventory, but also appears to be an obligate microaerophilic nitrate reducer capable of co-denitrification.[44] We are convinced that its high genetic redundancy conveys metabolic versatility to the organism and enables various ecological strategies in the soil matrix, in analogy to rRNA copy numbers.[45]

15.3 N_2O Emission by Non-Denitrifying Bacilli

15.3.1 N_2O Emission Related to Nitrite Accumulation and DNRA

More than three decades ago, Smith and Zimmerman[46] were the first to report that N_2O could also be anaerobically produced by non-denitrifiers. They found that 163 of 209 N_2O producers isolated from soil were not respiratory denitrifiers, but rather nitrite accumulators and ammonium producers. These non-denitrifying soil isolates evolved N_2O (up to 34% of nitrate), though the produced amount was always less than that of nitrite and/or ammonium. Most belonged to the genus *Bacillus* or were members of the family *Enterobacteriaceae*. Further characterisation of a *Bacillus* (and a *Citrobacter*) strain confirmed fermentative nitrite reduction to ammonium and N_2O production, while NO and N_2 were not produced in detectable amounts;[46] this was soon verified for another *Bacillus* and several other strains.[47] Additional ammonium did not inhibit N_2O or ammonium production, indicating that these processes were not assimilatory.[46,47] N_2O evolved slowly in batch cultures and mostly after apparent growth ceased.[46,47] We recently made similar observations for *Bacillus licheniformis*,[48] of which strains were previously described as denitrifiers.[19,49] Physiological experiments on three genotypically different strains (LMG 6934, LMG 7559 and LMG 17339) demonstrated that they reduced nitrate to nitrite, then switched to nitrite ammonification and produced ammonium. N_2O was produced in substantial but non-stoichiometric amounts (10.9–32.9% of the initial 10 mM of nitrate was recovered as N_2O) after growth had stopped, and important differences in the ratios of the end products nitrite, ammonium and N_2O were found between these closely related strains.

Work on *Citrobacter* isolate C48 suggested that N_2O production was unrelated to nitrite ammonification, but probably connected to intracellular nitrite accumulation, and both N conversions might occur in environmental conditions with contrasting C/N ratios.[50] Recently, the influence of carbon and nitrate availability on N_2O production from DNRA was verified and refined, again through the investigation of two fresh soil isolates affiliated with the genus *Bacillus* (and *Citrobacter*).[51] In defined media with lower C/N ratios (5- and 10-to-1) in which nitrite accumulated and no ammonium was produced, up to 2.7% of nitrate was reduced to N_2O by *Bacillus* sp. However, at higher C/N ratios (25- and 50-to-1), ammonium was produced and only 0.1% of nitrate was reduced to N_2O by *Bacillus* sp. These findings were corroborated with chemostat cultures under nitrate-sufficient and -limited conditions.[51] Further work was performed on *Bacillus vireti* LMG 21834^T, a versatile soil bacterium capable of fermentation, DNRA, N_2O production and its further reduction to N_2.[52] For nitrate concentrations equal to or higher than 15 mM, nitrite accumulated during nitrate reduction and growth (OD_{660}) ceased when nitrate was depleted. Only then did ammonium start to accumulate and N_2O and N_2 production was observed. In contrast, for lower nitrate concentrations, the nitrite peak was much lower and growth was still supported by nitrite reduction to ammonium. Also, nitrogen gases started to evolve during growth, which continued during the stationary phase. Confirming previous observations discussed above, the portion of the reduced nitrate recovered as gases in the headspace (*i.e.*, NO, N_2O and N_2) increased from 6% with 5 mM nitrate to between 49% and 55% with 20 mM or higher nitrate concentrations. Interesting to note is that the ratio of N_2O/N_2 in nitrogen gas end products increased with increasing nitrate concentrations.

So, from the currently available data on *Bacillus* (and *Enterobacteriaceae*) described here and *S. loihica*,[40] it can be concluded that indeed DNRA is favoured at higher C/N ratios or nitrate limitation. This fits with the traditional C/N hypothesis based on Gibbs free energy calculations that more energy can be conserved per mole of nitrate *via* DNRA and confirms findings from the soil environment[53] and enrichment cultures from a wastewater treatment plant (WWTP)[41] or marine environments.[42] In contrast, N_2O is a more significant product of DNRA at low C/N ratios (note that exact ratios are dependent on the type of carbon source used as the electron donor and the initial concentration of nitrate), under nitrate sufficiency or at high pH. N_2O production is probably a consequence of the (transient) accumulation of nitrite under these conditions. Indeed, it makes bioenergetic sense to maximise nitrate reduction to nitrite and minimise the use of scarce electrons to reduce nitrite to ammonium.[51]

15.3.2 Mechanisms for N_2O Production

It is currently unclear if all organisms capable of DNRA can produce N_2O. For example, *B. subtilis* is the second most studied model organism next to *E. coli*, and both are known to perform DNRA, but there are almost no reports of N_2O production from *B. subtilis*, and in those that exist, only very small amounts of N_2O were observed, being in the nanomolar range (compared to micromolar ranges for the microorganisms described in Section 15.3.1).[51] Nevertheless, this is within the same order of magnitude as *E. coli*.[51] Because *B. subtilis* was long considered not to produce N_2O, no experimental evidence is available on the mechanisms of its formation. Therefore, below we will combine the knowledge of gene inventories of the N_2O-producing, non-denitrifying *B. subtilis*, *B. vireti* and *B. licheniformis* (Table 15.1) with the experimental evidence for *E. coli* and *Salmonella enterica* serovar *Typhimurium* to propose several hypothetical pathways for NO and subsequent N_2O formation by non-denitrifying bacilli.

As in denitrification, the first step of DNRA is nitrate reduction. Subsequent nitrite reduction to ammonium can be catalysed by the cytoplasmic NADH-dependent nitrite reductase NirB or its to two-subunit variant NirBD and/or the periplasmic pentahaem cytochrome *c* nitrite reductase NrfA, depending on the organism and growth conditions. *B. vireti*, like *E. coli*, harbours and expresses genes for both enzymes,[35,52] with differential expression under high and low nitrite concentrations,[54] while *B. subtilis*[55] and *B. licheniformis*[48] only contain *nirB* (Table 15.1). All lack a nitrite reductase to NO for nitrite detoxification during its accumulation, but do encode a NO synthase and a flavohaemoglobin Hmp that can convert cytotoxic NO to N_2O. In addition, *B. licheniformis* and *B. vireti*, in contrast to *B. subtilis* (and *E. coli*), also encode a dedicated enzyme for the reduction of NO to N_2O (Table 15.1). Like *Wolinella succinogenes* DSM 1740,[56] *B. vireti* is also capable of converting N_2O to N_2 using the NosZ-type reductase.[52]

Early work on *E. coli* demonstrated that the nitrite conversion to NO occurred only after nitrate was depleted, required molybdate (the cofactor of NarGHI)

Table 15.1 Summarised overview of genes involved in dissimilatory nitrate reduction to ammonium and potential nitrous oxide formation and consumption based on genome analyses.

Gene	Probable function of product	*Bacillus subtilis*[52]	*Bacillus licheniformis*[48]	*Bacillus vireti*[49]
narG	Respiratory nitrate reductase subunit α	X	X	X
narH	Respiratory nitrate reductase subunit β	X	X	X
narJ	Respiratory nitrate reductase subunit δ	X	X	X
narI	Respiratory nitrate reductase subunit γ	X	X	X
nasC	Assimilatory nitrate reductase	X	X	X
nirB	Assimilatory NADH-dependent nitrite reductase large subunit	X	X	X
nirD	Assimilatory NADH-dependent nitrite reductase small subunit	X	X	X
nrfA	Cytochrome *c* nitrite reductase large subunit	—	—	X
nrfH	Cytochrome *c* nitrite reductase small subunit	—	—	X
amtB	Ammonium transport	X	X	—
narK	Nitrate/nitrite transporter	X	X	X
nirC	Formate/nitrite transporter	X	X	X
hmp	Flavohaemoglobin	X	X	X
nos	Nitric oxide synthase	X	X	X
qnorB	Quinol-dependent nitric oxide reductase	—	X	—
cbaA	Cu_A-dependent nitric oxide reductase subunit I	—	—	X
cbaB	Cu_A-dependent nitric oxide reductase subunit II	—	—	X
nosZ	Nitrous oxide reductase	—	—	X

and continued in *nirB* mutants,[50] but was absent in *narG* mutants.[57] These data suggested that NarGHI had a double function and converted nitrite to NO in the absence of nitrate. Also, in *S. typhimurium* strain 14028s, NarGHI appeared to be unequivocally responsible for NO generation from nitrite, which was completely eliminated in a *narGHI* mutant.[58] Later mutagenesis experiments in *E. coli* could not confirm the involvement of the cytoplasmic nitrate reductase in NO evolution, probably because the experiments were conducted in the absence of the nitrate that might be necessary to induce NO reduction.[59,60] Rather, NirB and NrfA appeared to be involved in NO production,[59] with their relative importance dependent on the nitrite concentration.[54,61] The activity of NO synthase, which produces NO from arginine as a defence mechanism against oxidative stress under aerobic conditions in *B. subtilis*,[62–64] is unlikely under anoxic, nitrate-reducing conditions.

Next, NO homeostasis is crucial to limiting the toxicity of NO, which is a reactive nitrogen species that is capable of damaging bacterial proteins, lipids and DNA and of binding to metal centres. NO can freely diffuse through the membrane and can be converted to N_2O in the periplasmic-like space by the quinol-dependent NO reductase in *B. licheniformis* or the Cu_A-dependent NO reductase in *B. vireti*. These reductases are known to be present both in denitrifiers and in non-denitrifiers,[21,22,29,52,65–67] including in pathogenic bacteria, where it is part of their defence mechanism against nitrosative stress. The periplasmic NrfA can also anaerobically reduce NO to ammonium and/or N_2O.[68,69] In addition, Hmp, which can be found both in the cytoplasm and the periplasm, is known to convert NO to nitrate aerobically and to N_2O anaerobically.[70,71] However, the latter conversion is at greatly reduced activity.[72] In addition, *hmp* appeared not to be upregulated in anaerobic conditions at high levels of nitrite in *B. vireti* with a Cu_ANor.[73] Furthermore, the cytoplasmic flavorubredoxin NorV and its associated oxidoreductase NorW[74,75] can both anaerobically reduce NO to ammonium and/or N_2O, but the encoding genes have thus far not been found in the genomes of non-denitrifying bacilli.

Taken together, several routes for NO and N_2O production are plausible in *B. subtilis*, *B. vireti* and *B. licheniformis*, depending on the organism, its genetic capacity (Table 15.1) and the growth conditions. These have been summarised and visualised in Figure 15.3. Mutagenic and transcriptomic studies are of course necessary in order to confirm these hypothetical pathways. Interestingly, the most striking difference in gene inventories for anaerobic nitrate and nitrite metabolism between *B. subtilis* on the one hand and *B. licheniformis* and *B. vireti* on the other hand is the presence of a dedicated NO reductase in the latter organisms (qNorB and Cu_ANor, respectively). These NO reductases could well be the reason for the three orders of magnitude difference in N_2O production between the organisms as described earlier and might be indicative of a higher tolerance to nitrosative stress. How environmentally relevant these N_2O emissions from non-denitrifiers are remains difficult to establish, though we lack ways to differentiate them from denitrification; in isotope pairing experiments, non-denitrifiers will also produce $^{30}N–N_2O$. In addition, even with mechanistic understanding,

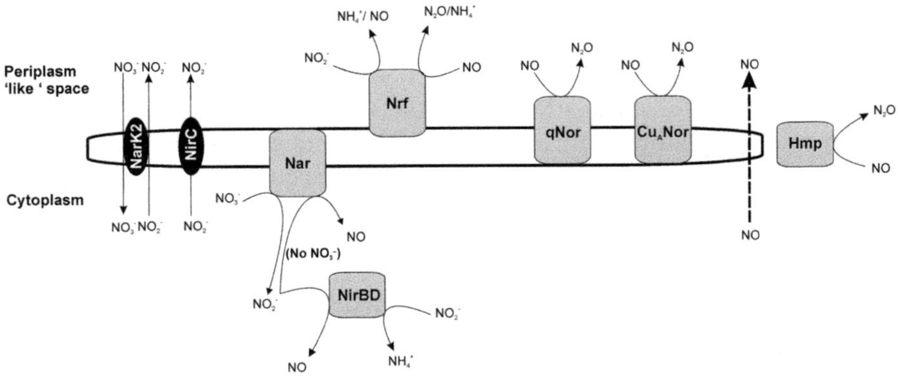

Figure 15.3 Proposed hypothetical pathways for nitric oxide (NO) and nitrous oxide (N_2O) production in non-denitrifying bacilli (see text for more explanation). Schematic representations of enzymes are given in grey and transporters are in black. Diffusion of gaseous NO through the cytoplasmic membrane is indicated by the dashed arrow. Soluble Hmp can be located in the cytoplasm and periplasm.

deducing specific target genes for molecular surveys will be nearly impossible, as different mechanisms are possible within a single microorganism and the genes involved have dual functions (*e.g.*, NarG, NirB and NrfA).

15.4 Environmental Relevance of Bacilli and Related Methodological Issues

15.4.1 Their Ubiquitous Nature

Bacilli are widely distributed in the natural environment. Their habitats range from all kinds of soils to the water columns and sediments of fresh and marine waters. They are found in the rhizosphere of various plants and crops, can be associated with sea weeds, are part of the gut microbiota of humans and other higher organisms, can be major soil-borne food contaminants, can be either food-borne pathogens or cause microbial food spoilage and also have numerous commercial and agricultural uses (*e.g.*, production of peptide antibiotics, chemicals and proteases and mitigation of fungal pathogens). Despite their ubiquitous nature, bacilli are generally considered to be soil-related microorganisms. This dates back to cultivation-based the qualitative and quantitative microbial diversity studies of soils, in which their numerical dominance in isolates was assumed to reflect their *in situ* relative abundance. However, this was disproved by Peter Janssen's landmark meta-data analyses of 3240 16S rRNA gene sequences from 32 clone libraries from a variety of bulk soil samples.[76] Soil bacterial communities appeared to be dominated by *Acidobacteria*, *Actinobacteria*, *Bacteroidetes*, *Proteobacteria* and *Verrucomicrobia*, and bacilli only contributed a mean of 2% (range of 0–8%), while in contrast, bacilli comprised up to 45% of isolates from traditional cultivation-based studies. These findings were later confirmed by many studies, and the most elaborate to date was a Illumina-based 16S rRNA gene amplicon sequencing survey targeting 16 soils selected to span a wide range of ecologically distinct biomes.[77] Bacilli were found in nearly all soils, but their relative abundances were highly variable and typically represented less than 5% of the 16S rRNA reads in any individual soil. However, molecular surveys using DNA extracted from environmental samples are confounded by the lack of information on contributions of endospores, which can persist for a long time in the environment, and vegetative cells. In addition, it is possible that inappropriate DNA extraction protocols might cause insufficient lyses of Gram-positive vegetative cells and definitely of their endospores, making them under-represented in DNA-based molecular surveys.[78] Besides the bias in DNA extraction, there can also be an uncoupling between specific activity and the abundance of microorganisms in the environment,[79,80] although this has not been reported for bacilli. Nevertheless, a series of 16S rRNA studies based on ribosome isolation,[81–83] later validated through fluorescent whole-cell *in situ* hybridisation,[84] demonstrated that bacilli can predominate in the active bacterial population in grassland soil, next to *Acidobacteria* and *Alphaproteobacteria*. In addition to soil, bacilli have been found to make up a considerable part of the bacterial community of other ecosystems with important nitrate removal activities and high N_2O emissions. Various DNA-based studies using denaturing gradient gel electrophoresis demonstrated that bacilli are abundant in the thermophilic stage of animal manure composting.[85] They are also major constituents of the microbiota in wastewater treatment systems, with, for example, 25% of the 16S rRNA gene sequences attributed to bacilli in the biofilm of a constructed wetland for enhanced nitrate removal,[86] and more than 30% in a modified rotating biological contactor wastewater treatment process.[87]

15.4.2 Their Contribution to Denitrification

Comprehensive understanding of the identities and activities of microorganisms, as well as the cellular mechanisms involved in nitrate removal, are crucial for improving models that predict fluxes of nitrate, nitrite and N_2O.[88] Denitrification is a facultative trait and thus is believed to be a weak selector of the microbial community,[6,89] meaning that the occurrence of these functional guilds is mainly determined by their aerobic, mostly heterotrophic, metabolism. So, the ubiquitous nature and the abundance of bacilli in specific environments can indeed be informative of their potential relevance as anaerobic nitrate reducers. This of course needs to be verified through specific determination of denitrifier community composition, their abundance and their activity using functional genes as biomarkers.[90] For denitrification, frequent targets are either *nirK* or *nirS* for copper- or cytochrome cd_1-dependent nitrite reductase to NO, as these encode the key step of converting fixed nitrogen into a gaseous form, or *nosZ* for the N_2O reductase, converting the potent greenhouse gas into harmless N_2. Unfortunately, until recently, most commonly used polymerase chain reaction (PCR) primers or molecular probes for these functional genes were designed using almost exclusively reference sequences from *Proteobacteria*,[17,91–95] thus producing consistently negative results when tested on physiologically confirmed denitrifying members of the genera *Bacillus* (Heylen, unpublished results), *Paenibacillus*[14] or *Geobacillus*.[96] In the exceptional case that amplicons were obtained from Gram-positive denitrifiers, their sequence phylogeny was highly related to that of sequences derived from Gram-negative denitrifiers,[97] suggesting horizontal gene transmission. It has now become generally accepted that widely used primers are not broad range,[98] and we believe that, as a consequence, the ecological relevance of bacilli, but also of other non-targeted taxa such as *Bacteroidetes* or *Actinobacteria*, for anaerobic nitrate removal and N_2O emission has been minimalised over the past two decades.

In the last few years, whole-genome mining and detailed functional sequence analyses have demonstrated that *Firmicutes* and other unaccounted denitrifiers and ammonifiers have very divergent biomarker sequences and form clades that are distinct from commonly detected *Alpha-*, *Beta-* and *Gammaproteobacteria* for *nosZ*,[37,99] *nirK* and *nirS* (Decleyre and Heylen, unpublished).[100] Using newly developed primers, the *nosZ* clade II organisms were detected in a range of geographically diverse environmental samples, including various soil types, wetlands, lake sediments and activated sludge from WWTPs, and appeared to be at least as abundant as the commonly targeted clade I.[99] Given the relative coherence of the *nosZ* and 16S rRNA phylogenies,[101,102] the phylogenetic placement of environmental sequences in combination with the quantitative PCR results indicate that organisms within the *Bacteroidetes*, *Gemmatimonadetes* and *Deltaproteobacteria* make up a significant proportion of the N_2O-reducing communities in different environments. So, unfortunately, the use of improved primers for *nosZ* did not result in the identification of bacilli as a major component of the targeted functional guild. Indeed, of the 403 cloned *nosZ* clade II sequences, only two sequences (from activated sludge from WWTPs) were most similar to those from included Firmicute representatives *Geobacillus thermophilus* and *Desulfitobacterium hafniense*. We again refer to the above-mentioned potential difficulties with cell lyses in DNA extraction protocols, as we noticed that the DNA of two out of the four included WWTP samples was extracted with a modified protocol for Gram-positive bacteria. In addition, the validation of the *nosZ* clade II primers was least strong for the Firmicute strains, with weak amplification from pure cultures and non-specific amplification products of considerable size (500 bp).[99] Of course, we cannot be certain that these issues have caused an under-detection of *in situ*-present bacilli within *nosZ* clade II; they might just have been absent from the investigated environmental samples or might have had truncated denitrification pathways (as suggested in Section 15.1). Another very recent effort to target unaccounted clades of *nirS* and *nirK* again demonstrated that the previously undetected denitrifiers are highly diverse and two- to six-times more abundant *in situ* than the commonly targeted clades.[100] Unfortunately, this study did not include any sequences from bacilli in the analysed data set and an *in silico* analysis indeed confirmed that the newly developed primers do not target bacilli-derived *nir* genes (Decleyre and Heylen, unpublished). Because of the high divergence of denitrification genes from bacilli, attempts to design primers specifically targeting those genes have thus far either been unsuccessful (Decleyre and Heylen, unpublished) or have resulted in primers with a coverage limited to those sequences included for the primer design.[96,103]

Without efficient molecular tools to detect denitrifying bacilli and other Gram-positive bacteria *in situ*, culture-dependent methods, albeit extremely biased, can provide some insight into their environmental relevance. Cultivation studies suggest that bacilli and other Gram-positive bacteria capable of denitrification can be found in various ecosystems, from WWTPs,[104–106] many types of soil and plant rhizospheres,[16,107,108] mangrove roots[17] to lagoon sediments,[94] and thus might actively contribute to denitrification in these systems.

Methodological issues for PCR-based community structure analyses and abundance assessment might become less relevant as shotgun sequence data analyses becomes more accessible to non-expert users, but are until then very relevant for assessing which organisms contribute to nitrate removal and denitrification. As has become clear for other previously not considered groups of denitrifiers, bacilli and other Gram-positive bacteria might be relevant and abundant denitrifiers is specific environments, but we currently lack the tools to specifically detect and quantify them. In addition, we know too little of their ecological lifestyle, besides their ubiquitous nature, to predict in what kinds of habitats they could thrive as denitrifiers. Nevertheless, their persistence in nature through endospore formation is an important feature for resisting environmental stresses and may also contribute to the metabolic resilience of denitrifying communities under perturbation. Furthermore, their high level of modularity of the dissimilatory nitrate reduction pathway, as outlined in this chapter both for denitrifiers and for non-denitrifiers, will definitely determine their niche specialisation and relevance in nature.

Acknowledgements

YS was funded by the Chinese Scholarship Council (file number 201206330054) and BOF CSC co-funding from Ghent University (grant 01SC2713). KH was funded by the Fund for Scientific Research (FWO), Flanders, for a position as a postdoctoral research fellow (grants FWO11/PDO/0840 and FWO15/PDOH1/084).

References

1. I. Mahne and J. M. Tiedje, *Appl. Environ. Microbiol.*, 1995, **61**, 1110.
2. J. P. Shapleigh, in *The Prokaryotes*, ed. E. Rosenberg et al., Springer-Verlag, Berlin Heidelberg, 4th edn, 2013, vol. 10, p. 405.
3. D. R. Graf, C. M. Jones and S. Hallin, *PLoS One*, 2014, **9**, e114118.

4. I. Verbaendert, P. De Vos, N. Boon and K. Heylen, *Biochem. Soc. Trans.*, 2011, **39**, 254.
5. W. J. Payne, *Denitrification*, John Wiley & Sons, New York, 1981.
6. J. M. Tiedje, in *Biology of anaerobic microorganisms*, ed. A. J. B. Zehnder, John Wiley & Sons, New York, 1988, vol. 717, (4), p. 179.
7. W. G. Zumft, in *The Prokaryotes: a handbook on the biology of bacteria: ecophysiology, isolation, identification, applications*, ed. A. Balows et al., Springer-Verlag, New York, 2nd edn, 1992, vol. 1, ch. 23, pp. 554–582.
8. L. Philippot, S. Hallin and M. Schloter, *Adv. Agron.*, 2007, **96**, 135.
9. F. Pichinoty, M. Durand and C. Job, *Can. J. Microbiol.*, 1978, **24**, 608.
10. P. L. Manachini, D. Mora, G. Nicastro, C. Parini, E. Stackebrandt, R. Pukall and M. G. Fortina, *Int. J. Syst. Evol. Microbiol.*, 2000, **50**, 1331.
11. G. Denariaz, W. J. Payne and J. L. Gall, *Int. J. Syst. Bacteriol.*, 1989, **39**, 145.
12. G. Denariaz, W. J. Payne and J. LeGall, *Biochim. Biophys. Acta*, 1991, **1056**, 225.
13. H. de Barjac and A. Bonnefoi, *Ann. Inst. Pasteur*, 1972, **122**, 463.
14. U. Behrendt, P. Schumann, M. Stieglmeier, R. Pukall, J. Augustin, C. Spröer, P. Schwendner, C. Moissl-Eichinger and A. Ulrich, *Syst. Appl. Microbiol.*, 2010, **33**, 328.
15. H. Shoun, M. Kano, I. Baba, N. Takaya and M. Matsuo, *J. Bacteriol.*, 1998, **180**, 4413.
16. D. Cheneby, L. Philippot, A. Hartmann, C. Henault and J.-C. Germon, *FEMS Microbiol. Ecol.*, 2000, **34**, 121.
17. A. L. Flores-Mireles, S. C. Winans and G. Holguin, *Appl. Environ. Microbiol.*, 2007, **73**, 7308.
18. E. D. Renner and G. E. Becker, *J. Bacteriol.*, 1970, **101**, 821.
19. I. Verbaendert, N. Boon, P. De Vos and K. Heylen, *Syst. Appl. Microbiol.*, 2011, **34**, 385.
20. F. Pichinoty, H. de Barjac, M. Mandel and J. Asselineau, *Int. J. Syst. Bacteriol.*, 1983, **33**, 660.
21. Suharti and S. de Vries, *Biochem. Soc. Trans.*, 2005, **33**, 130.
22. K. Heylen and J. Keltjens, *Front. Microbiol.*, 2012, **3**, 371.
23. R. A. Rothery, G. J. Workun and J. H. Weiner, *Biochim. Biophys. Acta*, 2008, **1778**, 1897.
24. J. Simon and M. G. Klotz, *Biochim. Biophys. Acta*, 2013, **1827**, 114.
25. D. J. Richardson, B. C. Berks, D. A. Russell, D. Spiro and C. J. Taylor, *Cell. Mol. Life Sci.*, 2001, **58**, 165.
26. J. Simon, M. Sänger, C. Schuster and R. Gross, *Mol. Microbiol.*, 2003, **49**, 69.
27. M. Kern and J. Simon, *Mol. Microbiol.*, 2008, **69**, 1137.
28. S. Al-Attar and S. de Vries, *FEBS Lett.*, 2015, **589**, 2050.
29. Suharti, M. J. F. Strampaard, I. Schröder and S. de Vries, *Biochemistry*, 2001, **40**, 2632.
30. Suharti, H. A. Heering and S. de Vries, *Biochemistry*, 2004, **43**, 13487.
31. T. Hino, S. Nagano, H. Sugimoto, T. Tosha and Y. Shiro, *Biochim. Biophys. Acta*, 2011, **1817**, 680.

32. J. Reimann, U. Flock, H. Lepp, A. Honigmann and P. Adelroth, *Biochim. Biophys. Acta*, 2007, **1767**, 362.
33. J. ter Beek, N. Krause, J. Reimann, P. Lachmann and P. Adelroth, *J. Biol. Chem.*, 2013, **288**, 30626.
34. Y. Matsumoto, T. Tosha, A. V. Pisliakov, T. Hino, H. Sugimoto, S. Nagano and Y. Sugita, *Nat. Struct. Mol. Biol.*, 2012, **19**, 238.
35. J. Cole, *FEMS Microbiol. Lett.*, 1996, **136**, 1.
36. M. Simek and J. E. Cooper, *Eur. J. Soil Sci.*, 2002, **53**, 345.
37. B. Liu, P. T. Mørkved, Å. Frostegåard and L. R. Bakken, *FEMS Microbiol. Ecol.*, 2010, **72**, 407.
38. O. Einsle, A. Messerschmidt, P. Stach, G. P. Bourenkov, H. D. Bartunik, R. Huber and P. M. H. Kroneck, *Nature*, 1999, **400**, 476.
39. R. A. Sanford, D. D. Wagner, Q. Wu, J. C. Chee-Sanford, S. H. Thomas, C. Cruz-García, G. Rodríguez, A. Massol-Deyá, K. K. Krishnani, K. M. Ritalahti, S. Nissen, K. T. Konstantinidis and F. E. Löffler, *Proc. Natl. Acad. Sci. U. S. A.*, 2012, **109**, 19709.
40. S. Yoon, C. Cruz-Garcia, R. Sanford, K. M. Ritalahti and F. E. Loffler, *ISME J.*, 2015, **9**, 1093.
41. S. Yoon, R. A. Sanford and F. E. Loffler, *Appl. Environ. Microbiol.*, 2015, **81**, 3510.
42. B. Kraft, H. E. Tegetmeyer, R. Sharma, M. G. Klotz, T. G. Ferdelman, R. L. Hettich, J. S. Geelhoed and M. Strous, *Science*, 2014, **345**, 676.
43. E. M. van den Berg, U. van Dongen, B. Abbas and M. C. van Loosdrecht, *ISME J.*, 2015, **9**, 2153.
44. M. Nielsen, L. Schreiber, K. Finster and A. Schramm, *Stand. Genomic Sci.*, 2014, **10**, 4.
45. J. A. Klappenbach, J. M. Dunbar and T. M. Schmidt, *Appl. Environ. Microbiol.*, 2000, **66**, 1328.
46. M. S. Smith and K. Zimmerman, *Soil Sci. Soc. Am. J.*, 1981, **45**, 865.
47. B. H. Bleakley and J. M. Tiedje, *Appl. Environ. Microbiol.*, 1982, **44**, 1342.
48. Y. Sun, P. De Vos and K. Heylen, *BMC Genomics*, 2016, **17**, 68.
49. F. Pichinoty, J. L. Garcia, C. Job and M. Durand, *Can. J. Microbiol.*, 1978, **24**, 45.
50. M. S. Smith, *Appl. Environ. Microbiol.*, 1982, **43**, 854.
51. M. A. Streminíska, H. Felgate, G. Rowley, D. J. Richardson and E. M. Baggs, *Environ. Microbiol. Rep.*, 2012, **4**, 66.
52. D. Mania, K. Heylen, R. J. van Spanning and A. Frostegard, *Environ. Microbiol.*, 2014, **16**, 3196.
53. T. Rütting, P. Boeckx, C. Müller and L. Klemedtsson, *Biogeosci. Discuss*, 2011, **8**, 1169.
54. H. Wang, O. Robert, J. Hall and P. Gunsalus, *J. Bacteriol.*, 2000, **182**, 5813.
55. M. M. Nakano and P. Zuber, *Annu. Rev. Microbiol.*, 1998, **52**, 165.
56. J. Simon, *FEMS Microbiol. Rev.*, 2002, **26**, 285.
57. M. S. Smith, *Appl. Environ. Microbiol.*, 1983, **45**, 1545.
58. N. J. Gilberthorpe and R. K. Poole, *J. Biol. Chem.*, 2008, **283**, 11146.
59. H. Corker and R. K. Poole, *J. Biol. Chem.*, 2003, **278**, 31584.
60. C. E. Vine and J. A. Cole, *Biochem. Soc. Trans.*, 2011, **39**, 313.

61. G. Rowley, D. Hensen, H. Felgate, A. Arkenberg, C. Appia-Ayme, K. Prior, C. Harrington, S. J. Field, J. N. Butt, E. Baggs and D. J. Richardson, *Biochem. J.*, 2012, **441**, 755.
62. S. Adak, K. S. Aulak and D. J. Stuehr, *J. Biol. Chem.*, 2002, **277**, 16167.
63. I. Gusarov and E. Nudler, *Proc. Natl. Acad. Sci. U. S. A.*, 2005, **102**, 13855.
64. F. Schreiber, M. Beutler, D. Enning, M. Lamprecht-Grandio, O. Zafra, J. E. Gonzalez-Pastor and D. de Beer, *BMC Microbiol.*, 2011, **11**, 111.
65. A. Büsch, B. Friedrich and R. Cramm, *Appl. Environ. Microbiol.*, 2002, **68**, 668.
66. L. Philippot, *Biochim. Biophys. Acta*, 2002, **1577**, 355.
67. K. Heylen, B. Vanparys, D. Gevers, L. Wittebolle, N. Boon and P. De Vos, *Environ. Microbiol.*, 2007, **9**, 1072.
68. C. Costa, A. Macedo, I. Moura, J. J. G. Moura, J. Le Gall, Y. Berlier, M.-Y. Liu and W. J. Payne, *FEBS Lett.*, 1990, **276**, 67.
69. S. R. Poock, E. R. Leach, J. W. B. Moir, J. A. Cole and D. J. Richardson, *J. Biol. Chem.*, 2002, **277**, 23664.
70. P. R. Gardner, A. M. Gardner, L. A. Martin and A. L. Salzman, *Proc. Natl. Acad. Sci. U. S. A.*, 1998, **95**, 10378.
71. S. O. Kim, Y. Orii, D. Lloyd, M. N. Hughes and R. K. Poole, *FEBS Lett.*, 1999, **445**, 389.
72. C. E. Mills, S. Sedelnikova, B. Soballe, M. N. Hughes and R. K. Poole, *Biochem. J.*, 2001, **353**, 207.
73. D. Mania, K. Heylen, R. J. M. Van Spanning and Å. Frostegård, *Environ. Microbiol.*, 2016, in press.
74. C. M. Gomes, A. Giuffre, E. Forte, J. B. Vicente, L. M. Saraiva, M. Brunori and M. Teixeira, *J. Biol. Chem.*, 2002, **277**, 25273.
75. J. H. van Wonderen, B. Burlat, D. J. Richardson, M. R. Cheesman and J. N. Butt, *J. Biol. Chem.*, 2008, **283**, 9587.
76. P. H. Janssen, *Appl. Environ. Microbiol.*, 2006, **72**, 1719.
77. N. Fierer, J. W. Leff, B. J. Adams, U. N. Nielsen, S. T. Bates, C. L. Lauber, S. Owens, J. A. Gilbert, D. H. Wall and J. G. Caporaso, *Proc. Natl. Acad. Sci. U. S. A.*, 2012, **109**, 21390.
78. S. Filippidou, T. Junier, T. Wunderlin, C. C. Lo, P. E. Li, P. S. Chain and P. Junier, *Comput. Struct. Biotechnol. J.*, 2015, **13**, 299.
79. B. J. Campbell, L. Yu, J. F. Heidelberg and D. L. Kirchman, *Proc. Natl. Acad. Sci. U. S. A.*, 2011, **108**, 12776.
80. D. E. Hunt, Y. Lin, M. J. Church, D. M. Karl, S. G. Tringe, L. K. Izzo and Z. I. Johnson, *Appl. Environ. Microbiol.*, 2013, **79**, 177.
81. A. Felske, A. D. Akkermans and W. M. De Vos, *Appl. Environ. Microbiol.*, 1998, **64**, 4581.
82. A. Felske, A. Wolterink, R. Van Lis and A. D. Akkermans, *Appl. Environ. Microbiol.*, 1998, **64**, 871.
83. A. Felske, A. Wolterink, R. Van Lis, W. M. De Vos and A. D. Akkermans, *Appl. Environ. Microbiol.*, 2000, **66**, 3998.
84. A. Felske, A. D. L. Akkermans and W. M. De Vos, *Appl. Environ. Microbiol.*, 1998, **34**, 4588.

85. K. Maeda, D. Hanajima, S. Toyoda, N. Yoshida, R. Morioka and T. Osada, *Microb. Biotechnol.*, 2011, **4**, 700.
86. Z. Shen, Y. Zhou, J. Liu, Y. Xiao, R. Cao and F. Wu, *Bioresour. Technol.*, 2014, **175c**, 239.
87. S. J. Park, J. C. Yoon, K. S. Shin, E. H. Kim, S. Yim, Y. J. Cho, G. M. Sung, D. G. Lee, S. B. Kim, D. U. Lee, S. H. Woo and B. Koopman, *J. Microbiol.*, 2007, **45**, 113.
88. D. E. Canfield, A. N. Glazer and P. G. Falkowski, *Science*, 2010, **330**, 192.
89. L. Philippot, J. Cuhel, N. P. A. Saby, D. Chèneby, A. Chronáková, D. Bru, D. Arrouays, F. Martin-Laurent and M. Šimek, *Environ. Microbiol.*, 2009, **11**, 1518.
90. L. Philippot and S. Hallin, *Curr. Opin. Microbiol.*, 2005, **8**, 234.
91. G. Braker, A. Fesefeldt and K.-P. Witzel, *Appl. Environ. Microbiol.*, 1998, **64**, 3769.
92. S. Hallin and P.-E. Lindgren, *Appl. Environ. Microbiol.*, 1999, **65**, 1652.
93. I. N. Throback, K. Enwall, A. Jarvis and S. Hallin, *FEMS Microbiol. Ecol.*, 2004, **49**, 401.
94. C. M. Goregues, V. D. Michotey and P. C. Bonin, *Microb. Ecol.*, 2005, **49**, 198.
95. K. L. Casciotti and B. B. Ward, *Appl. Environ. Microbiol.*, 2001, **67**, 2213.
96. I. Verbaendert, S. Hoefman, P. Boeckx, N. Boon and P. De Vos, *FEMS Microbiol. Ecol.*, 2014, **89**, 162.
97. K. Heylen, D. Gevers, B. Vanparys, L. Wittebolle, J. Geets, N. Boon and P. De Vos, *Environ. Microbiol.*, 2006, **8**, 2012.
98. C. R. Penton, T. A. Johnson, J. F. Quensen, S. Iwai, J. R. Cole and J. M. Tiedje, *Front. Microbiol.*, 2013, **4**, 279.
99. C. M. Jones, D. R. H. Graf, D. Bru, L. Philippot and S. Hallin, *ISME J.*, 2013, **7**, 417.
100. X. Zhao, Z. Wei, Y. Zhao, B. Xi, X. Wang, T. Zhao, X. Zhang and Y. Wei, *Microb. Biotechnol.*, 2015, **8**, 541.
101. C. M. Jones, B. Stres, M. Rosenquist and S. Hallin, *Mol. Biol. Evol.*, 2008, **25**, 1955.
102. K. Palmer, H. L. Drake and M. A. Horn, *Appl. Environ. Microbiol.*, 2009, **75**, 5170.
103. C. M. Jones, A. Welsh, I. N. Throback, P. Dorsch, L. R. Bakken and S. Hallin, *FEMS Microbiol. Ecol.*, 2011, **76**, 541.
104. S. Y. Lee, S. H. Lee and Y. K. Park, *J. Microbiol. Biotechnol.*, 2005, **15**, 14.
105. B. Harbi, K. Chaieb, C. Jabeur, K. Mahdouani and A. Bakhrouf, *World J. Microbiol. Biotechnol. Bioeng.*, 2010, **26**, 473.
106. B. Zhao, Y. L. He and X. F. Zhang, *Environ. Technol.*, 2010, **31**, 409.
107. X. L. Dong and G. B. Reddy, *Bioresour. Technol.*, 2010, **101**, 1175.
108. G. McMullan, J. M. Christie, T. J. Rahman, I. M. Banat, N. G. Ternan and R. Marchant, *Biochem. Soc. Trans.*, 2004, **32**, 214.

CHAPTER 16

Denitrification Processes for Wastewater Treatment

BING-JIE NI[a], YUTING PAN[b], JIANHUA GUO[a], BERNARDINO VIRDIS[a], SHIHU HU[a], XUEMING CHEN[a] AND ZHIGUO YUAN*[a]

[a]Advanced Water Management Centre, The University of Queensland, St Lucia, Brisbane, Queensland 4072, Australia; [b]Department of Environmental Science and Engineering, School of Architecture and Environment, Sichuan University, Chengdu, Sichuan 610065, China
*E-mail: zhiguo@awmc.uq.edu.au

16.1 Introduction

Nitrogen is essential to life on earth. Nitrogen is an essential component of DNA, RNA and proteins, the building blocks of life. All organisms require nitrogen to live and grow. Nitrogen gas makes up 78% of the air we breathe. Nitrogen is an incredibly versatile element, existing in both inorganic and organic forms, as well as many different oxidation states. The predominant form is nitrogen gas. Nitrate (NO_3^-) and ammonia are the two forms used by plants. Organic nitrogen is mineralised by microorganisms to create these two forms that are found in our soils and water. Nitrogen, in various forms, passes from the air, to the soil, to all living things, and then back into the air. Microorganisms, particularly bacteria, play major roles in all of the principal nitrogen transformations. As microbe-mediated processes, these nitrogen transformations tend to occur faster than geological processes like plate

RSC Metallobiology Series No. 9
Metalloenzymes in Denitrification: Applications and Environmental Impacts
Edited by Isabel Moura, José J. G. Moura, Sofia R. Pauleta and Luisa B. Maia

Published by the Royal Society of Chemistry, www.rsc.org

motion, a very slow, purely physical process that is a part of the carbon cycle. Instead, rates are affected by environmental factors that influence microbial activity, such as temperature, moisture and resource availability.

Nitrogen becomes a concern to water quality when nitrogen in the soil is converted to the NO_3^- form. It is a concern because nitrate is very mobile and easily moves with water in the soil. The concern regarding nitrates and water quality is generally directed at groundwater. However, nitrates can also enter surface waters such as ponds, streams and rivers. The presence of nitrates in the soil is largely the result of natural biological processes associated with the decomposition of plant residues and organic matter. Nitrates can also come from rainfall, animal manure and nitrogen fertilisers. High levels of nitrates can be toxic to new-borns, causing anoxia, or internal suffocation. Alternative water sources are needed if nitrate levels exceed the health standard of 10 ppm nitrate-N. The most common symptom of nitrate poisoning in babies is a bluish colour to the skin, particularly around the baby's eyes and mouth. These symptoms of nitrate toxicity are commonly referred to as the "blue-baby" syndrome.

The U.S. Environmental Protection Agency has established a standard for nitrogen in drinking water of 10 mg nitrate-N per litre. Unfortunately, many systems (particularly in agricultural areas) already exceed this level. By comparison, nitrate levels in waters that have not been altered by human activity are rarely greater than 1 mg N L^{-1}. In surface waters, discharge of nitrogenous components to water bodies can lead to nutrient over-enrichment (*i.e.*, eutrophication), particularly in coastal waters receiving inflow from polluted rivers. Eutrophication has been blamed for increased frequencies of coastal fish-kill events, increased frequencies of harmful algal blooms and species shifts within coastal ecosystems. Reactive nitrogen such as NO_3^- and ammonium (NH_4^+) present in surface waters and soils can also enter the atmosphere as the smog-component nitric oxide (NO) and the greenhouse gas nitrous oxide (N_2O). Eventually, this atmospheric nitrogen can be blown into nitrogen-sensitive terrestrial environments, causing long-term changes.

Denitrification is one of the key processes in biological nitrogen removal (BNR) from water and wastewater. Through denitrification, oxidised forms of nitrogen, such as NO_3^- and nitrite (NO_2^-), are converted to dinitrogen (N_2) and, to a lesser extent, N_2O gas. Denitrification is an anaerobic process that is carried out by denitrifying bacteria, which convert nitrate to dinitrogen in the following sequence:

$$NO_3^- \rightarrow NO_2^- \rightarrow NO \rightarrow N_2O \rightarrow N_2 \qquad (16.1)$$

Once converted to dinitrogen, nitrogen is unlikely to be reconverted to a biologically available form because it is a gas and is rapidly lost to the atmosphere. Denitrification is the only nitrogen biotransformation process that removes nitrogen from ecosystems (essentially irreversibly). In this chapter, heterotrophic denitrification, anaerobic ammonium oxidation (AnAmmOx) coupled to nitrite reduction, denitrifying anaerobic methane oxidation (DAMO), autotrophic denitrification and microbial fuel cells (MFCs) with denitrification processes are delineated. An overview of the technological application of these denitrification processes for wastewater treatment is also presented. The aim is to provide knowledge of denitrification in order to facilitate the engineering and use of denitrification-based technology for nitrogen removal from wastewater.

16.1.1 Nitrogen Removal from Wastewater

Nitrogen compounds are among the most important pollutants of wastewater because of their role in eutrophication, their effect on the oxygen content of receiving waters and their toxicity to aquatic invertebrate and vertebrate species, including human beings. However, these compounds are also of interest because of the beneficial role that they can play in augmenting plant growth, which in turn stimulates the production of wildlife. Wastewater originating from many other sources such as tanneries, food processing, fertiliser manufacturing, slaughter houses and landfill leachate contains greater amounts of nitrogen, which should be treated before discharge into the surface water body. Wastewater containing huge amounts of nitrogen compounds is not allowed to be released into the surface water as it has ecological impacts and can affect human health.

Chemical, physicochemical and biological methods are broadly used for the treatment of wastewater loaded with highly concentrated nitrogen. In considering criteria such as a cost–benefit analysis, the requirement of energy and chemical doses, familiarity with operational procedures and environmental sustainability, a particular treatment for a specific pollutant is usually selected.[1] Still, the tradition is that depending on the concentration of nitrogen load presenting in the collected wastewater specimen, either physicochemical or biological treatment methods are decided. According to Mulder,[1] three concentration ranges could be differentiated:

- Nitrogen concentration less than 100 mg N L^{-1}: in this range, biological N removal is the preferred process based on cost-effectiveness. Domestic wastewater is within this range.
- Nitrogen concentrations in the range 100–5000 mg N L^{-1}: a typical example is sludge digestion liquor, for which biological treatment is preferred after extensive investigation.[2] Although ammonia stripping and the production of $MgNH_4PO_4$ were identified as interesting alternatives for resource recovery, these options are generally not cost effective.[2]
- Nitrogen concentrations greater than 5000 mg N L^{-1}: in this range, physicochemical methods are technically and economically feasible. A successful example is the steam stripping of wastewater with an ammonium concentration of 1.5% followed by ammonia recovery, which has been in operation on an industrial scale since 1985.[3]

The BNR process is cheaper and is the most widely practiced approach for nitrogen control in wastewater treatment. For many years, the traditional method for nitrogen removal from wastewater has been the combination of nitrification–denitrification processes. Frequently, space limitations or economic constrains do not allow the capacity of the existing treatment plants, especially for wastewaters with a high ammonium load, to be expanded. Fortunately, with the discovery of AnAmmOx and DAMO processes, new possibilities and alternatives have opened up.

16.1.2 Overview of Denitrification Processes

Heterotrophic denitrification converts nitrate generated from autotrophic nitrification to nitrogen gas, thus removing nitrogen from wastewater. Heterotrophic denitrification consists of four consecutive reductions, which produce nitrite, NO and N_2O as three obligatory intermediates. Each reduction step is catalysed by one or more specific reductase enzymes. The heterotrophic denitrification is accomplished with a variety of electron donors, including methanol, acetate, ethanol, lactate and glucose.[4] During denitrification, the requirement of organic carbon is significant. For example, 2.47 g of methanol is required per gram of nitrate nitrogen for complete denitrification.[4] Because the organic carbon present naturally in the wastewater is quite limited, the complete removal of nitrogen from wastewaters that contain high nitrogen concentrations requires a large amount of an added carbon source for denitrification.[5]

The AnAmmOx process could offer an alternative for the treatment of nitrogen. In the AnAmmOx process, ammonium is oxidised under anoxic conditions with nitrite as the electron acceptor. AnAmmOx metabolism also plays an important role in the global nitrogen cycle. Microbial activities, such as denitrification and AnAmmOx, are the major mechanisms that convert combined nitrogen to dinitrogen gas, thereby completing the nitrogen cycle. If the AnAmmOx process is combined with a preceding nitrification step, only part of the ammonium needs to be nitrified to nitrite, while the AnAmmOx process combines the remaining ammonium with the nitrite to yield dinitrogen gas. This will reduce oxygen demand in nitrification and organic carbon in denitrification. The biomass yield is very low; consequently, little sludge is produced. This is another factor that contributes to a substantially lower operation cost of AnAmmOx compared to the conventional denitrification process. However, the low biomass yield also necessitates an efficient system for sludge retention, and long start-up times are required to obtain a sufficient biomass concentration.[6]

The discovery of the DAMO process, in which methane is oxidised anaerobically to provide electrons for denitrification,[7,8] forms an important link between two major global nutrient cycles (*i.e.*, the carbon and nitrogen cycles). It not only stimulated the appreciation of the ecological significance of DAMO microorganisms, but also further opened some avenues for developing more sustainable wastewater treatment processes (*e.g.*, achieving high levels of nitrogen removal from wastewater with a minimised carbon footprint through using methane as the electron donor).[9,10] DAMO archaea are able to reduce nitrate to nitrite with methane as the electron donor,[8] while DAMO bacteria are capable of converting nitrite to nitrogen gas using electrons derived from methane oxidation.[11] By coupling DAMO with AnAmmOx, DAMO microorganisms could reduce the nitrate produced by AnAmmOx with methane, thereby further contributing to the total nitrogen (TN) removal.[9,10]

In addition to the three microbial processes mentioned above, there are also other microbial processes for denitrification such as autotrophic denitrifications, which coexist with regular denitrification processes and might play an important role in BNR from wastewater. In autotrophic denitrification, autotrophic denitrifiers can use hydrogen or sulphur as terminal electron donors in their metabolic chains that are based on inorganic carbon sources. Sulphur, typically supplied as a granular material, forms the physical support for the biofilm in the denitrification reactor, while hydrogen gas can be inserted by common diffusers and membranes or directly produced in the reactor by chemical or electrochemical reactions. These autotrophic denitrification processes would be limited by carbon dioxide availability.

16.2 Heterotrophic Denitrification

16.2.1 Heterotrophic Denitrifying Organisms

There are numerous genera of bacteria that contain species of heterotrophic denitrifying bacteria identified in activated sludge (Table 16.1). Depending on the influent characteristics, treatment configurations and operation conditions, heterotrophic denitrifying bacteria account for approximately 20–80% of all bacteria flocculated and dispersed in the activated sludge.[12] Further, the abundances of dominant populations in denitrifying communities also vary significantly from one treatment plant to another. Most of these denitrifiers are facultative aerobic bacteria preferring free molecular oxygen to ionic nitrogen oxides as electron acceptors. Therefore, the absence of oxygen or the presence of an oxygen gradient within the activated sludge floc particles is necessary for heterotrophic denitrification to occur.

Not all heterotrophic denitrifiers can perform complete denitrification by reducing nitrate to molecular nitrogen. Some of them lack critical enzymes to reduce nitrate to nitrite, nitrite to NO or N_2O to nitrogen gas. In wastewater treatment plants (WWTPs), complete denitrification is likely achieved by the

Table 16.1 Genera of denitrifying species in activated sludge.[28]

Achromobacter	*Escherichia*	*Neisseria*
Acinetobacter	*Flavobacterium*	*Paracoccus*
Agrobacterium	*Gluconobacer*	*Propionibacterium*
Alcaligens	*Holobacterium*	*Pseudomonas*
Bacillus	*Hyphomicrobium*	*Rhizobium*
Chromobacterium	*Kingella*	*Rhodopseudomonas*
Corynebacterium	*Methanonas*	*Spirillum*
Denitrobacillus	*Moraxella*	*Thiobacillus*
Enterobacter	*Xanthomonas*	

cooperation of denitrifying bacteria with different nitrogen oxide reduction capabilities.

16.2.2 Stoichiometry and Kinetics of Heterotrophic Denitrification

Complete heterotrophic denitrification consists of sequential reductive reactions from NO_3^- to NO_2^-, NO, N_2O and finally to N_2 using organic carbon as the electron donor. NO_3^- is normally produced by the nitrification process in WWTPs and carbon sources are mostly from the wastewater influent. Additional external organic carbon supplementation (such as methanol or acetate) is often applied when there is an inadequate amount of organic carbon in wastewaters for removing ionic nitrogen oxides.

The stoichiometric relationship of nitrate reduction to nitrogen gas is generally coupled with a portion of carbon sources used for energy production, while another portion of carbon sources is utilised during denitrification for bacterial growth. There is a significant amount of biomass growth during denitrification. The ratio of the amount of biomass produced to the amount of the overall carbon source used is defined as the biomass yield (Y). The biomass yield largely depends on the nature of the carbon source and the microbial ecology of the denitrification process. Some reported yield values for different carbon sources, such as methanol, acetate, ethanol and glucose, range between 0.23 and 0.65 (mg biomass COD produced/mg COD removed).[13–18]

16.2.3 Factors Affecting Heterotrophic Denitrification

16.2.3.1 Carbon Source

The carbon sources for denitrification can be classified into three types based on the origin of carbon source: COD in the influent wastewater; COD generated by the biomass itself through the death and lysis of organisms; and external carbon sources dosed into anoxic zones to enhance denitrification, such as methanol or acetate. In wastewater influent, there are mainly two forms of carbon source: readily biodegradable organics (rbCODs) and slowly biodegradable organics (sbCODs). The rbCODs contain those small molecule compounds that are able to pass through the bacteria cell wall directly (*e.g.*, sugars and short chain fatty acids), while sbCODs refer to those large organic compounds requiring degradation before passing through the cell wall. The sbCODs can be used as electron donors only after being hydrolysed into rbCODs. Generally, the rbCODs utilisation is considerably faster (by seven to ten times) than the rate of sbCODs.[19] The decay of microorganisms produces some biodegradable organics, and these parts of COD can be utilised for denitrification as well. The denitrification rate of endogenous carbon sources is much slower than for rbCODs, but comparable to sbCODs.[20] When there is not enough COD in the influent serving as a carbon source, or the concentration limit of nitrogen compounds in the effluent is stringent (*e.g.*, less than 5 mg TN L^{-1}), the addition of an external carbon source is often necessary.

Simple organic compounds are usually used as additional carbon sources. In choosing the type of external carbon source, criteria such as the cost of the carbon source, denitrification rate, degree of carbon utilisation, sludge production, handling and storage and content of unfavourable toxic compounds are taken into consideration.[21] Methanol is the most commonly used external carbon source since it is cheaper compared to most other carbon sources (in terms of chemical cost per mass of N removed). It also results in lower biomass yield, and thus reduces the amount of sludge production and the related cost of residual sludge handling and disposal. Methanol is utilised by a specific group of facultative methylotroph organisms.[22] An acclimatisation period varying from a week to several weeks is required in WWTPs in order for the methylotroph organisms to establish themselves.[23] Similarly, an adaptation period may also be needed for other kinds of functional denitrifying groups to develop when a different type of external carbon source is used.

16.2.3.2 COD to N Ratio

In biological denitrification, it is important to estimate the mass of COD that is required to achieve nitrate removal. Generally, around 4 g of biodegradable COD is needed per gram of NO_3^- reduced.[12] However, this ratio largely depends on the nature of the carbon source and the microbial ecology of the denitrification process. Denitrifying bacteria growing on carbon sources with higher biomass yields (Y) require a higher COD to N ratio.

16.2.3.3 Dissolved Oxygen Concentration

Dissolved oxygen (DO) provides a better electron acceptor for denitrifiers to generate energy and has inhibitory effects on the activity of nearly all N reductases. In general, DO is not detectable in the anoxic zone/phase in WWTPs. However, a very limited amount of oxygen could be brought into the denitrification process through natural surface air–liquid oxygen transfer or through flow circulation, and is consumed by denitrifiers immediately. Denitrification can still proceed with a limited DO concentration, with its rate decreasing as the DO concentration increases. In an activated sludge system, the bacteria are clumped together, forming sludge flocs. Depending on the morphology of the sludge flocs, the concentration of substrates such as ammonia and organics in the bulk liquid, oxygen may not be able to diffuse through the entire floc depth, resulting in an oxygen gradient across the floc. This will allow denitrification to happen on the inner side of the flocs.

16.2.3.4 pH and Temperature

The optimum pH for denitrification was found to be in the range of 7–9, with the denitrification rate falling off sharply outside this region.[24] In most conventional WWTPs, the pH is stable at between 7 and 8, suggesting that the pH effect is expected to play a minor role in affecting denitrification performance.[25] The optimal temperature reported ranges from 20 to 35 °C in WWTPs.[12,26,27] The growth rate of denitrifiers decreases with a decrease in wastewater temperature, potentially leading to a reduced capacity for denitrification in the winter period. Denitrification is significantly inhibited at wastewater temperatures lower than 5 °C.[28]

16.2.4 Denitrification Intermediate Accumulation and Electron Competition

Although the targeted end product of denitrification in wastewater treatment is N_2, the accumulation of denitrification intermediates is an ever-present problem in some WWTPs. Nitrite that has accumulated in the denitrification tank would be oxidised to nitrate again in the nitrification tank when a pre-anoxic configuration is applied, leading to increasing oxygen demand in nitrification and extra carbon source consumption during denitrification.[29] In addition, nitrite is known to be toxic, which could suppress the activity of denitrifiers.[30,31] In recent years, N_2O escaping from the denitrification process has become a problem of increasing concern for WWTPs, because N_2O is a strong greenhouse gas, with its global warming potential approximately 300-times higher than CO_2.[32] It is reported that around 0.01–1.8% of the influent nitrogen load of a WWTP is emitted as N_2O, based on the results measured by relatively reliable online monitoring equipment.[33–36] Although nitrification is believed to be a key contributing process, denitrification is certainly a source of the N_2O that is emitted.

Some environmental conditions such as the availability of a carbon source, DO concentration, pH and sulphide have been reported to affect intermediate accumulation.[37–43] This is likely due to the fact that different denitrification enzymes have different sensitivities to environmental factors. Further, the electron competition between the four denitrification steps likely plays an important role in intermediate accumulation. Competition for electrons occurs when the electron supply rate from the oxidation process cannot meet the demand for electrons by nitrogen oxide reduction.[44] Therefore, it is likely that the four denitrification steps could exert influences on each other through electron competition, thus leading to unbalanced rates and intermediate accumulation. It is demonstrated that electron competition processes could contribute to N_2O accumulation. Using a methanol-utilising denitrifying sludge, Pan *et al.*[45] revealed that the culture did not allocate electrons according to the maximum turnover rates of the different nitrogen oxide reductases, with electron supply being the limiting step. Rather, the reduction of nitrite was prioritised over the other denitrification steps, consequently leading to N_2O accumulation.

16.2.5 Modelling the Heterotrophic Denitrification Processes

WWTPs are engineered systems designed to achieve the removal of pollutants, mainly organics (C), nitrogen (N) and phosphorus (P). Modelling of activated sludge processes has become a common part of the design and control of

WWTPs. The activated sludge models (ASMs) no. 1, 2 and 3,[46] published by the International Water Association (IWA) focusing on the prediction of pollutants (C, N and P) removal and related microorganisms growth, are well received by both the scientific community and wastewater industry. In ASMs, denitrification was modelled as a single-step process, with nitrate being directly reduced to N_2. The removal of COD and nitrate were coupled to denitrifying bacteria growth directly. The ASM model series provides a standard platform for wastewater modellers by introducing a nomenclature of a common language that wastewater modellers speak when using the concepts and the matrix notation that facilitates the communication of complex models.

To predict denitrification intermediate accumulation, denitrification needs to be modelled as a multiple-step process. Based on ASMs, many modifications have been developed. So far, denitrification has been modelled as two-step (from NO_3^- to N_2 *via* NO_2^-), three-step (from NO_3^- to N_2 *via* NO_2^- and N_2O) and four-step (from NO_3^- to N_2 *via* NO_2^-, NO and N_2O) processes in literature. Each denitrification step is modelled with individual, reaction-specific kinetics. Among these variations, two distinct approach in describing the electron allocation among different steps of denitrification, represented by the activated sludge model for nitrogen (ASMN)[47] and the activated sludge model with indirect coupling of electrons (ASM-ICE),[48] have been adopted. The key differences of the two modelling structures are illustrated in Figure 16.1. Pan *et al.*[48] evaluated the two types of models by comparing their ability to

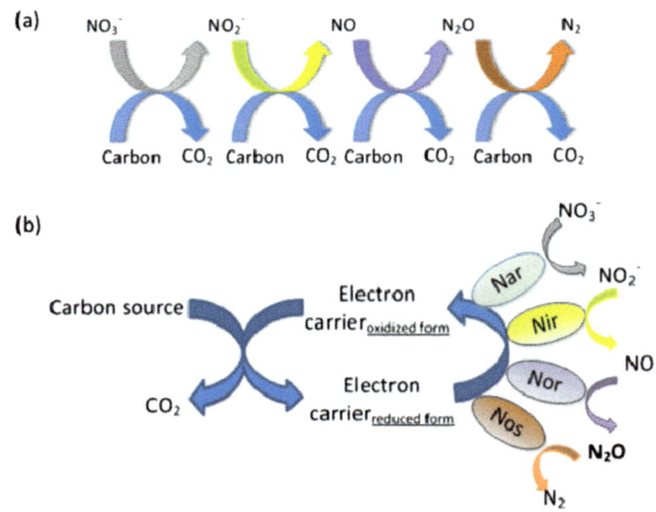

Figure 16.1 Simplified reaction schemes used in the activated sludge model for nitrogen (a) and the activated sludge model with indirect coupling of electrons (b).[48] Reproduced from *Water Research*, 71, Yuting Pan, Bing-Jie Ni, Huijie Lu, Kartik Chandran, David Richardson, Zhiguo Yuan, Evaluating two concepts for the modelling of intermediates accumulation during biological denitrification in wastewater treatment, 21–31, Copyright 2015 with permission from Elsevier.

predict nitrogen conversion dynamics in one pure culture and three mixed culture studies. It was demonstrated that the ASMN-type model is able to predict the overall nitrogen and COD removal performance with low-level accumulation of denitrification intermediates, but is inadequate for predicting the N_2O production with intensive electron competition. The ASM-ICE-type model has the potential to describe all of the N_2O data under different conditions, but requires information on both the carbon oxidation and the nitrogen reduction kinetics.

16.2.6 Nitrogen Removal Based on Heterotrophic Denitrification

The conventional BNR process, incorporating both the nitrification process to convert NH_4^+ to NO_3^- and the denitrification process to reduce NO_3^- to remove nitrogen pollutants from wastewater, is the most mature and widely applied nitrogen removal process in the world. In conventional BNR systems, the microbial community (activated sludge) is exposed to different environmental conditions to accomplish COD and nitrogen removal. Specifically, aerobic conditions are required for nitrification, whereas a sufficient amount of organic carbon compound and the absence of oxygen are required to support denitrification. To achieve this, biological reactors providing different conditions are usually linked in series in a WWTP, whereby the mixed liquor of wastewater and the activated sludge flows through. At the end of the biological treatment process, the mixed liquor is passed into the clarifier, where the sludge is settled and separated from the treated water. The latter is discharged as the effluent. Most of the settled sludge is returned to the bioreactor, while a small fraction of the sludge is removed and disposed of after several steps of sludge treatment. When complete denitrification is achieved, around 75% of the influent N is removed by emitting as N_2, and 20% of the influent N is incorporated into the sludge and removed as waste sludge, with only around 5% of the influent N remaining in the effluent.

16.2.6.1 Pre-Denitrification and Post-Denitrification

The conventional BNR processes can be classified into two basic types based on whether the anoxic zone is located before or after the aeration zone. (i) Pre-denitrification: the wastewater and the activated sludge contact and mix in an anoxic zone before an aeration zone, such as the modified Ludzack–Ettinger process. Nitrate produced in the aerobic zones by nitrification is recycled back to the pre-denitrification compartment for denitrification. This process enables the use of the organic carbon source that is available in the influent for denitrification. The denitrification rate is affected by the rbCOD concentration in the influent and other parameters, such as MLVSS concentration and temperature. Since nitrification is located after denitrification, nitrate is present in the effluent. The pre-denitrification process will typically produce effluent TKN concentrations of between 6 and 10 mg N L^{-1}. (ii) Post-denitrification: the first reactor is aerobic and the second is un-aerated. Self-generated endogenous organics and/or external carbon sources are used as the carbon source. A post-denitrification configuration on its own is not a commonly used BNR process. The influent COD, rather than being used as the denitrification carbon source, is consumed in the aerobic zone with an extra energy cost for aeration. In addition, the post-denitrification process may produce other problems, such as the release of ammonium in the anoxic zone if only self-generated endogenous organics are used as carbon sources. In practice, upgrading WWTPs by adding a post-denitrification compartment in some cases is necessary in order to meet the more stringent effluent requirement (<5 mg TN L^{-1}) of some countries in recent years. At this point, most of the rbCOD has been depleted in the previous steps; therefore, external carbon addition is usually necessary.

16.2.6.2 Modified Ludzack–Ettinger Configuration

In a typical modified Ludzack–Ettinger configuration, an anoxic compartment/zone (where DO is usually not detectable) precedes the aerobic compartment (where DO concentration is typically within the range of 0.5–2 mg L^{-1}) (Figure 16.2). At the end of the aerobic compartment, the mixed liquor of nitrified wastewater containing NO_3^- and the activated sludge is recirculated back to the anoxic compartment with a flow rate that is a few times that of the wastewater influent. Wastewater is also fed into the anoxic compartment, which provides the organic carbon for denitrification.

16.2.6.3 Oxidation Ditch

Oxidation ditches (Figure 16.3) are usually equipped with horizontal brush aerators to provide aeration and also to move the mixed liquor along the ditch at a relatively high velocity (0.25–0.35 m s^{-1}).[49] Each pass of mixed liquor in the ditch typically lasts for several minutes. A relatively high DO concentration is obtained at or close to the aerator, and anoxic conditions develop further away from the aerator. The DO is typically low (*e.g.*, around 0.5 mg L^{-1}) in oxidation ditches, favouring simultaneous nitrification and denitrification (SND).

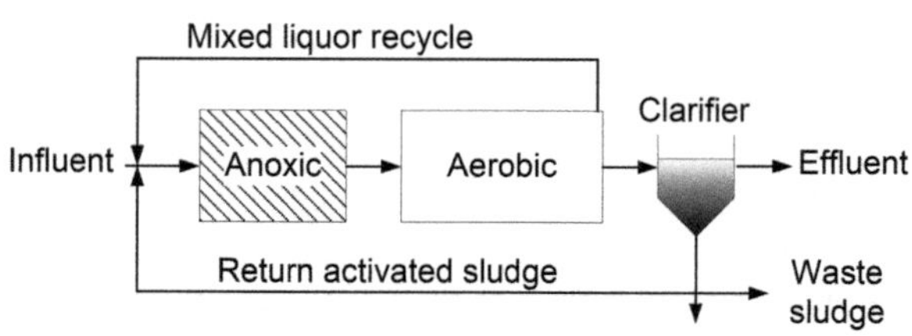

Figure 16.2 Diagram of a modified Ludzack–Ettinger system.

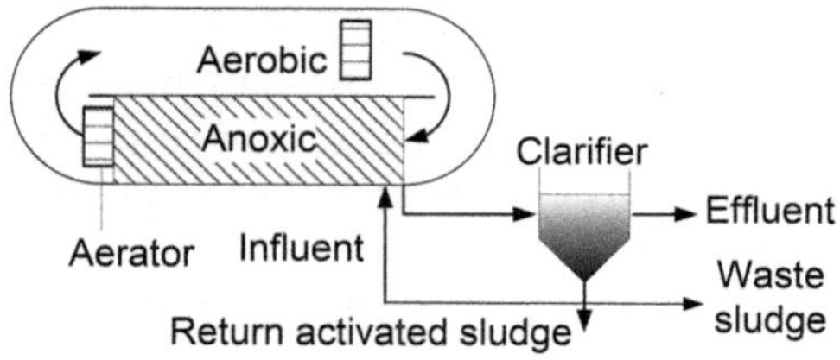

Figure 16.3 Diagram of an oxidation ditch.

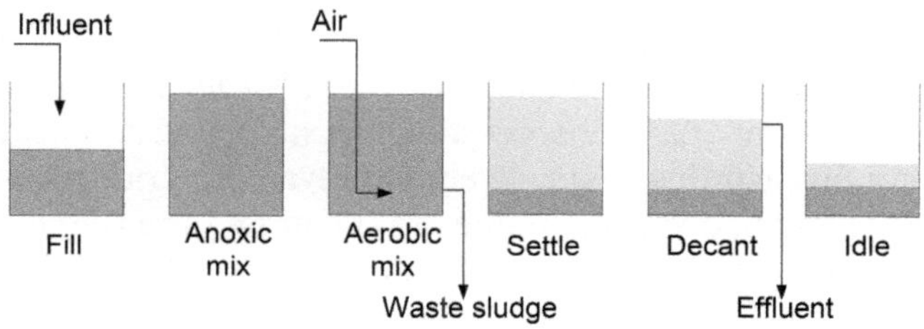

Figure 16.4 Diagram of a sequencing batch reactor system.

16.2.6.4 Sequencing Batch Reactors

Unlike the continuous flow systems outlined above, sequencing batch reactors (SBRs) can also be used to achieve the removal of nitrogen and organic carbon (Figure 16.4). Aerobic and anoxic conditions are separated by time instead of space. The sewage is introduced into a reaction tank, one batch at a time. Wastewater treatment is achieved by a timed sequence of operations that occurs in the same SBR tank, consisting of filling, reaction (aeration), settling, decanting, idling and sludge wasting.

16.3 Anaerobic Ammonia Oxidation

Understanding of the global nitrogen cycle has been altered fundamentally because of the discovery of the AnAmmOx bacteria. The identification of the responsible chemolithoautotrophic bacteria (*i.e.*, AnAmmOx bacteria)[50] not only stimulated the appreciation of their applied and ecological significance, but also provided economically attractive and environmentally friendly alternatives to current wastewater treatment processes.[51–53]

16.3.1 AnAmmOx Organisms

To date, five genera of AnAmmOx bacteria have been discovered in natural and engineered environments, including *Candidatus "Brocadia"*, *"Kuenenia"*, *"Scalindua"*, *"Anammoxoglobus"* and *"Jettenia"*.[54] *Candidatus "Brocadia anammoxidans"*, which was the first AnAmmOx bacterium identified,[55] was cultured by seeding sludge from a full-scale WWTP. All AnAmmOx organisms are related to the phylum *Planctomycetes* and show no distinct differences in physiology, metabolism or ultrastructure. Furthermore, all five currently recognised genera of AnAmmOx bacteria share unique physiological and morphological features, such as the presence of an anammoxosome and the biosynthesis of ladderane lipids.

Candidatus "*Scalindua*" has frequently been detected in marine sediments and marine oxygen minimum zones.[56–58] The other four AnAmmOx genera have been identified and enriched in bioreactors treating nitrogenous wastewater, including "*Kuenenia*",[59,60] "*Brocadia*",[50,61] "*Anammoxoglobus*"[62] and "*Jettenia*".[63] Until now, draft genomes of four AnAmmOx species have been assembled and reported, including *Candidatus* "*Kuenenia stuttgartiensis*",[60] "*Jettenia caeni*",[64] "*Brocadia fulgida*"[65] and "*Scalindua profunda*".[66] The genome assemblies of "*Ca. Kuenenia stuttgartiensis*" and "*Ca. Jettenia caeni*" are in five and four contigs, respectively. However, the draft genome of "*Ca. Brocadia fulgida*" is relatively fragmented (411 contigs) and the "*Ca. Scalindua profunda*" draft genome is highly fragmented (1580 contigs).[67]

16.3.2 Metabolic Pathways of AnAmmOx

AnAmmOx bacteria have the unique metabolic ability to utilise ammonium and nitrite to form nitrogen gas under anaerobic conditions.[55] The metabolic model of AnAmmOx has been postulated and gradually revealed from genome analysis, ^{15}N-labelling experiments and thermodynamic calculations.[60,68] A biochemical model has been proposed according to the genome of the AnAmmOx bacterium *Ca.* "*Kuenenia stuttgartiensis*".[60] In the model, nitrite is first reduced to NO with the uptake of one low-energy electron, after which NO oxidises ammonium to form hydrazine *via* catalysis by hydrazine hydrolase using three low-energy electrons. The hydrazine is then oxidised to nitrogen gas *via* a reaction catalysed by hydrazine dehydrogenase/oxidase while simultaneously releasing four high-energy electrons. These electrons are transferred to a respiratory chain at the level of ubiquinone *via* the cytochrome bc_1 complex (complex III) and ubiquinone oxidoreductase.[69,70] The bc_1 complex shuttles these electrons toward nitrite reduction and hydrazine synthesis. Simultaneously, protons are transported across a membrane system, thereby establishing a proton motive force, which could be utilised by membrane-bound ATPases to generate ATP. The production of ATP over the anammoxosomal membrane is utilised for the autotrophic growth of AnAmmOx bacteria (Figure 16.5).

16.3.3 Physiological Characteristics and Enrichment of AnAmmOx

The competitive advantage of a microorganism will mainly depend on the kinetic parameters, including maximum growth rate, substrate affinity constants (*i.e.*, half-saturation constant, K_m) and biomass yields (Y). Table 16.2

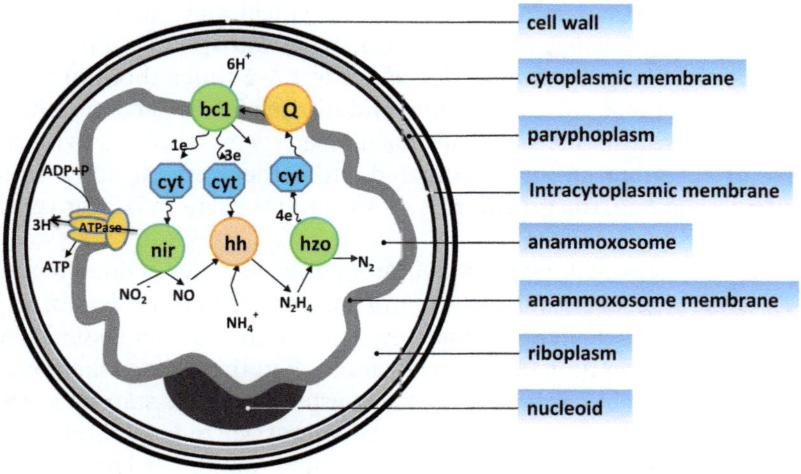

Figure 16.5 Schematic diagram of the ultrastructure of anaerobic ammonium oxidation (AnAmmOx) bacteria and the metabolic pathway of the AnAmmOx process based on literature.[59,240] The background picture is the granular biomass from an AnAmmOx reactor treating domestic wastewater. nir: nitrite reductase (cytochrome cd_1); hh: hydrazine hydrolase; hzo: hydrazine/hydroxylamine oxidoreductase; cyt: mono- or di-haem cytochrome *c* electron carriers; bc_1: cytochrome bc_1 complex (complex III); Q: coenzyme Q (ubiquinone).

Table 16.2 Physiological characteristics and kinetics of anaerobic ammonium oxidation bacteria.[a]

Parameters	*Brocadia anammoxidans*	*Brocadia sinica*	*Kuenenia stuttgartiensis*	*Scalindua* sp.	*Scalindua profunda*	*Jettenia caeni*
Growth temperature (°C)	20–43	25–45	25–37	10–30	15–45	20–42.5
Growth pH	6.7–8.3	7.0–8.8	6.5–9.0	6.0–8.5	7.4	6.5–8.5
Maximum growth rate (μ_{max}, day^{-1})	0.065	0.098	0.062–0.084	0.048	ND	0.048
Doubling time (days)	10.7	7	8–11	ND	ND	ND
Biomass yield (Y, mmol C mmol N^{-1})	0.07	0.063	ND	0.03	ND	0.056
K_m for ammonia (μM)	<5	28 ± 4	ND	3	ND	17.1 ± 4.3
K_m for nitrite (μM)	<5	34 ± 21	0.2–3	0.45	ND	35.6 ± 0.92
References	50 and 79	241	73, 83, 242	243	244	245

[a]ND: no data.

compares the kinetic parameters among six typical AnAmmOx bacteria, *K. stuttgartiensis*, *B. anammoxidans*, *B. sinica*, *Scalindua* sp., *S. profunda* and *J. caeni*. Specially, the growth rate of *B. sinica* is 1.5–2-times those of the others. In contrast, the K_m for ammonium and nitrite of *B. sinica* is significantly higher, which indicated *B. sinica* has maximised its metabolic activity at the expense of substrate affinity. Compared with other general heterotrophic denitrifying bacteria, the growth rate of AnAmmOx bacteria is relatively low, in the range of 0.002–0.0041 h^{-1}.

Due to the slow growth rate of the AnAmmOx bacteria (with doubling times of 7–11 days), the successful culturing of AnAmmOx bacteria has been recognised as a major challenge. Generally, AnAmmOx reactors should have a long sludge retention time (SRT; usually more than three times the doubling time) to keep a high biomass concentration, especially at the start-up stage. To address the longer start-up period required, extensive studies have been conducted to evaluate different reactors for enriching AnAmmOx bacteria. At the laboratory-scale level, AnAmmOx enrichment has been successfully achieved in fixed biofilm bed, fluidized bed (analogous to up-flow anaerobic sludge blanket [UASB]), SBR, gas-lift and well-mixed reactors, as well as in rotating biological contactors; however, the maximum loading rate and the operation costs differ among these reactor configurations.[71,72] Specially, a culture with high purity of the AnAmmOx enrichment (97.6%) was achieved in a membrane bioreactor (MBR), despite of a short SRT of 12 days. It seems that the MBR is a promising tool for the cultivation of AnAmmOx bacteria.[73] The large amounts of active biomass harvested by existing AnAmmOx reactors can also be used for the inoculation of other reactors, which may distinctly shorten the start-up periods. Furthermore, studies have revealed that biofilm systems[74,75] and granular bioreactors[76] are feasible options for enriching AnAmmOx bacteria. However, biofilm or granular AnAmmOx systems would need extremely long start-up times (*e.g.*, the first full-scale AnAmmOx reactor [in Rotterdam, The Netherlands] has achieved a rate of 10 kg N m^{-3} day^{-1} after a start-up period of 3 years).[77]

16.3.4 Factors Affecting AnAmmOx

The activity and distribution of AnAmmOx are affected by various environmental conditions, including DO, ammonia and nitrite levels, temperature, salinity and metal concentrations.

16.3.4.1 Dissolved Oxygen

Initial batch experiments with *B. anammoxidans* enrichments showed that oxygen levels as low as 2 μM completely inhibited the AnAmmOx activity.[78] However, this inhibitory effect is reversible, and the rate of AnAmmOx is the same before and after aeration in an intermittent aeration (2 hours oxic and 2 hours anoxic) reactor for 20 days.[79] AnAmmOx and nitrification can form a stable symbiotic relationship under oxygen-limited conditions ($DO < 0.5 mg L^{-1}$), which is particularly beneficial for the one-stage partial nitrification and AnAmmOx processes.

16.3.4.2 Ammonia and Nitrite Levels

Ammonia and nitrite, which are substrates for the growth of AnAmmOx bacteria, both affect the activity, growth and distribution of AnAmmOx bacteria. Ammonia has little inhibitory effect on AnAmmOx at concentrations lower than 1000 mg N L^{-1}.[80,81] Nitrite exhibits a distinct inhibitory effect against AnAmmOx bacteria at levels above 50–150 mg N L^{-1}.[50,82,83] With nitrite at concentrations higher than 7 mM at pH 7.0 (correspondingly, a free nitrous acid [FNA] concentration of 0.020 mg L^{-1}), the AnAmmOx activity was completely lost in a SBR with 74% AnAmmOx bacteria. However, this nitrite inhibition could be overcome by the addition of trace amounts of either hydrazine or hydroxylamine.[50] Moreover, in batch tests with a *K. stuttgartiensis* enrichment culture, nitrite of 18 mM at pH 7.0 (correspondingly, a FNA of 0.14 mg L^{-1}) completely inactivated the AnAmmOx bacteria. However, the inhibitory effect of 12 mM nitrite at pH 7.0 was temporary and the AnAmmOx activity resumed at the same rate, with nitrite being reduced to 5 mM after 3 days.[83]

16.3.4.3 Temperature

The AnAmmOx process commonly occurs in mesophilic environments with temperatures ranging from 6 °C to 43 °C, while the AnAmmOx activity drops rapidly at temperatures lower than 15 °C or higher than 40 °C. However, there is increasing evidence showing that AnAmmOx bacteria can tolerate a relatively wide temperature range of −2 °C to 85 °C.[84,85] The optimal temperature for the AnAmmOx process is considered to be around 30–40 °C; therefore, most studies of the process are conducted at temperatures greater than 30 °C, even though many studies have shown that the AnAmmOx process could also be successfully conducted at around 20 °C.[86,87]

16.3.4.4 Salinity

The adaptation of a freshwater AnAmmOx population to wastewater with high salinity demonstrated that AnAmmOx bacteria had good tolerance to salinity. The activity of AnAmmOx was not affected by exposure to salinity levels up to 6 g L^{-1} NaCl. They could adapt to NaCl as high as 30 g L^{-1}, provided that the increase was gradual.[88] AnAmmOx's high resistance to salinity would be favourable for removing wastewater with a high salt concentration.[65] However, the variations in salinity may affect the abundance, diversity and distribution of AnAmmOx bacteria in marine or estuarine systems.[89]

16.3.4.5 Fe(II) Concentration

As the essential substrate element that is required for culturing AnAmmOx sludge, Fe(II) is expected to affect AnAmmOx bacterial growth. Several studies have demonstrated that iron could affect the metabolism of AnAmmOx bacteria.[90,91] It was found that AnAmmOx bacteria would store iron inside the anammoxosome compartment in order to have an excess supply of iron for further haem *c* synthesis.[92] Currently, the Fe(II) concentration in the feeding solution of almost all of the enriched AnAmmOx sludge systems has been constantly set as 0.03 mM since the discovery of AnAmmOx process. Little effort has been dedicated to understanding the impacts of different amounts of Fe(II) addition on AnAmmOx activity.[93,94]

16.3.5 Modelling the AnAmmOx Processes

Mathematical modelling has been widely used to describe the AnAmmOx process and evaluate the effects of operational conditions on the AnAmmOx performance. Most modelling work has focused on biofilm-based or granule-based AnAmmOx reactors, since these two types of AnAmmOx reactors are widely used in laboratory- or full-scale applications. For example, biofilm models were developed to simulate the nitrification, heterotrophic denitrification and AnAmmOx processes in a completely autotrophic ammonium removal over nitrite (CANON) reactor, as well as the influences of factors on reactor performance. Modelling results showed that COD in the influent has no significant effects on the trends in the partial nitrification–AnAmmOx biofilm process. AnAmmOx plays a major role in the TN removal in stable biofilm systems and conventional denitrification only takes a share of <20% in the TN removal.[95] The activities of AnAmmOx bacteria would decrease with the decreasing of temperature. For CANON processes, thicker biofilms are required, or else the ammonium surface load to the biofilm should be decreased in order to maintain full N removal at lower temperatures.[96]

In the case of granular sludge reactors, the granule size would affect the distribution of substrates (nitrite, ammonium and oxygen) in a single granule, resulting in different AnAmmOx reactor performance. For one-stage partial nitrification–AnAmmOx reactors, the AnAmmOx process is favoured in relatively large granules, while smaller granules tend to accumulate nitrite or even nitrate, since nitrite-oxidising bacteria (NOB) are present in the smallest granules, while they are outcompeted by AnAmmOx bacteria in larger granules.[97] In addition, based on an AnAmmOx granule model of UASB treating high-strength ammonium wastewater, the optimum granule diameter for the maximum N removal should be between 1.0 and 1.3 mm and the optimum N loading rate should be 0.8 kg N m^{-3} day^{-1}.[86]

16.3.6 Autotrophic Nitrogen Removal by AnAmmOx

Compared to conventional nitrification and denitrification, the AnAmmOx process exhibits a series of advantages, including less sludge production, decreased oxygen supply, less space required and reduced CO_2 and N_2O emissions.[53,71,98,99] So far, over 40 full-scale AnAmmOx plants have been made operational or are under design or construction, mainly in Europe, China and the United States.[100] Table 16.3 shows the operation performance of some typical full-scale AnAmmOx plants around the world. AnAmmOx has been successfully used to treat the ammonium-rich anaerobic sludge digestion liquor (so-called

Table 16.3 Some typical full-scale anaerobic ammonium oxidation plants in the world.[a]

Location	Volume (m^3)	Seeding sludge	One- or two-step	Influent	Reactor type	Specific max. conversion (kg N m^{-3} day^{-1})	Nitrogen removal efficiency	Reference
Rotterdam (The Netherlands)	70	Nitrifying sludge	Two-step	Reject water	GSR	10	85%	77
Strass (Austria)	500	AnAmmOx + nitrifying sludge	One-step	Reject water	SBR	0.6	>85%	246
Köllikon (Switzerland)	33	Nitrifying sludge	One-step	Landfill leachate	RDC	0.5	70%	247
Lichtenvoorde (The Netherlands)	100	AnAmmOx sludge	Two-step	Tannery	GSR	1	85%	77
Olburgen (The Netherlands)	600	AnAmmOx sludge	One-step	Potato processing	Bubble column	1.2	>70%	248
Zürich (Switzerland)	1400	AnAmmOx + nitrifying sludge	One-step	Reject water	SBR	2	90%	249
Hattingen (German)	102	AnAmmOx sludge	One-step	Reject water	Moving bed	1	70–80%	250
Mie Prefecture (Japan)	58	AnAmmOx sludge	Two-step	Semiconductor	GSR	3	ND	251
Glarnerland (Switzerland)	400	AnAmmOx sludge	One-step	Reject water	SBR	0.4	>90%	252
Pitsea (England)	240	AnAmmOx + nitrifying sludge	One-step	Landfill leachate	RDC	1.7	>90%	253
Tongliao (China)	6700	AnAmmOx sludge	One-step	Monosodium glutamate	GSR	1.6	ND	
Jiang Su (China)	1600	AnAmmOx sludge	One-step	Sweetener	GSR	2.18	ND	
Xinjiang (China)	5400	AnAmmOx sludge	One-step	Monosodium glutamate	GSR	10.7	ND	
Shaoxing	560	AnAmmOx sludge	One-step	Distillery	GSR	0.9	ND	

[a]AnAmmOx: anaerobic ammonium oxidation; GSR: granular sludge reactor; ND: no data; RDC: rotating biological contactor; SBR: sequencing batch reactor.

side-stream autotrophic N removal). Recently, the revolutionary view that the AnAmmOx process should be implemented at the mainstream for autotrophic N removal following organic C separation has been formed.[53,101,102] Like in the side-stream process, the mainstream autotrophic N removal process involves ammonia-oxidising bacteria (AOB) converting approximately half of the ammonium to nitrite, with the AnAmmOx bacteria converting the nitrite formed by AOB and the remaining ammonium to N_2 (and nitrate).

Based on the AnAmmOx reaction, the most appropriate nitrite : ammonium molar ratio should be 1.32 : 1. To provide an appropriate influent, the SHARON process has often been employed to produce the 50 : 50 mixture of ammonium and nitrite that is required in the AnAmmOx process.[77] If AnAmmOx is combined with partial nitrification in a one-step process (often referred to as a one-stage AnAmmOx process), it is a promising new technology for wastewater treatment. However, a critical requirement of the one-stage autotrophic N removal process is the suppression of NOB growth, as otherwise nitrate would be the final product of nitrification, which could not be removed by the AnAmmOx process. Several strategies were developed in side-stream autotrophic N removal systems for the successful washout of NOB. However, these strategies either require particular influent characteristics (*e.g.*, high ammonium concentrations and an approximately equal molar ratio between ammonium and bicarbonate) or unique growth conditions (*e.g.*, at a high temperature of ~35 °C). These conditions do not exist in mainstream wastewater, and indeed the selection for AOB against NOB has proven very difficult for domestic wastewater. The strategies developed to date, including the use of low DO, intermittent aeration and the augmentation of AOB using AOB from side-stream systems, have been unable to reliably suppress NOB. Recently, it has been reported that FNA (or HNO_2 at ppm levels) can selectively suppress NOB.[103] FNA treatment might potentially be used to manipulate the nitrifying community structure, in particular to suppress the activity of NOB to achieve stable nitrification, thus providing an appropriate nitrite : ammonium ratio for the AnAmmOx conversion.

16.4 Denitrifying Anaerobic Methane Oxidation

16.4.1 Stoichiometry of DAMO

Under anoxic conditions, methane is oxidised to carbon dioxide in a process called anaerobic oxidation of methane (AOM) while coupled to different electron acceptors such as sulphate and metal oxides. AOM is estimated to consume up to 90% of the methane produced in anaerobic marine sediments,[104] representing an important process regulating the release of this greenhouse gas into the atmosphere.

In contrast to the numerous studies of AOM coupled to sulphate reduction, there have only been a few publications focusing on the DAMO process. Thermodynamically, oxidised nitrogenous compounds are more favourable electron acceptors than sulphate, where the energy yield of AOM coupling to sulphate reduction is only −20 to −40 kJ mol^{-1} CH_4.[105] The reactions may be described as follows:

$$5CH_4 + 8NO_3^- + 8H^+ \rightarrow 4N_2 + 14H_2O + 5CO_2; \Delta G^{0\prime} = -765 \text{ kJ mol}^{-1} \text{ CH}_4 \quad (16.2)$$

$$3CH_4 + 8NO_2^- + 8H^+ \rightarrow 4N_2 + 10H_2O + 3CO_2; \Delta G^{0\prime} = -928 \text{ kJ mol}^{-1} \text{ CH}_4 \quad (16.3)$$

16.4.2 Enrichment and Characteristics of DAMO

Process engineers consider methane to be a potentially inexpensive external electron donor for denitrification for the treatment of wastewater or landfill leachate where methane is generated on site.[106,107] However, several attempts to enrich the DAMO microorganisms in the laboratory have failed.[108–110] The slow growth of these microorganisms is a likely explanation for persistent failures to obtain DAMO cultures.

DAMO microorganisms were first enriched in 2006 by Raghoebarsing *et al.*[7] from freshwater canal sediments. The enrichment culture was dominated by *Candidatus Methylomirabilis oxyfera* (~80%); however, a group of anaerobic methanotrophic (ANME-2D) archaea were also detected (~10%). Substrate utilisation studies using ^{13}C-labelled methane showed that both members of this consortia were able to incorporate ^{13}C into their lipids.[7] ANME had previously been shown to play an important role in AOM coupled to sulphate reduction or manganese and iron reduction.[111,112] It was initially hypothesised that the ANME-2D worked with *M. oxyfera* to perform AOM coupled with denitrification.[7] However, the ANME-2D later disappeared from the culture without influencing DAMO performance, indicating that *M. oxyfera* alone could carry out the DAMO process.[113]

Later, Hu *et al.*[114] reported the successful enrichment of two DAMO cultures from a mixture of activated sludge and digester sludge from a WWTP and anoxic sediments from a lake. One enrichment culture (enriched at 35 °C) comprised organisms that were closely related to *M. oxyfera* and ANME-2D. The second enrichment culture (enriched at 22 °C) contained *M. oxyfera*, but no archaea. Based on the relative performance of all reported DAMO cultures, Hu *et al.*[114] suggested that the archaea found in these cultures might play an important role in nitrate reduction.

16.4.3 Proposed Mechanisms of DAMO Microorganisms

16.4.3.1 *DAMO Bacteria*—Candidatus Methylomirabilis oxyfera

The first co-culture of DAMO bacteria and DAMO archaea was enriched at a ratio of 8 : 1.[7] After subsequent cultivation, a bacteria-dominant culture without archaea was obtained, suggesting that bacteria alone could couple anaerobic methane oxidation to denitrification.[113] FISH analysis (Figure 16.6) showed that the DAMO bacteria had thin rod shapes, about 1 µm

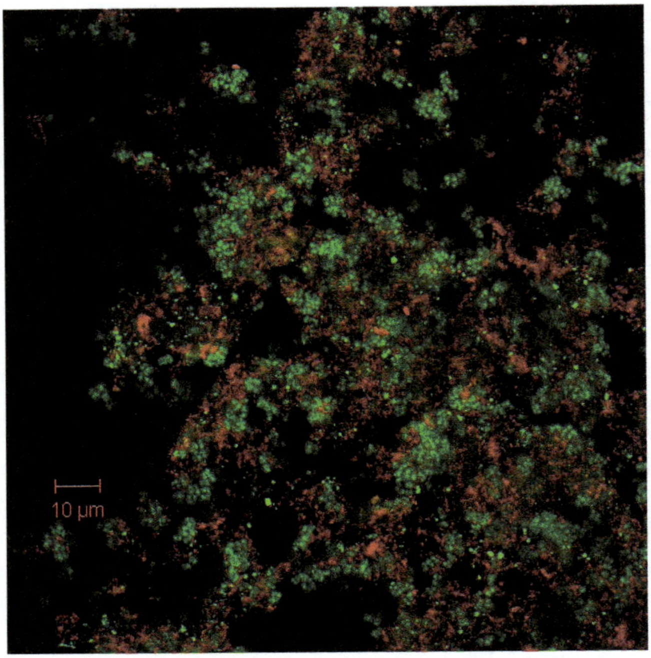

Figure 16.6 Fluorescence *in situ* hybridisation of denitrifying anaerobic methane oxidation organisms. Epifluorescence micrograph taken after hybridisation with the archaeal probe DARCH-872 (green) and specific bacterial probe NC10-1162 (red).

long and 0.3 µm wide, with DNA concentrated in the centre of the cell.[113] A new enriched culture monitored by quantitative polymerase chain reaction revealed that the enriched bacteria belonged to a phylum NC10.[115]

Using metagenomics, Ettwig *et al.*[11] assembled the genome of *M. oxyfera*. Genes for the reduction of nitrate to nitrite (*narFHJI* and *napAB*), nitrite to NO (*nirSJFD*/*GH*/*L*) and NO to N_2O (*norZ* = *qnor*), as well as genes encoding the complete pathway for aerobic methane, were present in its genome. However, the gene for reducing N_2O to dinitrogen gas was missing. So a new pathway coupling anaerobic methane oxidation to denitrification was proposed, where oxygen is generated from NO in the denitrification process and used for aerobic methane oxidation.[11] This pathway is the fourth biological pathway for oxygen production in nature besides photosynthesis, chlorate respiration and the detoxification of reactive oxygen species. Since then, the "*M. oxyfera*" bacteria have been discovered in many different places all over the world.[116–119]

16.4.3.2 DAMO Archaea—Candidatus Methanoperedens nitroreducens

Since DAMO cultures with or without archaea were both fostered, there was some debate as to the function of DAMO archaea. The fact that the cultures containing DAMO archaea displayed a nitrate reduction rate 30-times higher than those of DAMO cultures without archaea suggested that archaea likely play a significant role in reducing nitrate.

Using metagenomic and metatranscriptomic approaches, Haroon *et al.*[8] confirmed that these archaea, named as "*Candidatus Methanoperedens nitroreducens*", can oxidise methane to carbon dioxide through reverse methanogenesis and reduce nitrate to nitrite. Meta-omics analyses combined with bioreactor performance and ^{13}C- and ^{15}N-labelling experiments showed that ANME-2D is capable of independent AOM *via* reverse methanogenesis using nitrate as the terminal electron acceptor. Comparative analyses revealed that the genes for nitrate reduction have been laterally transferred from a bacterial donor, suggesting selection for this novel process within ANME-2D. Nitrite produced by ANME-2D was reduced to dinitrogen gas through a syntrophic relationship with an AnAmmOx bacterium, effectively outcompeting "*M. oxyfera*" in the system.[120]

16.4.4 Factors Affecting DAMO

Temperature is a significant factor affecting the growth of DAMO microorganisms. The influence of temperature on the selection between DAMO bacteria and DAMO archaea could be demonstrated by comparing DAMO cultures enriched at 22 °C and cultures enriched at 35 °C.[114] The cultures at 22 °C contained 15% DAMO bacteria without DAMO archaea, while the cultures at 35 °C consisted of 35% DAMO bacteria and 40% DAMO archaea, indicating that DAMO archaea might be more readily activated at higher temperatures.

pH and the presence of oxygen could also affect the DAMO activities. Although the DAMO organisms were enriched under slightly different pH conditions (pH for different enrichments ranged from 6.9 to 7.5), the optimum pH for DAMO was not reported. Furthermore, anaerobic conditions were applied to all of the enrichment reactors, yet the hypothesised pathway showed that DAMO bacteria could generate their own oxygen for the oxidation of methane. The effects of extra oxygen on DAMO bacteria were recently tested.[121] The results showed that the addition of 2–8% oxygen resulted in the genes encoding the methane oxidation pathway being expressed, but this repressed the genes expression that encodes the denitrification pathway.

16.4.5 Emerging Technologies Based on DAMO Processes

The discovery of DAMO processes, in which methane is oxidised anaerobically to provide electrons for denitrification,[7,8,11,114] provides some new opportunities for achieving complete nitrogen removal from wastewater by utilising *in situ*-produced methane as the electron donor under anaerobic conditions.[9,10] One possible way to apply this process for wastewater treatment is to combine DAMO with the AnAmmOx process.

The AnAmmOx process is an economically attractive and environmentally friendly alternative to current wastewater treatment in terms of less sludge production, decreased oxygen supply and reduced N_2O emissions.[51–53] To date, AnAmmOx-related processes have been widely installed and operated in full-scale for side-stream wastewater treatment. Several laboratory studies showed that nitrogen removal from mainstream wastewater could also be achieved by AnAmmOx process, although it was difficult to control the level of TN in the effluent to below 10 mg N L^{-1}.[122–124] However, the AnAmmOx process presents some other limitations. Even with an optimal ammonium : nitrite ratio of 1 : 1.32 in the feed, the AnAmmOx process can only remove 89% of the TN theoretically, with 11% of the nitrogen being converted to nitrate. The nitrogen removal efficiency reported in literature was mostly lower than 85%, since the effluent from the partial nitrification reactor cannot ensure the ideal ratio of 1 : 1.32.[125]

The combination of DAMO and AnAmmOx processes (Figure 16.7) enables us to achieve complete nitrogen removal in wastewater treatment. Nitrite, either from the influent or produced by DAMO archaea, can be removed by DAMO bacteria using methane as the electron donor or AnAmmOx bacteria using ammonium from the influent as the electron donor. DAMO archaea will further polish the TN removal by using the nitrate produced by AnAmmOx, which presents a significant advantage over the single AnAmmOx process. A co-culture of DAMO bacteria and AnAmmOx bacteria was enriched previously.[126] In this culture, AnAmmOx consumed 77% of the nitrite fed with ammonium as the electron donor, and DAMO bacteria consumed the rest of the nitrite with methane as the electron donor.[126] In Australia, a co-culture of DAMO archaea and AnAmmOx bacteria was enriched by feeding methane, nitrate and ammonium into a suspended culture.[8] In another study, in a mixed culture consisting of AnAmmOx bacteria, DAMO archaea and DAMO bacteria fed with methane, nitrate and ammonium, the DAMO bacteria were

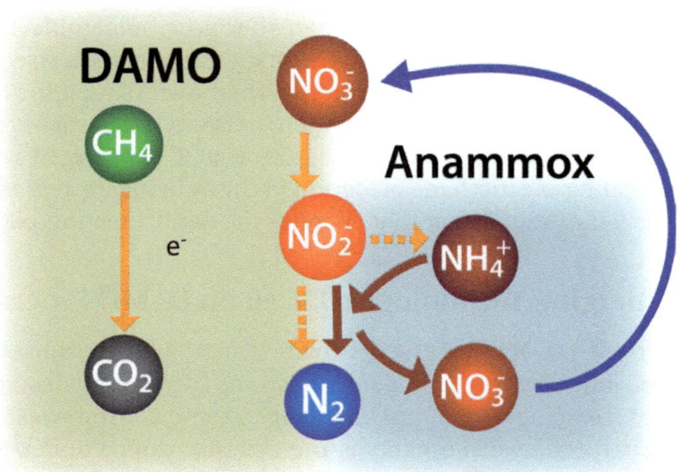

Figure 16.7 A conceptual model for the interactions between the denitrifying anaerobic methane oxidation (DAMO) and anaerobic ammonium oxidation (AnAmmOx) processes when treating wastewater containing methane, ammonium, nitrate and nitrite.

outcompeted, and the DAMO archaea and AnAmmOx bacteria became the dominant groups in the culture.[120] The possible reason for this could be that the DAMO archaea are more competitive than the DAMO bacteria for nitrate reduction to nitrite, and the AnAmmOx bacteria are more competitive than the DAMO bacteria for nitrite reduction under ammonium-present condition.

Membrane biofilm reactors (MBfRs) have been developed to remove various contaminants by using gas (*e.g.*, H_2 or O_2) delivered through hollow fibre membranes.[127] Since the SRT can be uncoupled from the hydraulic retention time (HRT) in a MBfR, this is a powerful tool for developing biofilms consisting of slow-growing microorganisms. Moreover, the methane supply through a gas-permeable membrane leads to enhanced methane transfer through the biofilm, which will provide a huge benefit to the DAMO process. Recently, a novel MBfR seeded with a co-culture of DAMO and AnAmmOx microorganisms was developed (as shown in Figure 16.8).[9] The results demonstrated the important synergistic relationship between the AnAmmOx and DAMO organisms. The stratification of DAMO and AnAmmOx microorganisms in the biofilm was successfully established *via* methane delivery from inside the

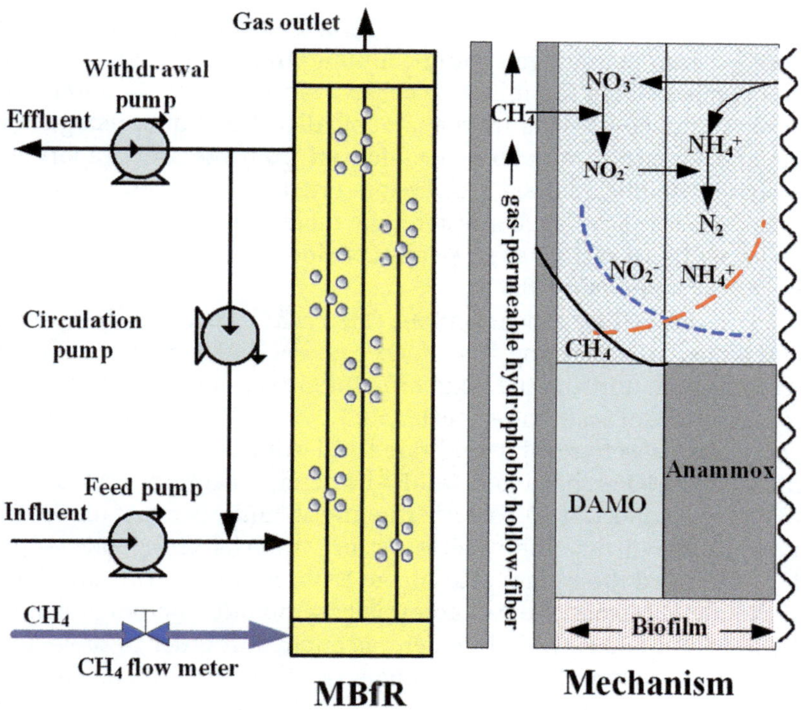

Figure 16.8 Schematic diagram of complete nitrogen removal in membrane biofilm reactor (MBfR) driven by anaerobic ammonium oxidation (AnAmmOx) and denitrifying anaerobic methane oxidation (DAMO) organisms.

membrane and nitrogenous substrate delivery from bulk liquid outside the membrane. A total NRR of 250 mg N L^{-1} day^{-1} with a TN removal efficiency of 85% was obtained in this MBfR.[9] The AnAmmOx organisms removed ammonium and most of the nitrite that was fed in, and DAMO organisms reduced the nitrate formed by the AnAmmOx reactions and also the residual nitrite. The nitrate reduction rate achieved by the DAMO process in this promising approach has been recently demonstrated to be high enough for removing nitrate produced by AnAmmOx processes, which would enable complete nitrogen removal from wastewater.[128]

16.4.6 Modelling DAMO Processes

Mathematical modelling is a useful tool for studying new processes such as DAMO and provides strong support for the understanding and optimisation of emerging technologies. However, only limited modelling research on DAMO processes has been reported so far. He *et al.*[129] applied Monod-type kinetics in order to describe the activity of DAMO bacteria in a SBR and acquired some key kinetic parameters of DAMO bacteria. Nevertheless, the model structure and the obtained parameters require further assessment in view of the obvious discrepancy between the model prediction and the experimental data at the later stage of the long-term operation. Winkler *et al.*[130] proposed a granular sludge system coupling AnAmmOx and DAMO bacteria, and evaluated its feasibility to achieve simultaneous ammonium and dissolved methane removal from anaerobic digestion liquor using stoichiometric and kinetic parameter values derived from literature-reported data and thermodynamic analysis. However, in order to offer a solid basis for the practical application of such a system, the model could be further improved by including DAMO archaea, especially considering the high nitrate concentration in the simulated effluent.

Based on the well-established ASMs, Chen *et al.*[10] successfully developed a biofilm model incorporating AnAmmOx bacteria, DAMO archaea and DAMO bacteria through utilising the long-term, intensively monitored data and the batch test results of a laboratory-scale MBfR. Several key parameters of DAMO microorganisms were reliably calibrated and validated, including yield coefficients for DAMO archaea and DAMO bacteria, maximum growth rates of DAMO archaea and DAMO bacteria, the nitrate affinity constant for DAMO archaea and the nitrite affinity constant for DAMO bacteria. Follow-up studies[131,132] extended the model with nitrite/DO inhibition terms and incorporated the metabolisms of other potentially symbiotic microorganisms (*e.g.*, AOB and NOB). The extended model was applied in order to prove the feasibility and to explore the optimal ranges of process parameters (*e.g.*, influent loadings, gas supply and biofilm thickness) to achieve complete nitrogen and dissolved methane removal in MBfRs taking on DAMO-based technologies from the modelling perspective.

All of these modelling studies provide significant insights and useful information for the design and operation of the new DAMO-based technologies. With the increasing availability of DAMO microorganisms, more modelling research on DAMO cultures and on the key factors would advance the knowledge of DAMO processes, which in turn would facilitate their future applications.

16.5 Autotrophic Denitrification

16.5.1 Autotrophic Denitrifying Organisms

Autotrophic denitrification refers to the biological reduction of nitrogen oxides (*e.g.*, NO_3^- or NO_2^-) to nitrogen gas, with hydrogen (H_2) or reduced sulphur compounds (*e.g.*, S^{2-}, S^0 or $S_2O_3^{2-}$) as electron donors. The energy derived from the inorganic redox reactions facilitates the microbial growth through assimilating inorganic carbon compounds (*i.e.*, CO_2 or HCO_3^-) as the carbon source. A variety of studies has been conducted to isolate and characterise the microbial populations in their natural habitats, as well as man-made environments where autotrophic denitrification processes take place. It has been found that both hydrogen-oxidising and sulphur-oxidising denitrifiers belong to the phylum of *Proteobacteria*.[133,134] Tables 16.4 and 16.5 list some of the reported genera of hydrogen-oxidising and sulphur-oxidising denitrifiers, respectively. Although autotrophic denitrifying bacteria are chemolithotrophic, many of them are capable of adapting to autotrophic as well as heterotrophic and mixotrophic growth. Several species, such as *Paracoccus denitrificans*,[135] *Thiomicrospira* CVO and *Sulfurimonas paralvinellae*,[134] can use both hydrogen and reduced sulphur compounds as electron donors for autotrophic denitrification.

Metabolic diversity exists among different species of sulphur-oxidising denitrifiers. For example, a few species, such as *Thiobacillus thioparus*, can only reduce nitrate to nitrite, while others, such as *Thiobacillus denitrificans* and *Thiomicrospira denitrificans*, are able to carry out the complete reduction of nitrate to nitrogen gas.[136] There has been no elemental sulphur formation reported for *S. denitrificans* due to its lack of a *dsr* gene cluster. In contrast, sulphur accumulation has been observed for a strain that is closely related to *Thiomicrospira* CVO.[134]

Table 16.4 The genera of hydrogen-oxidising denitrifiers.[133]

Pseudomonas	*Acinetobacter*	*Aeromonas*
Ochrobactrum	*Paracoccus*	*Acidovorax*
Rhodocyclus	*Shewanella*	*Alcaligenes*

Table 16.5 The genera of sulphur-oxidising denitrifiers.[99,134]

Paracoccus	*Thioalkalivibrio*	*Thiohalorhabdus*
Thiobacillus	*Thialkalivibrio*	*Thiohalophilus*
Thermothrix	*Thiohalomonas*	*Thioalkalispira*
Thioploca	*Thiosphaera*	*Sulfurimonas*
Thiomicrospira	*Beggiatoa*	

16.5.2 Stoichiometry of Autotrophic Denitrification

Similarly to heterotrophic denitrification, hydrogen-based autotrophic denitrification has been deemed to comprise sequential reduction from NO_3^- to NO_2^-, NO, N_2O and finally to N_2.[137] Without considering the cell synthesis, 2.5 mol of H_2 is used when 1 mol of NO_3^- is fully converted to N_2. However, the biomass growth can be linked to the substrate consumption through the biomass yield. The biomass yield for hydrogen-oxidising denitrifiers was determined to be 0.22–0.37 g VSS g^{-1} NO_3^--N.[137–140]

For sulphur-based autotrophic denitrification, the complete reduction of nitrate to nitrogen gas utilises electrons derived from the oxidation of sulphide, elemental sulphur and thiosulphate to sulphate. However, the final redox products rely on the N:S molar ratio. If nitrate is stoichiometrically excessive, nitrite would accumulate. By contrast, the formation of elemental sulphur would be expected in the case of stoichiometric deficiency of nitrate. In addition to nitrate, nitrite can also serve as the electron acceptor for sulphur-oxidising denitrifiers. The biomass yield of sulphur-oxidising denitrifiers is highly dependent on the electron donor/acceptor properties and microbial species involved. For example, the biomass yield with elemental sulphur as the electron donor was found to be 0.85–1.11 g VSS g^{-1} NO_3^--N,[141] while it was 0.17–0.50 g VSS g^{-1} NO_3^--N with thiosulphate as the electron donor.[142,143] The biomass yield was 0.22 g VSS g^{-1} S^{2-} for *Thiomicrospira* CVO,[144] while it was 0.38 g VSS g^{-1} S^{2-} for *Thiobacillus denitrificans*.[145]

16.5.3 Key Factors Affecting Autotrophic Denitrification

16.5.3.1 pH

The optimum pH for hydrogen-based autotrophic denitrification ranges between 7.6 and 8.6.[138,146–148] Nevertheless, it might vary among microbial species under different operational conditions. Generally, the increase of pH above the optimum range will cause nitrite accumulation, while a pH below the optimum value can inhibit the denitrification rate.[133] Due to the alkalinity released during hydrogen-based autotrophic denitrification, pH control strategies should be applied in order to avoid pH upheaval and precipitation of mineral deposits and to maintain the denitrification capacity. Injection of a chemical buffer[137,149] or carbon dioxide[133,150–152] have been proposed as effective solutions. pH also plays a significant role in sulphur-based autotrophic denitrification. The optimum pH for *Thiobacillus denitrificans*, the most commonly studied sulphur-based denitrifying species, was found to be between 6.8 and 8.2, and the specific denitrification rate would be inhibited as pH decreased below 6.8 and would approach zero at a pH of 5.5.[153] The alkalinity consumption represents one major drawback related to autotrophic denitrification with elemental sulphur as the electron donor. This drawback is commonly handled by supplementing limestone ($CaCO_3$), considering its low cost and broad availability.[154,155] Limestone can also provide an inorganic carbon source for sulphur-oxidising denitrifiers. However, the autotrophic denitrification rate is to some extent restricted by the inefficient alkalinity supply of slowly dissolving limestone. The hardness of the treated water will also increase due to the Ca^{2+} released during limestone dissolution. Alternatively, autotrophic denitrification can be combined with heterotrophic denitrification to form mixotrophic process, which can not only control the sulphate formation,[156] but also alleviate the requirement for external alkalinity due to the alkalinity supplementation during heterotrophic denitrification.[157,158]

16.5.3.2 Temperature

Temperature has been found to impact denitrification processes by affecting bacteria behaviour. Hydrogen-based autotrophic denitrification has been commonly applied at temperatures between 10 and 30 °C, with high temperatures allowing the growth and good performance of the hydrogen-oxidising denitrifying cultures.[140,159] The majority of chemolithotrophic sulphide-oxidising denitrifiers are mesophilic.[136] The optimum growth and maximum denitrification rate were obtained at 30 °C for sulphur-based autotrophic denitrification processes.[134,160] Higher temperatures might adversely influence the denitrification performance *via* enzymatic inactivation.[161]

16.5.3.3 Electron Acceptor

Electron acceptors (*i.e.*, nitrate and nitrite) have significant effects on autotrophic denitrification. While the roles of nitrate and nitrite in hydrogen-based autotrophic denitrifying systems varied, the nitrate removal rate in some studies increased with the increasing initial nitrate concentration/loading, despite the intermediate nitrite accumulation.[162,163] However, Zhou *et al.*[164] observed nitrite inhibition when the initial nitrate concentration exceeded 30 mg NO_3^--N L^{-1}. Reduced denitrification rates were also found even with low nitrite accumulation at a high initial nitrate concentration.[165] The performance of sulphur-based autotrophic denitrification highly depends on influent nitrate concentration/loading.[166] In general, the specific nitrate and nitrite reduction rates increase almost linearly with the increasing initial nitrate and nitrite concentrations. However, high nitrite concentrations (*e.g.*, 15–16 mM) could have strong inhibitory effects on autotrophic denitrification.[167]

16.5.3.4 Electron Donor

Hydrogen provides the electrons that are fundamental to the reduction of nitrogen compounds, as well as cell synthesis. Complete nitrate nitrogen removal was achieved with a dissolved hydrogen concentration varying from 0.4 to 0.8 mg L^{-1},[133] while incomplete denitrification occurred when the hydrogen concentration fell below 0.2 mg L^{-1}.[168] The crucial role of hydrogen in autotrophic denitrification represents a potential control strategy over the hydrogen supply in order to regulate the process rates as well as biomass growth, which is of great interest to biofilm-based systems taking on hydrogen-based autotrophic denitrification. Sulphide, elemental sulphur and thiosulphate have all been used as electron donors to achieve autotrophic denitrification. Compared to sulphide and elemental sulphur, thiosulphate is most readily utilised with nitrate degradation rates that are 4.6- and 9.5-times higher than those with sulphide and elemental sulphur, respectively.[169] However, the production of a higher amount of sulphate compared to autotrophic denitrification with sulphide and elemental sulphur as the electron donor represents the main disadvantage, which hinders its application. The denitrification rate was found to increase linearly with the initial sulphide concentration up to 10.7 mM. However, further increases in the sulphide concentration to 16.3 mM resulted in a significant decrease in the removal rate, owing to the inhibitory effect of sulphide on autotrophic denitrifying bacteria.[170] The particle size of elemental sulphur determines the specific surface area and thus affects the denitrification rate.[171] For the sulphur–limestone autotrophic denitrification (SLAD) systems, the denitrification performance increased in the order of the sulphur : limestone volumetric ratio of 3 : 1, 2 : 1 and 1 : 1,[166] and the optimum sulphur : limestone ratio was identified to be 1 : 1 on weight basis.[172]

16.5.4 Application of the Autotrophic Denitrification Processes

Significant efforts have been devoted to the application of the autotrophic denitrification processes in environment technology, due to their advantages over the conventional heterotrophic denitrification processes: (i) autotrophic denitrification processes do not require organic carbon, which not only reduces the operational cost under the *in situ* condition of deficient organic carbon, but also eliminates the potential problems associated with the imprecise addition of external organic matter (*e.g.*, nitrite accumulation when organic carbon is stoichiometrically insufficient and residual organics in the effluent in the case of excessive supply);[155,173,174] and (ii) much less sludge is produced in the autotrophic processes,[133,158] which in turn lowers the costs arising from sludge treatment.

So far, autotrophic denitrification has been extensively investigated and applied to treat groundwater,[151,175–178] drinking water,[137,142,159,173,179] wastewater[150,180–182] and landfill leachate,[183] which are polluted with nitrate/nitrite but lacking biodegradable organic matter, making heterotrophic denitrification biologically unfavourable. Hydrogen makes an excellent electron donor due to its clean nature, and hydrogen-based autotrophic denitrification was applied in a full-scale groundwater treatment facility with the capacity of 50 m^3 h^{-1} at Rasseln near Mönchengladbach, Germany.[184] Elemental sulphur has been regarded as a particularly suitable electron donor for autotrophic denitrification and has been widely explored as it is non-toxic, water insoluble, stable under normal conditions and readily available. A full-scale autotrophic denitrifying bioreactor with elemental sulphur as the electron donor was set up in Turkey with an average flow rate of 40 m^3 day^{-1}.[185]

Due to the low biomass yield of autotrophic denitrifiers, autotrophic denitrification has mostly been studied and applied in attached growth (*i.e.*, biofilm) systems. Reactor configuration is crucial to ensuring the biomass retention and hence the successful application of autotrophic denitrification. In particular for hydrogen-based autotrophic denitrification, a proper reactor design offering efficient hydrogen delivery is of great importance considering the low solubility (1.6 mg L^{-1} at 1 atm and 20 °C) and the explosive nature of hydrogen gas. Packed-bed reactors (PBRs) (Figure 16.9a) use support media to achieve the attached growth of hydrogen-oxidising denitrifiers as a biofilm. The characteristics of support media such as size and shape, which determine the porosity and the specific surface area, have great influence on the reactor performance: the porosity concerns biofilm thickness and pore clogging, while the specific surface area determines the available surface for bacteria attachment and growth. Hydrogen is usually supplied through direct sparging or an external hydrogen absorption tank. However, the difficulty in biofilm control, the restricted mass transfer and the decreasing biomass activity resulting from thick biofilm formation greatly limit the use of PBRs.

Fluidised-bed reactors can overcome the problems associated with PBRs (*i.e.*, clogging and channelling) and ensure stable operation of reactors. However, in order to achieve fluidisation of the bed, high up-flow velocities must be applied, which will result in a short retention time. To guarantee sufficient nitrate elimination, the recirculation of effluent is often practiced, making it complicated to control the process performance.

Gas-permeable membranes can act as both hydrogen diffusers and biofilm carriers. MBfRs have been widely considered to be the most effective configuration for enhancing hydrogen delivery efficiency and limiting explosion risks *via* the bubble-less introduction of hydrogen. In such a MBfR (Figure 16.9b), a biofilm develops on the outside of the membranes, and as the bacteria oxidise H_2 to reduce nitrate, the concentration gradient drives more H_2 across the membrane wall. Almost 100% utilisation of hydrogen is possible,[186] which minimises the costs of electron donor supply. Moreover, counter-diffusion of nitrate and hydrogen gas also renders a flexible control strategy. Among all of the types of gas-permeable membranes, hollow fibre membranes are most commonly used due to their low space requirements and high specific surface areas. The denitrification capacity could reach up to 14.2 g N m^{-2} day^{-1}.[187] Although offering highly specific surface areas and nitrogen removal efficiencies, membranes require high costs associated with energy consumption as well as the cleaning of clogged and calcified membranes due to biomass growth and inorganic precipitation. Alternatively, the combined system of a biofilm

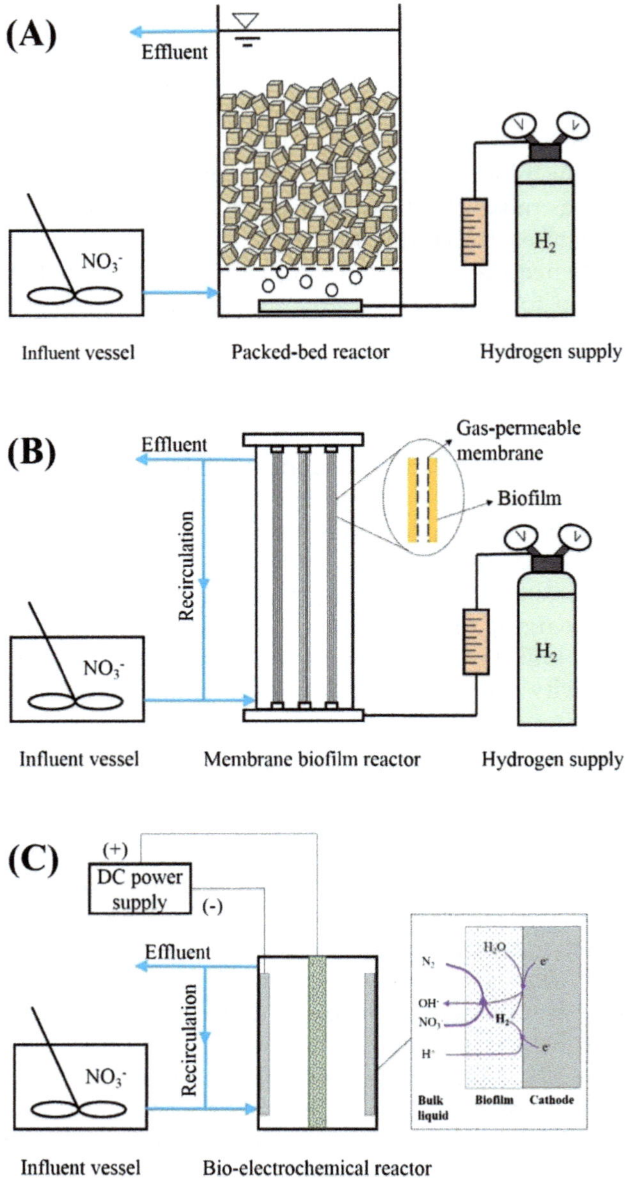

Figure 16.9 Schematic diagrams of (a) packed-bed reactor, (b) membrane biofilm reactor and (c) bioelectrochemical reactor systems performing hydrogen-based autotrophic denitrification.

reactor and an independent gas-permeable membrane unit for hydrogen supply was employed for hydrogen-based autotrophic denitrification.[188] This setup not only avoids the biomass lapse by cleaning or replacing the membrane, but also makes full use of the dissolved hydrogen in the bulk liquid.

Bio-electrochemical reactors are able to minimise the hydrogen supply and wastage through *in situ* generation and consumption of hydrogen with the passing of an electric current. In a BER (Figure 16.9c), hydrogen is produced by electrolysis of water at the cathode, on the surface of which denitrifying bacteria are directly immobilised. The advantage of this process lies in its easy operation and maintenance; however, the denitrification rate is slow.[133] Longer hydraulic retention times are therefore needed in order to achieve complete denitrification.[189–191]

With elemental sulphur as the electron donor as well as support media, PBRs have been commonly selected in order to implement the SLAD technology.[173,192] However, the running resistance and head loss will significantly increase due to the interception of influent-suspended solid or aged biofilms by sulphur particles. Frequent and sufficient backwashing is therefore indispensable for PBRs.[193]

MBRs facilitate complete biomass retention and possess good mass transfer characteristics, and therefore offer a more integrated approach. Zhang *et al.*[175] developed a novel anaerobic fluidised bed MBR (AFB-MBR) system, which integrates hollow fibre membranes and a FBR with elemental sulphur as the electron donor to treat nitrate-contaminated groundwater. Relatively high nitrate loading rates were achieved, and the nitrate removal rates obtained (1.25–4.0 g NO_3^--N L^{-1} day^{-1}) were higher than those observed in PBRs. Also, the AFB-MBR prevents secondary contamination by bacteria, which are emitted with the treated water in the conventional processes.

Sulphide-based autotrophic denitrification has been applied to develop the sulphate reduction, autotrophic denitrification and nitrification integrated (SANI) process. The system to carry out this SANI process usually comprises: (i) an anaerobic reactor to remove COD by sulphate-reducing bacteria (SRB); (ii) an anoxic reactor for autotrophic denitrification of nitrate with dissolved sulphide generated from the previous sulphate reduction; and (iii) an aerobic reactor to nitrify ammonia as well as recirculate nitrate to the anoxic reactor for denitrification.[194] The merit of this process lies in the effective reduction of excess sludge production due to its manipulation of three major functional microbial populations with low growth yields (*i.e.*, SRB, autotrophic denitrifiers and autotrophic nitrifiers).[194,195] The SANI process is particularly suitable for places like Hong Kong, where the household toilet flushing using sulphate-containing seawater is widely implemented and the disposal of excess sludge from sewage treatment remains a big challenge due to the restricted landfill and/or incineration capacity.[196] Both laboratory-scale and pilot-scale investigations have verified the feasibility of the SANI process to treat saline sewage, without the need for excess sludge withdrawal from the system.[194,196–198] Sulphide-based autotrophic denitrification was found to mainly contribute to the ultimate nitrogen removal in the SANI system and required a minimum influent sulphide : nitrate ratio of 1.6 : 1 for high-level (>90%) TN removal.[194] The recirculation ratio between the anoxic reactor and aerobic reactor was also shown to play a significant role in the TN removal.[196]

16.5.5 Modelling the Autotrophic Denitrification Processes

A number of mathematical models, such as zero-order kinetic models,[146] first- or second-order reactions,[178] double Monod functions[188,199,200] and kinetics of substitutable substrates,[159,165] have been applied to describe hydrogen-based autotrophic denitrification in both pure and mixed cultures. In general, hydrogen-based autotrophic denitrification is modelled as a two-step process (*i.e.*, the sequential reduction of nitrate to nitrite and then to nitrogen gas). The bacterial conversions of nitrate, nitrite and hydrogen are most commonly described using explicit Monod-type expressions incorporating the concentrations of all substrates and biomasses involved. The inhibition of nitrate and nitrite could be included in different forms.[133] Kinetic parameter values differ among studies of hydrogen-based autotrophic denitrification due to the different conditions and microbial species used. For example, the maximum specific growth rates with nitrate and nitrite as the electron acceptor vary between 0.0023[201] and 0.155 h^{-1},[202] and between 0.00813[203] and 0.917 h^{-1},[159] respectively. In addition, both high and low values of half-saturation constants for nitrate, nitrite and hydrogen have been reported.[133] To reliably model biofilm-based autotrophic denitrification systems, the diffusion of nitrate/nitrite, hydrogen and carbon dioxide from bulk/membranes into the biofilm, as well as biofilm growth and biomass detachment, should be considered, which greatly affect the dynamic conversions of substrates and biomass. The diffusion coefficients of nitrate, hydrogen and carbon dioxide were reported to be 0.0683, 0.2104 and 0.0691 $cm^2\ h^{-1}$.[199] Also, due to the changes in bacterial activity during fixation, the values of kinetics in a biofilm-based reactor are different from those of a suspended growth system.[165]

Due to the low half-saturation constant and less than fully effective penetration of substrate into the pores of biofilm, autotrophic denitrification with elemental sulphur as the electron donor is commonly described using a half-order kinetic model.[192,204] The half-order reaction rate constant is related to the specific surface area of the reactor media (*i.e.*, elemental sulphur and limestone).[204] In contrast, modelling sulphide-oxidising autotrophic denitrifiers as well as their competition against heterotrophic denitrifiers in mixotrophic processes is usually based on the framework of ASM1,[205,206] with the inhibition of sulphide and nitrite being included through the incorporation of inhibition functions.

16.6 Bioelectrochemical Denitrification Processes

16.6.1 Bioelectrochemical Systems

Bioelectrochemical systems are transducers that are capable of transforming chemical energy into electrical energy and *vice versa*. In their most simplistic form, BESs consist of anode and cathode compartments, usually separated by an ion-selective membrane. Redox reactions in BESs are split between the two compartments: while oxidation reactions take place at the anode, reduction reactions occur at the cathode. Figure 16.10 provides an overview of the typical anodic and cathodic half-reactions occurring in BESs. In order to maintain charge neutrality, and since electrons travel from the anode to the cathode along an external circuitry, the protons liberated by the oxidation reaction at the anode need to migrate to the cathode through the ion-permeable membrane (usually proton or, more generally, cation selective). Depending on the overall thermodynamics of the processes, BESs can be divided into two main typologies. If the overall Gibbs free energy change (ΔG) associated with the anodic and cathodic reactions is negative, then electric power is produced and the BES is referred to as a MFC. When the overall process is thermodynamically unfavourable (*i.e.*, $\Delta G > 0$), power needs to be externally supplied to the system, and the BES would be referred to as a microbial electrolysis cell, since the electrolysis of water would be the main cathodic reaction. The process of bioelectrochemical conversion of CO_2 into organic acids or alcohols is generally referred to as microbial electrosynthesis.[207] This subsection concerns on the bioelectrochemical reduction of nitrate at biocathodes, coupled with the oxidation of organics at bioanodes, primarily in MFC configurations.

A common characteristic of all BESs is that one or both electrode reactions are catalysed by microorganisms, referred to as electrochemically active (EA) microorganisms, since they possess the capability to exchange electrons extracellularly with solid electrode surfaces.[208] While extracellular electron transfer (EET) ability is widespread in nature, the strategies adopted by microorganisms to enable EET are still the subject of intense research

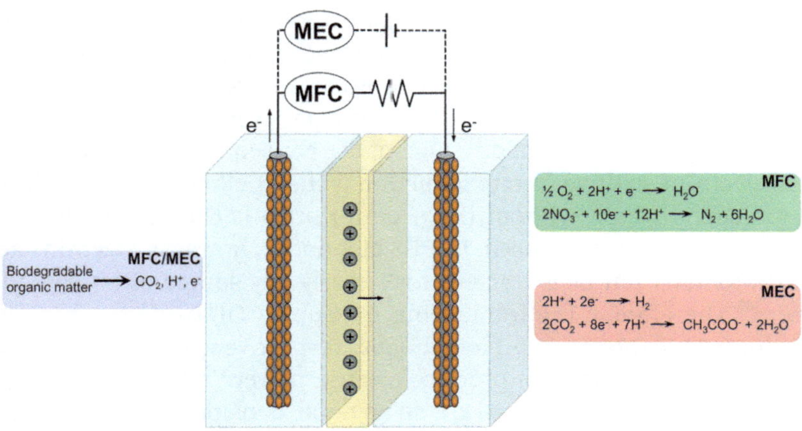

Figure 16.10 Schematic overview of anodic and cathodic reactions in bioelectrochemical systems. Microbial electrolysis cells (MECs) require additional energy input to overcome thermodynamic limitations, whereas in microbial fuel cells (MFCs), electric energy is produced. To maintain electroneutrality, protons or other cations are transported across a separating ion-permeable membrane.

efforts. Nevertheless, it is widely accepted that EA microorganisms use two strategies to achieve EET: indirect and direct mechanisms. Whereas indirect mechanisms include the use of soluble redox components (artificially added or naturally occurring) that *shuttle* electrons between the bacterial cell and electrodes,[209] direct electron transfer relies on the use of short-range physical contact using redox-active molecules located on the cell exterior,[210] or by long-range electron transfer *via* conductive filaments known as bacterial *nanowires*.[211]

16.6.2 Bioelectrochemical Denitrification

One of the major drawbacks affecting conventional technologies for nitrogen removal *via* heterotrophic denitrification from wastewater relates to the additional supply of carbonaceous organic matter to wastewater streams characterised by low COD : N ratios.[12] In spite of optimal process configurations that try to maximise the use of organics already present in the wastewater stream, a considerable fraction of the influent COD is consumed aerobically in the nitrification stage as a result of the leakage of soluble and particulate COD from the anoxic to the aerobic stages caused by the high recirculation flows typically required to achieve low nitrate levels in the effluent.[212] This leads to a significant increase in the treatment costs due to the higher oxygen demand for aerobic COD oxidation, the larger sludge production and the increase in COD demand per mass unit of nitrogen removed.[213] While this may not be an issue for wastewater streams that are rich in organic matter, as is the case in food-processing streams, wastewater with low organic contents often requires carbon supplements to achieve COD : N ratios of between 7 and 10 g COD g^{-1} N or higher.

In contrast, bioelectrochemical denitrification relies on the use of denitrifying organisms that use electrodes as the sole source of electrons instead of organic matter. While the concept was proposed first by Lewis in 1966,[214] it was only recently that the presence of nitrogen cycling in BES biocathodes was experimentally confirmed.[215] One of the consequences of the direct use of free electrons to drive nitrate reduction is that it allows the denitrification step to occur virtually without the need for organic carbon supply, as long as a source of electrons is used. In BES bioanodes, however, electrons at low redox potentials can be extracted from nearly any source of biodegradable matter.[207] Therefore, it is advantageous to couple COD oxidation and nitrate reduction at bioanodes and biocathodes, respectively, to obtain an overall exergonic reaction that can produce electric power.[216] The carbon requirements of such configurations are in general considerably lower than in traditional heterotrophic denitrification because COD oxidation and denitrification are kept physically separated. As such, the microbial competition for the available organics is minimised and the process can occur at COD : N ratios that are close to the stoichiometric requirement for the conversion of nitrate to dinitrogen without taking into account microbial growth (*i.e.*, 2.86 g COD g^{-1} N).

16.6.3 Bioelectrochemical System Configurations for Nitrogen Removal

Due to the nature of the processes involved, the removal of nitrogen from wastewater requires the sequential combination of oxidative and reductive processes. In fact, although nitrate can be bioelectrochemically reduced in biocathodes, most wastewaters contain nitrogen as ammonium rather than nitrate.[12] In a typical BES configuration, the anodic oxidation of organics should occur under anaerobic conditions to maximise the electron recovery; on the other hand, the conversion of ammonium into nitrate (*i.e.*, nitrification) needs oxygen. Therefore, for maximum efficiency in the use of the available organics, the nitrification step should ideally follow the oxidation process at the anode and precede the cathodic denitrification. This was realised for the first time by Virdis *et al.*[216] through incorporating a separate nitrification stage in a three-stage system (Figure 16.11a), whereby the synthetic wastewater was treated firstly at the anode, where the oxidation of the organics occurred, providing electrons for the cathodic denitrification stage; then the anode effluent was diverted into a passively aerated, biofilm-based nitrification stage, where ammonia was biologically oxidised to nitrate; and finally, the effluent of this intermediate stage was fed into the cathode compartment, where the reduction of nitrate to dinitrogen took place. Due to the fact that the anodic effluent was recycled to the cathodic compartment, this configuration is also referred to as loop-system.[216]

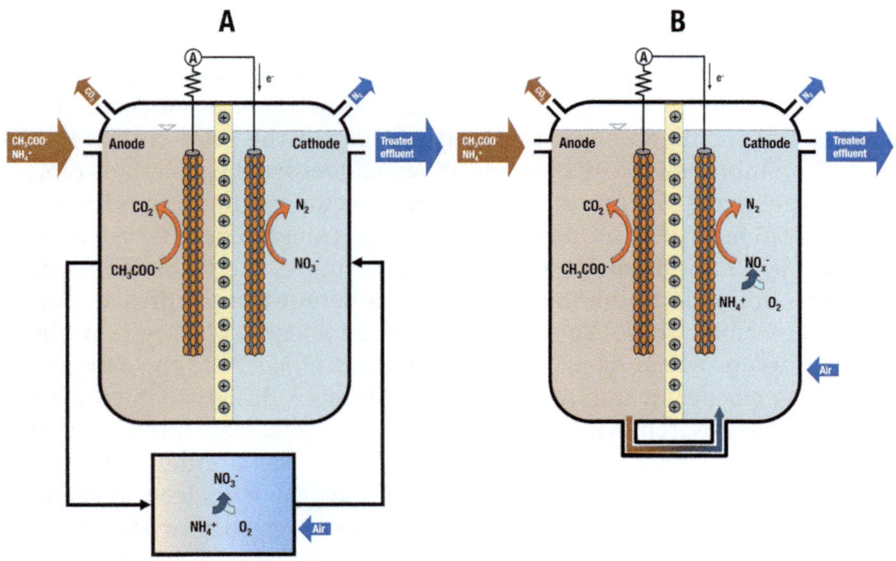

Figure 16.11 Schematic overview of proposed BES configurations for the combined removal of organics and nitrogen from wastewater. (a) Three-stage loop configuration. (b) Two-stage loop configuration.

The system was capable of removing COD and nitrogen at rates of 2 kg COD m^{-3} day^{-1} (net compartment volume) and 0.41 kg m^{-3} day^{-1}, respectively, and yielded a maximum power output of 34.6 W m^{-3}.[216] While this configuration allowed for a very high organic removal efficiency, since any carbon carried over from the anodic process was further oxidised under aerobic conditions in the nitrification stage, the nitrogen removal efficiency was affected by significant transport of NH_4^+ ions to the cathode compartment through the cation-exchange membrane. This NH_4^+ bypassed the nitrification stage and resulted in N removal efficiencies of lower than 70%. Therefore, the following technological advancement that was able to tackle this significant drawback in the design was to integrate the nitrification process into the cathode, thus achieving SND, as illustrated in Figure 16.11b.[217] The rationale behind this configuration was that denitrifying organisms would benefit from large concentration gradients developing across the biofilm, protecting them from oxidative stresses due to the presence of aerobic conditions in the surrounding catholyte. Analysis of the microbial community confirmed the presence of stratified biofilms whereby nitrifying bacteria occupied the outer layers and putative denitrifying organisms were concentrated in the layers in proximity to the electrode.[218] This configuration was capable of a higher nitrogen removal efficiency (up to 94%) compared to the three-stage system. Importantly, the organics requirement could also be considerably reduced, since values as low as 2.4 g COD per g N were still sufficient to remove over 77% of the influent N. This is a specific advantage of bioelectrochemical denitrification, where the microbial competition for the available substrates (organics and electron acceptors) is limited by the fact that the oxidative and reductive processes are kept physically separated (*vide supra*).

With the intent to simplify reactor design and operations, alternative configurations were also investigated. For instance, since ion-selective membranes that are typically used in BESs to separate the anode and cathode compartments account for about 20–40% of the total cost of a BES unit,[219] membrane-less BESs would be considerably cheaper than typical compartmented BESs. Accordingly, various reactor configurations for nitrogen removal have been proposed, including an integrated MFC system with a three-stage rotating biological contactor acting as the cathode,[220] where SND was obtained without the need for active aeration or the presence of an ion-selective membrane. Yu *et al.*[221] proposed instead a MFC system where aeration was provided by placing a gas diffuser adjacent to the cathode or by using a gas-permeable membrane. While both designs achieved high COD and NH_4^+-N removal efficiencies, the low observed nitrate removal efficiency suggested that the denitrification step was limited by the presence of DO in the cathode chamber. Also, the very low electron recoveries that were observed in these two studies suggested that the absence of a physical separation between the anode and cathode might promote competitive biological processes due to the significant cross-over of electron donors (organic matter) and electron acceptors (oxygen and nitrate) between compartments.

16.6.4 Factors Affecting Bioelectrochemical Denitrification

Several factors influence nitrogen removal efficiencies in bioelectrochemical denitrification, including the presence of oxygen or other competitive electron acceptors, the carbon source as well as abovementioned COD : N ratio, the electrode potential, the electrolyte buffering capacity and the pH. Understanding the specific and synergistic role of these parameters is critical for the operation of a robust bioelectrochemical nitrogen removal process.

The establishment of anoxic conditions is typically critical for efficient denitrification in wastewater treatment processes.[12] For instance, levels as low as 0.8 mg DO L^{-1} were reported as detrimental.[222] Similarly, it is important that the DO in a biocathode is maintained below levels that would be detrimental to the development of bioelectrochemical denitrification activity, particularly in the early phases of biofilm development.[216,223] Particularly crucial is the control of the DO in BESs performing SND, where it is necessary to find an appropriate balance between nitrification and denitrification; that is, above levels that would result in incomplete nitrification, but below levels that would be detrimental to denitrification. Biofilm systems are in general more resilient to high DO levels in the bulk liquid due to the development of concentration gradients that prevent high DO from reaching the inner portions of the biofilm (see, for example, Kuroda *et al.*[224]). In the biocathode, there is the additional competitive advantage for denitrifying organisms to occupy the portions closer to the electrodes, represented by the high availability of electrons in those areas, while nitrifying organisms can dominate the outer layers where oxygen availability is higher instead.[218] This mutually beneficial relationship results in the efficient bioelectrochemical SND (N removal >80%) reported at DO levels as high as 5 mg L^{-1}.[217,218] Another study by Yu *et al.*[221] reported a much lower optimal DO for SND of 0.5 mg L^{-1}, while a higher level resulted in suboptimal denitrification and a lower nitrogen removal efficiency.

As discussed above, while bioelectrochemical denitrification relies on electrons being delivered from the cathode electrode, for a BES operating in MFC mode, these electrons are generated at the anode from the oxidation of organics; therefore, it is important that the C : N ratio is appropriately balanced to maximise the electron recovery at the anode. While it was shown that efficient COD and N removal can be achieved from synthetic wastewaters with low C : N ratios (in some cases close to the stoichiometric balance of COD oxidation balanced with denitrification[216,225]), an oversupply of organics to the anode may trigger processes other than substrate oxidation to electrons, such as fermentation and methanogenesis, which compete with the anode as the sole electron sink.[226–228] While this might not represent a serious problem in configurations where the aerobic nitrification stage precedes the denitrifying biocathode, thus polishing the stream from residual organic matter before the last step of the process,[217 229] in BES configurations whereby the biocathode is directly fed with the effluent of the bioanode (such as SND-BES configurations), an oversupply of organics to the cathode can be detrimental, since it might promote heterotrophic processes over the bioelectrochemical process, and hence prevent the BES from generating electricity.[230]

The electrolyte pH is a problem that affects BESs in general and is not confined to bioelectrochemical denitrification. Acidic pH is produced as the result of microbial oxidation at bioanodes, while alkaline conditions are produced at biocathodes as a result of the microbial reduction of the final electron acceptor (*e.g.*, oxygen or nitrate). Systems removing nitrogen are also affected by the additional acidity produced during the nitrification step. The problem is exacerbated by the often non-ideal transport of protons through the ion-selective membrane, which leads cations other than protons to be preferentially transported across BES compartments.[219] High pH can be detrimental to biocathodes as it can limit bacterial activity in various ways.[231] Appropriate pH control can instead enhance the activity of bioelectrochemical denitrification.[232] However, the use of highly buffered synthetic wastewater or chemical dosing has become a serious challenge for practical BES implementations, which should operate at low HRT and with minimal need for the addition of chemicals.[219,231]

The use of membrane-less BESs can prevent pH gradients to some extent, but at the expense of significant losses in efficiency due to the contact between the organic-rich stream with the nitrate-rich stream, as discussed earlier. The use of membranes characterised by proton fluxes higher than traditional ion-selective membranes, as well as the use of the loop configurations whereby the acidic anode solution is directed to the cathode (*vide supra*), seems to be promising strategies for limiting pH variations between BES compartments.

16.6.5 Community Analyses in Denitrifying Biocathodes

Microbial communities in denitrifying biocathodes are very diverse and comprise often of species typically observed in natural and engineered denitrifying environments, together with other microbes whose roles in biocathodes have yet to be clarified.

One of the earliest accounts of a community analysis in nitrate-reducing biocathodes was conducted by Gregory *et al.*[215] 16S rRNA gene sequencing revealed a significant enrichment in δ-proteobacteria, together with a relatively small increase in γ-proteobacteria. While the δ-proteobacteria sequences were closely related to the genus *Geobacteraceae*, whose members are often observed in BES environments,[233] the γ-proteobacteria sequences were mostly similar to *Thermomonas* sp., whose members have been isolated from reactors performing heterotrophic denitrification of drinking water.[215,234] Puig *et al.*[235] analysed the biocathodic community using both nitrate and nitrite as electron acceptors. *Oligotropha carboxidovorans*, a member of the α-proteobacteria that is capable of utilising carbon monoxide, carbon dioxide and hydrogen as energy sources, was found to be dominant at their biocathode, and it was suggested to be the major contributor to denitrification in their system. In a more recent study by the same group, the authors used cytometric fingerprinting combined with 16S rDNA sequencing to derive the structure–function relationships of the biocathode microbiome under different stress tests.[236] Both analyses suggested that a major sub-community composed of β-proteobacteria *Thiobacillus* sp. contributed to the initial reduction step of nitrate to nitrite, while the consequent reduction steps were performed by a wider number of microbial sub-communities. These results confirmed the key role of *Thiobacillus* sp. and its respiratory nitrate reductase (NarGHI) in bioelectrochemical denitrification.[237]

A different approach was proposed by Wrighton *et al.*,[238] who compared the communities resulting from two different operational approaches: loop configuration (whereby the anodic effluent is fed to the cathode compartment; *vide supra*) *versus* a non-loop design (whereby the anodic and cathodic streams are kept separated). Their study identified *Proteobacteria*, *Firmicutes* and *Chloroflexi* members as dominant in denitrifying cathodes. However, they concluded that the generally better performance of loop configurations compared to non-loop approaches in terms of both current outputs and nitrate reduction rates was related to a greater bacterial richness and evenness, rather than a phylogenetic affiliation of dominant bacteria.

Recently, Gregoire *et al.*[239] performed a thorough electrochemical characterisation of highly performing denitrifying biocathodic microbial communities. Results of 16S rDNA sequencing revealed a high predominance of β-proteobacteria (>78% of all *Proteobacteria*) and included sequences identified as belonging to the orders *Rhodocyclales* and *Burkholderiales*, which are known environmental nitrogen cyclers and have been suggested as the main drivers of denitrification in this system.

16.7 Concluding Remarks

Over the past years, denitrification processes have been widely recognised as the key processes in BNR from wastewater. In this chapter, the traditional and emerging denitrification processes in wastewater treatment were reviewed in order to illuminate their stoichiometry, microbial communities, kinetics, affecting factors, mathematical models and technological applications, including heterotrophic denitrification, AnAmmOx, DAMO, autotrophic denitrification and bioelectrochemical denitrification. Although existing technological denitrification processes still have limitations, their applications will undoubtedly increase in the near future because of increasing attention that is being paid to high-rate, cost-effective nitrogen removal from wastewater. The increasing interest in elucidating the denitrifying mechanisms and microbial community diversity will likely also stimulate research activities on denitrification.

For further improvements in denitrification for wastewater treatment, we need to expand our knowledge of the involved nitrogen conversion pathways and the conditions determining the shifts and competitions of dominating pathways, which are all important elements in the denitrification processes. With this knowledge, more comprehensive mathematical models to take all of these parameters into consideration should be developed in the future in order to simulate the nitrogen dynamics under different conditions across different WWTPs. In addition, experiment designs should be optimised to provide more information on the nitrogen/carbon oxidation and the nitrogen reduction processes in the emerging denitrification processes under different conditions, such as DAMO and bioelectrochemical denitrification. Future studies addressing these new denitrification processes are likely to give us clearer insights into the nitrogen conversion mechanisms and facilitate their wider application for nitrogen removal.

Acknowledgements

Dr Bing-Jie Ni acknowledges the support of Australian Research Council Discovery Early Career Researcher Award DE130100451. Zhiguo Yuan acknowledges the funding support from the Australian Research Council through Discovery Project DP120100163.

References

1. A. Mulder, *Water Sci. Technol.*, 2003, **48**, 67.
2. H. Janus and H. Vanderroest, *Water Sci. Technol.*, 1997, **35**, 27.
3. L. Harmsen, P. Lourens and H. Leeuwen, *PT-Procestech.*, 1986, **41**, 27.
4. T. Khin and A. P. Annachhatre, *Biotechnol. Adv.*, 2004, **22**, 519.
5. U. van Dongen, M. S. Jetten and M. C. van Loosdrecht, *Water Sci. Technol.*, 2001, **44**, 153.
6. M. S. M. Jetten, S. J. Horn and M. C. M. vanLoosdrecht, *Water Sci. Technol.*, 1997, **35**, 171.
7. A. A. Raghoebarsing, A. Pol, K. T. van de Pas-Schoonen, A. J. Smolders, K. F. Ettwig, W. I. Rijpstra, S. Schouten, J. S. Damste, H. J. Op den Camp, M. S. Jetten and M. Strous, *Nature*, 2006, **440**, 918.
8. M. F. Haroon, S. Hu, Y. Shi, M. Imelfort, J. Keller, P. Hugenholtz, Z. Yuan and G. W. Tyson, *Nature*, 2013, **500**, 567.
9. Y. Shi, S. Hu, J. Lou, P. Lu, J. Keller and Z. Yuan, *Environ. Sci. Technol.*, 2013, **47**, 11577.
10. X. Chen, J. Guo, Y. Shi, S. Hu, Z. Yuan and B. J. Ni, *Environ. Sci. Technol.*, 2014, **48**, 9540.
11. K. F. Ettwig, M. K. Butler, D. Le Paslier, E. Pelletier, S. Mangenot, M. M. Kuypers, F. Schreiber, B. E. Dutilh, J. Zedelius, D. de Beer, J. Gloerich, H. J. Wessels, T. van Alen, F. Luesken, M. L. Wu, K. T. van de Pas-Schoonen, H. J. Op den Camp, E. M. Janssen-Megens, K. J. Francoijs, H. Stunnenberg, J. Weissenbach, M. S. Jetten and M. Strous, *Nature*, 2010, **464**, 543.
12. G. Tchobanoglous, F. Burton and H. D. Stensel, *Wastewater engineering: treatment and reuse*, Metcalf & Eddy, Inc., McGraw Hill Education, New York, 4th edn, 2003.
13. J. C. Akunna, C. Bizeau and R. Moletta, *Water Res.*, 1993, **27**, 1303.

14. M. Christensson, E. Lie and T. Welander, *Water Sci. Technol.*, 1994, **30**, 83.
15. Y. Mokhayeri, A. Nichols, S. Murthy, R. Riffat, P. Dold and I. Takacs, *Water Sci. Technol.*, 2006, **54**, 155.
16. Y. Z. Peng, Y. Ma and S. Y. Wang, *J. Environ. Sci. (Beijing, China)*, 2007, **19**, 284.
17. A. Muller, M. C. Wentzel, R. E. Loewenthal and G. A. Ekama, *Water Res.*, 2003, **37**, 2435.
18. K. Kujawa and B. Klapwijk, *Water Res.*, 1999, **33**, 2291.
19. M. Henze, *Biological wastewater treatment : principles, modelling and design*, IWA Publishing, London, 2008.
20. M. Henze, W. Gujer, T. Mino and M. C. M. van Loosdrecht, *Activated sludge models ASM1, ASM2, ASM2d and ASM3*, IWA Publishing, London, 2000.
21. M. P. Ginige, J. C. Bowyer, L. Foley, J. Keller and Z. Yuan, *Biodegradation*, 2009, **20**, 221.
22. V. Baytshtok, H. Lu, H. Park, S. Kim, R. Yu and K. Chandran, *Biotechnol. Bioeng.*, 2009, **102**, 1527.
23. K. Bilyk, T. Bruton, J. Rohrbacher, R. Latimer, P. Pitt, R. Dodson and J. Dodson, *Water Environ. Technol.*, 2011, **23**, 56.
24. C. L. Grady Jr, G. T. Daigger, N. G. Love and C. D. Filipe, *Biological wastewater treatment*, CRC Press, 2012.
25. M. J. Kampschreur, H. Temmink, R. Kleerebezem, M. S. Jetten and M. C. van Loosdrecht, *Water Res.*, 2009, **43**, 4093.
26. R. Dawson and K. Murphy, *Water Res.*, 1972, **6**, 71.
27. R. Knowles, *Microbiol. Rev.*, 1982, **46**, 43.
28. M. H. Gerardi, *Wastewater Bacteria*, Wiley-Interscience, Hoboken, New Jersey, 2006.
29. J. Oh and J. Silverstein, *J. Environ. Eng. Div. (Am. Soc. Civ. Eng.)*, 1999, **125**, 234.
30. B. J. Ni and H. Q. Yu, *Chem. Eng. Sci.*, 2008, **63**, 1449.
31. W. G. Zumft, *Microbiol. Mol. Biol. Rev.*, 1997, **61**, 533.
32. IPCC, *IPCC fourth assessment peport: climate change 2007*, Cambridge University Press, Cambridge, United Kingdom and New York, NY, USA, 2007.
33. J. H. Ahn, S. Kim, H. Park, B. Rahm, K. Pagilla and K. Chandran, *Environ. Sci. Technol.*, 2010, **44**, 4505.
34. L. Ye, B. J. Ni, Y. Law, C. Byers and Z. Yuan, *Water Res.*, 2014, **48**, 257.
35. A. Aboobakar, E. Cartmell, T. Stephenson, M. Jones, P. Vale and G. Dotro, *Water Res.*, 2013, **47**, 524.
36. A. Rodriguez-Caballero, I. Aymerich, M. Poch and M. Pijuan, *Sci. Total Environ.*, 2014, **493**, 384.
37. S. Otte, N. G. Grobben, L. A. Robertson, M. S. Jetten and J. G. Kuenen, *Appl. Environ. Microbiol.*, 1996, **62**, 2421.
38. G. Tallec, J. Garnier, G. Billen and M. Gousailles, *Water Res.*, 2006, **40**, 2972.

39. G. Tallec, J. Garnier, G. Billen and M. Gousailles, *Bioresour. Technol.*, 2008, **99**, 2200.
40. Y. Pan, L. Ye and Z. Yuan, *Environ. Sci. Technol.*, 2013, **47**, 8408.
41. Y. Pan, L. Ye, B. J. Ni and Z. Yuan, *Water Res.*, 2012, **46**, 4832.
42. K. Hanaki, Z. Hong and T. Matsuo, *Water Sci. Technol.*, 1992, **26**, 1027.
43. M. Thorn and F. Sorensson, *Water Res.*, 1996, **30**, 1543.
44. D. Richardson, H. Felgate, N. Watmough, A. Thomson and E. Baggs, *Trends Biotechnol.*, 2009, **27**, 388.
45. Y. Pan, B. J. Ni, P. L. Bond, L. Ye and Z. Yuan, *Water Res.*, 2013, **47**, 3273.
46. M. Henze, W. Gujer, T. Mino and M. C. M. van Loosdrecht, *IWA Scientific and Technical Report No. 9*, IWA Publishing, London, UK, 2000.
47. W. C. Hiatt and C. P. Grady, Jr, *Water Environ. Res.*, 2008, **80**, 2145.
48. Y. Pan, B. J. Ni, H. Lu, K. Chandran, D. Richardson and Z. Yuan, *Water Res.*, 2015, **71**, 21.
49. R. J. Seviour, K. C. Lindrea, P. C. Griffiths and L. L. Blackall, in *The Microbiology of Activated Sludge*, ed. R. J. Seviour and L. L. Blackall, Kluwer Publishing, Boston, 1999.
50. M. Strous, J. G. Kuenen and M. S. Jetten, *Appl. Environ. Microbiol.*, 1999, **65**, 3248.
51. Z. Hu, T. Lotti, M. van Loosdrecht and B. Kartal, *Biotechnol. Lett.*, 2013, **35**, 1145.
52. B. Kartal, N. C. Tan, E. Van de Biezen, M. J. Kampschreur, M. C. Van Loosdrecht and M. S. Jetten, *Appl. Environ. Microbiol.*, 2010, **76**, 6304.
53. B. Kartal, J. G. Kuenen and M. C. van Loosdrecht, *Science*, 2010, **328**, 702.
54. F. Gori, S. G. Tringe, B. Kartal, E. Marchiori and M. S. M. Jetten, *Biochem. Soc. Trans.*, 2012, **40**, 295.
55. M. Strous, J. A. Fuerst, E. H. M. Kramer, S. Logemann, G. Muyzer, K. T. van de Pas-Schoonen, R. Webb, J. G. Kuenen and M. S. M. Jetten, *Nature*, 1999, **400**, 446.
56. P. Lam, M. M. Jensen, G. Lavik, D. F. McGinnis, B. Muller, C. J. Schubert, R. Amann, B. Thamdrup and M. M. Kuypers, *Proc. Natl. Acad. Sci. U. S. A.*, 2007, **104**, 7104.
57. N. Risgaard-Petersen, R. L. Meyer, M. Schmid, M. S. M. Jetten, A. Enrich-Prast, S. Rysgaard and N. P. Revsbech, *Aquat. Microb. Ecol.*, 2004, **36**, 293.
58. D. Woebken, P. Lam, M. M. Kuypers, S. W. Naqvi, B. Kartal, M. Strous, M. S. Jetten, B. M. Fuchs and R. Amann, *Environ. Microbiol.*, 2008, **10**, 3106.
59. J. H. Guo, Y. Z. Peng, S. Y. Wang, B. Ma, S. J. Ge, Z. W. Wang, H. J. Huang, J. R. Zhang and L. Zhang, *Crit. Rev. Environ. Sci. Technol.*, 2013, **43**, 2213.
60. M. Strous, E. Pelletier, S. Mangenot, T. Rattei, A. Lehner, M. W. Taylor, M. Horn, H. Daims, D. Bartol-Mavel, P. Wincker, V. Barbe, N. Fonknechten, D. Vallenet, B. Segurens, C. Schenowitz-Truong, C. Medigue, A. Collingro, B. Snel, B. E. Dutilh, H. J. Op den Camp, C. van der Drift, I. Cirpus, K. T. van de Pas-Schoonen, H. R. Harhangi, L. van Niftrik, M. Schmid, J. Keltjens, J. van de Vossenberg, B. Kartal, H. Meier, D. Frishman, M. A. Huynen, H. W. Mewes, J. Weissenbach, M. S. Jetten, M. Wagner and D. Le Paslier, *Nature*, 2006, **440**, 790.

61. B. Kartal, L. van Niftrik, J. Rattray, J. L. van de Vossenberg, M. C. Schmid, J. Sinninghe Damste, M. S. Jetten and M. Strous, *FEMS Microbiol. Ecol.*, 2008, **63**, 46.
62. B. Kartal, J. Rattray, L. A. van Niftrik, J. van de Vossenberg, M. C. Schmid, R. I. Webb, S. Schouten, J. A. Fuerst, J. S. Damste, M. S. Jetten and M. Strous, *Syst. Appl. Microbiol.*, 2007, **30**, 39.
63. Z. X. Quan, S. K. Rhee, J. E. Zuo, Y. Yang, J. W. Bae, J. R. Park, S. T. Lee and Y. H. Park, *Environ. Microbiol.*, 2008, **10**, 3130.
64. D. Hira, H. Toh, C. T. Migita, H. Okubo, T. Nishiyama, M. Hattori, K. Furukawa and T. Fujii, *FEBS Lett.*, 2012, **586**, 1658.
65. C. Ferousi, D. R. Speth, J. Reimann, H. J. Op den Camp, J. W. Allen, J. T. Keltjens and M. S. Jetten, *BMC Microbiol.*, 2013, **13**, 265.
66. J. van de Vossenberg, D. Woebken, W. J. Maalcke, H. J. Wessels, B. E. Dutilh, B. Kartal, E. M. Janssen-Megens, G. Roeselers, J. Yan, D. Speth, J. Gloerich, W. Geerts, E. van der Biezen, W. Pluk, K. J. Francoijs, L. Russ, P. Lam, S. A. Malfatti, S. G. Tringe, S. C. Haaijer, H. J. Op den Camp, H. G. Stunnenberg, R. Amann, M. M. Kuypers and M. S. Jetten, *Environ. Microbiol.*, 2013, **15**, 1275.
67. D. R. Speth, L. Russ, B. Kartal, H. J. Op den Camp, B. E. Dutilh and M. S. Jetten, *Genome Announc.*, 2015, **3**, e01415.
68. A. A. vandeGraaf, P. deBruijn, L. A. Robertson, M. S. M. Jetten and J. G. Kuenen, *Microbiology*, 1997, **143**, 2415.
69. I. E. Cirpus, M. de Been, H. J. den Camp, M. Strous, D. Le Paslier, G. J. Kuenen and M. S. Jetten, *FEMS Microbiol. Lett.*, 2005, **252**, 273.
70. W. M. Huston, H. R. Harhangi, A. P. Leech, C. S. Butler, M. S. M. Jetten, H. J. M. O. den Camp and J. W. B. Moir, *Protein Expression Purif.*, 2007, **51**, 28.
71. I. Schmidt, O. Sliekers, M. Schmid, E. Bock, J. Fuerst, J. G. Kuenen, M. S. Jetten and M. Strous, *FEMS Microbiol. Rev.*, 2003, **27**, 481.
72. M. Strous, E. VanGerven, P. Zheng, J. G. Kuenen and M. S. M. Jetten, *Water Res.*, 1997, **31**, 1955.
73. W. R. van der Star, A. I. Miclea, U. G. van Dongen, G. Muyzer, C. Picioreanu and M. C. van Loosdrecht, *Biotechnol. Bioeng.*, 2008, **101**, 286.
74. A. Dapena-Mora, J. L. Campos, A. Mosquera-Corral, M. S. Jetten and R. Mendez, *J. Biotechnol.*, 2004, **110**, 159.
75. A. O. Sliekers, K. A. Third, W. Abma, J. G. Kuenen and M. S. Jetten, *FEMS Microbiol. Lett.*, 2003, **218**, 339.
76. B. J. Ni, B. L. Hu, F. Fang, W. M. Xie, B. Kartal, X. W. Liu, G. P. Sheng, M. Jetten, P. Zheng and H. Q. Yu, *Appl. Environ. Microbiol.*, 2010, **76**, 2652.
77. W. R. van der Star, W. R. Abma, D. Blommers, J. W. Mulder, T. Tokutomi, M. Strous, C. Picioreanu and M. C. van Loosdrecht, *Water Res.*, 2007, **41**, 4149.
78. M. S. Jetten, S. Logemann, G. Muyzer, L. A. Robertson, S. de Vries, M. C. van Loosdrecht and J. G. Kuenen, *Antonie van Leeuwenhoek*, 1997, **71**, 75.
79. M. Strous, E. Van Gerven, J. G. Kuenen and M. Jetten, *Appl. Environ. Microbiol.*, 1997, **63**, 2446.
80. R. Ganigue, H. Lopez, M. D. Balaguer and J. Colprim, *Water Res.*, 2007, **41**, 3317.

81. M. S. Jetten, M. Strous, K. T. van de Pas-Schoonen, J. Schalk, U. G. van Dongen, A. A. van de Graaf, S. Logemann, G. Muyzer, M. C. van Loosdrecht and J. G. Kuenen, *FEMS Microbiol. Rev.*, 1998, **22**, 421.
82. A. Dapena-Mora, I. Fernandez, J. L. Campos, A. Mosquera-Corral, R. Mendez and M. S. M. Jetten, *Enzyme Microb. Technol.*, 2007, **40**, 859.
83. K. Egli, U. Fanger, P. J. Alvarez, H. Siegrist, J. R. van der Meer and A. J. Zehnder, *Arch. Microbiol.*, 2001, **175**, 198.
84. S. Rysgaard, R. N. Glud, N. Risgaard-Petersen and T. Dalsgaard, *Limnol. Oceanogr.*, 2004, **49**, 1493.
85. A. Jaeschke, H. J. Op den Camp, H. Harhangi, A. Klimiuk, E. C. Hopmans, M. S. Jetten, S. Schouten and J. S. Sinninghe Damste, *FEMS Microbiol. Ecol.*, 2009, **67**, 343.
86. B. J. Ni, Y. P. Chen, S. Y. Liu, F. Fang, W. M. Xie and H. Q. Yu, *Biotechnol. Bioeng.*, 2009, **103**, 490.
87. M. Oshiki, K. Shinyako-Hata, H. Satoh and S. Okabe, *Genome Announc.*, 2015, **3**, e00267.
88. B. Kartal, M. Koleva, R. Arsov, W. van der Star, M. S. Jetten and M. Strous, *J. Biotechnol.*, 2006, **126**, 546.
89. M. Nielsen, A. Bollmann, O. Sliekers, M. Jetten, M. Schmid, M. Strous, I. Schmidt, L. H. Larsen, L. P. Nielsen and N. P. Revsbech, *FEMS Microbiol. Ecol.*, 2005, **51**, 247.
90. J. Zhang, Y. Zhang, Y. Li, L. Zhang, S. Qiao, F. Yang and X. Quan, *Bioresour. Technol.*, 2012, **114**, 102.
91. R. Zhao, H. Zhang, Y. Li, T. Jiang and F. Yang, *Curr. Microbiol.*, 2014, **69**, 880.
92. L. van Niftrik, W. J. Geerts, E. G. van Donselaar, B. M. Humbel, A. Yakushevska, A. J. Verkleij, M. S. Jetten and M. Strous, *J. Struct. Biol.*, 2008, **161**, 401.
93. S. Qiao, Z. Bi, J. Zhou, Y. Cheng and J. Zhang, *Bioresour. Technol.*, 2013, **142**, 490.
94. Y. Liu and B. J. Ni, *Sci. Rep.*, 2015, **5**, 8204.
95. X. D. Hao and M. C. M. van Loosdrecht, *5th IWA International Conference on Biofilm Systems*, Cape Town, South Africa, 2003.
96. X. Hao, J. J. Heijnen and M. C. Van Loosdrecht, *Water Res.*, 2002, **36**, 4839.
97. E. I. Volcke, C. Picioreanu, B. De Baets and M. C. van Loosdrecht, *Biotechnol. Bioeng.*, 2012, **109**, 1629.
98. M. S. M. Jetten, M. Wagner, J. Fuerst, M. van Loosdrecht, G. Kuenen and M. Strous, *Curr. Opin. Biotechnol.*, 2001, **12**, 283.
99. Y. H. Ahn, *Process Biochem.*, 2006, **41**, 1709.
100. B. Wett, A. Omari, S. M. Podmirseg, M. Han, O. Akintayo, M. Gomez Brandon, S. Murthy, C. Bott, M. Hell, I. Takacs, G. Nyhuis and M. O'Shaughnessy, *Water Sci. Technol.*, 2013, **68**, 283.
101. W. Verstraete and S. E. Vlaeminck, *Int. J. Sustainable Dev. World Ecol.*, 2011, **18**, 253.

102. P. L. McCarty, J. Bae and J. Kim, *Environ. Sci. Technol.*, 2011, **45**, 7100.
103. Q. Wang, L. Ye, G. Jiang, S. Hu and Z. Yuan, *Water Res.*, 2014, **55**, 245.
104. W. S. Reeburgh, *Chem. Rev.*, 2007, **107**, 486.
105. M. Strous and M. S. Jetten, *Annu. Rev. Microbiol.*, 2004, **58**, 99.
106. S. Islas-Lima, F. Thalasso and J. Gomez-Hernandez, *Water Res.*, 2004, **38**, 13.
107. O. Modin, K. Fukushi and K. Yamamoto, *Water Res.*, 2007, **41**, 2726.
108. F. Thalasso, A. Vallecillo, P. GarciaEncina and F. FdzPolanco, *Water Res.*, 1997, **31**, 55.
109. I. Mason, *J.–Water Pollut. Control Fed.*, 1977, **49**, 855.
110. A. Eisentraeger, P. Klag, B. Vansbotter, E. Heymann and W. Dott, *Water Res.*, 2001, **35**, 2261.
111. E. J. Beal, C. H. House and V. J. Orphan, *Science*, 2009, **325**, 184.
112. V. J. Orphan, K. U. Hinrichs, W. Ussler 3rd, C. K. Paull, L. T. Taylor, S. P. Sylva, J. M. Hayes and E. F. Delong, *Appl. Environ. Microbiol.*, 2001, **67**, 1922.
113. K. F. Ettwig, S. Shima, K. T. van de Pas-Schoonen, J. Kahnt, M. H. Medema, H. J. Op den Camp, M. S. Jetten and M. Strous, *Environ. Microbiol.*, 2008, **10**, 3164.
114. S. Hu, R. J. Zeng, L. C. Burow, P. Lant, J. Keller and Z. Yuan, *Environ. Microbiol. Rep.*, 2009, **1**, 377.
115. K. F. Ettwig, T. van Alen, K. T. van de Pas-Schoonen, M. S. Jetten and M. Strous, *Appl. Environ. Microbiol.*, 2009, **75**, 3656.
116. C. Bakermans and E. L. Madsen, *Microb. Ecol.*, 2002, **44**, 95.
117. A. López-Archilla, D. Moreira, S. Velasco and P. López-García, *Aquat. Microb. Ecol.*, 2007, **47**, 123.
118. M. G. Kalyuzhnaya, M. E. Lidstrom and L. Chistoserdova, *ISME J.*, 2008, **2**, 696.
119. Y. Koizumi, H. Kojima and M. Fukui, *FEMS Microbiol. Ecol.*, 2003, **46**, 147.
120. S. Hu, R. J. Zeng, M. F. Haroon, J. Keller, P. A. Lant, G. W. Tyson and Z. Yuan, *Sci. Rep.*, 2015, **5**, 8706.
121. F. A. Luesken, M. L. Wu, H. J. Op den Camp, J. T. Keltjens, H. Stunnenberg, K. J. Francoijs, M. Strous and M. S. Jetten, *Environ. Microbiol.*, 2012, **14**, 1024.
122. T. L. Hendrickx, Y. Wang, C. Kampman, G. Zeeman, H. Temmink and C. J. Buisman, *Water Res.*, 2012, **46**, 2187.
123. J. R. Vazquez-Padin, I. Fernandez, N. Morales, J. L. Campos, A. Mosquera-Corral and R. Mendez, *Water Sci. Technol.*, 2011, **63**, 1282.
124. K. Isaka, Y. Date, Y. Kimura, T. Sumino and S. Tsuneda, *FEMS Microbiol. Lett.*, 2008, **282**, 32.
125. S. W. H. Van Hulle, H. J. P. Vandeweyer, B. D. Meesschaert, P. A. Vanrolleghem, P. Dejans and A. Dumoulin, *Chem. Eng. J.*, 2010, **162**, 1.
126. F. A. Luesken, J. Sanchez, T. A. van Alen, J. Sanabria, H. J. Op den Camp, M. S. Jetten and B. Kartal, *Appl. Environ. Microbiol.*, 2011, **77**, 6802.

127. B. E. Rittmann, *Water Sci. Technol.*, 2006, **53**, 219.
128. C. Cai, S. Hu, J. Guo, Y. Shi, G. J. Xie and Z. Yuan, *Water Res.*, 2015, **87**, 211.
129. Z. He, C. Cai, S. Geng, L. Lou, X. Xu, P. Zheng and B. Hu, *Bioresour. Technol.*, 2013, **147**, 315.
130. M. K. Winkler, K. F. Ettwig, T. P. Vannecke, K. Stultiens, A. Bogdan, B. Kartal and E. I. Volcke, *Water Res.*, 2015, **73**, 323.
131. X. Chen, J. Guo, G. J. Xie, Y. Liu, Z. Yuan and B. J. Ni, *Water Res.*, 2015, **85**, 295.
132. X. Chen, J. Guo, G. J. Xie, Z. Yuan and B. J. Ni, *Biotechnol. Bioeng.*, 2016, 1035.
133. K. A. Karanasios, I. A. Vasiliadou, S. Pavlou and D. V. Vayenas, *J. Hazard. Mater.*, 2010, **180**, 20.
134. M. F. Shao, T. Zhang and H. H. Fang, *Appl. Microbiol. Biotechnol.*, 2010, **88**, 1027.
135. D. Pokorna and J. Zabranska, *Biotechnol. Adv.*, 2015, **33**, 1246.
136. K. Tang, V. Baskaran and M. Nemati, *Biochem. Eng. J.*, 2009, **44**, 73.
137. K. C. Lee and B. E. Rittmann, *Water Res.*, 2002, **36**, 2040.
138. K. C. Lee and B. E. Rittmann, *Water Res.*, 2003, **37**, 1551.
139. S. Ghafari, M. Hasan and M. K. Aroua, *J. Hazard. Mater.*, 2009, **162**, 1507.
140. B. O. Mansell and E. D. Schroeder, *Water Res.*, 2002, **36**, 4683.
141. H. Zeng and T. C. Zhang, *Water Res.*, 2005, **39**, 4941.
142. I. Manconi, A. Carucci and P. Lens, *Biotechnol. Bioeng.*, 2007, **98**, 551.
143. S. E. Oh, K. S. Kim, H. C. Choi, J. Cho and I. S. Kim, *Water Sci. Technol.*, 2000, **42**, 59.
144. C. McComas, K. L. Sublette, G. Jenneman and G. Bala, *Biotechnol. Prog.*, 2001, **17**, 439.
145. K. L. Sublette and N. D. Sylvester, *Biotechnol. Bioeng.*, 1987, **29**, 249.
146. B. Rezania, N. Cicek and J. A. Oleszkiewicz, *Biotechnol. Bioeng.*, 2005, **92**, 900.
147. S. Ghafari, M. Hasan and M. K. Aroua, *Bioresour. Technol.*, 2010, **101**, 2236.
148. S. Ghafari, M. Hasan and M. K. Aroua, *J. Biosci. Bioeng.*, 2009, **107**, 275.
149. K. C. Lee and B. E. Rittmann, *Water Sci. Technol.*, 2000, **41**, 219.
150. K. Y. Park, S. K. Maeng, K. H. Ahn, K. Lee, O. K. Choi and J. W. Lee, *Biotechnol. Bioprocess Eng.*, 2013, **18**, 326.
151. M. R. Schnobrich, B. P. Chaplin, M. J. Semmens and P. J. Novak, *Water Res.*, 2007, **41**, 1869.
152. D. Celmer, J. Oleszkiewicz, N. Cicek and H. Husain, *Water Sci. Technol.*, 2006, **54**, 165.
153. A. Koenig and L. H. Liu, *J. Biotechnol.*, 2002, **99**, 161.
154. E. Sahinkaya, A. Yurtsever, O. Aktas, D. Ucar and Z. W. Wang, *Chem. Eng. J.*, 2015, **268**, 180.
155. R. Sierra-Alvarez, R. Beristain-Cardoso, M. Salazar, J. Gomez, E. Razo-Flores and J. A. Field, *Water Res.*, 2007, **41**, 1253.
156. E. Sahinkaya, N. Dursun, A. Kilic, S. Demirel, S. Uyanik and O. Cinar, *Water Res.*, 2011, **45**, 6661.

157. D. U. Lee, I. S. Lee, Y. D. Choi and J. H. Bae, *Process Biochem.*, 2001, **36**, 1215.
158. S. E. Oh, Y. B. Yoo, J. C. Young and I. S. Kim, *J. Biotechnol.*, 2001, **92**, 1.
159. I. A. Vasiliadou, K. A. Karanasios, S. Pavlou and D. V. Vayenas, *J. Hazard. Mater.*, 2009, **165**, 812.
160. A. Koenig and L. H. Liu, *Water Environ. Res.*, 2004, **76**, 37.
161. C. Fajardo, M. Mora, I. Fernandez, A. Mosquera-Corral, J. L. Campos and R. Mendez, *Chemosphere*, 2014, **97**, 10.
162. H. I. Park, Y. J. Choi and D. Pak, *Biotechnol. Lett.*, 2005, **27**, 949.
163. H. I. Park, D. K. Kim, Y. J. Choi and D. Pak, *Process Biochem.*, 2005, **40**, 3383.
164. M. H. Zhou, W. J. Fu, H. Y. Gu and L. C. Lei, *Electrochim. Acta*, 2007, **52**, 6052.
165. I. A. Vasiliadou, S. Pavlou and D. V. Vayenas, *Process Biochem.*, 2006, **41**, 1401.
166. A. Kilic, E. Sahinkaya and O. Cinar, *Environ. Technol.*, 2014, **35**, 2796.
167. Y. Sun and M. Nemati, *Bioresour. Technol.*, 2012, **114**, 207.
168. C. C. Chang, S. K. Tseng and H. K. Huang, *Bioresour. Technol.*, 1999, **69**, 53.
169. R. B. Cardoso, R. Sierra-Alvarez, P. Rowlette, E. R. Flores, J. Gomez and J. A. Field, *Biotechnol. Bioeng.*, 2006, **95**, 1148.
170. S. An, K. Tang and M. Nemati, *Water Res.*, 2010, **44**, 1531.
171. H. S. Moon, S. W. Chang, K. Nam, J. Choe and J. Y. Kim, *Environ. Pollut.*, 2006, **144**, 802.
172. I. Zhu and T. Getting, *Environ. Technol. Rev.*, 2012, **1**, 46.
173. H. Liu, W. Jiang, D. Wan and J. Qu, *J. Hazard. Mater.*, 2009, **169**, 23.
174. E. Sahinkaya and A. Kilic, *Water Res.*, 2014, **50**, 278.
175. L. Zhang, C. Zhang, C. Hu, H. Liu and J. Qu, *Appl. Microbiol. Biotechnol.*, 2015, **99**, 2815.
176. M. I. Soares, *Water Res.*, 2002, **36**, 1392.
177. H. S. Moon, K. Nam and J. Y. Kim, *J. Environ. Eng. Div. (Am. Soc. Civ. Eng.)*, 2006, **132**, 971.
178. K. S. Haugen, M. J. Semmens and P. J. Novak, *Water Res.*, 2002, **36**, 3497.
179. H. Wang and J. Qu, *Water Res.*, 2003, **37**, 3767.
180. T. C. Zhang, *J. Environ. Eng. Div. (Am. Soc. Civ. Eng.)*, 2002, **128**, 73.
181. I. S. Kim, S. E. Oh, M. S. Bum, J. L. Lee and S. T. Lee, *Appl. Microbiol. Biotechnol.*, 2002, **59**, 91.
182. P. Li, W. Xing, J. Zuo, L. Tang, Y. Wang and J. Lin, *Bioresour. Technol.*, 2013, **144**, 452.
183. A. Koenig and L. H. Liu, *Water Sci. Technol.*, 1996, **34**, 469.
184. H. Gros, G. Schnoor and P. Rutten, *Water Supply*, 1988, **6**, 193.
185. E. Sahinkaya, A. Kilic and B. Duygulu, *Water Res.*, 2014, **60**, 210.
186. B. E. Rittmann, *Environ. Eng. Res.*, 2007, **12**, 157.
187. B. Rezania, J. A. Oleszkiewicz, N. Cicek and H. Mo, *Water Sci. Technol.*, 2005, **51**, 403.
188. C. Lu, P. Gu, P. He, G. Zhang and C. Song, *J. Hazard. Mater.*, 2009, **168**, 1581.
189. Z. Feleke and Y. Sakakibara, *Water Res.*, 2002, **36**, 3092.

190. S. Islam and M. T. Suidan, *Water Res.*, 1998, **32**, 528.
191. Z. Feleke, K. Araki, Y. Sakakibara, T. Watanabe and M. Kuroda, *Water Res.*, 1998, **32**, 2728.
192. A. Koenig and L. H. Liu, *Water Res.*, 2001, **35**, 1969.
193. X. Lv, M. Shao, J. Li and C. Xie, *Environ. Technol.*, 2014, **35**, 2692.
194. J. Wang, H. Lu, G. H. Chen, G. N. Lau, W. L. Tsang and M. C. van Loosdrecht, *Water Res.*, 2009, **43**, 2363.
195. G. N. Lau, K. R. Sharma, G. H. Chen and M. C. van Loosdrecht, *Water Sci. Technol.*, 2006, **53**, 227.
196. H. Lu, J. Wang, S. Li, G. H. Chen, M. C. van Loosdrecht and G. A. Ekama, *Water Res.*, 2009, **43**, 3613.
197. H. Lu, G. A. Ekama, D. Wu, J. Feng, M. C. van Loosdrecht and G. H. Chen, *Water Res.*, 2012, **46**, 475.
198. H. Lu, D. Wu, D. T. Tang, G. H. Chen, M. C. van Loosdrecht and G. Ekama, *Water Sci. Technol.*, 2011, **63**, 2149.
199. Y. Sakakibara, J. R. V. Flora, M. T. Suidan and M. Kuroda, *Water Res.*, 1994, **28**, 1077.
200. M. Kurt, I. J. Dunn and J. R. Bourne, *Biotechnol. Bioeng.*, 1987, **29**, 493.
201. C. Visvanathan, N. Q. Hung and V. Jegatheesan, *Process Biochem.*, 2008, **43**, 673.
202. I. A. Vasiliadou, S. Siozios, I. T. Papadas, K. Bourtzis, S. Pavlou and D. V. Vayenas, *Biotechnol. Bioeng.*, 2006, **95**, 513.
203. A. Tiemeyer, H. Link and D. Weuster-Botz, *Appl. Microbiol. Biotechnol.*, 2007, **76**, 75.
204. A. Darbi and T. Viraraghavan, *Water Qual. Res. J. Can.*, 2003, **38**, 183.
205. A. Wang, C. Liu, N. Ren, H. Han and D. Lee, *J. Hazard. Mater.*, 2010, **178**, 35.
206. X. Xu, C. Chen, A. Wang, W. Guo, X. Zhou, D. J. Lee, N. Ren and J. S. Chang, *J. Hazard. Mater.*, 2014, **264**, 16.
207. B. E. Logan and K. Rabaey, *Science*, 2012, **337**, 686.
208. S. Freguia, B. Virdis, F. Harnisch and J. Keller, *Electrochim. Acta*, 2012, **82**, 165.
209. E. Marsili, D. B. Baron, I. D. Shikhare, D. Coursolle, J. A. Gralnick and D. R. Bond, *Proc. Natl. Acad. Sci. U. S. A.*, 2008, **105**, 3968.
210. C. R. Myers and J. M. Myers, *J. Bacteriol.*, 1992, **174**, 3429.
211. G. Reguera, K. D. McCarthy, T. Mehta, J. S. Nicoll, M. T. Tuominen and D. R. Lovley, *Nature*, 2005, **435**, 1098.
212. Z. Yuan, J. Keller and P. Lant, *Biotechnology for the Environment: Wastewater Treatment and Modeling, Waste Gas Handling, Springer*, 2003, pp. 187–227.
213. C. Fux and H. Siegrist, *Water Sci. Technol.*, 2004, **50**, 19.
214. K. Lewis, *Bacteriol. Rev.*, 1966, **30**, 101.
215. K. B. Gregory, D. R. Bond and D. R. Lovley, *Environ. Microbiol.*, 2004, **6**, 596.
216. B. Virdis, K. Rabaey, Z. Yuan and J. Keller, *Water Res.*, 2008, **42**, 3013.
217. B. Virdis, K. Rabaey, R. A. Rozendal, Z. Yuan and J. Keller, *Water Res.*, 2010, **44**, 2970.

218. B. Virdis, S. T. Read, K. Rabaey, R. A. Rozendal, Z. Yuan and J. Keller, *Bioresour. Technol.*, 2011, **102**, 334.
219. R. A. Rozendal, H. V. Hamelers, K. Rabaey, J. Keller and C. J. Buisman, *Trends Biotechnol.*, 2008, **26**, 450.
220. R. R. Sayess, P. E. Saikaly, M. El-Fadel, D. Li and L. Semerjian, *Water Res.*, 2013, **47**, 881.
221. C. P. Yu, Z. Liang, A. Das and Z. Hu, *Water Res.*, 2011, **45**, 1157.
222. K. Pochana and J. Keller, *Water Sci. Technol.*, 1999, **39**, 61.
223. P. Clauwaert, K. Rabaey, P. Aelterman, L. de Schamphelaire, T. H. Pham, P. Boeckx, N. Boon and W. Verstraete, *Environ. Sci. Technol.*, 2007, **41**, 3354.
224. M. Kuroda, T. Watanabe and Y. Umedu, *Water Sci. Technol.*, 1996, **34**, 101.
225. F. Zhang and Z. He, *J. Chem. Technol. Biotechnol.*, 2012, **87**, 153.
226. B. Virdis, K. Rabaey, Z. Yuan, R. A. Rozendal and J. Keller, *Environ. Sci. Technol.*, 2009, **43**, 5144.
227. S. Freguia, K. Rabaey, Z. Yuan and J. Keller, *Environ. Sci. Technol.*, 2008, **42**, 7937.
228. S. Freguia, K. Rabaey, Z. Yuan and J. Keller, *Environ. Sci. Technol.*, 2007, **41**, 2915.
229. S. Xie, P. Liang, Y. Chen, X. Xia and X. Huang, *Bioresour. Technol.*, 2011, **102**, 348.
230. S. Freguia, K. Rabaey, Z. Yuan and J. Keller, *Water Res.*, 2008, **42**, 1387.
231. L. Huang, J. M. Regan and X. Quan, *Bioresour. Technol.*, 2011, **102**, 316.
232. P. Clauwaert, J. Desloover, C. Shea, R. Nerenberg, N. Boon and W. Verstraete, *Biotechnol. Lett.*, 2009, **31**, 1537.
233. D. R. Lovley, *Annu. Rev. Microbiol.*, 2012, **66**, 391.
234. J. Mergaert, M. C. Cnockaert and J. Swings, *Int. J. Syst. Evol. Microbiol.*, 2003, **53**, 1961.
235. S. Puig, M. Serra, A. Vilar-Sanz, M. Cabre, L. Baneras, J. Colprim and M. D. Balaguer, *Bioresour. Technol.*, 2011, **102**, 4462.
236. N. Pous, C. Koch, A. Vila-Rovira, M. D. Balaguer, J. Colprim, J. Muhlenberg, S. Muller, F. Harnisch and S. Puig, *RSC Adv.*, 2015, **5**, 68326.
237. N. Pous, C. Koch, J. Colprim, S. Puig and F. Harnisch, *Electrochem. Commun.*, 2014, **49**, 93.
238. K. C. Wrighton, B. Virdis, P. Clauwaert, S. T. Read, R. A. Daly, N. Boon, Y. Piceno, G. L. Andersen, J. D. Coates and K. Rabaey, *ISME J.*, 2010, **4**, 1443.
239. K. P. Gregoire, S. M. Glaven, J. Hervey, B. C. Lin and L. M. Tender, *J. Electrochem. Soc.*, 2014, **161**, 3049.
240. L. van Niftrik, W. J. Geerts, E. G. van Donselaar, B. M. Humbel, R. I. Webb, J. A. Fuerst, A. J. Verkleij, M. S. Jetten and M. Strous, *J. Bacteriol.*, 2008, **190**, 708.
241. M. Oshiki, M. Shimokawa, N. Fujii, H. Satoh and S. Okabe, *Microbiology*, 2011, **157**, 1706.
242. L. Russ, D. R. Speth, M. S. Jetten, H. J. Op den Camp and B. Kartal, *Environ. Microbiol.*, 2014, **16**, 3487.

243. T. Awata, M. Oshiki, T. Kindaichi, N. Ozaki, A. Ohashi and S. Okabe, *Appl. Environ. Microbiol.*, 2013, **79**, 4145.
244. J. van de Vossenberg, J. E. Rattray, W. Geerts, B. Kartal, L. van Niftrik, E. G. van Donselaar, J. S. Sinninghe Damste, M. Strous and M. S. Jetten, *Environ. Microbiol.*, 2008, **10**, 3120.
245. M. Ali, M. Oshiki, T. Awata, K. Isobe, Z. Kimura, H. Yoshikawa, D. Hira, T. Kindaichi, H. Satoh, T. Fujii and S. Okabe, *Environ. Microbiol.*, 2015, **17**, 2172.
246. B. Wett, *Water Sci. Technol.*, 2006, **53**, 121.
247. H. Siegrist, S. Reithaar and P. Lais, *Water Sci. Technol.*, 1998, **37**, 589.
248. W. R. Abma, W. Driessen, R. Haarhuis and M. C. van Loosdrecht, *Water Sci. Technol.*, 2010, **61**, 1715.
249. A. Joss, D. Salzgeber, J. Eugster, R. Konig, K. Rottermann, S. Burger, P. Fabijan, S. Leumann, J. Mohn and H. Siegrist, *Environ. Sci. Technol.*, 2009, **43**, 5301.
250. A. Hippen, K. H. Rosenwinkel, G. Baumgarten and C. F. Seyfried, *Water Sci. Technol.*, 1997, **35**, 111.
251. W. R. Abma, C. E. Schultz, J. W. Mulder, M. C. M. van Loosdrecht, W. R. L. van der Star, M. Strous and T. Tokutomi, *Water*, 2007, **21**, 36.
252. G. Nyhuis, V. Stadler and B. Wett, *Proc. 6th Aachen Conf. on N-return Load*, Aachen, Germany, 2006.
253. M. Schmid, K. Walsh, R. Webb, W. I. C. Rijpstra, K. van de Pas-Schoonen, M. J. Verbruggen, T. Hill, B. Moffett, J. Fuerst, S. Schouten, J. S. S. Damste, J. Harris, P. Shaw, M. Jetten and M. Strous, *Syst. Appl. Microbiol.*, 2003, **26**, 529.

CHAPTER 17

Lessons from Denitrification for the Human Metabolism of Signalling Nitric Oxide

LUÍSA B. MAIA*[a] AND JOSÉ J. G. MOURA[a]

[a]UCIBIO, REQUIMTE, Departamento de Química, Faculdade de Ciências e Tecnologia, Campus da Caparica, 2829-516 Caparica, Portugal
*E-mail: luisa.maia@fct.unl.pt

17.1 "Classic" Metabolism of Signalling Nitric Oxide

The nitric oxide radical, ${}^{\bullet}NO$ (herein abbreviated as NO), is a signalling molecule that is involved in several physiological processes in animals, plants and also in prokaryotes.[1–8] In humans, NO controls a plethora of functions, including vasodilation (through the well-known activation of guanylate cyclase), neurotransmission, immune response, platelet aggregation, apoptosis and gene expression, and it mediates a wide range of both anti-tumour and anti-microbial activities.[1] Conversely, NO is also involved in several pathological conditions, either when it is not produced at sufficient concentrations or when it is overproduced, as during chronic inflammation, infection, septic shock syndrome, diabetes or Parkinson's and Alzheimer's diseases.[9–11]

Humans produce NO form the guanidinium nitrogen atom of L-arginine and one of the oxygen atoms of dioxygen (eqn (17.1) and (17.2)) in a reaction

RSC Metallobiology Series No. 9
Metalloenzymes in Denitrification: Applications and Environmental Impacts
Edited by Isabel Moura, José J. G. Moura, Sofia R. Pauleta and Luisa B. Maia

Published by the Royal Society of Chemistry, www.rsc.org

that is catalysed by specific NO synthases (NOSs).[12–15] These enzymes are encoded by three distinct genes and are expressed in a tissue-dependent manner: (i) neuronal NOS, present in nervous tissue and skeletal muscle type II and involved in cell communication; (ii) inducible NOS, present in immune and cardiovascular systems and involved in immune defence; and (iii) endothelial NOS, present in the endothelium and involved in vasodilation.

NOSs are complex homodimeric enzymes that are composed of one flavinic reductase C-terminal domain (that accepts electrons from NADPH) and one haemic oxygenase N-terminal domain, where L-arginine is hydroxylated (eqn (17.1)) and the N$^{\omega}$-hydroxy-L-arginine is oxidised and cleaved to yield NO (eqn (17.2)).

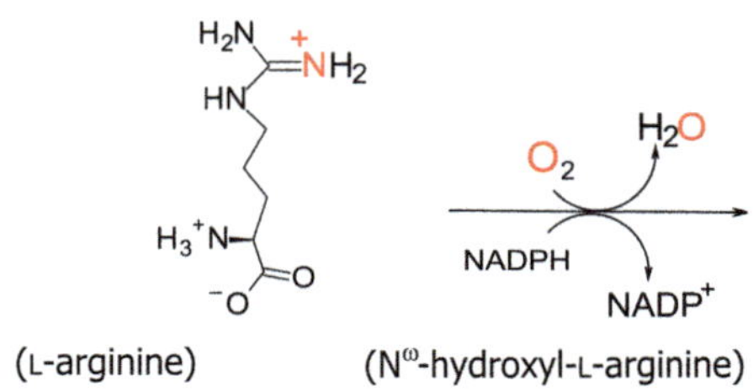

(17.1)

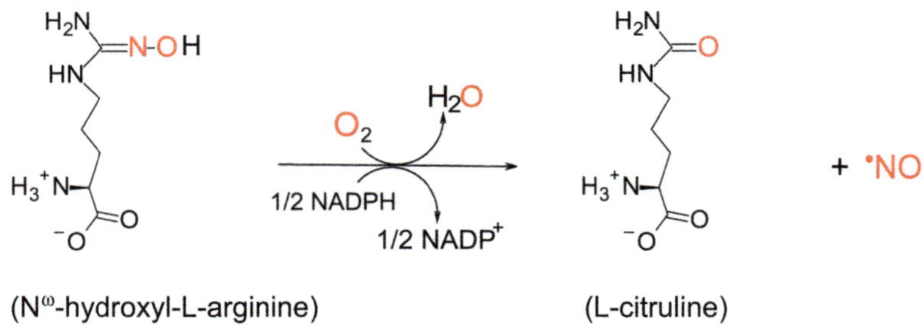

(17.2)

The biological effects of NO are largely accomplished by post-translational modification of transition metal centres (mostly haems and labile [4Fe–4S] centres) and of cysteine residues and other thiols, which would become converted into the respective nitrosyl (–metal–N = O) and S-nitrosothiol (–S–N = O) derivates.[7,9,10,16–29] To manage the *timing* and specificity of the signal, the cell tightly controls the NO lifetime, regulating its formation (the NOS activity) and promoting its timely oxidation to nitrite and nitrate.[1,7,30–44] Yet, upon an hypoxic event, when the decreased (or abolished) dioxygen concentration[†] compromises the NOS activity, how does the cell cope with insufficient NOS-dependent NO formation?

[†]It is generally accepted that hypoxia and anoxia correspond to dioxygen concentrations lower than 20 and 2 μM, respectively.

In this Chapter, we will try to answer the following highly intriguing and motivating key questions: can signalling NO be produced from nitrite? How is nitrite homeostasis regulated in humans? "Who" reduces nitrite to NO in humans?

17.2 Nitrite-Dependent NO Formation

17.2.1 A New Concept Emerged

While our knowledge about the physiological roles and metabolism of NO was growing exponentially, nitrate and nitrite were thought of as *useless* end products of NO metabolism and were essentially forgotten. *Science* magazine elected NO as the "Molecule of the Year" in 1992[45] and the Nobel Prize in Physiology or Medicine was awarded in 1998 for the identification of the "endothelium-derived relaxing factor" as the NO molecule. Only later, at the end of the 1990s, did it became clear that nitrite can be reduced back to NO under hypoxic/anoxic conditions (eqn (17.3)) and it was realised that nitrite can exert significant protective action *in vivo* during ischaemia and other pathological conditions,[46–111] "re-discovering" a beneficial role that was already known in 1880.[112] Since then, a new concept has begun, and presently, blood and tissue nitrite are thought of as NO "storage forms" that can be made available for maintaining NO formation and ensuring cell functioning under conditions of hypoxia/anoxia, precisely when the oxygen-dependent NOS activity is impaired. The physiological relevance of this nitrite-derived NO during hypoxia goes well beyond the obvious vasodilation[50,54,60,77,84] and includes angiogenesis,[77] smooth muscle proliferation,[92] regulation of gene expression[53] and the most important regulation of mitochondrial "respiration"/energy production.[59,68,69,72,90,91,107] Hence, through this nitrite/NO "recycling" pathway, an organ under ischaemia can maintain (or even increase) its blood flow, modulate the dioxygen distribution and the reactive oxygen species (ROS) formation and, at the same time, maintain an anti-inflammatory and anti-apoptotic environment.

$$\mathbf{NO_2^- + 1e^- + 2H^+ \rightarrow {}^{\bullet}NO + H_2O} \qquad (17.3)$$

This new concept of *nitrite-derived NO* is already being explored in animal models and preclinical assays of novel therapeutic approaches for the management of several pathological conditions,[74,78,96–98,113–118] including hypertension, cardiovascular dysfunctions, myocardial infarction, stroke, pulmonary hypertension and ischaemic injury.[48,50,52,54,61–63,66,75,80,82,83,85,88,89,95,98–104] Nitrite is also being explored as a bactericide/fungicide in the stomach and airways.[47,119–121] Nitration of the virulence factors of bacterial pathogens such as *Pseudomonas aeruginosa* decreases the ability of the factors to induce airway inflammation and cytotoxicity *in vivo* in mice,[122] and acidified nitrite is being examined as a therapy for highly refractory infections by mucoid *P. aeruginosa* in patients with cystic fibrosis.[123,124] Certainly, future research of human nitrite metabolism will create other innovative and improved therapeutic approaches.

17.2.2 Present Key Questions

While the experimental evidence for the cytoprotective role of nitrite is building up and the scientific community is embracing the new concept of NO formation, new questions are being raised.

17.2.2.1 Nitrite Homeostasis

One key aspect regards the control of nitrite concentration: if it is a key molecule for cellular homeostasis, how is the nitrite concentration regulated to respond to daily changes?

Nitrite is obtained from dietary sources either directly from vegetables (*e.g.*, spinach, lettuce or beetroot),[125] curried meat[125–127] or drinking water,[128] or indirectly through nitrate ingestion and subsequent reduction to nitrite by commensal bacteria in the mouth and gastrointestinal tract (Figure 17.1).[129–134] Nitrate also originates in skeletal muscle[135] and the circulation, from where up to 25% is actively taken up by the salivary glands and secreted into the saliva (where it can be reduced to nitrite).[136–139] The exogenous nitrite can be converted into NO in the stomach (where it plays a role in vasodilation,[140–142] gastric mucosa formation[143] and host defence[130,131,138,139,144–150]) or it can diffuse into the systemic circulation, where it is transported to resistance vessels and tissues.[130] The second main source of nitrite is NO oxidation itself (endogenous NO),[1,151,152] which is believed to account for as much as 70% of the plasma nitrite.[153,154]

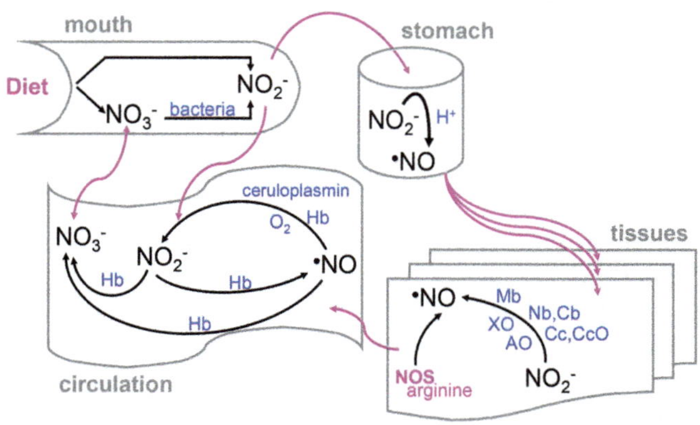

Figure 17.1 Nitrate/nitrite sources and metabolism in humans. AO: aldehyde oxidase; Cb: cytogloblin; *Cc*: cytochrome *c*; CcO: cytochrome *c* oxidase; Hb: haemoglobin; Mb: myoglobin; Nb: neuroglobin; NOS: nitric oxide synthase; XO: xanthine oxidase. Adapted with permission from ref. 8. Copyright (2014) American Chemical Society.

All of the nitrite sources are thus potentially subjected to great daily variations (Figure 17.1). Presently, the mechanisms of nitrite transport and intracellular accumulation/export are poorly understood and mainly restricted to the erythrocytes, where the anion exchanger AE1 of band 3 is probably involved.[155–160] It is clear that much more work has to be done in this area.

17.2.2.2 "Non-Dedicated" Nitrite Reductases

Another key question regards "who" is reducing nitrite to NO in humans?

To date, no "dedicated" mammalian nitrite reductase (NiR) has been identified. On the contrary, the specific nitrite reduction to NO has been ascribed only to prokaryotic organisms that use the d1 haem-containing NiRs and copper-containing NiRs (see Chapters 4 and 5 in this book) to derive energy *via* denitrification, anaerobic ammonium oxidation and other related respiratory pathways (see Chapter 1 in this book, and also ref. 8 and [161–168]). Mammals do not have these enzymes! In addition, while the absence of a "dedicated" NiR was interpreted as "mammals do not need to metabolise nitrite", the correct question was long disregarded: are other mammalian proteins reducing nitrite?

In recent years, several mammalian metalloproteins that are present in cells in order to carry out other functions were shown to be able to reduce nitrite to NO, constituting what we named "non-dedicated" NiRs.[8] Nitrite has been shown to be reduced to NO by metalloproteins of rather different cellular roles, structural organisation, tissue and subcellular localisations by enzymes, metabolite transporters and electron transfer proteins (Table 17.1). The list of "non-dedicated" NiRs includes the four mammalian molybdenum-containing enzymes,[169] xanthine oxidase (XO),[170–176] aldehyde oxidase,[175–177] sulphite oxidase[178] and mitochondrial amidoxime-reducing component,[179] a growing number of haem-containing proteins, like haemoglobin (Hb) and myoglobin (Mb), cytochrome *c*, cytochrome *c* oxidase, cytochrome P_{450}[59,72,180–187] and other diverse metalloproteins.[8,188]

The physiological relevance of each individual pathway with respect to NO formation is currently being debated, and the debate will surely continue over the coming years.[8] So far, only XO and Mb have been identified as crucial for the *ex vivo* and *in vivo* cytoprotective action of nitrite.

Mammalian XO is a cytoplasmic molybdoenzyme belonging to the so-called XO family that is present in various tissues.[189–195] Physiologically, XO is a key enzyme in purine catabolism, where it catalyses the hydroxylation of both hypoxanthine and xanthine to the terminal metabolite, urate, with the simultaneous reduction of dioxygen. In addition, this enzyme also catalyses the oxidation of a wide variety of substituted pyridines, purines, pteridines and related compounds, including the NADH and aldehydes, contributing in this way to the beneficial activation of pro-drugs,[196–200] but also to the deleterious activation of xenobiotics.[201] Its ability to reduce dioxygen to a superoxide radical anion and hydrogen peroxide (two ROS) suggests its

Table 17.1 Proteins involved in nitrite-derived nitric oxide formation in mammals.[a,b]

Protein	**"Classic" reaction** Functions	Site of nitrite reduction **Reaction of nitrite reduction** × Major effect(s) counteracting nitrite reduction and/or release of bioactive NO[c]
XO, xanthine dehydrogenase (liver, intestine, heart, mammal gland, small vessels and endothelial and epithelial cells)	**Xanthine → urate** Catabolism of purines "Non-classic" functions include metabolism of xenobiotics and ROS-mediated signalling cascades	Molybdenum centre of the XO family type $NO_2^- + 2H^+ + (XO)Mo^{4+} \rightarrow (XO)Mo^{5+} + {}^{\bullet}NO + H_2O$ $NO_2^- + 2H^+ + (XO)Mo^{5+} \rightarrow (XO)Mo^{6+} + {}^{\bullet}NO + H_2O$ × Competition with O_2
AO (liver, heart, lung, kidney, brain and eye)	**Aldehyde → carboxylate** Metabolism of xenobiotics, neurotransmitters and retinoic acid and ROS-mediated signalling cascades	Molybdenum centre of the XO family type $\mathbf{NO_2^- + 2H^+ + (AO)Mo^{4+} \rightarrow (AO)Mo^{5+} + {}^{\bullet}NO + H_2O}$ $\mathbf{NO_2^- + 2H^+ + (AO)Mo^{5+} \rightarrow (AO)Mo^{6+} + {}^{\bullet}NO + H_2O}$ × Competition with O_2
SO (all tissues)	$\mathbf{SO_3^{2-} \rightarrow SO_4^{2-}}$ Catabolism of sulphur-containing amino acids and other compounds	Molybdenum centre of the SO family type $\mathbf{NO_2^- + 2H^+ + (SO)Mo^{4+} \rightarrow (SO)Mo^{5+} + {}^{\bullet}NO + H_2O}$ × Competition with Cc; need of an appropriate one-electron donor to overcome Mo^{5+} dead-end formation
mARC (all tissues)	$\mathbf{R_2(N){-}OH \rightarrow R_2(N){-}OH}$ **and** $\mathbf{R_2(S){-}OH \rightarrow R_2(S){-}OH}$ No "classic" functions It is a *new* protein, probably involved in the detoxification of mutagenic and toxic aromatic hydroxyl-amines, among other possible roles	Molybdenum centre of the SO family type $\mathbf{NO_2^- + 2H^+ + (mARC)Mo^{4+} \rightarrow (mARC)Mo^{5+} + {}^{\bullet}NO + H_2O}$ $\mathbf{NO_2^- + 2H^+ + (mARC)Mo^{5+} \rightarrow (mARC)Mo^{6+} + {}^{\bullet}NO + H_2O}$
Mb (cardiac, skeletal and smooth muscle)	Dioxygen transport Recently, several *novel* functions were suggested: cardiac NO homeostasis, O_2 sensing, ROS scavenging and intracellular fatty acid transport	*b* haem, penta-coordinated by a histidine residue $\mathbf{NO_2^- + 2H^+ + (Mb)Fe^{2+} \rightarrow (Mb)Fe^{3+} + {}^{\bullet}NO + H_2O}$ × Competition with O_2; nitrite and NO oxidation by $Mb{-}O_2$ (eqn (17.4) and (17.6)); NO scavenging by haem (eqn 17.5)
Hb (blood)	Dioxygen transport	*b* haem, penta-coordinated by a histidine residue $\mathbf{NO_2^- + 2H^+ + (Hb)Fe^{2+} \rightarrow (Hb)Fe^{3+} + {}^{\bullet}NO + H_2O}$ × Competition with O_2; nitrite and NO oxidation by $Hb{-}O_2$ (eqn (17.4) and (17.6)); NO scavenging by haem (eqn 17.5))

Nb (nerve tissues)	No "classic" functions It is a *new* protein, whose functions are still controversial Cellular redox state sensing, scavenging/sensing of ROS/RNS, O_2, CO, NO and inhibition of Cc-induced apoptosis	*b* haem, reversible hexa-coordination (two histidine residues) $\mathbf{NO_2^- + 2H^+ + (Nb)Fe^{2+} \rightarrow (Nb)Fe^{3+} + {}^{\bullet}NO + H_2O}$ × NO scavenging by haem (eqn (17.5)); conversion of hexa- into penta-coordination under hypoxia/anoxia
Cb (apparently all tissue types)	No "classic" functions It is a *new* protein, whose functions are still controversial: NADH oxidase, O_2 sensor and NO scavenger/formation	*b* haem, reversible hexa-coordination (two histidine residues) $\mathbf{NO_2^- + 2H^+ + (Cb)Fe^{2+} \rightarrow (Cb)Fe^{3+} + {}^{\bullet}NO + H_2O}$ × NO scavenging by haem (eqn (17.5)); conversion of hexa- into penta-coordination under hypoxia/anoxia
Cc (all tissue types)	**Electron transfer complex III → complex IV** Mitochondrial oxidative phosphorylation "Non-classic" functions include apoptotic signalling molecule, $O_2^{\bullet-}$ scavenger, lipid peroxidation catalyst, peroxidase-like enzyme and nitrating agent	*c* haem, reversible hexa-coordination (histidine [proximal] and methionine [distal] residues) $\mathbf{NO_2^- + 2H^+ + (\mathit{Cc})Fe^{2+} \rightarrow (\mathit{Cc})Fe^{3+} + {}^{\bullet}NO + H_2O}$ × NO scavenging by haem (similar to eqn (17.5)); conversion of hexa- into penta-coordination under hypoxia/anoxia; mechanism to trigger release from mitochondria (apoptosis)
Carbonic anhydrase	$\mathbf{CO_2 + H_2O \rightleftharpoons H_2CO_3 \rightleftharpoons HCO_3^- + H^+}$ Acid–base homeostasis and electrolyte balance	Zinc (Zn^{2+}) $\mathbf{2NO_2^- + 2H^+ \rightleftharpoons 2HNO_2 \rightleftharpoons H_2O + N_2O_3, N_2O_3 \rightleftharpoons {}^{\bullet}NO + {}^{\bullet}NO_2}$ × Competition with CO_2; reaction mechanism in the absence of a redox-active metal is controversial; N_2O_3 dismutation to yield NO is controversial

[a] AO: aldehyde oxidase; Cb: cytoglobin; *Cc*: cytochrome c; Hb: haemoglobin; mARC: mitochondrial amidoxime reducing component; Mb: myoglobin; Nb: neuroglobin; NO: nitric oxide; RNS, reactive nitrogen species; ROS: reactive oxygen species; SO: sulphite oxidase; XO: xanthine oxidase.
[b] Adapted with permission from ref. 8. Copyright (2014) American Chemical Society.
[c] Some of which are relevant to control the "moonlighting" nitrite reductase activity of the protein.

involvement in signalling pathways and, most importantly, in some ROS-mediated diseases,[202–210] including ischaemia–reperfusion injury[211–218] and ethanol hepatotoxicity.[219–225] The demonstration that XO can also catalyse nitrite reduction with NO formation further stimulated and increased the medical interest in this enzyme.

In vitro, XO catalyses nitrite reduction to NO in a reaction that is highly favoured in acidic and hypoxic conditions.[170–177,226,227] As the pH decreases from 7.4 to 6.3, the XO specificity constant for nitrite (k_{cat}/K_m) increases eight-fold, to $2.3 \times 10^3\ M^{-1}\ s^{-1}$, while the K_m value decreases six-fold, to 250 μM, showing that XO's ability to trigger NO formation would be greatly enhanced under the acidic conditions that are characteristic of ischaemia.[176] Dioxygen, as would be expected, acts as a strong competitive inhibitor of nitrite reduction, "stealing" the electrons that are needed to reduce nitrite.[173,174,176,228] However, the inhibition constant, within the values of cellular dioxygen concentration from normoxia to hypoxia (25–50 μM), shows that *in vivo* NO formation would not be abolished by dioxygen.[176] On the contrary, it suggests that the NO generation can be fine-tuned by the dioxygen availability, which is precisely what is needed to regulate the NiR/NOS activity of this enzyme: as the dioxygen concentration starts to decrease and hypoxic conditions become settled, the XO NiR activity is no longer hampered and becomes less and less inhibited as dioxygen decreases, thus leading to fine-tuned NO generation. Supporting the involvement of XO in the *in vivo* NO formation, there are presently several *in situ* and *in vivo* studies in models of ischaemic (and other) injuries in the heart, liver, lung, kidney and vessels.[52,54–56,64,71,81,92,93,174,175,177,229–234] Particularly relevant is the demonstration of the XO/nitrite protective role within the context of cardiac ischaemia in an isolated heart model.[52]

Mb is presently thought as a "multi-task" protein. It has a "textbook" defined function as a dioxygen transporter (dioxygen-facilitated diffusion[235–239]) in smooth[235,240] skeletal and cardiac muscles.[235] However, intriguingly, Mb knockout mouse models do not have compromised respiration or dioxygen consumption,[241,242] and the slow Mb diffusion seems to oppose a predominant role in facilitating oxygen transport in myocardium.[243,244] These studies suggest that the Mb function, which was thought to be fully understood half a century ago, has to be reassessed! One important novel function of Mb is certainly the control of cardiac NO homeostasis, where it can function as a NiR/NOS (deoxy-Mb) and as a NO scavenger (oxy-Mb).[35,37,38,40,41,245–247] Other "unconventional" functions of Mb include as an oxygen sensor, ROS scavenger[248] and intracellular fatty acid transporter.[249]

To reduce nitrite to NO, the Mb haem must be reduced and penta-coordinated, in this way providing the necessary electrons and having a vacant coordination position to bind nitrite and carry out the reduction to NO.[8,250] Hence, the NiR/NOS activity of Mb is also dependent on the dioxygen concentration, as only ferrous deoxy (penta-coordinated)-Mb reacts with nitrite to yield NO, and sufficient Mb deoxygenation only occurs when the dioxygen concentration decreases to values lower than 3–4 μM. The reaction is also pH dependent, with the rate constant increasing from 0.1–1 to 6–12 $M^{-1}\,s^{-1}$ when the pH is decreased from 7.4 to 6.5.[50,69,251–257] It is noteworthy that nitrite also reacts with ferrous oxy-Mb, but is oxidised to nitrate (eqn (17.4)); this reaction is essential to prevent nitrite accumulation *in vivo* under normoxic conditions (Mb- and Hb-mediated reactions).[1,30,258]

$$\mathbf{NO_2^-} + 1H^+ + (Hb/Mb)Fe^{2+}\text{–}\mathbf{O_2} \rightarrow (Hb/Mb)Fe^{3+} + \mathbf{NO_3^-} + 1/2H_2O_2 \quad (17.4)$$

These two opposing reactions—deoxy-Mb/nitrite reduction to NO *versus* oxy-Mb/nitrite oxidation to nitrate—can operate to control NO formation throughout the entire physiological dioxygen gradient, from normoxia to anoxia. Under normoxic conditions, NO is synthesised by NOS and nitrite is oxidised to nitrate; as the dioxygen concentration decreases, the balance between the two reactions is deviated towards nitrite reduction to NO and the impairment of the NOS activity is compensated for.

This nitrite-derived NO pathway has, however, found opposition, because deoxy-Mb (and deoxy-Hb) should rapidly ($k_{on} \approx 10^7\ M^{-1}\ s^{-1}$)[31,259–267] trap the NO that is formed (eqn (17.5)) in a stable complex ($K_d \approx 10^{-12}$–10^{-10} M)[253,255,259,260,265,267–272] that would not allow NO release. In addition, the oxy-Mb (and oxy-Hb) molecules that would be present under non-anoxic conditions should rapidly oxidise the NO formed to nitrate (eqn (17.6)),[1,30,31,34,36,39,43,259,266,273–275] thus consuming the NO that is generated.

$$^{\bullet}\mathbf{NO} + (Hb/Mb)Fe^{2+} \rightleftharpoons (Hb/Mb)Fe^{2+}\text{–}\mathbf{NO} \quad (17.5)$$

$$^{\bullet}\mathbf{NO} + (Hb/Mb)Fe^{2+}\text{–}\mathbf{O_2} \rightleftharpoons (Hb/Mb)Fe^{3+} + \mathbf{NO_3^-} \quad (17.6)$$

Because of this well-documented *in vitro* haem chemistry, the *in vivo* physiological role of haemic proteins in the formation of bioactive NO has been greatly questioned! Nevertheless, several *in vivo*, *in situ* and *in vitro* studies suggest that isolated hearts and purified Mb, as well as erythrocytes and purified Hb, can contribute to NO formation and stimulate NO signalling in the presence of nitrite under hypoxic conditions.[48,50–52,54,58,60,62,65–67,69,73,76,139,235,240,255,276–298] Particularly relevant are the studies showing that the *in vivo* administration of nitrite generates NO, nitrosylates cardiomyocyte iron-containing proteins and reduces (by 60%) myocardial infarction in mice; by contrast, in Mb knockout mice, nitrite has no protective effects and there is a decrease in NO formation.[69,76,283,296] These studies suggest that nitrite cytoprotection is exerted through its reduction to NO, implying a cytoprotective role for the Mb/nitrite pair in myocardial ischaemia–reperfusion injury.

17.2.3 Human Nitrite Reduction in the Cellular Context

As in XO and Mb, in all other "non-dedicated" NiRs, the chemistry behind mammalian nitrite-derived NO formation is quite simple, being strikingly more simple than the complex reaction catalysed by NOS (eqn (17.1) and

(17.2)). It involves the one-electron reduction of nitrite by a redox-active metalloprotein, with the concomitant abstraction of one oxygen atom, and requires only a low dioxygen concentration, protons (acidosis) and an electron donor (reducing equivalents) to reduce the protein metal (eqn (17.3)).

From a chemical point of view, such cells are just doing redox chemistry with an available redox-active system (a metalloprotein) and an "adapted" substrate (nitrite) in order to obtain a different and specific product (NO). From a physiological point of view, such cells are "reusing" the available metalloproteins that are normally present in order to accomplish other cellular functions (Table 17.1) and are "switching" their activity (to a NiR/NOS) when it is necessary to generate NO and when NOS activity is impaired. This "reuse of proteins" for diverse functions is not an eccentricity! The biological use of a single protein to accomplish more than one function is a well-recognised and common phenomenon called moonlighting, which has important implications for systems biology and, in particular, for human physiology and pathology.[299,300] Moreover, the existence of different NO formation pathways seems very reasonable. It is not probable that nature developed only one specialised enzyme (NOS) to generate such a vital molecule. Certainly, it would be an advantage to have some "rescue" pathways in order to ensure the formation of NO. These would be particularly relevant for mammals in order for them to be able to avoid deleterious hypoxic/anoxic conditions, in which oxygen-dependent NOS activity would be impaired.

17.3 Nitrate Reduction to Signalling NO in the Context of the Nitrogen Cycle

Nitrate, nitrite and NO have long been known as players in the biogeochemical cycle of nitrogen, participating in key pathways that are crucial to life on Earth and to the planetary "recycling" of nitrogen (see Figure 1.1 from Chapter 1). More recently, nitrite has also being recognised as a key molecule for cell signalling and survival, contributing to NO formation not only in humans (as is described here), but also in other animals and plants, as well as in prokaryotes.[1–8] When this virtually universal signalling pathway is set in the scenario of the nitrogen cycle (Figure 17.2), it becomes evident why it is thought to be a "vestige" of the pre-aerobic pathways of the nitrogen cycle.[8,301–303]

To carry out the conversion of nitrate into NO, nature developed several strategies, exploring different metals, chemical strategies and protein structures, but using a similar "design" (Figure 17.2). To reduce nitrate to nitrite, only molybdenum is used, as far as is presently known (see Chapter 3 in this book). The unique chemistry of molybdenum makes enzymatic molybdenum active sites excellent "oxygen atom exchangers",[169,194,195] which is precisely what is needed to interconvert nitrate and nitrite and to reduce nitrite to NO. The utilisation of molybdenum was so successful that the organisms developed four types of molybdenum-containing nitrate reductases (one of which

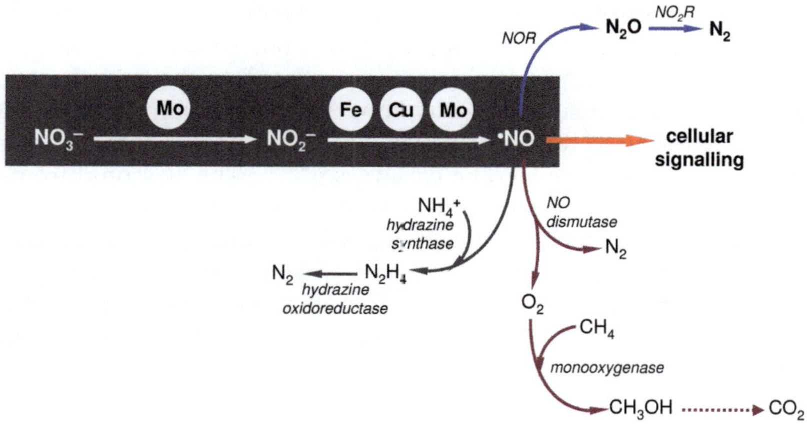

Figure 17.2 Signalling nitrate/nitrite/nitric oxide (NO) pathway in the nitrogen cycle context. The nitrate conversion into NO is highlighted, along with the metals responsible for the catalysis of each step. The NO formed can be subsequently reduced to: (i) nitrous oxide (by NO reductase) and dinitrogen (by nitrous oxide reductase [N_2OR]) in denitrification (blue arrows); and (ii) hydrazine (by hydrazine synthase) and dinitrogen (by hydrazine oxidoreductase) in anaerobic ammonium oxidation (grey arrows). The NO can also be dismutated to dinitrogen plus dioxygen in the "denitrification/intra-aerobic methane oxidation" pathway (violet arrows); the dioxygen formed is used to oxidise methane and eventually to yield carbon dioxide. In humans, as well as in other animals, plants and also prokaryotes, the nitrate/nitrite-derived NO is used in signalling pathways.

is used by higher eukaryotes) and two types of nitrite oxidoreductases (these are involved in nitrification; see Figure 1.1, black arrows). Moreover, it is now becoming evident that virtually all forms of life are also able to use molybdenum to generate signalling NO, with different types of molybdoenzymes (with different molybdenum centre structures, as exemplified in Table 17.1) being shown to reduce nitrite to NO.[169] Yet, to reduce nitrite, the most widely employed metal is probably iron, more specifically haem iron. Prokaryotes reduce nitrite for assimilatory and dissimilatory purposes, using different haem types: *c*-type haem and sirohaem to reduce it to ammonium (in dissimilatory nitrate reduction to ammonium [Figure 1.1, green arrows] and assimilatory ammonification [Figure 1.1, orange arrows], respectively) and d_1-type to reduce it to NO (in denitrification, anaerobic ammonium oxidation and others [Figure 1.1, blue and grey arrows, respectively]) (see Chapter 4 in this book). Haem iron is also able to reduce nitrite in order to generate signalling NO, with numerous haemic proteins being suggested to fulfil this role (as exemplified in Table 17.1). Organisms also use copper to reduce nitrite to NO (denitrification [Figure 1.1, blue arrows]) (see Chapter 5 in this book), although, comparatively, its utilisation is believed not to be so diversified.

Hence, once nature had found a "design" (Figure 17.2) that worked for nitrate conversion into NO, this approach was reused several times, introducing only *minor* adjustments in order to respond to the specific cellular demands. To generate signalling NO, modern mammalian and plant cells are just using the "old" redox chemistry, which has already been thoroughly "tested" in the earlier, pre-aerobic pathways.[8,301–303] The "lessons" from the ancestral, pre-aerobic world become even more evident when it is realised that the chemistry behind an anaerobic activity is the perfect solution in order for contemporary higher organisms to generate NO under hypoxic/anoxic conditions. In conclusion, nitrate reduction to NO is a ubiquitous pathway that came from a distant pre-aerobic past and has been "reused" ever since.[8,302]

Acknowledgements

This work was supported by the Unidade de Ciências Biomoleculares Aplicadas-UCIBIO, which is financed by national funds from FCT/MEC (UID/Multi/04378/2013) and co-financed by the ERDF under the PT2020 Partnership Agreement (POCI-01-0145-FEDER-007728). LBM wishes to thank to Fundacão para a Ciência e a Tecnologia, MEC, for a fellowship grant (SFRH/BPD/111404/2015, which is financed by national funds and co-financed by FSE).

References

1. S. Moncada, R. M. J. Palmer and E. A. Higgs, *Pharmacol. Rev.*, 1991, **43**, 109.
2. I. Gusarov and E. Nudler, *Proc. Natl. Acad. Sci. U. S. A.*, 2005, **102**, 13855.
3. S. Neill, J. Bright, R. Desikan, J. Hancock, J. Harrison and I. Wilson, *J. Exp. Bot.*, 2008, **59**, 25.
4. I. D. Wilson, S. J. Neill and J. T. Hancock, *Plant, Cell Environ.*, 2008, **31**, 622.
5. K. J. Gupta and W. M. Kaiser, *Plant Cell Physiol.*, 2010, **51**, 576.
6. M. Moreau, C. Lindermayr, J. Durner and D. F. Klessig, *Physiol. Plant.*, 2010, **138**, 372.
7. J. C. Toledo and O. Augusto, *Chem. Res. Toxicol.*, 2012, **25**, 975.
8. L. Maia and J. J. G. Moura, *Chem. Rev.*, 2014, **114**, 5273.
9. J. S. Beckman and W. H. Koppenol, *Am. J. Physiol.*, 1996, **271**, C1424.
10. J. P. Eiserich, R. P. Patel and V. B. O'Donnell, *Mol. Aspects Med.*, 1998, **19**, 221.
11. W. H. Koppenol, *Free Radical Biol. Med.*, 1998, **25**, 385.
12. S. Pfeiffer, B. Mayer and B. Hemmens, *Angew. Chem., Int. Ed.*, 1999, **38**, 1714.
13. D. J. Stuehr, *Biochim. Biophys. Acta*, 1999, **1411**, 217.
14. W. K. Alderton, C. E. Cooper and R. G. Knowles, *Biochem. J.*, 2001, **357**, 593.
15. N. Lehnert, T. C. Berto, M. G. I. Galinato and L. E. Goodrich, in *The Handbook of Porphyrin Science*, ed K. M. Kadish, K. M. Smith and R. Guilard, World Scientific, NJ, 2011, vol. 14, pp. 1–200.

16. G. J. Carr and S. J. Ferguson, *Biochim. Biophys. Acta*, 1990, **1017**, 57.
17. A. Jansen, J. Drazen, J. A. Osborne, R. Brown, J. Loscalzo and J. S. J. Stamler, *Pharmacol. Exp. Ther.*, 1992, **261**, 154.
18. J. S. Stamler, O. Jaraki, J. Osborne, D. I. Simon, J. Keaney, J. Vita, D. Singel, C. R. Valeri and J. Loscalzo, *Proc. Natl. Acad. Sci. U. S. A.*, 1992, **89**, 7674.
19. J. S. Stamler, D. I. Simon, J. A. Osborne, M. E. Mullins, O. Jaraki, T. Michel, D. J. Singel and J. Loscalzo, *Proc. Natl. Acad. Sci. U. S. A.*, 1992, **89**, 444.
20. G. C. Brown and C. E. Cooper, *FEBS Lett.*, 1994, **356**, 295.
21. M. W. Cleeter, J. M. Cooper, V. M. Darley-Usmar, S. Moncada and A. H. Schapira, *FEBS Lett.*, 1994, **345**, 50.
22. V. G. Kharitonov, A. R. Sundquist and V. S. Sharma, *J. Biol. Chem.*, 1995, **270**, 28158.
23. E. Clementi, G. C. Brown, M. Feelisch and S. Moncada, *Proc. Natl. Acad. Sci. U. S. A.*, 1998, **95**, 7631.
24. M. Feelisch, T. Rassaf, S. Mnaimneh, N. Singh, N. S. Bryan, D. Jourd'Heuil and M. Kelm, *FASEB J.*, 2002, **16**, 1775.
25. B. M. Gaston, J. Carver, A. Doctor and L. A. Palmer, *Mol. Interventions*, 2003, **3**, 253.
26. N. S. Bryan, T. Rassaf, R. E. Maloney, C. M. Rodriguez, F. Saijo, J. R. Rodriguez and M. Feelisch, *Proc. Natl. Acad. Sci. U. S. A.*, 2004, **101**, 4308.
27. L. Liu, Y. Yan, M. Zeng, J. Zhang, M. A. Hanes, G. Ahearn, T. J. McMahon, T. Dickfeld, H. E. Marshall, L. G. Que and J. S. Stamler, *Cell*, 2004, **116**, 617.
28. B. Lima, G. K. Lam, L. Xie, D. L. Diesen, N. Villamizar, J. Nienaber, E. Messina, D. Bowles, C. D. Kontos, J. M. Hare, J. S. Stamler and H. A. Rockman, *Proc. Natl. Acad. Sci. U. S. A.*, 2009, **106**, 6297.
29. P. C. Sanghani, W. I. Davis, S. L. Fears, S. L. Green, L. Zhai, Y. Tang, E. Martin, N. S. Bryan and S. P. Sanghani, *J. Biol. Chem.*, 2009, **284**, 24354.
30. M. P. Doyle and J. W. Hoekstra, *J. Inorg. Biochem.*, 1981, **14**, 351.
31. V. S. Sharma, T. G. Taylor and R. Gardiner, *Biochemistry*, 1987, **26**, 3837.
32. D. A. Wink, J. F. Darbyshire, R. W. Nims, J. E. Saavedra and P. C. Ford, *Chem. Res. Toxicol.*, 1993, **6**, 23.
33. S. Goldstein and G. Czapski, *J. Am. Chem. Soc.*, 1995, **117**, 12078.
34. R. F. Eich, T. Li, D. D. Lemon, D. H. Doherty, S. R. Curry, J. F. Aitken, A. J. Mathews, K. A. Johnson, R. D. Smith Jr and J. S. Olson, *Biochemistry*, 1996, **35**, 6976.
35. X. Liu, M. J. S. Miller, M. S. Joshi, H. Sadowska-Krowicka, D. A. Clark and J. R. Lancaster, *J. Biol. Chem.*, 1998, **273**, 18709.
36. A. J. Gow, B. P. Luchsinger, J. R. Pawloski, D. J. Singel and J. S. Stamler, *Proc. Natl. Acad. Sci. U. S. A.*, 1999, **96**, 9027.
37. M. Brunori, *Trends Biochem. Sci.*, 2001, **26**, 21.
38. U. Flögel, M. W. Merx, A. Gödecke, U. K. M. Decking and J. Schrader, *Proc. Natl. Acad. Sci. U. S. A.*, 2001, **98**, 735.
39. S. Herold, M. Exner and T. Nauser, *Biochemistry*, 2001, **40**, 3385.
40. K.-T. Huang, T. H. Han, D. R. Hyduke, M. W. Vaughn, H. V. Herle, T. W. Hein, C. Zhang, L. Kuo and J. C. Liao, *Proc. Natl. Acad. Sci. U. S. A.*, 2001, **98**, 11771.

41. P. K. Witting, D. J. Douglas and A. G. Mauk, *J. Biol. Chem.*, 2001, **276**, 3991.
42. X. Liu, Q. Liu, E. Gupta, N. Zorko, E. Brownlee and J. L. Zweier, *Nitric Oxide*, 2005, **13**, 68.
43. P. R. Gardner, A. M. Gardner, W. T. Brashear, T. Suzuki, A. N. Hvitved, K. D. R. Setchell and J. S. Olson, *J. Inorg. Biochem.*, 2006, **100**, 542.
44. S. Shiva, X. Wang, L. A. Ringwood, X. Xu, S. Yuditskaya, V. Annavajjhala, H. Miyajima, N. Hogg, Z. L. Harris and M. T. Gladwin, *Nat. Chem. Biol.*, 2006, **2**, 486.
45. D. E. Koshland Jr, *Science*, 1992, **258**, 1861.
46. G. Johnson III, P. S. Tsao, D. Mulloy and A. M. Lefer, *J. Pharmacol. Exp. Ther.*, 1990, **252**, 35.
47. R. S. Dykhuizen, A. Fraser, H. McKenzie, M. Golden, C. Leifert and N. Benjamin, *Gut*, 1998, **42**, 334.
48. M. T. Gladwin, J. H. Shelhamer, A. N. Schechter, M. E. Pease-Fye, M. A. Waclawiw, J. A. Panza, F. P. Ognibene and R. O. Cannon III, *Proc. Natl. Acad. Sci. U. S. A.*, 2000, **97**, 11482.
49. E. A. Demoncheaux, T. W. Higenbottam, P. J. Foster, C. D. Borland, A. P. Smith, H. M. Marriott, D. Bee, S. Akamine and M. B. Davies, *Clin. Sci.*, 2002, **102**, 77.
50. K. Cosby, K. S. Partovi, J. H. Crawford, R. P. Patel, C. D. Reiter, S. Martyr, B. K. Yang, M. A. Waclawiw, G. Zalos, X. Xu, K. T. Huang, H. Shields, D. B. Kim-Shapiro, A. N. Schechter, R. O. Cannon III and M. T. Gladwin, *Nat. Med.*, 2003, **9**, 1498.
51. C. J. Hunter, A. Dejam, A. B. Blood, H. Shields, D. B. Kim-Shapiro, R. F. Machado, S. Tarekegn, N. Mulla, A. O. Hopper, A. N. Schechter, G. G. Power and M. T. Gladwin, *Nat. Med.*, 2004, **10**, 1122.
52. A. Webb, R. Bond, P. McLean, R. Uppal, N. Benjamin and A. Ahluwalia, *Proc. Natl. Acad. Sci. U. S. A.*, 2004, **101**, 13683.
53. N. S. Bryan, B. O. Fernandez, S. M. Bauer, M. F. Garcia-Saura, A. B. Milsom, T. Rassaf, R. E. Maloney, A. Bharti, J. Rodriguez and M. Feelisch, *Nat. Chem. Biol.*, 2005, **1**, 290.
54. M. R. Duranski, J. J. Greer, A. Dejam, S. Jaganmohan, N. Hogg, W. Langston, R. P. Patel, S. F. Yet, X. Wang, C. G. Kevil, M. T. Gladwin and D. J. Lefer, *J. Clin. Invest.*, 2005, **115**, 1232.
55. P. Lu, F. Liu, Z. Yao, C. Y. Wang, D. D. Chen, Y. Tian, J. H. Zhang and Y. H. Wu, *Hepatobiliary Pancreatic Dis. Int.*, 2005, **4**, 350.
56. J. O. Lundberg and E. Weitzberg, *Arterioscler., Thromb., Vasc. Biol.*, 2005, **25**, 915.
57. R. M. Pluta, A. Dejam, G. Grimes, M. T. Gladwin and E. H. Oldfield, *JAMA, J. Am. Med. Assoc.*, 2005, **293**, 1477.
58. K. Tsuchiya, Y. Kanematsu, M. Yoshizumi, H. Ohnishi, K. Kirima, Y. Izawa, M. Shikishima, T. Ishida, S. Kondo, S. Kagami, Y. Takiguchi and T. Tamaki, *Am. J. Physiol.: Heart Circ. Physiol.*, 2005, **288**, H2163.
59. P. R. Castello, P. S. David, T. McClure, Z. Crook and R. O. Poyton, *Cell Metab.*, 2006, **3**, 277.

60. J. H. Crawford, T. S. Isbell, Z. Huang, S. Shiva, B. K. Chacko, A. N. Schechter, V. M. Darley-Usmar, J. D. Kerby, J. D. Lang Jr, D. Kraus, C. Ho, M. T. Gladwin and R. P. Patel, *Blood*, 2006, **107**, 566.
61. K. H. Jung, K. Chu, S. Y. Ko, S. T. Lee, D. I. Sinn, D. K. Park, J. M. Kim, E. C. Song, M. Kim and J. K. Roh, *Stroke*, 2006, **37**, 2744.
62. F. J. Larsen, B. Ekblom, K. Sahlin, J. O. Lundberg and E. Weitzberg, *N. Engl. J. Med.*, 2006, **355**, 2792.
63. D. Lefer, *Am. J. Physiol. Renal, Physiol.*, 2006, **290**, F777.
64. J. E. Baker, J. Su, X. Fu, A. Hsu, G. J. Gross, J. S. Tweddell and N. Hogg, *J. Mol. Cell. Cardiol.*, 2007, **43**, 437.
65. N. S. Bryan, J. W. Calvert, J. W. Elrod, S. Gundewar, S. Y. Ji and D. Lefer, *Proc. Natl. Acad. Sci. U. S. A.*, 2007, **104**, 19144.
66. C. Dezfulian, N. Raat, S. Shiva and M. T. Gladwin, *Cardiovasc. Res.*, 2007, **75**, 327.
67. A. Dejam, C. J. Hunter, C. Tremonti, R. M. Pluta, Y. Y. Hon, G. Grimes, K. Partovi, M. M. Pelletier, E. H. Oldfield, R. O. Cannon III, A. N. Schechter and M. T. Gladwin, *Circulation*, 2007, **116**, 1821.
68. F. J. Larsen, E. Weitzberg, J. O. Lundberg and B. Ekblom, *Acta Physiol.*, 2007, **191**, 59.
69. S. Shiva, Z. Huang, R. Grubina, J. Sun, L. A. Ringwood, P. H. MacArthur, X. Xu, E. Murphy, V. M. Darley-Usmar and M. T. Gladwin, *Circ. Res.*, 2007, **100**, 654.
70. S. Shiva, M. N. Sack, J. J. Greer, M. Duranski, L. A. Ringwood, L. Burwell, X. Wang, P. H. MacArthur, A. Shoja, N. Raghavachari, J. W. Calvert, P. S. Brookes, D. Lefer and M. T. Gladwin, *J. Exp. Med.*, 2007, **204**, 2089.
71. P. Tripatara, N. S. Patel, A. Webb, K. Rathod, F. M. Lecomte, E. Mazzon, S. Cuzzocrea, M. M. Yaqoob, A. Ahluwalia and C. Thiemermann, *J. Am. Soc. Nephrol.*, 2007, **18**, 570.
72. A. Benamar, H. Rolletschek, L. Borisjuk, M. H. Avelange-Macherel, G. Curien, H. A. Mostefai, R. Andriantsitohaina and D. Macherel, *Biochim. Biophys. Acta*, 2008, **1777**, 1268.
73. N. S. Bryan, J. W. Calvert, S. Gundewar and D. J. Lefer, *Free Radical Biol. Med.*, 2008, **45**, 468.
74. A. R. Butler and M. Feelisch, *Circulation*, 2008, **117**, 2151.
75. F. M. Gonzalez, S. Shiva, P. S. Vincent, L. A. Ringwood, L. Y. Hsu, Y. Y. Hon, A. H. Aletras, R. O. Cannon 3rd, M. T. Gladwin and A. E. Arai, *Circulation*, 2008, **117**, 2986.
76. U. B. Hendgen-Cotta, M. W. Merx, S. Shiva, J. Schmitz, S. Becher, J. P. Klare, H. Steinhoff, A. Goedecke, J. Schrader, M. T. Gladwin, M. Kelm and T. Rassaf, *Proc. Natl. Acad. Sci. U. S. A.*, 2008, **105**, 10256.
77. D. Kumar, B. G. Branch, C. B. Pattillo, J. Hood, S. Thoma, S. Simpson, S. Illum, N. Arora, J. H. Chidlow, W. Langston, X. Teng, D. J. Lefer, R. P. Patel and C. G. Kevil, *Proc. Natl. Acad. Sci. U. S. A.*, 2008, **105**, 7540.
78. J. O. Lundberg, E. Weitzberg and M. T. Gladwin, *Nat. Rev. Drug Discovery*, 2008, **7**, 156.

79. A. R. Maher, A. B. Milsom, P. Gunaruwan, K. Abozguia, I. Ahmed, R. A. Weaver, P. Thomas, H. Ashrafian, G. V. Born, P. E. James and M. P. Frenneaux, *Circulation*, 2008, **117**, 670.
80. S. S. Sinha, S. Shiva and M. T. Gladwin, *Trends Cardiovasc. Med.*, 2008, **18**, 163.
81. A. J. Webb, A. B. Milsom, K. S. Rathod, W. L. Chu, S. Qureshi, M. J. Lovell, F. M. Lecomte, D. Perrett, C. Raimondo, E. Khoshbin, Z. Ahmed, R. Uppal, N. Benjamin, A. J. Hobbs and A. Ahluwalia, *Circ. Res.*, 2008, **103**, 957.
82. J. W. Calvert and D. J. Lefer, *Cardiovasc. Res.*, 2009, **83**, 195.
83. D. B. Casey, A. M. Badejo, J. S. Dhaliwal, S. N. Murthy, A. L. Hyman, B. D. Nossaman and P. J. Kadowitz, *Am. J. Physiol. Heart. Circ. Physiol.*, 2009, **296**, H524.
84. A. Cauwels, E. S. Buys, R. Thoonen, L. Geary, J. Delanghe, S. Shiva and P. Brouckaert, *J. Exp. Med.*, 2009, **206**, 2915.
85. D. J. Lefer, *Arch. Pharm. Res.*, 2009, **32**, 1127.
86. J. O. Lundberg, M. T. Gladwin, A. Ahluwalia, N. Benjamin, N. S. Bryan, A. Butler, P. Cabrales, A. Fago, M. Feelisch, P. C. Ford, B. A. Freeman, M. Frenneaux, J. Friedman, M. Kelm, C. G. Kevil, D. B. Kim-Shapiro, A. V. Kozlov, J. R. Lancaster Jr, D. J. Lefer, K. McColl, K. McCurry, R. P. Patel, J. Petersson, T. Rassaf, V. P. Reutov, J. B. Richter-Addo, A. Schechter, S. Shiva, K. Tsuchiya, E. E. Faassen, A. J. Webb, B. S. Zuckerbraun, J. L. Zweier and E. Weitzberg, *Nat. Chem. Biol.*, 2009, **5**, 865.
87. N. J. Raat, S. Shiva and M. T. Gladwin, *Adv. Drug Delivery Rev.*, 2009, **61**, 339.
88. J. W. Calvert and D. J. Lefer, *Nitric Oxide*, 2010, **22**, 91.
89. T. E. Ingram, A. G. Pinder, D. M. Bailey, A. G. Fraser and P. E. James, *Am. J. Physiol. Heart. Circ. Physiol.*, 2010, **298**, H331.
90. F. J. Larsen, E. Weitzberg, J. O. Lundberg and B. Ekblom, *Free Radical Biol. Med.*, 2010, **48**, 342.
91. S. Shiva, *Nitric Oxide*, 2010, **22**, 64.
92. B. S. Zuckerbraun, S. Shiva, E. Ifedigbo, M. A. Mathier, K. P. Mollen, J. Rao, P. M. Bauer, J. J. Choi, E. Curtis, A. M. Choi and M. T. Gladwin, *Circulation*, 2010, **121**, 98.
93. M. J. Alef, R. Vallabhaneni, E. Carchman, S. M. Morris Jr, S. Shiva, Y. Wang, E. E. Kelley, M. M. Tarpey, M. T. Gladwin, E. Tzeng and B. S. Zuckerbraun, *J. Clin. Invest.*, 2011, **121**, 1646.
94. A. B. Blood, H. J. Schroeder, M. H. Terry, J. Merrill-Henry, S. L. Bragg, K. Vrancken, T. Liu, J. Herring, L. C. Sowers, S. M. Wilson and G. G. Power, *Circulation*, 2011, **123**, 605.
95. M. Gilchrist, A. C. Shore and N. Benjamin, *Cardiovasc. Res.*, 2011, **89**, 492.
96. C. G. Kevil, G. K. Kolluru, C. B. Pattillo and T. Giordano, *Free Radical Biol. Med.*, 2011, **51**, 576.
97. F. J. Larsen, T. A. Schiffer, S. Borniquel, K. Sahlin, B. Ekblom, J. O. Lundberg and E. Weitzberg, *Cell Metab.*, 2011, **13**, 149.

98. J. O. Lundberg, M. Carlstrom, F. J. Larsen and E. Weitzberg, *Cardiovasc. Res.*, 2011, **89**, 525.
99. A. Machha and A. N. Schechter, *Eur. J. Nutr.*, 2011, **50**, 293.
100. D. Murillo, C. Kamga, L. Mo and S. Shiva, *Nitric Oxide*, 2011, **25**, 70.
101. C. B. Pattillo, S. Bir, V. Rajaram and C. G. Kevil, *Cardiovasc. Res.*, 2011, **89**, 533.
102. A. L. Sindler, B. S. Fleenor, J. W. Calvert, K. D. Marshall, M. L. Zigler, D. J. Lefer and D. R. Seals, *Aging Cell*, 2011, **10**, 429.
103. R. S. Baliga, A. B. Milsom, S. M. Ghosh, S. L. Trinder, R. J. Macallister, A. Ahluwalia and A. J. Hobbs, *Circulation*, 2012, **125**, 2922.
104. C. E. Sparacino-Watkins, Y. C. Lai and M. T. Gladwin, *Ther. Circ.*, 2012, **125**, 2824.
105. M. Bueno, J. Wang, A. L. Mora and M. T. Gladwin, *Antioxid. Redox Signal.*, 2013, **18**, 1797.
106. N. Cantu-Medellin and E. E. Kelley, *Nitric Oxide*, 2013, **34**, 19.
107. E. T Chouchani, C. Methner, S. M. Nadtochiy, A. Logan, V. R. Pell, S. Ding, A. M. James, H. M. Cochemé, J. Reinhold, K. S. Lilley, L. Partridge, I. M. Fearnley, A. J. Robinson, R. C. Hartley, R. A. J. Smith, T. Krieg, P. S. Brookes and M. P. Murphy, *Nature Med.*, 2013, **19**, 753.
108. S. M. Ghosh, V. Kapil, I. Fuentes-Calvo, K. J. Bubb, V. Pearl, A. B. Milsom, R. Khambata, S. Maleki-Toyserkani, M. Yousuf, N. Benjamin, A. J. Webb, M. J. Caulfield, A. J. Hobbs and A. Ahluwalia, *Hypertension*, 2013, **61**, 1091.
109. C. Helms and D. B. Kim-Shapiro, *Free Radical Biol. Med.*, 2013, **61**, 464.
110. D. B. Kim-Shapiro and M. T. Gladwin, *Nitric Oxide*, 2014, **38**, 58.
111. S. A. Omar and A. Webb, *J. Mol. Cell. Cardiol.*, 2014, **73**, 57.
112. E. T. Reichert and S. W. Mitchell, *Am. J. Med. Sci.*, 1880, **159**, 158.
113. C. G. Kevil, G. K. Kolluru, C. B. Pattillo and T. Giordano, *Free Radical Biol. Med.*, 2011, **51**, 576.
114. M. Delmastro-Greenwood, K. S. Hughan, D. A. Vitturi, S. R. Salvatore, G. Grimes, G. Potti, S. Shiva, F. J. Schopfer, M. T. Gladwin, B. A. Freeman and S. Gelhaus-Wendell, *Free Radical Biol. Med.*, 2015, **89**, 333.
115. D. N. Granger and P. R. Kvietys, *Redox Biol.*, 2015, **6**, 524.
116. B. Kautza, H. Gomez, D. Escobar, C. Corey, B. Ataya, J. Luciano, A. M. Botero, L. Gordon, J. Brumfield, S. Martinez, A. Holder, O. Ogundele, M. Pinsky, S. Shiva and B. S. Zuckerbraun, *Nitric Oxide*, 2015, **51**, 7.
117. M. C. Madigan, R. M. McEnaney, A. J. Shukla, G. Hong, E. Kelley, M. M. Tarpey, M. Gladwin, B. S. Zuckerbraun and E. Tzeng, *Mol. Med.*, 2015, **21**, 313.
118. J. Eggebeen, D. B. Kim-Shapiro, M. Haykowsky, T. M. Morgan, S. Basu, P. Brubaker, J. Rejeski and D. W. Kitzman, *J.A.C.C. Heart Fail*, 2016, **4**, 428.
119. R. Phillips, S. Kuijper, N. Benjamin, M. Wansbrough-Jones, M. Wilks and A. H. Kolk, *Antimicrob. Agents Chemother.*, 2004, **48**, 3130.
120. M. Anyim, N. Benjamin and M. Wilks, *Int. J. Antimicrob. Agents*, 2005, **26**, 85.
121. A. Rao, R. L. Jump, N. J. Pultz, M. J. Pultz and C. J. Donskey, *Antimicrob. Agents Chemother.*, 2006, **50**, 3901.

122. K. J. Reszka, Y. Xiong, L. Sallans, R. Pasula, O. Olakanmi, D. J. Hassett and B. E. Britigan, *Am. J. Physiol.: Lung Cell. Mol. Physiol.*, 2012, **302**, L1044.
123. S. S. Yoon, R. Coakley, G. W. Lau, S. V. Lymar, B. Gaston, A. C. Karabulut, R. F. Hennigan, S. H. Hwang, G. Buettner, M. J. Schurr, J. E. Mortensen, J. L. Burns, D. Speert, R. C. Boucher and D. J. Hassett, *J. Clin. Invest.*, 2006, **116**, 436.
124. T. A. Major, W. Panmanee, J. E. Mortensen, L. D. Gray, N. Hoglen and D. J. Hassett, *Antimicrob. Agents Chemother.*, 2010, **54**, 4671.
125. J. A. Pennington, *Food Control*, 1998, **9**, 385.
126. R. G. Cassens, I. Ito, M. Lee and D. Buege, *Bioscience*, 1978, **28**, 633.
127. B. Bartholomew and M. Hill, *J. Food Chem. Toxicol.*, 1984, **22**, 789.
128. N. S. Bryan, *Free Radical Biol. Med.*, 2006, **41**, 691.
129. T. Sasaki and K. Matano, *J. Food Hyg. Soc. Jpn.*, 1979, **20**, 363.
130. J. O. Lundberg and M. Govoni, *Free Radical Biol. Med.*, 2004, **37**, 395.
131. M. Govoni, E. A. Jansson, E. Weitzberg and J. O. Lundberg, *Nitric Oxide*, 2008, **19**, 333.
132. E. Weitzberg, M. Hezel and J. O. Lundberg, *Anesthesiology*, 2010, **113**, 1460.
133. V. Kapil, S. M. Haydar, V. Pearl, J. O. Lundberg, E. Weitzberg and A. Ahluwalia, *Free Radical Biol. Med.*, 2012, **55**, 93.
134. M. Woessner, J. M. Smoliga, B. Tarzia, T. Stabler, M. Van Bruggen and J. D. Allen, *Nitric Oxide*, 2016, **54**, 1.
135. B. Piknova, J. W. Park, K. M. Swanson, S. Dey, C. T. Noguchi and A. N. Schechter, *Nitric Oxide*, 2015, **47**, 10.
136. B. Spiegelhalder, G. Eisenbrand and R. Preussman, *Food Cosmet. Toxicol.*, 1976, **14**, 545.
137. S. R. Tannenbaum, M. Weisman and D. Fett, *Food Cosmet. Toxicol.*, 1976, **14**, 549.
138. C. Duncan, H. Douglas, P. Johnston, S. Green, R. Brogan, C. Leifert, L. Smith, M. Golden and N. Benjamin, *Nat. Med.*, 1995, **1**, 546.
139. A. J. Webb, N. Patel, S. Loukogeorgakis, M. Okorie, Z. Aboud, S. Misra, R. Rashid, P. Miall, J. Deanfield, N. Benjamin, R. MacAllister, A. J. Hobbs and A. Ahluwalia, *Hypertension*, 2008, **51**, 784.
140. H. Kitagawa, F. Takeda and H. Kohei, *J. Pharmacol. Exp. Ther.*, 1990, **253**, 1133.
141. B. J. R. Whittle, *Br. J. Pharmacol.*, 1993, **110**, 3.
142. A. Mohammad, N. Ali, B. Reza and K. Ali, *Indian J. Pharmacol.*, 2010, **42**, 78.
143. J. F. Brown, P. J. Hanson and B. J. R. Whittle, *Eur. J. Pharmacol.*, 1992, **223**, 103.
144. N. Benjamin, F. O'Driscoll, H. Dougall, C. Duncan, L. Smith and M. Golden, *Nature*, 1994, **368**, 502.
145. Y. Kono, H. Shibata, K. Adachi and K. Tanaka, *Arch. Biochem. Biophys.*, 1994, **311**, 153.
146. R. Dykhuizen, R. Frazer, C. Duncan, C. Smith, M. Golden, N. Benjamin and C. Leifert, *Antimicrob. Agents Chemother.*, 1996, **40**, 1422.

147. S. Kaplan, J. Lancaster, R. Basford and R. Simmons, *Infect. Immun.*, 1996, **64**, 69.
148. N. Benjamin, S. Pattullo, R. Weller, L. Smith and A. Ormerod, *Lancet*, 1997, **349**, 1776.
149. J. O. Lundberg, S. Carlsson, L. Engstrand, E. Morcos, N. P. Wiklund and E. Weitzberg, *Urology*, 1997, **50**, 189.
150. J. Petersson, M. Carlstrom, O. Schreiber, M. Phillipson, G. Christoffersson, A. Jagare, S. Roos, E. A. Jansson, E. A. Persson, J. O. Lundberg and L. Holm, *Free Radical Biol. Med.*, 2009, **46**, 1068.
151. L. C. Green, K. R. Luzuriaga, D. A. Wagner, W. Rand, N. Istfan, V. R. Young and S. R. Tannenbaum, *Proc. Natl. Acad. Sci. U. S. A.*, 1981, **78**, 7764.
152. P. M. Rhodes, A. M. Leone, P. L. Francis, A. D. Struthers and S. Moncada, *Biochem. Biophys. Res. Commun.*, 1995, **209**, 590.
153. T. Lauer, M. Preik, T. Rassaf, B. E. Strauer, A. Deussen, M. Feelisch and M. Kelm, *Proc. Natl. Acad. Sci. U. S. A.*, 2001, **98**, 12814.
154. P. Kleinbongard, A. Dejam, T. Lauer, T. Rassaf, A. Schindler, O. Picker, T. Scheeren, A. Godecke, J. Schrader, R. Schulz, G. Heusch, G. A. Schaub, N. S. Bryan, M. Feelisch and M. Kelm, *Free Radical Biol. Med.*, 2003, **35**, 790.
155. R. Shingles, M. H. Roh and R. E. McCarty, *J. Bioenerg. Biomembr.*, 1997, **29**, 611.
156. F. B. Jensen, *Acta Physiol. Scand.*, 2005, **184**, 243.
157. A. Samouilov, Y. Y. Woldman, J. L. Zweier and V. V. Khramtsov, *Nitric Oxide*, 2007, **16**, 362.
158. D. A. Vitturi, X. Teng, J. C. Toledo, S. Matalon, J. R. Lancaster and R. P. Patel, *Am. J. Physiol.: Heart Circ. Physiol.*, 2009, **296**, H1398.
159. F. B. Jensen and S. Rohde, *Am. J. Physiol.: Regul., Integr. Comp. Physiol.*, 2010, **298**, R972.
160. T. Arakawa, T. Kobayashi-Yurugi, Y. Alguel, H. Iwanari, H. Hatae, M. Iwata, Y. Abe, T. Hino, C. Ikeda-Suno, H. Kuma, D. Kang, T. Murata, T. Hamakubo, A. D. Cameron, T. Kobayashi, N. Hamasaki and S. Iwata, *Science*, 2015, **350**, 680.
161. D. J. Richardson and N. J. Watmough, *Curr. Opin. Chem. Biol.*, 1999, **3**, 207.
162. J. N. Galloway, F. J. Dentener, D. G. Capone, E. W. Boyer, R. W. Howarth, S. P. Seitzinger, G. P. Asner, C. C. Cleveland, P. A. Green, E. A. Holland, D. M. Karl, A. F. Michaels, J. H. Porter, A. R. Townsend and C. J. Vorosmarty, *Biogeochemistry*, 2004, **70**, 153.
163. H. Bothe, S. J. Ferguson and W. E. Newton, *Biology of the Nitrogen Cycle*, Elsevier, Amsterdam, 2007.
164. M. S. M. Jetten, *Environ. Microbiol.*, 2006, **10**, 2903.
165. D. E. Canfield, A. N. Glazer and P. G. Falkowski, *Science*, 2010, **330**, 192.
166. R. M. Martínez-Espinosa, J. A. Cole, D. J. Richardson and N. J. Watmough, *Biochem. Soc. Trans.*, 2011, **39**, 175.
167. D. Fowler, M. Coyle, U. Skiba, M. A. Sutton, J. N. Cape, S. Reis, L. J. Sheppard, A. Jenkins, B. Grizzetti, J. N. Galloway, P. Vitousek, A. Leach, A. F. Bouwman, K. Butterbach-Bahl, F. Dentener, D. Stevenson, V. Amann and M. Voss, *Philos. Trans. R. Soc., B*, 2013, **368**, 164.

168. J. N. Galloway, A. M. Leach, A. Bleeker and J. W. Erisman, *Philos. Trans. R. Soc., B*, 2013, **368**, 120.
169. L. Maia and J. J. G. Moura, *J. Biol. Inorg. Chem.*, 2015, **20**, 403.
170. T. M. Millar, C. R. Stevens, N. Benjamin, R. Eisenthal, R. Harrison and D. R. Blake, *FEBS Lett.*, 1998, **427**, 225.
171. Z. Zhang, D. Naughton, P. G. Winyard, N. Benjamin, D. T. Blake and M. C. Symons, *Biochem. Biophys. Res. Commun.*, 1998, **249**, 767.
172. H. L. J. Godber, J. J. Doel, G. P. Sapkota, D. R. Blake, C. R. Stevens, R. Eisenthal and R. Harrison, *J. Biol. Chem.*, 2000, **275**, 7757.
173. H. Li, A. Samouilov, X. Liu and J. L. Zweier, *J. Biol. Chem.*, 2001, **276**, 24482.
174. H. Li, A. Samouilov, X. Liu and J. L. Zweier, *J. Biol. Chem.*, 2004, **279**, 16939.
175. H. Li, H. Cui, T. K. Kundu, W. Alzawahra and J. L. Zweier, *J. Biol. Chem.*, 2008, **283**, 17855.
176. L. Maia, V. Pereira, L. Mira and J. J. G. Moura, *Biochemistry*, 2015, **54**, 685.
177. H. Li, T. K. Kundu and J. L. Zweier, *J. Biol. Chem.*, 2009, **284**, 33850.
178. J. Wang, S. Krizowski, K. Fischer-Schrader, D. Niks, J. Tejero, C. Sparacino-Watkins, L. Wang, V. Ragireddy, S. Frizzell, E. E. Kelley, Y. Zhang, P. Basu, R. Hille, G. Schwarz and M. T. Gladwin, *Antioxid. Redox. Signal.*, 2015, **23**, 283.
179. C. E. Sparacino-Watkins, J. Tejero, B. Sun, M. C. Gauthier, J. Thomas, V. Ragireddy, B. A. Merchant, J. Wang, I. Azarov, P. Basu and M. T. Gladwin, *J. Biol. Chem.*, 2014, **289**, 10345.
180. A. V. Kozlov, K. Staniek and H. Nohl, *FEBS Lett.*, 1999, **454**, 127.
181. H. Li, X. Liu, H. Cui, Y. R. Chen, A. J. Cardounel and J. L. Zweier, *J. Biol. Chem.*, 2006, **281**, 12546.
182. P. R. Castello, D. K. Woo, K. Ball, J. Wojcik, L. Liu and R. O. Poyton, *Proc. Natl. Acad. Sci. U. S. A.*, 2008, **105**, 8203.
183. A. M. Badejo, C. Hodnette, J. S. Dhaliwal, D. B. Casey, E. Pankey, S. N. Murthy, B. D. Nossaman, A. L. Hyman and P. Kadowitz, *Am. J. Physiol.: Heart Circ. Physiol.*, 2010, **299**, H819.
184. S. Hematian, I. Garcia-Bosch and K. D. Karlin, *Acc. Chem. Res.*, 2015, **48**, 2462.
185. S. Hematian, I. Kenkel, T. E. Shubina, M. Dürr, J. J. Liu, M. A. Siegler, I. Ivanovic-Burmazovic and K. D. Karlin, *J. Am. Chem. Soc.*, 2015, **137**, 6602.
186. J. Tejero, C. E. Sparacino-Watkins, V. Ragireddy, S. Frizzell and M. T. Gladwin, *Biochemistry*, 2015, **54**, 722.
187. M. G. I. Galinato, R. S. Fogle III, A. Stetz and J. F. Galanb, *J. Inorg. Biochem.*, 2016, **154**, 7.
188. D. A. Vitturi, L. Minarrieta, S. R. Ssalvatore, E. M. Ppostlethwait, M. Fazzari, G. Ferrer-Sueta, J. Lancaster Jr, B. Freeman and F. J. Sschopfer, *Nat. Chem. Biol.*, 2015, **11**, 504.
189. R. Hille, *Arch. Biochem. Biophys.*, 2005, **433**, 107.

190. R. Hille, *Eur. J. Inorg. Chem.*, 2006, 1913.
191. T. Nishino, K. Okamoto, B. T. Eger, E. F. Pai and T. Nishino, *FEBS J.*, 2008, **275**, 3278.
192. R. Hille, T. Nishino and F. Bittner, *Coord. Chem. Rev.*, 2011, **255**, 1179.
193. K. Okamoto, T. Kusano and T. Nishino, *Curr. Pharm. Des.*, 2013, **19**, 2606.
194. R. Hille, J. Hall and P. Basu, *Chem. Rev.*, 2014, **114**, 3963.
195. L. Maia, I. Moura and J. J. G. Moura, in *Molybdenum and Tungsten Enzymes - Metallobiochemistry Series of the Royal Society of Chemistry*, ed R. Hille, M. Kirk and C. Schulzke, Royal Society of Chemistry, 2016, in press.
196. E. Garattini and M. Terao, *Drug Metab. Rev.*, 2011, **43**, 374.
197. D. C. Pryde, D. Dalvie, Q. Hu, P. Jones, R. S. Obach and T. D. Tran, *J. Med. Chem.*, 2010, **53**, 8441.
198. G. Fabre, R. Seither and D. Goldman, *Biochem. Pharmacol.*, 1986, **35**, 1325.
199. S. B. Yee and C. A. Pritsos, *Arch. Biochem. Biophys.*, 1997, **347**, 235.
200. M. Dambrova, S. Uhlén, C. J. Welch and J. E. S. Wikberg, *Eur. J. Biochem.*, 1998, **257**, 178.
201. A. M. Stoddart and W. G. Levine, *Biochem. Pharmacol.*, 1992, **43**, 2227.
202. R. Harrison, *Biochem. Soc. Trans.*, 1997, **25**, 786.
203. R. M. Wright and J. E. Repine, *Biochem. Soc. Trans.*, 1997, **25**, 799.
204. K. B. Beckman and B. N. Ames, *Physiol. Rev.*, 1998, **78**, 547.
205. H. Suzuki, F. A. Delano, D. A. Parks, N. Jamshidi, D. N. Granger, H. Ishii, M. Suematsu, B. W. Zweifach and G. W. Schmid-Schonbein, *Proc. Natl. Acad. Sci. U. S. A.*, 1998, **95**, 4754.
206. R. Harrison, *Free Radical Biol. Med.*, 2002, **33**, 774.
207. F. Stirpe, M. Ravaioli, M. G. Battelli, S. Musiani and G. L. Grazi, *Am. J. Gastroenterol.*, 2002, **97**, 2079.
208. D. M. Small, J. S. Coombes, N. Bennett, D. W. Johnson and G. C. Gobe, *Nephrology*, 2012, **17**, 311.
209. M. M. Bachschmid, S. Schildknecht, R. Matsui, R. Zee, D. Haeussler, R. A. Cohen, D. Pimental and B. Loo, *Ann. Med.*, 2013, **45**, 17.
210. N. R. Madamanchi and M. S. Runge, *Free Radical Biol. Med.*, 2013, **61**, 473.
211. D. N. Granger, G. Rutili and J. M. McCord, *Gastroenterology*, 1981, **81**, 22.
212. J. M. McCord, *N. Engl. J. Med.*, 1985, **312**, 159.
213. J. L. Zweier, P. Kuppusamy and G. A. Lutty, *Proc. Natl. Acad. Sci. U. S. A.*, 1988, **85**, 4046.
214. J. L. Zweier, R. Broderick, P. Kuppusamy, S. Thompson-Gorman and G. A. Lutty, *J. Biol. Chem.*, 1994, **269**, 24156.
215. T. Nishino, S. Nakanishi, K. Okamoto, J. Mizushima, H. Hori, T. Iwasaki, T. Nishino, K. Ichimori and H. Nakazawa, *Biochem. Soc. Trans.*, 1997, **25**, 783.
216. C. E. Berry and J. M. Hare, *J. Physiol.*, 2004, **555**, 589.
217. L. Maia, A. Vala and L. Mira, *Free Radical Res.*, 2005, **39**, 979.
218. L. Maia, R. O. Duarte, A. Ponces-Freire, J. J. G. Moura and L. Mira, *J. Biol. Inorg. Chem.*, 2007, **12**, 777.

219. C. S. Lieber, *N. Engl. J. Med.*, 1988, **319**, 1639.
220. A. I. Cederbaum, *Free Radical Biol. Med.*, 1989, **7**, 537.
221. S. Kato, T. Kawase, J. Alderman, N. Inatomi and C. Lieber, *Gastroenterology*, 1990, **98**, 203.
222. R. Nordmann, C. Ribière and H. Rouach, *Free Radical Biol. Med.*, 1992, **12**, 219.
223. L. Mira, L. Maia, L. Barreira and C. F. Manso, *Arch. Biochem. Biophys.*, 1995, **318**, 53.
224. R. M. Wright, J. L. McManaman and J. E. Repine, *Free Radical Biol. Med.*, 1999, **26**, 348.
225. D. Wu and A. I. Cederbaum, *Alcohol Res. Health*, 2003, **27**, 277.
226. Z. A. Alikulov, N. P. L'vov and V. L. Kretovich, *Biokhimiia*, 1980, **45**, 1714.
227. H. Li, A. Samouilov, X. Liu and J. L. Zweier, *Biochemistry*, 2003, **42**, 1150.
228. T. M. Millar, *FEBS Lett.*, 2004, **562**, 129.
229. Z. Zhang, D. Naughton, P. G. Winyard, N. Benjamin, D. R. Blake and M. C. Symons, *Biochem. Biophys. Res. Commun.*, 1998, **249**, 767.
230. H. Li, C. Hemann, T. M. Abdelghany, M. A. El-Mahdy and J. L. Zweier, *J. Biol. Chem.*, 2012, **287**, 36623.
231. P. A. Pickerodt, M. J. Emery, R. Zarndt, W. Martin, R. C. Francis, W. Boemke and E. R. Swenson, *Anesthesiology*, 2012, **117**, 592.
232. A. A. Samal, J. Honavar, A. Brandon, K. M. Bradley, S. Doran, Y. Liu, C. Dunaway, C. Steele, E. M. Postlethwait, G. L. Squadrito, M. V. Fanucchi, S. Matalon and R. P. Patel, *Free Radical Biol. Med.*, 2012, **53**, 1431.
233. R. Sugimoto, T. Okamoto, A. Nakao, J. Zhan, Y. Wang, J. Kohmoto, D. Tokita, C. F. Farver, M. M. Tarpey, T. R. Billiar, M. T. Gladwin and K. R. McCurry, *Am. J. Transplant.*, 2012, **12**, 2938.
234. E. E. Kelley, *Pharmacol. Rep.*, 2015, **67**, 669.
235. B. A. Wittenberg and J. B. Wittenberg, *Annu. Rev. Physiol.*, 1989, **51**, 857.
236. G. A. Millikan, *Proc. R. Soc. London, Ser. B*, 1937, **123**, 218.
237. J. B. Wittenberg, *Physiol. Rev.*, 1970, **50**, 559.
238. J. B. Wittenberg, *J. Exp. Biol.*, 2003, **206**, 2011.
239. D. J. Garry and P. P. Mammen, *Adv. Exp. Med. Biol.*, 2007, **618**, 181.
240. Y. Qiu, L. Sutton and A. F. Riggs, *J. Biol. Chem.*, 1998, **273**, 23426.
241. D. J. Garry, G. A. Ordway, J. N. Lorenz, N. B. Radford, E. R. Chin, R. W. Grange, R. Bassel-Duby and R. S. Williams, *Nature*, 1998, **395**, 905.
242. A. Godecke, U. Flogel, K. Zanger, Z. Ding, J. Hirchenhain, U. K. Decking and J. Schrader, *Proc. Natl. Acad. Sci. U. S. A.*, 1999, **96**, 10495.
243. P. C. Lin, U. Kreutzer and T. Jue, *Biophys. J.*, 2007, **92**, 2608.
244. P. C. Lin, U. Kreutzer and T. Jue, *J. Physiol.*, 2007, **578**, 595.
245. M. Kelm and J. Schrader, *Circ. Res.*, 1990, **66**, 1561.
246. J. R. Lancaster, *Proc. Natl. Acad. Sci. U. S. A.*, 1994, **91**, 8137.
247. D. D. Thomas, X. Liu, S. P. Kantrow and J. R. Lancaster, *Proc. Natl. Acad. Sci. U. S. A.*, 2001, **98**, 355.
248. U. Flogel, A. Godecke, L. O. Klotz and J. Schrader, *FASEB J.*, 2004, **18**, 1156.
249. R. Sriram, U. Kreutzer, L. Shih and T. Jue, *FEBS Lett.*, 2008, **582**, 3643.

250. M. Sundararajan and F. Neese, *Inorg. Chem.*, 2015, **54**, 7209.
251. J. Haldane, *J. Hyg.*, 1901, **1**, 115.
252. J. Brooks, *Proc. R. Soc. London, Ser. B*, 1937, **123**, 368.
253. M. P. Doyle, R. A. Pickering, T. M. DeWeert, J. W. Hoekstra and D. Pater, *J. Biol. Chem.*, 1981, **256**, 12393.
254. F. Sulc, C. E. Immoos, D. Pervitsky and P. J. Farmer, *J. Am. Chem. Soc.*, 2004, **126**, 1096.
255. Z. Huang, S. Shiva, D. B. Kim-Shapiro, R. P. Patel, L. A. Ringwood, C. E. Irby, K. T. Huang, C. Ho, N. Hogg, A. N. Schechter and M. T. Gladwin, *J. Clin. Invest.*, 2005, **115**, 2099.
256. M. T. Gladwin and D. B. Kim-Shapiro, *Blood*, 2008, **112**, 2636.
257. J. M. Salhany, *Biochemistry*, 2008, **47**, 6059.
258. N. J. Parks, K. J. Krohn, C. A. Mathis, J. H. Chasko, K. R. Geiger, M. E. Gregor and N. F. Peek, *Science*, 1981, **212**, 58.
259. R. Cassoly and Q. H. Gibson, *J. Mol. Biol.*, 1975, **91**, 301.
260. E. G. Moore and Q. H. Gibson, *J. Biol. Chem.*, 1976, **251**, 2788.
261. E. J. Rose and B. M. Hoffman, *J. Am. Chem. Soc.*, 1983, **105**, 2866.
262. T. G. Traylor and V. S. Sharma, *Biochemistry*, 1992, **31**, 2847.
263. M. Hoshino, K. Ozawa, H. Seki and P. C. Ford, *J. Am. Chem. Soc.*, 1993, **115**, 9568.
264. R. Radi, *Chem. Res. Toxicol.*, 1996, **9**, 828.
265. V. G. Kharitonov, V. S. Sharma, D. Magde and D. Koesling, *Biochemistry*, 1997, **36**, 6814.
266. C. E. Cooper, *Biochim. Biophys. Acta*, 1999, **1411**, 290.
267. F. Azizi, J. E. Kielbasa, A. M. Adeyiga, R. D. Maree, M. Frazier, M. Yakubu, H. Shields, S. B. King and D. B. Kim-Shapiro, *Free Radical Biol. Med.*, 2005, **39**, 145.
268. V. S. Sharma and H. M. Ranney, *J. Biol. Chem.*, 1978, **253**, 6467.
269. J. S. Olson and G. N. Phillips, *J. Biol. Chem.*, 1996, **271**, 17593.
270. D. Ionascu, F. Gruia, X. Ye, A. Yu, F. Rosca, C. Beck, A. Demidov, J. S. Olson and P. M. Champion, *J. Am. Chem. Soc.*, 2005, **127**, 16921.
271. B. P. Luchsinger, E. N. Rich, Y. Yan, E. M. Williams, J. S. Stamler and D. J. Singel, *J. Inorg. Biochem.*, 2005, **99**, 912.
272. K. Nienhaus, P. Palladino and G. U. Nienhaus, *Biochemistry*, 2008, **47**, 935.
273. R. J. Morris and Q. H. Gibson, *J. Biol. Chem.*, 1980, **255**, 8050.
274. M. P. Doyle, R. A. Pickering and B. R. Cook, *J. Inorg. Biochem.*, 1983, **19**, 329.
275. M. S. Joshi, T. B. Ferguson, T. H. Han, D. R. Hyduke, J. C. Liao, T. Rassaf, N. Bryan, M. Feelisch and J. R. Lancaster, *Proc. Natl. Acad. Sci. U. S. A.*, 2002, **99**, 10341.
276. E. Nagababu, S. Ramasamy, D. R. Abernethy and J. M. Rifkind, *J. Biol. Chem.*, 2003, **278**, 46349.
277. A. Jeffers, X. Xu, K. T. Huang, M. Cho, N. Hogg, R. P. Patel and D. B. Kim-Shapiro, *Comp. Biochem. Physiol., Part A*, 2005, **142**, 130.
278. A. V. Kozlov, G. Costantino, B. Sobhian, L. Szalay, F. Umar, H. Nohl, S. Bahrami and H. Redl, *Antioxid. Redox Signaling*, 2005, **7**, 515.

279. M. T. Gladwin, N. J. Raat, S. Shiva, C. Dezfulian, N. Hogg, D. B. Kim-Shapiro and R. P. Patel, *Am. J. Physiol.: Heart Circ. Physiol.*, 2006, **291**, H2026.
280. R. Hataishi, A. C. Rodrigues, T. G. Neilan, J. G. Morgan, E. Buys, S. Shiva, R. Tambouret, D. S. Jassal, M. J. Raher, E. Furutani, F. Ichinose, M. T. Gladwin, A. Rosenzweig, W. M. Zapol, M. H. Picard, K. D. Bloch and M. Scherrer-Crosbie, *Am. J. Physiol.: Heart Circ. Physiol.*, 2006, **291**, H379.
281. T. Dalsgaard, U. Simonsen and A. Fago, *Am. J. Physiol.: Heart Circ. Physiol.*, 2007, **292**, H3072.
282. T. S. Isbell, M. T. Gladwin and R. P. Patel, *Am. J. Physiol.: Heart Circ. Physiol.*, 2007, **293**, H2565.
283. T. Rassaf, U. Flogel, C. Drexhage, U. Hendgen-Cotta, M. Kelm and J. Schrader, *Circ. Res.*, 2007, **100**, 1749.
284. W. F. Alzawahra, M. A. H. Talukder, X. Liu, A. Samouilov and J. L. Zweier, *Am. J. Physiol.: Heart Circ. Physiol.*, 2008, **295**, H499.
285. A. Cossins and M. Berenbrink, *Nature*, 2008, **454**, 416.
286. P. C. Minneci, K. J. Deans, S. Shiva, H. Zhi, S. M. Banks, S. Kern, C. Natanson, S. B. Solomon and M. T. Gladwin, *Am. J. Physiol.: Heart Circ. Physiol.*, 2008, **295**, H743.
287. B. Yu, M. J. Raher, G. P. Volpato, K. D. Bloch, F. Ichinose and W. M. Zapol, *Circulation*, 2008, **117**, 1982.
288. N. J. Raat, A. C. Noguchi, V. B. Liu, N. Raghavachari, D. Liu, X. Xu, S. Shiva, P. J. Munson and M. T. Gladwin, *Free Radical Biol. Med.*, 2009, **47**, 510.
289. J. M. Rifkind, E. Nagababu, Z. Cao, E. Barbiro-Michaely and A. Mayevsky, *Adv. Exp. Med. Biol.*, 2009, **645**, 27.
290. S. P. Dufour, R. P. Patel, A. Brandon, X. Teng, J. Pearson, H. Barker, L. Ali, A. H. Yuen, R. T. Smolenski and J. Gonzalez-Alonso, *Am. J. Physiol.: Heart. Circ. Physiol.*, 2010, **299**, H1936.
291. U. B. Hendgen-Cotta, M. Kelm and T. Rassaf, *Nitric Oxide*, 2010, **22**, 75.
292. V. Kapil, A. B. Milsom, M. Okorie, S. Maleki-Toyserkani, F. Akram, F. Rehman, S. Arghandawi, V. Pearl, N. Benjamin, S. Loukogeorgakis, R. Macallister, A. J. Hobbs, A. J. Webb and A. Ahluwalia, *Hypertension*, 2010, **56**, 274.
293. N. Cantu-Medellin, D. A. Vitturi, C. Rodriguez, S. Murphy, S. Dorman, S. Shiva, Y. Zhou, Y. Jia, A. F. Palmer and R. P. Patel, *Nitric Oxide*, 2011, **25**, 59.
294. J. O. M. Ormerod, H. Ashrafian, A. R. Maher, S. Arif, V. Steeples, G. V. R. Born, S. Egginton, M. Feelisch, H. Watkins and M. P. Frenneaux, *Cardiovasc. Res.*, 2011, **89**, 560.
295. R. P. Patel, N. Hogg and D. B. Kim-Shapiro, *Cardiovasc. Res.*, 2011, **89**, 507.
296. S. Shiva, T. Rassaf, R. P. Patel and M. T. Gladwin, *Cardiovasc. Res.*, 2011, **89**, 566.
297. C. Liu, N. Wajih, X. Liu, S. Basu, J. Janes, M. Marvel, C. Keggi, C. C. Helms, A. N. Lee, A. M. Belanger, D. I. Diz, P. J. Laurienti, D. L. Caudell, J. Wang, M. T. Gladwin and D. B. Kim-Shapiro, *J. Biol. Chem.*, 2015, **290**, 1281.

298. M. M. Rahaman, F. G. Reinders, D. Koes, A. T. Nguyen, S. M. Mutchler, C. Sparacino-Watkins, R. A. Alvarez, M. P. Miller, D. Cheng, B. B. Chen, E. K. Jackson, C. J. Camacho and A. C. Straub, *J. Biol. Chem.*, 2015, **290**, 16861.
299. O. Khersonsky and D. S. Tawfik, *Annu. Rev. Biochem.*, 2010, **79**, 471.
300. S. D. Copley, *BioEssays*, 2012, **34**, 578.
301. M. Feelisch and J. F. Martin, *Trends Ecol. Evol.*, 1995, **10**, 496.
302. F. Cutruzzolá, S. Rinaldo, N. Castiglione, G. Giardina, I. Pecht and M. Brunori, *BioEssays*, 2009, **31**, 885.
303. O. Khersonsky and D. S. Tawfik, *Annu. Rev. Biochem.*, 2010, **79**, 471.

Subject Index